Oil in the Environment
Legacies and Lessons of the *Exxon Valdez* Oil Spill

What light does nearly 25 years of scientific study of the *Exxon Valdez* oil spill shed on the fate and effects of a spill? How can the results help in assessing future spills? How can ecological risks be assessed and quantified?

In this, the first book on the effects of *Exxon Valdez* in 15 years, scientists directly involved in studying the spill provide a comprehensive perspective on, and synthesis of, scientific information on long-term spill effects. The coverage is multidisciplinary, with chapters discussing a range of issues including effects on biota; successes and failures of postspill studies and techniques; and areas of continuing disagreement. An even-handed and critical examination of more than two decades of scientific study, this is an invaluable guide for studying future oil spills and, more broadly, for unraveling the consequences of any large environmental disruption.

A full bibliography of related literature is available online at www.cambridge.org/9781107027176.

John A. Wiens is Chief Scientist at PRBO Conservation Science in California, USA, and Winthrop Research Professor in the School of Plant Biology at the University of Western Australia. He has served on the faculties of Oregon State University, the University of New Mexico, and Colorado State University, where he was a University Distinguished Professor. From 2002 to 2008 he was Lead Scientist with The Nature Conservancy. His work emphasizes landscape ecology, conservation, and the ecology of birds.

Frontispiece. Southeastern Herring Bay, Knight Island, Prince William Sound, Alaska, July 2008. Herring Bay was one of the areas most heavily oiled by the *Exxon Valdez* oil spill in 1989. Photo: John A. Wiens.

Oil in the Environment

Legacies and Lessons of the *Exxon Valdez* Oil Spill

Edited by

John A. Wiens

PRBO CONSERVATION SCIENCE, CALIFORNIA
and THE UNIVERSITY OF WESTERN AUSTRALIA, PERTH

CAMBRIDGE
UNIVERSITY PRESS

University Printing House, Cambridge CB2 8BS, United Kingdom

Published in the United States of America by Cambridge University Press, New York

Cambridge University Press is part of the University of Cambridge.

It furthers the University's mission by disseminating knowledge in the pursuit of education, learning and research at the highest international levels of excellence.

www.cambridge.org
Information on this title: www.cambridge.org/9781107027176

© Cambridge University Press 2013

This publication is in copyright. Subject to statutory exception
and to the provisions of relevant collective licensing agreements,
no reproduction of any part may take place without the written
permission of Cambridge University Press.

First published 2013

Printed and bound in Europe by Grafos S. A.

A catalog record for this publication is available from the British Library

Library of Congress Cataloging in Publication data

Oil in the environment: legacies and lessons of the Exxon Valdez oil spill / edited by John A. Wiens, PRBO Conservation Science, California and The University of Western Australia, Perth.
    pages cm
  Includes bibliographical references and index.
  ISBN 978-1-107-02717-6 (Hardback) – ISBN 978-1-107-61469-7 (Paperback)
    1. Petroleum–Environmental aspects.   2. Oil spills–Cleanup.   3. Oil pollution of soils.
4. Shore protection.   5. Environmental disturbance–Analysis.   6. Oil pollution of the sea.
7. Oil spills–Cleanup–Alaska–Prince William Sound Region.   8. Exxon Valdez Oil Spill, Alaska, 1989.   I. Wiens, John A.

TD196.P4O386 2013
363.738′2097983–dc23

2012041990

ISBN 978-1-107-02717-6 Hardback
ISBN 978-1-107-61469-7 Paperback

Additional resources for this publication at www.cambridge.org/9781107027176

Cambridge University Press has no responsibility for the persistence or
accuracy of URLs for external or third-party internet websites referred to in
this publication, and does not guarantee that any content on such websites is,
or will remain, accurate or appropriate.

A subvention of £12 000 toward the production of this book was provided
by Exxon Mobil Corporation. The publishers wish to make clear that the
subvention was not provided with any conditions constraining the editorial
scope or content of the book, nor the freedom of the editor and contributors
to express their views.

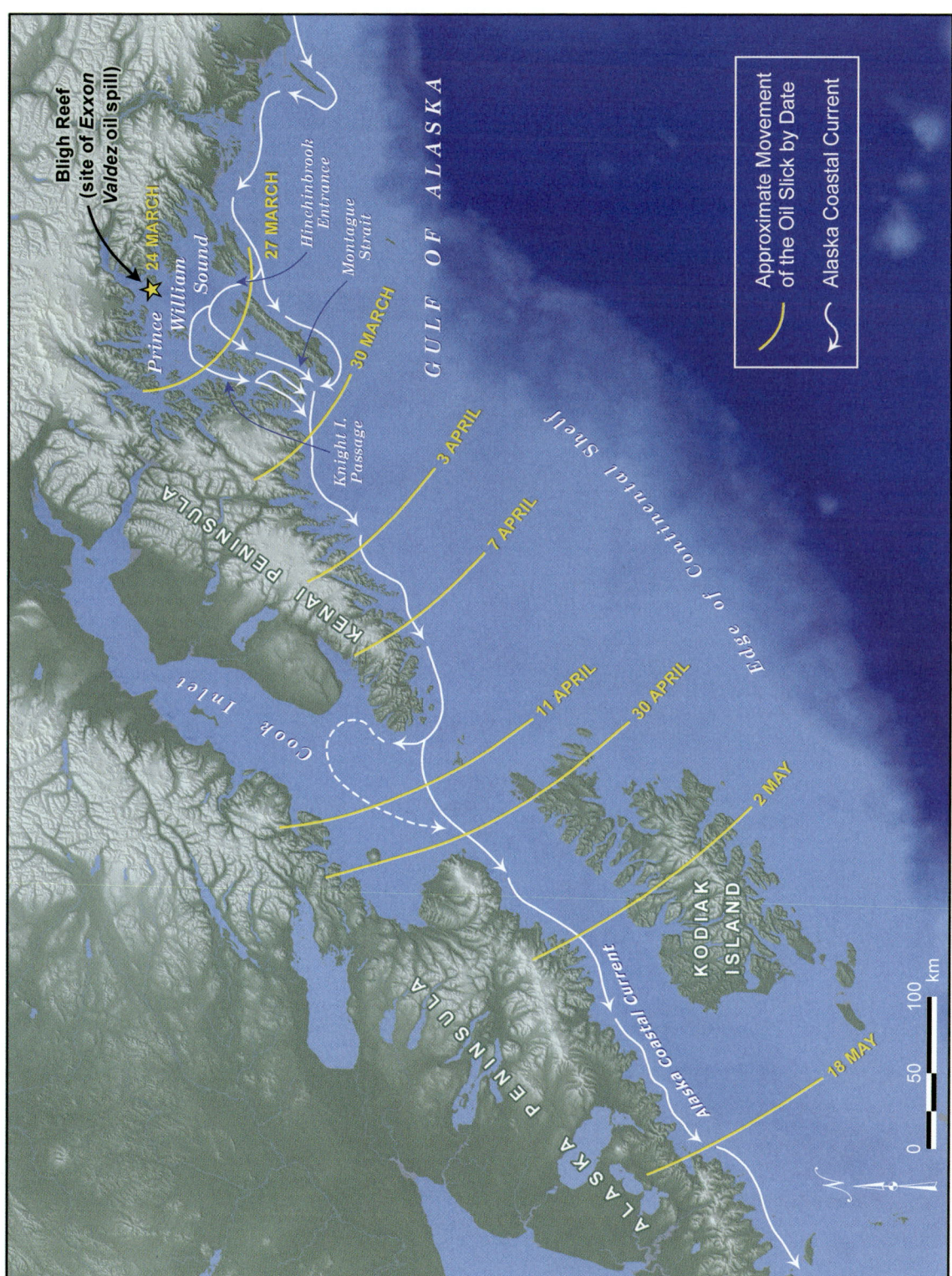

Map 1 The general path of the Alaska coastal current and the approximate progression of the oil spill from its source at Bligh Reef in Prince William Sound on March 23, 1989. (Source: Exxon Corporation).

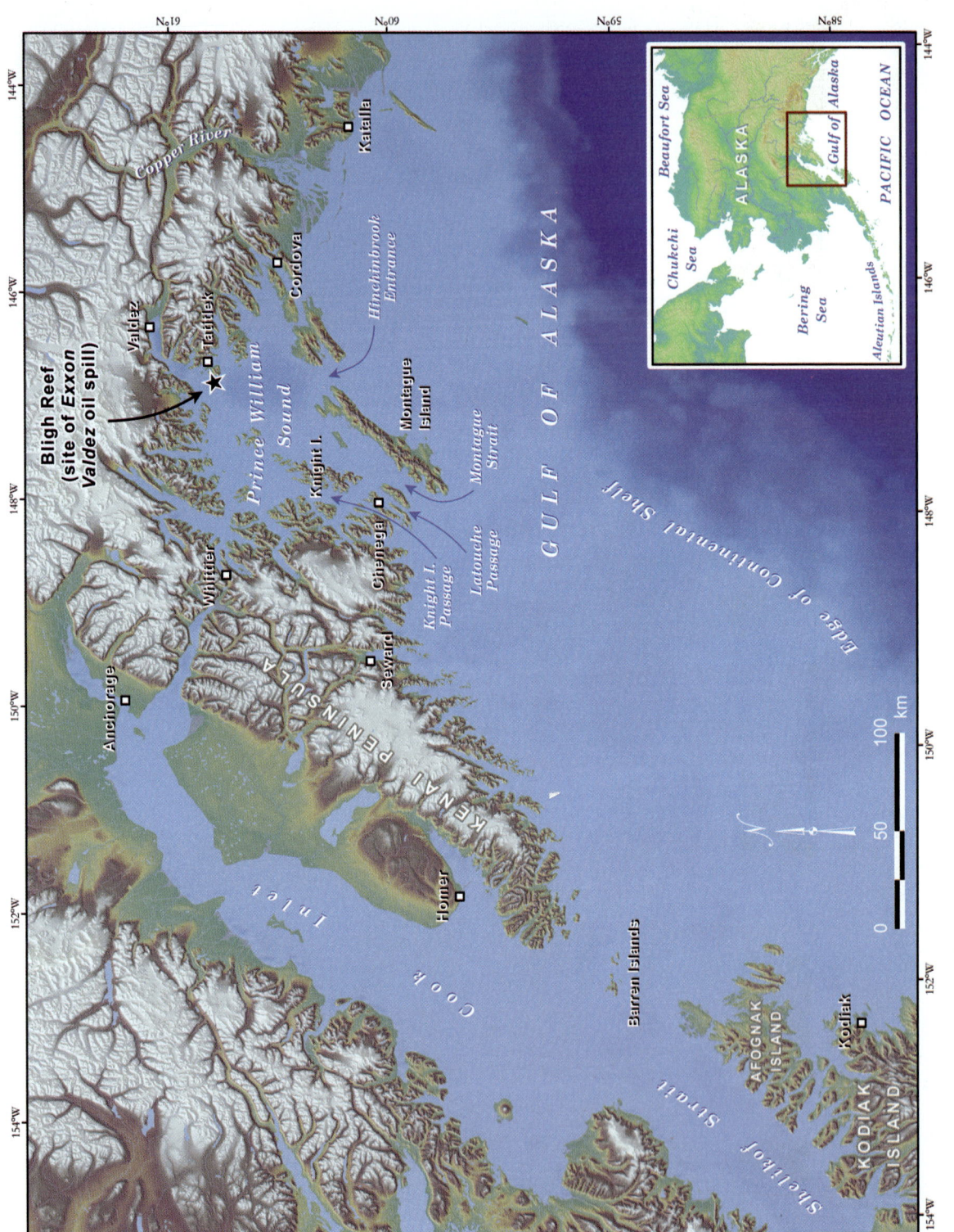

Map 2 Important features of the northern Gulf of Alaska where the *Exxon Valdez* oil spill occurred.

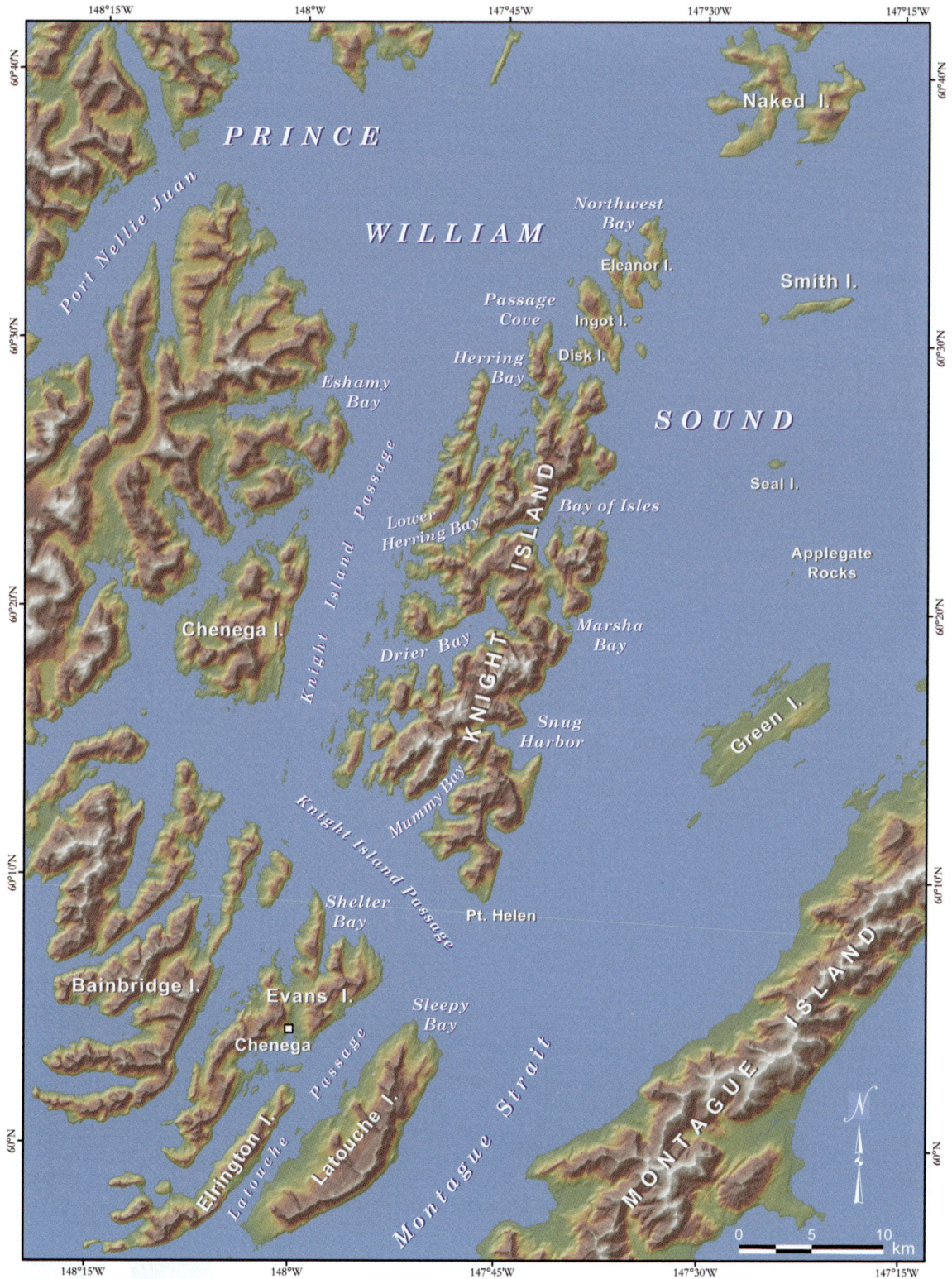

Map 3 An expanded map of western Prince William Sound, identifying important locations mentioned throughout this book.

CONTENTS

List of contributors xi

Use of acronyms xvii

Acknowledgments xviii

A bibliographic note xx
   David K. Johnson and Laura R. Rustin

Prologue xxii

PART I INTRODUCTION AND BACKGROUND

1. Introduction and background   3
   John A. Wiens
2. The phases of an oil spill and scientific studies of spill effects   37
   Paul D. Boehm, Erich R. Gundlach, and David S. Page

PART II OIL IN THE ENVIRONMENT

3. Oil in the water column   57
   Paul D. Boehm, Jerry M. Neff, and David S. Page
4. Surveying oil on the shoreline   78
   Edward H. Owens and P. Douglas Reimer
5. Ancient sites and emergency response: cultural resource protection   98
   Chris B. Wooley and James C. Haggarty
6. Fate of oil on shorelines   116
   David S. Page, Paul D. Boehm, John S. Brown, Erich R. Gundlach, and Jerry M. Neff
7. Understanding subsurface contamination using conceptual and mathematical models   144
   Gary A. Pope, Kimberly D. Gordon, and James R. Bragg
8. Removal of oil from shorelines: biodegradation and bioremediation   176
   Ronald M. Atlas and James R. Bragg

PART III BIOLOGICAL EFFECTS

9. Cytochrome P450 1A (CYP1A) as a biomarker in oil spill assessments   201
   James T. Oris and Aaron P. Roberts
10. Assessing effects and recovery from environmental accidents   220
    Keith R. Parker, John A. Wiens, Robert H. Day, and Stephen M. Murphy
11. Shoreline biota   241
    Erich R. Gundlach, David S. Page, Jerry M. Neff, and Paul D. Boehm
12. Oiling effects on pink salmon   263
    Ernest L. Brannon, Matthew A. Cronin, Alan W. Maki, Larry L. Moulton, and Keith R. Parker
13. Pacific herring   292
    Walter H. Pearson, Ralph A. Elston, Karen Humphrey, and Richard B. Deriso
14. Oil and marine birds in a variable environment   318
    John A. Wiens, Robert H. Day, and Stephen M. Murphy
15. Sea otters: trying to see the forest for the trees since the *Exxon Valdez*   348
    David L. Garshelis and Charles B. Johnson

PART IV ASSESSING OIL SPILL EFFECTS AND ECOLOGICAL RECOVERY

16. Characterizing ecological risks, significance, and recovery   383
    Mark A. Harwell, John H. Gentile, and Keith R. Parker

PART V CONCLUSIONS

17. Science and oil spills: the broad picture   423
    John A. Wiens

Index   446

CONTRIBUTORS

John A. Wiens, the editor of this volume as well as a contributor, has conducted research on birds in semiarid environments, marine birds, landscape ecology, and conservation in three continents, producing over 250 scientific publications and several books. He held faculty positions at Oregon State University, the University of New Mexico, and Colorado State University, where he is a Distinguished Professor Emeritus. John has also been a visiting professor or research scientist in Norway, Canada, and Australia. He was Chief Scientist for The Nature Conservancy, and currently holds appointments as Chief Scientist at PRBO Conservation Science in California and is a Winthrop Research Professor in the School of Plant Biology at the University of Western Australia. He worked on the *Exxon Valdez* spill from 1989 through the editing of this book. He received M.S. and Ph.D. degrees in zoology from the University of Wisconsin–Madison after his B.S. degree (zoology) at the University of Oklahoma. John has served as the president of the International Association for Landscape Ecology and has received a Fulbright Senior Scholar Award, the Elliott Coues Award (American Ornithologists' Union), the Distinguished Landscape Ecologist Award and Distinguished Service Award (International Association for Landscape Ecology), and the Loy and Alden Miller Research Award (Cooper Ornithological Society). He is the Editor in Chief of Cambridge University Press's Series on Landscape Ecology and serves or has been a member of the editorial boards for *The Wildlife Professional, Landscape Ecology, Ecosystems, Avian Conservation and Ecology,* and *Ecological Applications*.

Ronald M. Atlas is Professor of Biology at the University of Louisville. He has studied petroleum biodegradation for over 40 years and pioneered the field of bioremediation. He has conducted studies on hydrocarbon biodegradation in Prince William Sound since 1973 and has served as a consultant to Exxon and the US Environmental Protection Agency on the use of bioremediation following the *Exxon Valdez* oil spill. He received a B.S. from the State University of New York at Stony Brook in biology and M.S. and Ph.D. degrees from Rutgers University in microbiology.

Paul D. Boehm is Environmental Sciences Group Vice-president and Principal Scientist at Exponent, Inc., and is an internationally recognized expert in the field of environmental chemistry, marine pollution, and oil-spill scientific assessment. He has conducted investigations of the chemistry and effects of oil spills in the United States, Europe, and the Middle East, and has published extensively on oil spills. Paul has served on several National Research Council panels and other groups reviewing environmental and marine pollution and oil-spill assessments. He received a B.S. in chemical engineering from the University of Rochester and M.S. and Ph.D. degrees in chemical oceanography from the Graduate School of Oceanography of the University of Rhode Island.

James R. Bragg is a consultant with Creative Petroleum Solutions LLC, in Houston, Texas. He has conducted research and consulting in reservoir and environmental engineering for over 43 years in the areas of enhanced oil recovery and the chemistry and physics of interactions among hydrocarbons, water, and minerals. Following the *Exxon Valdez* spill, he conducted research on bioremediation and the fate of the spilled hydrocarbons. He has published extensively and holds numerous US and foreign patents. He holds degrees in chemical engineering from Virginia Polytechnic Institute and State University (B.S.) and the University of Illinois (M.S., Ph.D.).

Ernest L. Brannon is a Professor Emeritus and Distinguished Research Professor at the University of Idaho, Moscow. He received his B.S. in fisheries and his Ph.D. in salmon life history and genetics from the University of Washington. He was the Chief Biologist for the International Pacific Salmon Fisheries Commission from 1970 to 1972, was a member of the College of Fisheries faculty, University of Washington, from 1973 to 1989, and joined the College of Natural Resources faculty, University of Idaho, in 1989.

John S. Brown is a Senior Managing Scientist with Exponent Environmental Sciences in Maynard, Massachusetts. He has over 25 years of experience in the design, conduct, and synthesis of oil-spill studies and marine environmental monitoring programs. His studies of the fate of oil in the environment have included work in the United States, Europe, Asia, and Africa. He received a B.S. degree in biology from Hobart College.

Matthew A. Cronin is a scientist at Northwest Biotechnology Company, Anchorage, Alaska, and a research professor at the University of Alaska–Fairbanks School of Natural Resources and Agricultural Sciences. He received a B.S. in forest biology from the State University of New York, College of Environmental Science and Forestry, Syracuse, an M.S. in biology from Montana State University, and a Ph.D. in biology from Yale University. He conducts research in population genetics with applications in natural resource management and agriculture.

Robert H. Day is a Senior Scientist and Principal at ABR, Inc.–Environmental Research & Services, a scientific consulting firm based in Fairbanks, Alaska. He received his M.S. in zoology and his Ph.D. in oceanography from the University of Alaska–Fairbanks, where he is also affiliate faculty in the Institute of Marine Sciences. He has studied the ecology of marine organisms, especially seabirds and marine mammals, and marine pollution in Alaska and the North Pacific since 1975.

Richard B. Deriso is currently Chief Scientist of the Inter-American Tropical Tuna Commission (IATTC). He received his Ph.D. in biomathematics from the University of Washington. His research interests include population dynamics, quantitative ecology, and fishery stock assessment.

Ralph A. Elston has over 30 years of experience in aquatic animal health management and disease analysis and currently practices as chief scientist and founder of AquaTechnics, Inc., in Sequim, Washington. He has a Ph.D. in veterinary medicine from Cornell University and B.S. (biology) and M.S. (ecology) degrees from the University of California at Davis.

David L. Garshelis is a wildlife research scientist with the Minnesota Department of Natural Resources and an adjunct associate professor of wildlife conservation at the University of Minnesota. He studied the ecology of sea otters in Prince William Sound for his Ph.D. from the University of Minnesota and continued sea otter studies after the *Exxon Valdez* oil spill. His principal research and conservation efforts, however, are on the world's bears, where he focuses on improved methods of population monitoring.

John (Jack) H. Gentile served as a senior scientist with the US Environmental Protection Agency for 30 years, directing programs in marine ecotoxicology, ocean disposal of hazardous wastes, and marine water-quality criteria. Jack co-directed the development of the EPA Framework and Guidelines for Ecological Risk Assessments and has facilitated the development and application of risk-based conceptual ecosystem models for National Estuarine Research Reserves and the Everglades Restoration. He has conducted research to assess marine ecosystem risks in Biscayne and Tampa Bays, the Bay of Fundy, and Prince William Sound. He received his B.S. degree in biology and chemistry from Northeastern University and M.S. and Ph.D. degrees in ecology from the University of New Hampshire.

Kimberly D. Gordon is a Consulting Engineer with Platt, Sparks & Associates in Austin, Texas. Her professional experience encompasses both the oil and gas industry and environmental and natural resource studies. Her work focuses on chemical spills, including site characterization and fate of contaminants, simulation studies, and oil-field development studies with an emphasis on enhanced recovery techniques. She graduated from the University of Texas at Austin with a B.S. in geosystems engineering and hydrogeology. She is a licensed professional engineer and a member of the Society of Petroleum Engineers.

Erich R. Gundlach has been involved with the scientific investigation of oil spills and environmental issues since 1975, specializing in environmental assessments, contingency planning, and emergency-response management. He is a co-developer of the Environmental Sensitivity Index (ESI) and has authored over 200 publications and technical reports on oil spills, environmental impacts, contingency planning, and spill-response management. He received degrees from Stony Brook University (B.S., earth sciences), the University of New Hampshire (M.S., earth sciences and marine biology), and the University of South Carolina (Ph.D., coastal geomorphology).

James (Jim) C. Haggarty is the owner of Shoreline Archaeological Services, Inc. in Victoria, British Columbia, Canada. He has directed numerous coastal multidisciplinary research projects in British Columbia since 1971 and was the Assistant Director (1989) and Director (1990–93) of the Exxon Cultural Resource Program. Jim has published extensively, including several oil-spill response atlases for coastal British Columbia. Jim earned a B.Sc. in zoology from the University of Victoria, an M.A. in anthropology from the University of Victoria, and a Ph.D. in anthropology from Washington State University.

Mark A. Harwell specializes in ecological risk assessments, ecological modeling, and ecosystem management. He co-led the development of the US Environmental Protection Agency ecological risk assessment framework, the US Man and the Biosphere ecosystem management principles, and a series of conceptual and quantitative ecological models applied to risk assessments, including on the Everglades, Biscayne Bay, Tampa Bay, Apalachicola Bay, Prince William Sound, and the Bay of Fundy. He earned his Ph.D. in ecology at Emory University and was elected Fellow of the American Association for the Advancement of Science for his scientific service to governments.

Karen Humphrey is a laboratory specialist and research analyst for AquaTechnics, Inc., in Sequim, Washington. A registered pharmacist, Karen is skilled in the development of research protocols, experimental design, and data collection and analysis. She has a Bachelor of Pharmacy degree from Washington State University.

Charles B. (Rick) Johnson is a Senior Scientist at ABR, Inc.–Environmental Research & Services in Fairbanks, Alaska, where he has worked since 1987. Rick has studied marine and terrestrial mammals and birds in Alaska for over 30 years, focusing on the effects of human disturbance on habitat use, behavior, and productivity of wildlife. He participated in studies of marine birds, river otters, and mink after the *Exxon Valdez* oil spill, and was co-principal investigator on sea otter research. Rick has a B.S. in wildlife biology from the University of Montana and an M.S. in wildlife management from the University of Alaska–Fairbanks.

David K. Johnson is a Senior Staff Research Specialist with Exxon Mobil Corporation, where he has held various scientific and technical-information positions since 1976. His involvement with the *Exxon Valdez* oil spill began early Monday morning, March 27, 1989, doing literature searches on the effects of crude oil on biota. Since the summer of 1990, he has provided and managed information resources for the *Exxon Valdez* science programs and its consulting scientists. David has a B.S. in chemistry from the University of Toledo, an A.M. in organic chemistry from Princeton University, and studied information science and communications at Rutgers University's Graduate School of Communication, Information, and Library Studies.

Alan W. Maki served as Senior Environmental Scientist for Exxon in Alaska from 1985 to 2006. He was responsible for organizing the scientific assessment of ecological damage and recovery from the *Exxon Valdez* oil spill. He received his Ph.D. in wildlife and fisheries management from Michigan State University. He has authored and co-authored over 250 publications and reports and six books on various aspects of environmental quality. He is a former member of the US Environmental Protection Agency–Science Advisory Board and has served on many advisory panels for the EPA Office of Research and Development and on National Academy of Science panels concerned with the assessment and management of ecological risks. He is a former President of the Society of Environmental Toxicology and Chemistry.

Larry L. Moulton is a fisheries biologist and owner of MJM Research in Lopez Island, Washington. He has worked on fisheries issues and research throughout Alaska since 1978. He received his M.S. and Ph.D. degrees in fisheries science from the College of Fisheries at the University of Washington.

Stephen M. Murphy is a Senior Scientist and Principal at ABR, Inc.–Environmental Research & Services in Fairbanks, Alaska, and a Research Associate at the Institute of Arctic Biology at the University of Alaska–Fairbanks. He has over 30 years' experience studying the effects of anthropogenic disturbances on wildlife. He has a B.S. in wildlife biology from the University of Rhode Island and an M.S. in wildlife management from the University of Alaska–Fairbanks.

Jerry M. Neff is an internationally recognized authority on the fate and effects of petroleum hydrocarbons in marine, freshwater, and terrestrial environments. During the past 40 years, he has conducted research on the toxicity and ecological effects of petroleum and participated in 16 oil-spill environmental assessments worldwide, including the *Amoco Cadiz* and *Exxon Valdez* oil spills. He has published more than 200 scientific articles and three books, and co-edited two books dealing with the environmental fates and biological effects of petroleum in marine and freshwater ecosystems. Jerry received a B.S. in biology from Antioch College and a Ph.D. in zoology and biochemistry from Duke University.

James T. Oris is Professor of Zoology and the Associate Provost for Research and Scholarship and the Dean of the Graduate School at Miami University in Ohio. He has over 30 years' experience in the ecological risk assessment of fossil fuel components in freshwater and marine systems, with a focus on biochemical and population-level impacts of ultraviolet radiation and polycyclic aromatic hydrocarbons in fish. He earned a B.A. in biology from Wittenberg University and his Ph.D. in environmental toxicology and fisheries and wildlife from Michigan State University. He has served as president of the Society of Environmental Toxicology and Chemistry (SETAC) North America and received its Eugene Kenaga Membership Award. He was a member of the Ecological Processes and Effects Committee of the US Environmental Protection Agency's Science Advisory Board and is currently a member of the National Academies of Science Committee on Human and Environmental Exposure Science in the 21st Century.

Edward H. Owens is the Owner of Owens Coastal Consultants, Ltd., in Bainbridge Island, Washington. He is a coastal geomorphologist who has published extensively on the fate, behavior and cleanup of oil on shorelines. He has been involved with shoreline response operations worldwide since the *T/V Arrow* spill in 1970 and was the Shoreline Cleanup Technical Advisor on the *T/V Exxon Valdez*, on which he pioneered the Shoreline Cleanup Assessment Technique (SCAT) process, and the Deepwater Horizon operations. He earned degrees from the University College of Wales, Aberystwyth (B.Sc., Honours, physical geography), McMaster University (M.Sc., geography), and the University of South Carolina (Ph.D., geology).

David S. Page is the Charles Weston Pickard Professor of Chemistry and Biochemistry Emeritus at Bowdoin College in Brunswick, Maine. Since 1975, David has more than 125 publications, most on aspects of the fate and effects of petroleum and other contaminants on marine environments, including over 60 on the *Exxon Valdez* oil spill. He earned his B.S. in chemistry from Brown University and his Ph.D. in physical chemistry from Purdue University.

Keith R. Parker has 35 years of experience assessing environmental impacts to marine environments of North and South America, Africa, and Russia. He has published 40 peer-reviewed papers on anthropogenic effects on fisheries, marine mammals, seabirds, and shoreline and benthic ecology. He has organized sessions for assessing accidental impacts for the International Biometrics Society. Keith received an M.S. in biostatistics from the University of Washington.

Walter H. Pearson, the owner of Peapod Research in Allendale, Michigan, is Senior Environmental Consultant with Stantec Consulting Services, Inc., and previously conducted environmental research for 25 years at the Battelle Marine Sciences Laboratory in Sequim, Washington. He has led multidisciplinary, multi-organizational studies that address the effects of oil and other stressors on fisheries resources and marine and aquatic environments. He has studied five oil spills. He has a Ph.D. in oceanography from Oregon State University.

Gary A. Pope is the Director of the Center for Petroleum and Geosystems Engineering at the University of Texas at Austin, where he has taught since 1977 and holds the Texaco Centennial Chair in Petroleum Engineering. Previously he worked in production research at Shell Development Company. Gary earned a B.S. from Oklahoma State University and a Ph.D. from Rice University, both in chemical engineering. He was elected to the National Academy of Engineering in 1999 for his contributions to understanding multiphase flow and transport in porous media and applications of these principles to improved oil recovery and aquifer remediation. He has authored or co-authored more than 250 technical papers.

P. Douglas Reimer is a principal with EML Environmental Mapping, Ltd., In Saanichton, British Columbia, Canada. He has been involved with coastal resource analysis, data collection, sensitivity mapping, and oil-spill responses worldwide since 1980. He was a SCAT team leader and a major participant in the aerial videotape program on the *Exxon Valdez* spill. He has participated in multiple spill responses internationally, including the Deepwater Horizon as a team leader and SCAT data manager. He earned a B.Sc. in geology from the University of British Columbia and has worked extensively with governments and industry over the last 30 years.

Aaron P. Roberts is an Associate Professor of Environmental Science at the University of North Texas. His areas of expertise and research include the use of biomarkers in environmental assessment, photoinduced toxicity of environmental contaminants, and the impacts of mercury on fish species. He earned his B.S. (biological sciences) from the University of Missouri and M.S. and Ph.D. (both in zoology) from Miami University.

Laura R. Rustin has been an information consultant to ExxonMobil for more than 30 years. She has provided reference assistance, document delivery, information management, and database services in support of *Exxon Valdez* scientific work since the summer of 1989. Laura indexed the book Exxon Valdez *Oil Spill: Fate and Effects in Alaskan Waters*. Laura has a B.A. from Brooklyn College in history, an M.A. in library science from the University of Wisconsin, and an M.A. in history from New York University.

Chris B. Wooley is the owner of Chumis Cultural Resource Services in Anchorage, Alaska. He has been working on archaeological, anthropological, and cultural-resource-management projects in Alaska and British Columbia since 1982. He participated in the 1989 SCAT archaeological surveys during the *Exxon Valdez* response, was Assistant Director of the Exxon Cultural Resource Program, and has responded to other oil spills affecting remote Alaska shorelines. He has an M.A. in anthropology from Washington State University.

USE OF ACRONYMS

Acronyms are often mystical and cryptic, known only to those who work in a discipline. They interrupt the flow of a text, leading the reader to lurch from phrase to phrase. An acronym in a sentence is like a speed bump in a road. At the same time, some acronyms are widely understood and, once grasped, allow one to read a passage without getting bogged down in cumbersome terminology or labels. Here we list the acronyms that will appear frequently throughout this book; all will also be defined on their first mention in a chapter.

ADEC	Alaska Department of Environmental Conservation
ADFG	Alaska Department of Fish and Game
ASTM	American Society for Testing and Materials
BTEX	Benzene, toluene, ethylbenzene, xylenes
CERCLA	Comprehensive Environmental Response, Compensation, and Liability Act
CYP1A	Cytochrome P450 1A
EPA	United States Environmental Protection Agency
ERA	Ecological risk assessment
EROD	7-ethoxyresorufin-o-deethylase
GC-FID	Gas chromatography–flame ionization detection
GC-MS	Gas chromatography–mass spectrometry
GIS	Geographic Information System(s)
GOA	Gulf of Alaska
FAC	Fluorescent aromatic compounds
MDL	Method detection limit
MAYSAP	May 1991 shoreline assessment program
NMFS	National Marine Fisheries Service
NOAA	National Oceanographic and Atmospheric Administration
NRDA	Natural Resource Damage Assessment
OPA 90	Oil Pollution Act of 1990
PAH	Polycyclic aromatic hydrocarbon (singular or plural)
ppb	Part per billion; $ng \cdot g^{-1}$ or $\mu g \cdot L^{-1}$
ppm	Part per million; $\mu g \cdot g^{-1}$ or $mg \cdot L^{-1}$
PWS	Prince William Sound
SCAT	Shoreline Cleanup Assessment Technique
SEP	Shoreline ecology program
SOR	Surface oil residues
SSOR	Subsurface oil residues
TEH	Total extractable hydrocarbons
TPAH	Total PAH
TPH	Total petroleum hydrocarbons
USCG	United States Coast Guard
USFWS	United States Fish and Wildlife Service

ACKNOWLEDGMENTS

During the more than 20 years since the *Exxon Valdez* oil spill, hundreds of people have contributed in one way or another to the scientific studies that are the foundation of this book. Field teams surveyed oil on shorelines, collected water and hydrocarbon samples, inventoried cultural sites, and conducted counts and surveys of shoreline biota, fish, birds, and sea otters, all the time with extraordinary attention to proper quality control and data management. Ocean Explorer and its captains and crews (especially some awesome cooks!) of many vessels provided the logistical support that made the studies possible. They took the scientists and field crews to all sorts of places, in all kinds of weather, with a deep commitment to safety. Several laboratories, especially Battelle in Duxbury, Massachusetts, conducted sophisticated analyses of environmental and tissue samples for hydrocarbons, again with careful attention to precision, accuracy, and quality control. Statisticians, particularly John Skalski (University of Washington) and Jim Harner (West Virginia University), helped design ways of analyzing complex and uneven data sets and enhanced the statistical reliability of the results.

Placing the science and the data in context requires a wide variety of background information. The staff of the Alaska Resources Library and Information Services in Anchorage, especially Carrie Holba, have been unfailingly helpful. Valuable assistance was provided by personnel with The *Exxon Valdez* Oil Spill Trustee Council; Anchorage's Z. J. Loussac Public Library and its Alaska Collection; and the information and library staff of agencies involved with the *Exxon Valdez* spill, especially Nancy Tileston. Susan S. Howison (Exxon Mobil Corporation) provided all manner of support. For over 23 years, Laura Rustin (L. R. Rustin Research Consulting) has provided unexcelled library and information support, answering reference questions from the simple to the bizarre (often before they were asked), obtaining the most obscure documents, indexing, abstracting, building databases, and much more.

Bringing all the science together into a book rests on the contributions and support of many individuals and organizations other than those whose names appear on the chapter headings. The many Trustee scientists who also conducted studies – and those who supported them – have made important contributions to our knowledge of Prince William Sound, the *Exxon Valdez* oil spill, and its effects. Many of the authors of chapters in this book contributed to its development well beyond the domains of their individual chapters by sharing their findings and perspectives and participating in wide-ranging discussions. Many authors were supported by their home institutions or organizations during the development of their chapters. Steve Moffitt and Sherri Dressel of the Alaska Department of Fish and Game provided the data and models used to update herring recruitment figures in Chapter 13. Carla Christofferson, Dawn Sestito, and Reuben Wilson (O'Melveny & Myers, LLP) and Barat LaPorte (Patton Boggs, LLP) provided critical input on legal issues and their background.

Finally, several individuals deserve special mention, for without their efforts this book would not exist. Dominic Lewis (Cambridge University Press) helped guide us through the publication process. Mike Smith has been a steadfast supporter of science within Exxon Mobil Corporation and helped spearhead this project. David K. Johnson (Exxon Mobil Corporation) provided invaluable help to the editor and authors during the development of the book. Allison Zusi-Cobb (ABR, Inc.) and Betty Dowd (Exponent,

Inc.) applied their drafting and artistic talents to produce the maps and figures, respectively. Kyra Wiens brought a keen sense of style to bear on the prose through her detailed and sensitive editing, and she and Laura Rustin worked assiduously to create uniformity in referencing and ensure the accuracy of literature citations.

All of us who have contributed to this book, and have worked to learn the lessons of the *Exxon Valdez* oil spill, thank all of these individuals, teams, and organizations, and the many others too numerous to mention. This book is theirs as much as ours.

Financial support to conduct these studies, analyze and interpret the results, and publish and communicate the findings over the years has been provided by Exxon Mobil Corporation and the *Exxon Valdez* Oil Spill Trustee Council. The contents of this book, however, do not necessarily reflect the views of the funding sources.

A BIBLIOGRAPHIC NOTE

David K. Johnson and Laura R. Rustin

Over the last 23 years, we have obtained over 20 000 documents requested by the hundreds of Exxon-associated scientists, engineers, and others working on the *Exxon Valdez* oil spill. If nothing else, we have come to deeply appreciate accurate – and *useful* – literature citations that actually help you lay hands on a document. Consequently, we have paid particular attention to the references cited in this volume and have done our best to verify authors, titles, volumes, issues, years, journal names, publishers, agencies, page ranges, etc. We have tested all the URLs included in text and citations.[1]

The body of *Exxon Valdez* literature is large, and it continues to grow. In mid-March 2012, a very broad online literature search for *Exxon Valdez* oil spill citations carried out in 20 scientific and technical databases on the Chemical Abstract Service's STN® bibliographic database service retrieved roughly 2000 citations to scientific papers (after removing duplicates using STN®'s algorithm).[2] We estimate that 800–900 may be unique citations directly relevant to the *Exxon Valdez* oil spill. We say "may be" because "*Exxon Valdez*" has, among other things, become a metaphor and simile; a point in time ("not since the"); a byword for magnitude ("larger/smaller than the"); and a term of comparison ("like/unlike the"). Thus, one has to actually look at retrieved citations to determine their relevance. In addition to these papers, there are conference presentations, books, technical reports, dissertations, and theses.

A comprehensive bibliography of the *Exxon Valdez* literature, too large for inclusion in this printed volume, is available online at www.cambridge.org/9781107027176.

There are three important works that summarize *Exxon Valdez* scientific studies:

- **Wells *et al*.** (1995): papers from a symposium of the American Society for Testing and Materials reviewing the work of Exxon scientists.
- **Rice *et al*.** (1996): papers from a symposium of the American Fisheries Society reviewing the work of *Exxon Valdez* Oil Spill Trustees' ("Trustees") scientists.
- **Integral Consulting, Inc.** (2006): a critical review of both Exxon's and Trustees' work relating to the recovery of the spill area. Exxon and Trustee scientists were interviewed, and their relevant published papers were examined.

One of the goals of this book is to complement and update these three works.

Other key resources include:

- ***Exxon Valdez* Oil Spill Trustees' Website** (www.evostc.state.ak.us): includes Trustee annual and final reports, plus a great deal of other information. This is an important source because it includes significant information that does not appear in the peer-reviewed scientific literature. There is an online method for searching for specific Trustee projects by year, researchers, injured resource or service, project numbers, titles, or keywords.

---

[1] The internet URLs (Uniform Resource Locators) cited in the book were tested in January 2013.
[2] Books, dissertations and theses, patents, technical reports, and other non–journal-paper citations were removed. Also, social science, law, business, and humanities papers were not included in the search as STN® does not cover this literature.

- **The Alaska Resources Library and Information Service** (ARLIS; www.arlis.org): a consortium of Alaskan federal and state libraries in Anchorage, including: the *Exxon Valdez* Oil Spill Trustee Council; National Park Service; US Geological Survey; US Fish and Wildlife Service; US Bureau of Land Management; University of Alaska–Anchorage; Alaska Department of Fish & Game; and the US Bureau of Ocean Energy Management. ARLIS has a large collection of Trustee and governmental agency materials. It has a joint, online library catalog with the Anchorage Public Library, the University of Alaska–Anchorage/Alaska Pacific University Consortium Library, and the Anchorage Museum's Bob and Evangeline Atwood Alaska Resource Center. The staff are extremely knowledgeable and helpful.
- **The University of Alaska–Fairbank's Elmer E. Rasmuson Library's Alaska and Polar Regions Collection**: a repository of archaeological artifacts found during the spill cleanup and records from Exxon's *Exxon Valdez* Cultural Resource Program (Chapter 5).
- **Valdez Sciences Website** (www.valdezsciences.com): an Exxon website that has reports and data particularly relevant to this volume, including Prince William Sound polycyclic aromatic hydrocarbons (PAH) in water data (2001–08) (Chapters 3, 12, 13); PAH data particularly related to pink salmon (*Oncorhynchus gorbuscha*) and Pacific herring (*Clupea pallasii*) (Chapters 12, 13); and the results of Exxon's 1991 May Shoreline Assessment Program (MAYSAP), which are important for later studies of the persistence of subsurface oil residues (Chapters 4, 6). The site also includes brochures and reports issued in the early days of the spill that were intended for a more general, less scientific audience.
- **Scientific meetings and conferences**: a considerable body of *Exxon Valdez* work has been reported by all parties at three regular meetings:
  - *International Oil Spill Conference (IOSC)*, organized by the US Coast Guard, the US Environmental Protection Agency, and the American Petroleum Institute (with other occasional sponsoring agencies groups), publishes proceedings with full papers.
  - *Arctic and Marine Oilspill Program (AMOP) Technical Seminar*, organized by Environment Canada, publishes proceedings with full papers.
  - *Annual Meeting of the Society for Environmental Toxicology and Chemistry (SETAC)* produces an abstract book but does not publish full papers.

REFERENCES

Integral Consulting, Inc. (2006). *Information Synthesis and Recovery Recommendations for Resources and Services Injured by the* Exxon Valdez *Oil Spill*. Mercer Island, WA, USA: Integral Consulting, Inc.; *Exxon Valdez* Oil Spill Restoration Project 060783 Final Report. [http://www.evostc.state.ak.us/Files.cfm?doc=/Store/FinalReports/2006–060783-Final.pdf&]

Rice, S. D., R. B. Spies, D. A. Wolfe, and B. A. Wright, eds (1996). *Proceedings of the* Exxon Valdez *Oil Spill Symposium*. Bethesda, MD, USA: American Fisheries Society; Symposium 18; ISBN-10: 0913235954.

Wells, P. G., J. N. Butler, and J. S. Hughes, eds (1995). Exxon Valdez *Oil Spill: Fate and Effects in Alaskan Waters*. Philadelphia, PA, USA: American Society for Testing and Materials; ASTM Special Technical Publication 1219; ISBN-10: 0803118961.

PROLOGUE

Shortly after midnight on March 24, 1989, the *Tanker/Vessel Exxon Valdez*, fully loaded with its cargo of Alaska North Slope crude oil, grounded on Bligh Reef in Prince William Sound, Alaska. Eight of its 11 cargo tanks ruptured, releasing some 11 million gallons (40 million liters) of oil – about 20% of its cargo – into the icy waters of the Sound. The floating oil rapidly spread, pushed by strong coastal currents. Three days later, a severe winter storm moved in, widening and accelerating the spread of oil and thwarting attempts at containment. As the floating oil reached shorelines, it began coating beaches and intertidal algae and animals with layers of thick, black oil. Over 200 000 seabirds may have died. Commercial fisheries were closed. The spill eventually extended down the Kenai and Alaska peninsulas to beyond Kodiak Island, roughly the distance between Boston and Washington, District of Columbia.

It would be hard to imagine a worse place for an oil spill. Prince William Sound is widely regarded as pristine, and the remoteness and climate of the region present formidable challenges to mobilizing spill responses and cleanup. It supports large populations of charismatic wildlife: sea otters, seabirds, whales, bears, and seals. Tourism, cruises, recreation, and sport fishing are significant activities, and there are economically important commercial fisheries. Alaska Natives also rely heavily on the area for subsistence harvesting.

Because of the spill's magnitude and extent, the natural beauty of coastal Alaska and Prince William Sound, and the involvement of one of the world's largest corporations, the spill commanded intense public and media attention for years. Images of oiled sea otters and seabirds and of shorelines awash with oil were played and replayed. Speculations about long-term devastation of populations were rampant. Even now, more than 20 years later, the *Exxon Valdez* oil spill remains the benchmark against which other oil spills are often measured.

As the oil spread, there were urgent calls for action. Understanding and documenting how the environment and natural resources were affected by the spill demanded the rigor, objectivity, and clarity of science. But the normal scientific process – framing hypotheses, designing experiments, conducting tests, marshaling and managing data, developing models, and the like – takes time. It was not well-suited to responding to a rapidly unfolding environmental disaster in a wild and contentious setting. And communication in 1989 was not what it is now: there were no cell phones, no GPS, no internet access, little capacity for high-speed data processing, and no Facebook or Twitter. Sometimes we didn't know quite where we were – especially when the fog rolled in.

Despite these challenges, science-based studies were quickly implemented, often on the fly. In many cases, the studies involved using emerging technology that had never been applied to such problems or developing new study designs or techniques to disentangle the effects of the oil spill from everything else going on in a harsh and variable environment. The *Exxon Valdez* spill became the most intensively studied oil spill in history. Multiple scientific studies were conducted by multiple parties, at a cost of multiple millions of dollars, producing multiple (and sometimes contradictory)

Figure 1 The *Exxon* Valdez aground on Bligh Reef, Prince William Sound, Alaska, on March 26, 1989. This view, two days after the spill, is toward the northeast. Reef Island is in the immediate background, with Bligh Island behind that and Ellamar Mountain in the background. (Photo: Erich R. Gundlach).

conclusions. Most of this work focused on injuries to wildlife and the environment and their subsequent recovery from the spill. Although there were substantial initial effects of the spill, most vanished within a few years. Now, more than two decades later, oil from the *Exxon Valdez* has disappeared from all but a tiny portion of the shoreline. Debates continue, however, about whether animals are still exposed to oil residues and, if so, whether there are any harmful effects.

This book is about how scientists unraveled the consequences of a huge environmental disruption – the *Exxon Valdez* oil spill. How were studies designed and conducted? What are the limitations of science? What lessons were learned? The insights and lessons we share in this book are intended to provide guidance to those challenged with responding to other oil spills. But they can also apply to other large environmental disturbances, both natural and human-caused.

Our emphasis throughout this book is on how the tools and approaches of the physical and biological sciences have been applied to understand the dynamics of oil in the environment, oil's effects on the biota, and ecological recovery. With the exception of spill effects on archaeological and related historical/cultural sites, we do not consider studies in the domains of the economic, social, and legal sciences, nor do we include food safety or human health and safety. These aspects are covered extensively elsewhere (for example, Smith, 1992; Hausman, 1993; Jones *et al.*, 1994; Owen *et al.*, 1995; Picou *et al.*, 1997; Field *et al.*, 1999; Sunstein *et al.*, 2002; Carson *et al.*, 2003) and are included in the comprehensive bibliography of *Exxon Valdez* studies that is available for downloading from the publisher's website.[1]

The chapters in this book have been written by scientists from multiple disciplines, most of whom initiated studies shortly after the *Exxon Valdez* spill and who have been involved in assessing its effects ever since. As the studies developed, we shared findings, insights, and impressions with one another, expanding the scope of individual studies well beyond the boundaries of traditional disciplines. We also shared ship time, the beauty (and the weather!) of Prince William Sound, and, yes, fish stories. We developed the collegial, collaborative relationships that are as much a part of good science as statistics or study design.

Our investigations were funded by Exxon Corporation (and, since November 30, 1999, by Exxon Mobil Corporation).[2] From the outset, Exxon insisted on the highest standards of scientific integrity. The approach was to enlist respected, independent-minded scientists to design and implement research to determine the spill's effects and then document the extent of recovery with as much certainty as possible, separating fact from fiction and science from speculation. The association with Exxon, however, raised concerns by some of the public (and some scientists) that the work of "Exxon scientists" was biased or intended solely to cast doubt on the findings of other scientists. The position of Exxon as a defendant in civil litigation with nongovernmental plaintiffs further enhanced the image of

---

[1] At www.cambridge.org/9781107027176.
[2] So too, albeit indirectly, were the studies carried out by scientists engaged by the *Exxon Valdez* Oil Spill Trustee Council, whose work is funded by the $686.9 million paid by Exxon in its civil settlement with the federal and Alaska governments for environmental damages (http://www.evostc.state.ak.us/facts/settlement.cfm).

PART I

INTRODUCTION AND BACKGROUND

Owen, B. M., D. A. Argue, H. W. Furchgott-Roth, G. J. Hurdle, and G. Mosteller (1995). *The Economics of a Disaster: The* Exxon Valdez *Oil spill*. Westport, CT, USA: Quorum Books; ISBN-10: 0899309879.

Payerhin, M., ed. (2004). *Alaska after* Exxon Valdez. *A Collection of Papers by the Students of Political Science 280*. Alma, MI, USA: Alma College, Department of Political Science; Spring 2004.

Picou, J. S., D. A. Gill, and M. J. Cohen, eds (1997). *The* Exxon Valdez *Disaster: Readings on a Modern Social Problem*, 2nd edn. Dubuque, IA, USA: Kendall/Hunt Publishing Company; ISBN-10: 0787256854.

Reichman, D. (1992). *Tanker on the Rocks or The Great Alaskan Bad Friday Fish Spill of '89*. Valdez, AK, USA: Prince William Sound Books; ISBN-10: 1877900028; ISBN-13: 9781877900020.

Rice, S. D., R. B. Spies, D. A. Wolfe, and B. A. Wright, eds (1996). *Proceedings of the* Exxon Valdez *Oil Spill Symposium*. Bethesda, MD, USA: American Fisheries Society; Symposium 18; ISBN-10: 0913235954; ISSN: 08922284.

Robinson, R. (1997). *Light All Night*. Edmonton, AB, Canada: Commonwealth Publications; Commonwealth edition 1997; ISBN-10: 1551970155; (Also, New York, NY, USA: iUniverse, Inc.; 2000; ISBN-10: 0595138845; ISBN-13: 9780595138845).

Schultz, M. (1992). *The Sun, Split like Spun Glass. Three Poems of Marianne Moore for Soprano and Chamber Orchestra*. Austin, TX, USA: JOMAR Press.

Smith, C. (1992). *Media and Apocalypse: News Coverage of the Yellowstone Forest Fires,* Exxon Valdez *Oil Spill, and Loma Prieta Earthquake*. Westport, CT, USA: Greenwood Press; Contributions to the Study of Mass Media and Communications, Number 36; ISBN-10: 0313277257; ISSN: 07324456.

Spencer, P. (1990). *White Silk & Black Tar. A Journal of the Alaska Oil Spill*. Minneapolis, MN, USA: Bergamot Books; ISBN-10: 0943127041.

Sunstein, C. R., R. Hastie, J. W. Payne, D. A. Schkade, and W. K. Viscusi (2002). *Punitive Damages: How Juries Decide*. Chicago, IL, USA: The University of Chicago Press; ISBN-10: 0226780147 (cloth); ISBN-10: 0226780155 (paperback).

University of Nebraska (2001). *Sea Otter Biologist: Brenda Ballachey*. Lincoln, NE, USA: University of Nebraska; *Wonderwise, Women in Science Learning Series*, 2nd edn; DVD, CD, and Activity Book; distributed by Destination Education, Inc., Lincoln, NE, USA.

Wells, P. G., J. N. Butler, and J. S. Hughes, eds (1995). Exxon Valdez *Oil Spill: Fate and Effects in Alaskan Waters*. Philadelphia, PA, USA: American Society for Testing and Materials; ASTM Special Technical Publication 1219; ISBN-10: 0803118961.

REFERENCES

Anonymous (1990). *The Two Billion Dollar Cookbook: A Collection of Anecdotes and Treasured Recipes from the Hearts and Homes of the Alaskan Oil Spill Cleanup Workers, their Families and Friends.* Anchorage, AK, USA: EFFC, Inc., Ken Wray's Printing, Inc.; ISBN-10 0963044001; ISBN-13: 9780963044006.

Bushell, S. and S. Jones, eds (2009). *The Spill. Personal Stories from the* Exxon Valdez *Disaster.* Kenmore, WA, USA: Epicenter Press; ISBN-13: 9780980082586.

Carson, R. T., R. C. Mitchell, M. Hanemann, R. J. Kopp, S. Presser, and **P. A. Rudd** (2003). Contingent valuation and lost passive use: damages from the *Exxon Valdez* Oil Spill. *Environmental and Resource Economics* 25: 257–286.

Chandrasekhar, A. (1991). *Oliver and the Oil Spill.* Kansas City, MO, USA: Landmark Editions, Inc.; ISBN-10: 0933849338.

Erickson, L. (1991). *The Northern Light.* Toronto, ON, Canada: Harlequin Books; Harlequin Superromance Volume 439; ISBN-10: 0373704399; ISBN-13: 9780373704392.

Field, L. J., J. A. Fall, T. S. Nighswander, N. Peacock, and **U. Varanasi**, eds (1999). *Evaluating and Communicating Subsistence Seafood Safety in a Cross-Cultural Context: Lessons Learned from the* Exxon Valdez *Oil Spill.* Pensacola, FL, USA: Society of Environmental Toxicology and Chemistry, SETAC Press; ISBN-10: 1880611295.

Frost, H., ed. (1990). *Season of Dead Water.* Portland, OR, USA: Breitenbush Books, Inc.; ISBN-10: 0932576826 (hardcover); ISBN-10: 0932576834 (paper).

Garshelis, J. S. (2009). *The Otter Spotters: A Wildlife Adventure in Alaska.* New York, NY, USA: iUniverse, Inc.; ISBN-10: 1440161305; ISBN-13: 9781440161308.

Hausman, J. A., ed. (1993). *Contingent Valuation: A Critical Assessment.* Amsterdam, The Netherlands: Elsevier Science Publishers; Contributions to Economic Analysis 220; ISBN-10: 0444814698.

Holleman, M. (2004). *The Heart of the Sound. An Alaskan Paradise Found and Nearly Lost.* Salt Lake City, UT, USA: The University of Utah Press; ISBN-10: 0874807913.

Integral Consulting, Inc. (2006). *Information Synthesis and Recovery Recommendations for Resources and Services Injured by the* Exxon Valdez *Oil Spill.* Mercer Island, WA, USA: Integral Consulting, Inc.; *Exxon Valdez* Oil Spill Restoration Project 060783 Final Report. [http://www.evostc.state.ak.us/Files.cfm?doc=/Store/FinalReports/2006–060783-Final.pdf&]

Jones, J. D., C. L. Jones, and **F. Phillips-Patrick** (1994). Estimating the costs of the *Exxon Valdez* oil spill. In *Research in Law and Economics.* R. O. Zerbe, Jr., ed. Greenwich, CT, USA: JAI Press; Volume 16; ISBN-10: 1559385006, pp. 109–149.

Larson, W. (2002). *The Sitka Incident:* Exxon Valdez *Retold.* Port Orchard, WA, USA: Windstorm Creative; ISBN-10: 1590920511; ISBN-13: 9781590920510.

Lynn, R. (1989). *On the Rocks: The Great Alaska Oil Spill.* Valdez, AK, USA; On the Rock Enterprises; Board Game.

Markle, S. (1999). *After the Spill. The* Exxon Valdez *Disaster Then and Now.* New York, NY, USA: Walker and Company; ISBN-10: 0802786103.

scientists funded by Exxon pitted against those working for government agencies or testifying for the plaintiffs. We address this issue in the concluding chapter of the book.

Doubt, disagreement, and criticism are essential parts of the scientific process, of course – this is why it is so important that the findings of investigations be subjected to peer review. Over 800 peer-reviewed scientific papers have been published about the spill, and the work continues. In addition, there are at least a dozen books, hundreds of presentations at scientific meetings, hundreds of technical reports, and over 70 theses and dissertations that deal with the spill.[3] Integral Consulting, Inc. (2006) also produced a comprehensive and independent review, sponsored by the *Exxon Valdez* Oil Spill Trustee Council ("Trustees"), which was based on interviews with governmental and Exxon-supported scientists and an evaluation of the findings in published literature and reports authored by all parties. As a result of all these efforts, we have a much better understanding of the dynamics of oil in marine environments and of how to design and conduct investigations. These studies provide a wealth of knowledge on which responses to future oil spills will depend.

In some respects, this book provides an update of the work presented in the proceedings of two symposia that covered the studies of Trustee scientists (Rice *et al.*, 1996) and Exxon-supported scientists (Wells *et al.*, 1995). Unlike other books about the *Exxon Valdez* oil spill, however, a major goal of this book has been to consider the published work of all investigators – regardless of their funding sources, affiliations, employment, or other considerations. The chapter authors, each an expert in his or her own area of study, have aimed to examine the full breadth of scientific work related to their area and provide interpretations that rest on the firmest foundations of the science. The studies covered in the following chapters evolved over time, often responding to or building on previous studies. The differences in approaches, assumptions, and results provide exceptional insights – not just into what happened, but also into how studies of such large environmental disruptions can and should be conducted.

Much was learned in the process of investigating the many facets of the *Exxon Valdez* oil spill, not only about the dynamics of oil in the environment and the ways in which the spill affected (or did not affect) natural resources, but also about the value and limitations of science in assessing the consequences of environmental disruptions on such a scale. These lessons, which are the capstone elements of this book, are perhaps the most enduring legacy of the *Exxon Valdez* oil spill.

---

[3] In addition to the scientific literature (including the social science literature), there are also *Exxon Valdez* oil spill related poetry, fiction, children's books, essays, photography, oral history, class projects, and memoirs (for example, Frost, 1990; Spencer, 1990; Chandrasekhar, 1991; Markle, 1999; University of Nebraska, 2001; Larson, 2002; Holleman, 2004; Payerhin, 2004; Bushell and Jones, 2009; Garshelis, 2009), and at least one made-for-TV movie (*Dead Ahead: The* Exxon Valdez *Disaster*, http://www.imdb.com/title/tt0104060/), one composition for soprano and chamber orchestra (Schultz, 1992), one cookbook (Anonymous, 1990), one melodrama (Reichman, 1992), one mystery (Robinson, 1995), one Harlequin romance (Erickson, 1991), and one board game (Lynn, 1989).

INTRODUCTION

Every oil spill is different from the ones that preceded it. This is due in part to differences in the spill itself – the nature of the oil and the circumstances of the release – but it also stems from differences in the context of the spill – the setting, the resources at risk, and how people respond to the spill. All of these factors affect how science is brought to bear on investigating the spill and its consequences.

The two chapters in this section consider these factors in two ways. Chapter 1 (to which many of the authors contributed) introduces the environmental setting of the *Exxon Valdez* oil spill, what happened during the spill and to the oil that was released, the cleanup responses and the launch of scientific studies, and the regulatory and legal context of the spill responses and studies. Separate discussions consider: the characteristics of crude oil; the importance of assessing exposure of biota to the oil in gauging potential effects; the thorny issue of defining "impact," "effect," "injury," and "recovery"; and the legal issues that followed from the spill.

In Chapter 2, Paul Boehm, Erich Gundlach, and David Page take a broader view. When oil is released in a marine environment, it immediately begins to undergo transformations resulting from a variety of physical and chemical processes. Although these changes are continuous, they can be separated into three overlapping phases that align with shifts in the focus of scientific studies over time. Recognizing these phases can help to plan which studies to conduct, and when. This, in turn, can facilitate an efficient and effective allocation of research efforts.

CHAPTER ONE

Introduction and background

John A. Wiens

1.1 Introduction

In the aftermath of an oil spill, the effects on the environment and wildlife are often painful to see. After the initial emotional impact come the questions: What wildlife and environments are at risk, and when will they recover? How can the oil be removed without causing further harm? Is it safe to eat the seafood or to be on the beaches? What will happen to the oil? Science can offer objectivity, rigor, and focus in addressing such questions, helping to separate fact from fiction, evidence from conjecture. A science-based approach defines potential spill effects and then formulates testable hypotheses, follows an unbiased study design, collects and analyzes data using rigorous methods, and interprets the results with a mind open to alternative explanations that evolve during the investigations. This is how good science is done.

Conducting science following the *Exxon Valdez* spill was not always easy, however. Along with everyone else, the first scientists on the scene were distraught over what they saw – shorelines awash with oil, oiled seabirds and sea otters (*Enhydra lutris*) struggling to survive, and fisheries closed for fear of contamination. It was challenging to come up with good, objective study designs. The remote location of the spill and the wide variation among places in the spill zone complicated data collection. Studies conducted at different times or of different durations produced different results, and relationships documented at different spatial scales did not always match. Study designs often seemed to be confounded by other factors or uncontrolled sources of variation at every turn, making it difficult to separate changes in the environment due to the oil spill from changes due to other, unrelated factors.

This book is about how the investigations of the *Exxon Valdez* oil spill carried out by multiple parties met these challenges. Much was learned in the process. These lessons and scientific insights provide essential guidance for designing responses and assessing the consequences of future spills or, indeed, of any large environmental disruptions.

In this introductory chapter, we set the stage by describing the geographical, physical, environmental, and cultural settings of the *Exxon Valdez* spill. We discuss the event and what happened to the oil over time. We emphasize the importance of

---

Oil in the Environment: Legacies and Lessons of the Exxon Valdez *Oil Spill*, ed. J. A. Wiens. Published by Cambridge University Press. © Cambridge University Press 2013.

documenting the pathways by which organisms might be exposed to oil or its toxic[1] components and the value of clear, operational definitions of *impact*, *effect*, *injury*, and *recovery*. Finally, we touch very briefly on the regulatory and legal context of responses to the spill.

1.2 The setting: the northern Gulf of Alaska and Prince William Sound

The northern Gulf of Alaska (GOA), in the northeastern corner of the Pacific Ocean, is bounded to the east by southeastern Alaska and to the west by Kodiak Island and the Alaska Peninsula (see Map 1, p. v). It is a diverse, highly productive ecosystem that includes one of the world's largest fisheries (Spies, 2007; Gaichas *et al.*, 2009).[2] Because Prince William Sound (PWS) was the most heavily oiled part of the northern GOA and the vast majority of scientific studies of the effects of the *Exxon Valdez* oil spill were conducted there, we focus on the Sound throughout this book.

1.2.1 Geography and geology

Prince William Sound is a large, semi-enclosed estuary ecosystem east of the Kenai Peninsula and south of the Chugach Mountains of the Pacific Coast Ranges (see Map 2, p. vi). It is a place of stunning natural beauty surrounded by rugged mountains (see frontispiece). Tidewater glaciers calve icebergs into its northern reaches. Human settlements are sparse and scattered – only five were documented in the 2010 US census: Valdez (population 3976), Cordova (2239), Whittier (220), Tatitlek (88), and Chenega (76). There are some 4800 km of shoreline in an area of only 9000 km² (Adams *et al.*, 2002), with forests, high tides, and winter storms restricting most rocky beaches to narrow strands. Islands and shorelines on the western side of PWS are sharply dissected by numerous bays and deep fjords, whereas topography and bathymetry on the eastern side of PWS are more gradual, and the shoreline is less dissected (see http://www.charts.noaa.gov/OnLineViewer/16700.shtml).

Water from the Alaska Coastal Current enters PWS from the GOA through Hinchinbrook Entrance and leaves through Montague Strait and Knight Island Passage (see Map 1, p. v) 2 to 3 weeks later (Royer *et al.*, 1990). PWS is also fed by melting glaciers in the surrounding watershed and by coastal rain and snowfall (which can exceed 4 m in normal years and more than 10 m in exceptional years). Most of the glaciers, particularly large ones such as the Columbia Glacier, are in the mountainous areas bordering PWS to the north and west. They are responsible for most of the freshwater entering PWS.

The complex flows and mixing of fresh and marine water, combined with daily tidal fluctuations of up to 6 m, create the foundation for substantial marine productivity. This productivity varies considerably among locations, but it is higher where the Alaska

---

[1] To paraphrase Paracelsus, who lived in the sixteenth century and provided the basis for both modern medicine and ecotoxicology, there is nothing that is not a poison – depending on the dose. The word "toxic" is used in various chapters of this book with this reality in mind – the chemicals associated with spilled oil can be toxic or they may not be, depending on the dose.

[2] Additional details about the oceanography, geology, and biology of the GOA are provided in Mundy (2005) and Spies (2007).

Figure 1.1 Examples of shoreline types in PWS. (a) Exposed bedrock/rubble, north Smith Island, 2007. (b) Sheltered bedrock/rubble, southeastern Herring Bay, 2007. (c) Exposed boulder/cobble, northeastern Evans Island, 2002. (d) Mixed pebble/gravel, Foul Pass, Ingot Island, 2004 (Photos: David S. Page).

Coastal Current creates warmer and more saline waters to the east and south than in the glacial waters to the north and west (Royer et al., 1990).

The principal sources of subtidal sediments in PWS are streams and rivers to the east of PWS. Some of these sediments come from areas containing active natural petroleum seeps, such as those to the east of the Copper River delta near Katalla (Map 2, p. vi; Chapter 6). In addition, the Copper River discharges an estimated 107 million tons of sediment into the Alaska Coastal Current annually (Reimnitz, 1966). Suspended sediments are transported westward along the coastline, some entering PWS through Hinchinbrook Entrance (Royer et al., 1990), where they fall to the seafloor. Local sediment input from glaciers, by comparison, is small (Sharma, 1979).

Shorelines in PWS are generally narrow, although there are large tidal flats in some eastern parts of PWS. Metamorphic rocks form island landmasses in the central part of western PWS. The intertidal zone is often steep, particularly on islands bordered by deep fjords. Hillsides above the tidal zone are similarly restricted by rocky cliffs that abut dense forest. The underlying geology is composed of sedimentary rocks that are interbedded with volcanic rocks (both extrusive and intrusive; Wilson and Hults, 2008). Shoreline substrates in PWS are diverse, ranging from small pebbles, gravel, sand, and mudflats (in a few places) to large boulders, exposed rocky shelves, cliffs, and steep talus slopes (Fig. 1.1).

Shoreline substrates in western PWS differ from those of eastern PWS. The spill zone shoreline – located entirely in western PWS – consists largely of sheltered and exposed bedrock and bedrock/rubble, with boulder/cobble/gravel and mixed pebble/gravel beaches making up almost all of the remaining shoreline (Chapter 6). The sand/gravel beaches that dominate the eastern part of PWS are largely absent (Page et al., 1995).

Most shorelines are eroding; consequently, unless protected by surface bedrock/rubble or boulder/cobble/gravel armoring, shoreline sediments tend to be coarse.

1.2.2 The environment

Severe winter storms buffet the shorelines of PWS. Where wave exposure is low, the distribution of shoreline plants and invertebrates is determined largely by competition for space, predator–prey interactions, and resistance to physical stress. Where wave exposure is high, loose sediment particles – or even small boulders – can become missiles in high waves (Shanks and Wright, 1986). Logs, ice, and glacial bergs may further scour the shoreline, disrupting biotic communities and creating spaces open to colonization. The dominance of physical stress and disturbance results in a mosaic of species whose existence in a location is less a function of their competitive edge than their colonizing ability, the timing of their breeding cycle, and the vagaries of wave action and shoreline disturbance (Connell, 1961; Paine, 1966; see Chapter 11, Fig. 11.1).

Weather and ocean conditions in PWS are affected by broad-scale oceanographic phenomena, such as El Niño–Southern Oscillation events (Mantua et al., 1997; Mantua and Hare, 2002), as well as changes operating over longer periods (Finney et al., 2002). El Niño events have become more frequent, and La Niña events less frequent, since the 1970s (Trenberth and Hoar, 1996, 1997). Recent warming and freshening of waters in the northern GOA, possibly associated with increased coastal freshwater flows from melting glaciers and wind forcing associated with climate change, have led to a westward shift in oceanic isotherms that may have broad-scale, regional impacts (Royer and Grosch, 2006). Sudden shifts in atmospheric and broad-scale oceanographic conditions ("regime shifts") have occurred in the northern Pacific several times in the last 50 years (Peterson and Schwing, 2003; Litzow, 2006), leading to shifts in species' distributions, community composition, and food chains (Francis and Hare, 1994; Anderson and Piatt, 1999; Trites et al., 2006; Springer, 2007). In addition to the large regime shifts of 1976–77, major regime shifts occurred in 1989 (the year of the *Exxon Valdez* spill) and 1998 (Hare and Mantua, 2000; Overland et al., 2008). The Pacific Decadal Oscillation, which has broad-scale effects on sea-surface temperatures, has also undergone at least one phase shift since 1989 (Yatsu et al., 2008). This phase shift caused changes in freshwater input into the northern GOA, affecting salmon and other keystone species (Royer et al., 2001).

Spatial variation is also pronounced. At a broad scale, the eastern and western parts of PWS differ not only in topography and depth, but also in habitat, as exemplified by variation in the number and length of salmon spawning streams (the east has more, longer streams; Wiens et al., 2010; Alaska Department of Fish and Game, 2012). At a finer scale of tens of kilometers, areas such as parts of Knight Island (see Map 3, p. vii) differ in both physical features (e.g., number of islands and islets, shoreline substrate, steepness of the adjacent land) and biological attributes (e.g., cover of intertidal rockweed, *Fucus* spp.). At even finer scales, stretches of shoreline may have a heterogeneous mixture of habitats within even a few tens of meters (e.g., Fig. 1.1c).

Natural ecosystems are unstable, and not all changes are gradual: single, unpredictable events can have profound impacts. This is perhaps nowhere more evident than in portions of western PWS, where the 5-minute, 9.2-magnitude Great Alaska Earthquake on March 24, 1964, and the consequent tsunami fundamentally altered the geomorphology, bathymetry, and ecology of PWS (Plafker, 1965; Haven, 1971; Losey, 2005). Many parts of PWS were uplifted, some, as around Montague Island, by as much as 10 m

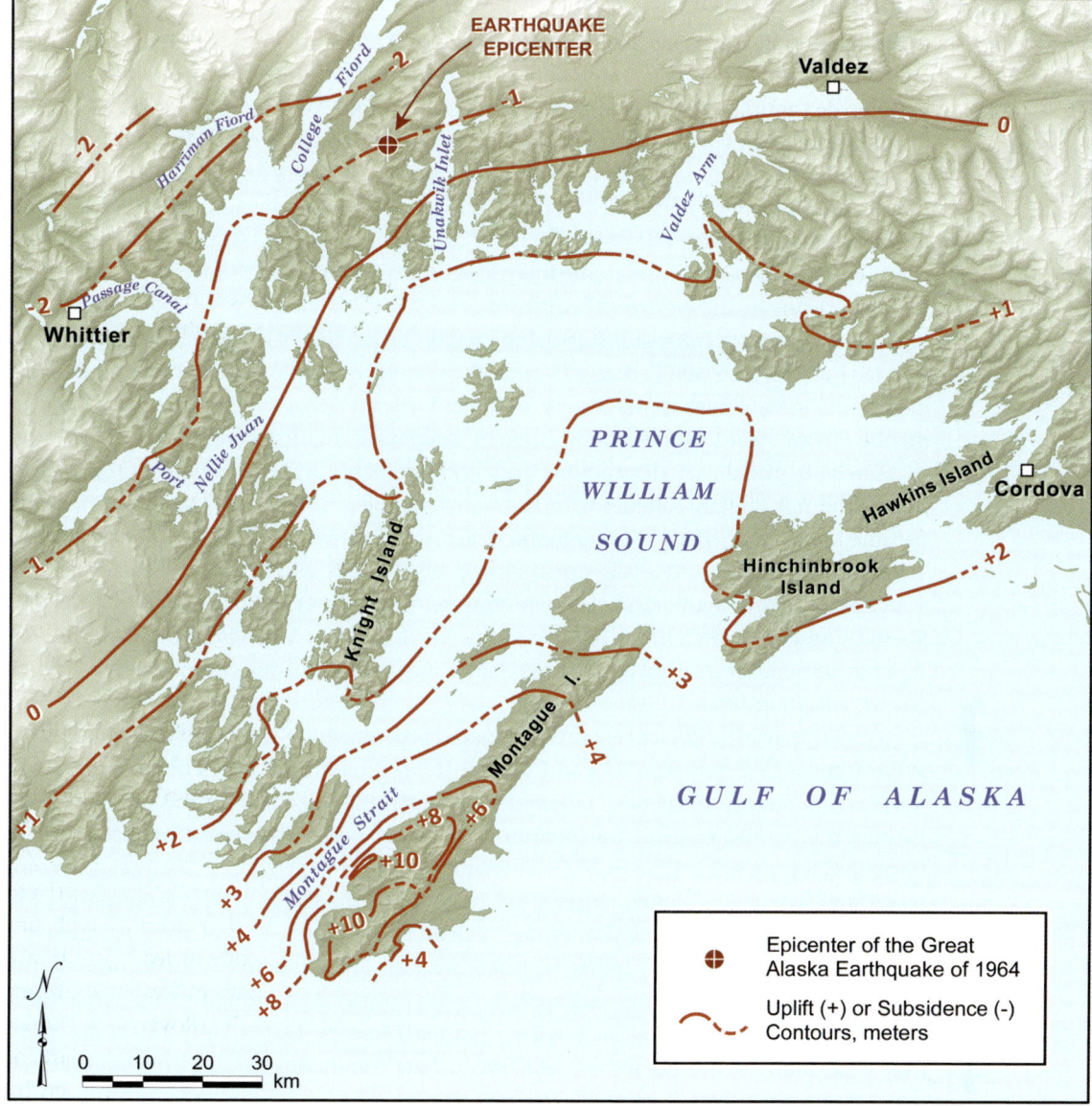

Figure 1.2 Ground deformation in Prince William Sound resulting from the 1964 Alaska earthquake (adapted from Plafker, 1969).

(Reimnitz and Marshall, 1965). The extent of uplift or subsidence varied substantially among different parts of PWS (Fig. 1.2; Haven, 1971). These changes had long-term effects on habitat suitability for a variety of organisms (Losey, 2005; Blanchard et al., 2010).

The earthquake and tsunami also destroyed settlements and structures throughout PWS, including much of the towns of Valdez and Old Chenega Village (which was abandoned as a result), as well as canneries, sawmills, mines, and other facilities. Ruptured fuel-storage tanks released significant quantities of petroleum products into intertidal shoreline and subtidal areas (Wilson and Torum, 1972). These processes continue even now (Chapters 5, 6).

1.2.3 Human history

Although people have occupied and altered coastal areas of PWS for millennia, human cultures of the ancient past are known from only a handful of archaeological excavations (de Laguna, 1956; Yarborough and Yarborough, 1996; Yarborough, 1997). These archaeological sites record eons of social adaptations to an ever-fluctuating maritime environment, including the remarkable skin-boat and sea-mammal hunting cultures developed by ancient maritime people (Fitzhugh and Chaussonnet, 1994; Yarborough and Yarborough, 1998). The oldest phase of occupation dates to 4400–3300 years before the present, and several other cultural phases followed before contact with Europeans and Americans in the eighteenth century.

The "postcontact" cultural history of the region began in 1741, when a Russian expedition led by Vitus Jonassen Bering, a Danish-born navigator in the Russian navy, landed just outside of PWS at Kayak Island. Georg Wilhelm Steller, a German naturalist and physician on the voyage, went ashore briefly (thereby becoming the first European to set foot on Alaskan soil), noted a small structure, and left trade beads and other items as token gifts, but he did not directly interact with the native people. (Steller also described what later became known as Steller's jay, *Cyanocitta stelleri*, and Steller's sea lion, *Eumetopias jubatus*.)

Russia's interest in PWS was not renewed until 1760, when sea otters had become scarce around the Aleutian Islands, which were much closer to Russia. In 1781, Grigory Shelikov and other merchants organized the North-Eastern America Company. The first Russian settlement in Alaska, at Three Saints Bay on the southeastern coast of Kodiak Island, was founded in 1784.

Meanwhile, fearing Russian expansion, Spain had been sending expeditions from Mexico as far north as PWS between 1774 and 1791. In 1790, explorer Salvador Fidalgo named Port Valdez. Also during this period, England sent Captain James Cook to find the Northwest Passage that would connect the Atlantic and Pacific oceans through the Arctic Ocean. Captain Cook arrived in PWS in 1778 and named many of its geographic features: Hinchinbrook and Montague islands, Bligh[3] Reef (the site of the *Exxon Valdez* grounding), and Sandwich Sound (later changed to Prince William Sound by the editors of Cook's maps).

The late 1700s to early 1800s featured warfare and epidemic disease among the indigenous people, creating widespread social upheaval. At roughly the same time, the maritime fur trade brought English, Spanish, French, Dutch, and American ships. The local, native cultures were fused into a new quasi-commercial, culturally mixed subsistence economy. Russian Orthodox communities of mixed Alutiiq, Aleut, and Euro-American ethnicity began to predominate, with the Alutiiq and Aleut members coming from areas to the west to hunt sea otters and seals.

Industrial-scale commercial whaling began in the mid-1800s, followed in the 1880s by the rapid development of commercial fishing and fish processing. Serbian, Italian, Scandinavian, and Midwestern American cultural influences affected life in PWS. Copper prospecting and mining, fur ranching, gold mining, fish salting and canning, infrastructural support, transportation services, and maritime supply (including petroleum products) transformed the region politically, economically, and culturally. This period saw considerable ecological and environmental impacts, and many industrial archaeological sites in the region date to this era (Wooley, 2002; see Chapter 5, Fig. 5.4).

---

[3] Named after Cook's sailing master, William Bligh, of *Mutiny on the Bounty* fame.

Salmon (*Oncorhynchus* spp.) and Pacific herring (*Clupea pallasii*) fisheries in PWS developed in the late 1800s and early 1900s, expanding rapidly with the demand for canned salmon during World War I. Canneries and other processing facilities were established at several locations throughout the western part of PWS. Overfishing caused fisheries to decline, and the last cannery in PWS ceased operating in 1959. In time, the fisheries recovered, and in 1988, the year before the *Exxon Valdez* spill, the PWS salmon fisheries harvested over 14.9 million fish (of all species), valued at approximately $80 million; at the same time, the Pacific herring fisheries harvested approximately 12 000 tons of herring roe and fish, valued at almost $12 million (Brady et al., 1990).

A new era of human activity in PWS began on March 13, 1968, when Atlantic Richfield Company and Humble Oil and Refining Company (an Exxon company) announced their discovery of oil in Alaska's Prudhoe Bay, 1300 km from PWS. Four months later, an Atlantic Richfield–Humble team began to investigate the feasibility of a pipeline to carry Prudhoe Bay crude oil to the Port of Valdez for tanker transport to the lower 48 states. The Trans Alaska Pipeline Act became law on November 16, 1973; the first pipe was laid on March 27, 1975; and the first oil moved through the pipeline on June 20, 1977. The 122-cm pipeline still carries crude oil from Prudhoe Bay to the Alyeska Pipeline Service Company's PWS tanker terminal in Valdez, currently North America's most northern ice-free port.

1.3 The event: the *Exxon Valdez* oil spill

At 9:21 p.m. on March 23, 1989, the T/V *Exxon Valdez*, loaded with 53 million gallons (~200 million liters) of Alaska North Slope crude oil, cleared the dock at Valdez bound for Long Beach, California. Shortly after midnight, it grounded on Bligh Reef in PWS, spilling some 11 million gallons (40 million liters) of its cargo (Leschine et al., 1993).[4] The spilled oil would affect some 2100 km of shoreline in PWS and the GOA (Neff et al., 1995) (see Map 1, p. v). The spill path included only areas in western PWS, leaving the eastern side untouched. An estimated 40% of the spilled oil was stranded on 783 km of shoreline (about 16% of the total shoreline of PWS; Wolfe et al., 1994; Neff et al., 1995; Chapter 4). The remainder evaporated or was carried out into the GOA (Chapter 3).

1.3.1 What is crude oil?

One reason that every oil spill is unique is because each crude oil is itself unique in its composition and physical properties. Crude oils are natural products, complex mixtures of thousands of compounds made up of carbon and hydrogen (hydrocarbons), plus other compounds containing sulfur, oxygen, nitrogen, and metals. Because the oils have physical and chemical properties that affect their persistence and toxicity in the environment, defining those properties and how they change with time is essential to assessments of environmental impacts. Box 1.1 highlights some fundamental properties of petroleum crude oils and how they can vary with respect to their physical characteristics, chemistry, and extent of weathering. This box also provides some important technical details specific to Alaska North Slope crude, to which several chapters in this book refer.

---

[4] The circumstances and consequences of the *Exxon Valdez* oil spill have been discussed and reviewed from a variety of perspectives in a large number of publications (see online bibliography) and several books (e.g., Davidson, 1990; Keeble, 1991; Smith, 1992; Loughlin, 1994; Wheelwright, 1994; Owen et al., 1995; Wells et al., 1995; Rice et al., 1996; Lebedoff, 1997; Leacock, 2005; Ott, 2005, 2008; Spies, 2007; Coll, 2012).

Box 1.1 The chemistry of petroleum

Crude oil is a complex mixture of thousands of organic (carbon-based) compounds formed by the natural breakdown of ancient organisms, mainly plants. By weight, it is 98% hydrocarbon molecules, which contain only hydrogen and carbon. The rest is made up of other organic compounds, containing sulfur, nitrogen, and oxygen, and nickel, vanadium, and other metals.

When a crude oil or refined product is exposed to the environment, its composition and physical properties begin to change as a result of chemical, physical, and biological processes collectively termed *weathering* (Fig. 1.1.1). These processes include evaporation, dissolution and dispersion in water, chemical reactions (such as oxidation, including photo-oxidation), the formation of water-in-oil emulsions (i.e., mousse), and biodegradation by microorganisms (Chapter 8). The composition of an oil and the spill environment determine which weathering processes dominate and at what rates they occur. These processes and their rates determine how the oil's specific gravity, viscosity, volatility, and other physical properties change over time.

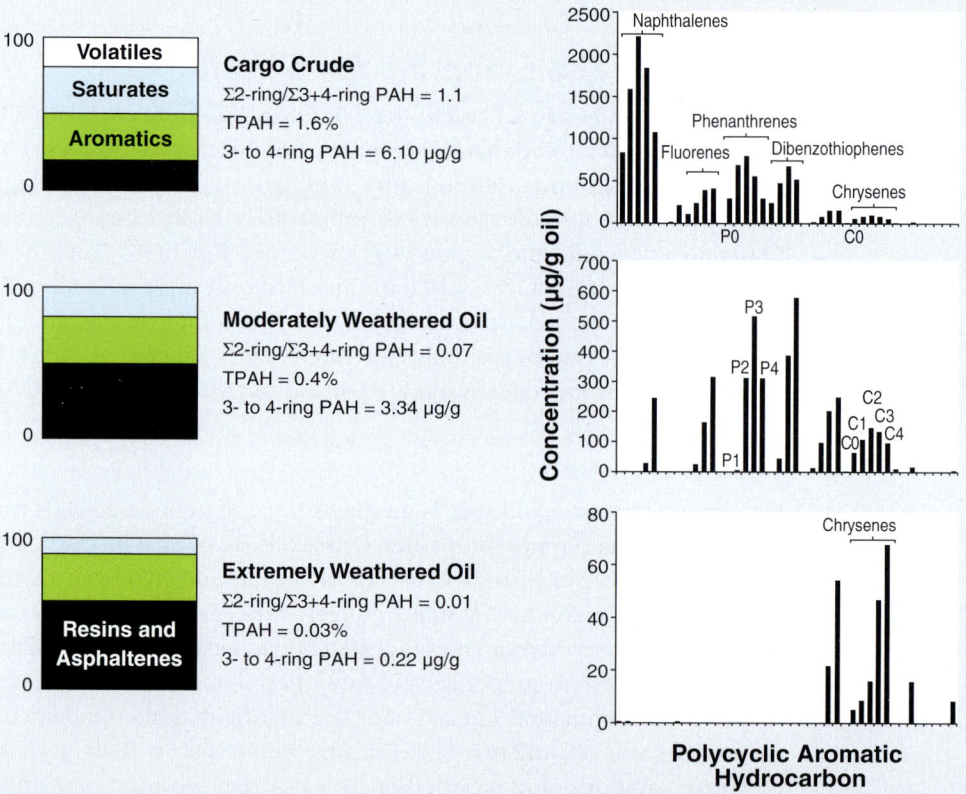

Figure 1.1.1 Bulk chemistry and PAH chemistry of *Exxon Valdez* cargo crude and weathered equivalents recovered from PWS shorelines. The PAH fingerprints, with the labeled series of naphthalenes, fluorenes, phenanthrenes, dibenzothiophenes, and chrysenes, illustrate the weathering behavior of the PAH (TPAH = total PAH).

Major classes of compounds

There are four major classes of compounds in crude oils (Fig. 1.1.1):[1]

SATURATED HYDROCARBONS (also called alkanes, aliphatics, saturates, or paraffins) have single carbon–carbon bonds and dominate the makeup of most crude oils (30–100% of total hydrocarbons). They include straight-chain alkanes (e.g., hexane and dodecane), branched-chain alkanes (e.g., isopentane, pristane, and phytane), and ring-structured cycloalkanes (also called naphthenes) (e.g., cyclopentane and cyclohexane). When released into the environment, the straight-chain alkanes are most rapidly degraded, followed by the branched alkanes.

Crude oils contain low concentrations of polycyclic alkanes (e.g., steranes and triterpanes), also called chemical biomarkers (see Peters et al., 2005). Each crude oil contains a suite of biomarkers that is unique to the petroleum source. Because of their resistance to biodegradation, these biomarkers are useful for both identifying and tracking a crude oil as it weathers in the environment. They help discriminate among multiple hydrocarbon sources (Chapter 6), and they help in the calculation of oil-mass losses due to weathering (Chapter 8, Box 8.1).

AROMATIC HYDROCARBONS (including monocyclic aromatics, polycyclic aromatics, and naphthenoaromatics) have one or more six-carbon benzene rings and are the class of greatest concern in an oil spill because of their persistence and potential toxicity to marine organisms. Two or more benzene rings are fused by sharing two carbons to form polycyclic aromatic hydrocarbons (PAH) (Fig. 1.1.1). The hydrogen on the aromatic carbons can be substituted by a methyl or longer-chain aliphatic structure to form alkylaromatic hydrocarbons. Generally, 1-, 2-, and 3-ring aromatic hydrocarbons – most containing one or more alkyl groups – account for at least 90% of the aromatic hydrocarbons in crude oil. The acutely toxic monoaromatics (which include benzene, toluene, ethylbenzene, and xylenes [BTEX]) are volatile and relatively water soluble and thus disappear rapidly from oil on the water surface or on the shore (Chapter 4; Neff, 1979).

PAH usually comprise a small component of crude oil (about 0.2–4.0% by weight); the *Exxon Valdez* crude oil contained about 1.3% PAH. The dominant PAH in crude oils are 2- and 3-ring (light PAH) (Fig. 1.1.1), including the 3-ring, sulfur-containing dibenzothiophenes. The high molecular weight PAH (4-rings and greater) are more toxic than the lower molecular weight PAH, although they occur in much lower concentrations in crude oils (Neff, 1979, 2002).

Monitoring of PAH concentrations and compositions in water, sediments, and biota is used to document exposure to spilled oil. The distribution of individual PAH (the PAH "fingerprint") varies for different oils, so PAH fingerprints are commonly used to identify oils from different sources (Chapters 3 and 6). These PAH fingerprints change in a predictable manner with weathering. Under aerobic (oxidizing) weathering conditions, there is early and rapid mass loss of all hydrocarbons, including PAH from crude oil. The absolute concentrations of each PAH, even the most stubborn, decrease through biodegradation. The rate of biodegradation decreases with increasing number of aromatic rings and with increasing numbers of attached alkyl groups. For example, parent 3-ring phenanthrene (P0) degrades more rapidly than parent 4-ring chrysene (C0). Within the phenanthrenes, the weathering sequence, as shown in Figure 1.1.1, is P0 > P1 > P2 > P3 (where P1, P2, and P3 indicate the number of alkyl substituents added to the parent phenanthrene, P0). As *Exxon Valdez* oil weathered, all PAH concentrations decreased, as did the ratio of low to high molecular weight PAH.

ASPHALTENES AND RESINS are the third and fourth classes of compounds in crude oil; they are generally considered together as they are what is left when the saturated and aromatic hydrocarbons are removed from the oil. Asphaltenes are the heaviest components of crude oil. They are large, complex molecules composed of condensed PAH and multiring cyclic alkanes with some nitrogen, sulfur, and oxygen heterocyclics and substituents and abundant alkyl side chains (Speight, 1991). They are operationally defined as the solid fraction of crude or refined oil that precipitates in n-pentane or n-heptane. Resins are in the pentane-soluble phase of the oil and are precipitated by addition of acids or alkalis. They have similar structures but lower molecular weights than the asphaltenes. The resin-asphaltene fraction of crude oil also contains high molecular weight, polar compounds with abundant carboxylic acid, phenol, and alcohol groups. Despite its polar nature, this fraction is not water soluble and is largely responsible for the density and viscosity of crude oil. Viscous oils with over 0.5% asphaltenes can form stable water-in-oil emulsions (mousse) that contain up to 80% water by weight.

Physical properties of crude oil

The relative proportions of the above four classes of compounds in a crude oil determine its physical properties. The physical properties most relevant to a marine spill are specific gravity, viscosity, and volatility.

SPECIFIC GRAVITY determines whether oil will float on water or sink. It is usually measured as:

$$\text{Specific gravity} = \frac{\text{oil density (g/mL)}}{\text{density of pure water at 15°C [1g/mL]}}$$

Fresh Alaska North Slope crude oil, similar to that from the *Exxon Valdez*, has a density of 0.87 g/mL and therefore floats. Heavy crude and fuel oils, such as bunker fuels, have densities above 0.95 g/mL and sometimes sink when they weather and lose some of their lighter components. Light crude and distillate oil, such as gasoline and diesel fuel, range from less than 0.8 to about 0.85 g/mL and float.

VISCOSITY is a measure of a fluid's resistance to flow. It is the primary factor in determining how readily oil will spread on the sea surface, disperse into the water column (Chapter 3), or penetrate porous beach sediments (Chapter 7). In general, viscosity decreases with increasing temperature and increases with weathering. Fresh Alaska North Slope crude has a dynamic viscosity of ~23.2 centipoise[2] at 0°C, decreasing to ~11.5 centipoise at 15°C. After artificially weathering to 70% of its original mass, the viscosity increases to about 4230 centipoise at 0°C (Wang *et al.*, 2003).

VOLATILITY determines how readily the oil or its components evaporate. Compounds in petroleum that boil at temperatures below 250°C, including alkanes (lighter than dodecane) and aromatics (from benzene to naphthalene), tend to evaporate early in a spill. The original *Exxon Valdez* oil spilled from the vessel contained about 20% volatile alkanes and aromatic hydrocarbons (Fig. 1.1.1). Most of these compounds evaporated in the first few days following the spill.

Notes

[1] For more detailed descriptions of these compound classes and structures, see Hunt (1995), Peters *et al.* (2005), and Chapters 2 and 3.

[2] Centipoise (cP), a measure of dynamic viscosity, has SI units of kg/m sec ($1\,\text{cP} = 10^{-3}$ kg/m sec). At 20°C, the (approximate) viscosities of water = 1 cP; motor oil = 300 cP; honey = 2000 cP; and peanut butter = 250 000 cP.

Hunt, J.M. (1995). *Petroleum Geochemistry and Geology (Second Edition)*. New York, NY, USA: W.H. Freeman; ISBN-10: 0716724413; ISBN-13: 9780716724414.

Neff, J.M. (1979). *Polycyclic Aromatic Hydrocarbons in the Aquatic Environment: Sources, Fates, and Biological Effects*. London, UK: Applied Science Publishers Ltd.; ISBN-10: 0853348324; ISBN-13: 9780853348320.

Neff, J.M. (2002). *Bioaccumulation in Marine Organisms: Effect of Contaminants from Oil Well Produced Water*. Oxford, UK: Elsevier; ISBN-10: 0080437168; ISBN-13: 9780080437163.

Peters, K.E., C.C. Walters, and J.M. Moldowan (2005). *The Biomarker Guide. Volume 1. Biomarkers in the Environment and Human History (Second Edition)*. Cambridge, UK: Cambridge University Press; ISBN-10: 0521781582; ISBN-13: 978–0521781589.

Speight, J.G. (1991). *The Chemistry and Technology of Petroleum. (Second Edition)*. New York, NY, USA: Marcel Dekker; ISBN-10: 0824784812; ISBN-13: 9780824784812.

Wang, Z., B.P. Hollebone, M. Fingas, B. Fieldhouse, L. Sigouin, M. Landriault, P. Smith, J. Noonan, G. Thouin, and J.W. Weaver (2003). *Characteristics of Spilled Oils, Fuels and Petroleum Products: 1. Composition and Properties of Selected Oils*. Research Triangle Park, NC, USA: US Environmental Protection Agency, Office of Research and Development, National Exposure Research Laboratory; EPA/600/R-03/072. [http://nepis.epa.gov/Adobe/ PDF../P1000AE6.pdf]

Changes in crude oil properties due to *weathering* – the collective term for a variety of chemical and physical changes that a crude oil undergoes when exposed to the environment – differ with the physical and chemical properties of the spilled oil, where it is released, and the environmental conditions during and after the spill. The most volatile components of the oil evaporate rapidly (usually within days), leaving the more persistent oil residues. Components can also be biodegraded by microorganisms or react with oxygen. The fresh oil may mix with water to form a stable, viscous emulsion ("mousse"), or it may weather into tar balls in the water or to asphalt pavements when they wash ashore. Some oil that comes ashore can percolate through coarse sediment layers into finer-grained sediments, where it becomes trapped and isolated from the weathering forces of the environment (Chapters 6, 7).

1.3.2 What happened to the oil?

Oil was not distributed evenly throughout the spill area. Because the oil drifted southwesterly with the prevailing surface currents, shorelines with a northern or northeastern exposure were generally oiled the most. Because northerly and northeasterly facing bays tended to funnel the oncoming oil, they particularly accumulated oil. Three days after the spill, a storm pushed the advancing oil into passages on both sides of Knight Island, so some shorelines on the western side of the island were also heavily oiled (Fig. 1.3a).

A massive cleanup effort began soon after the spill (Leschine *et al.*, 1993; Chapter 4). Directed by the US Coast Guard (USCG) Federal On-Scene Coordinator (FOSC) in consultation with state agencies, Exxon and a large number of contractor firms cleaned up offshore oil slicks during the spring and summer of 1989 and shorelines in PWS and the GOA during the summer of 1989. At the height of the cleanup effort in 1989, over 11 000 people and 1000 vessels were involved (Harrison, 1991).

(a)

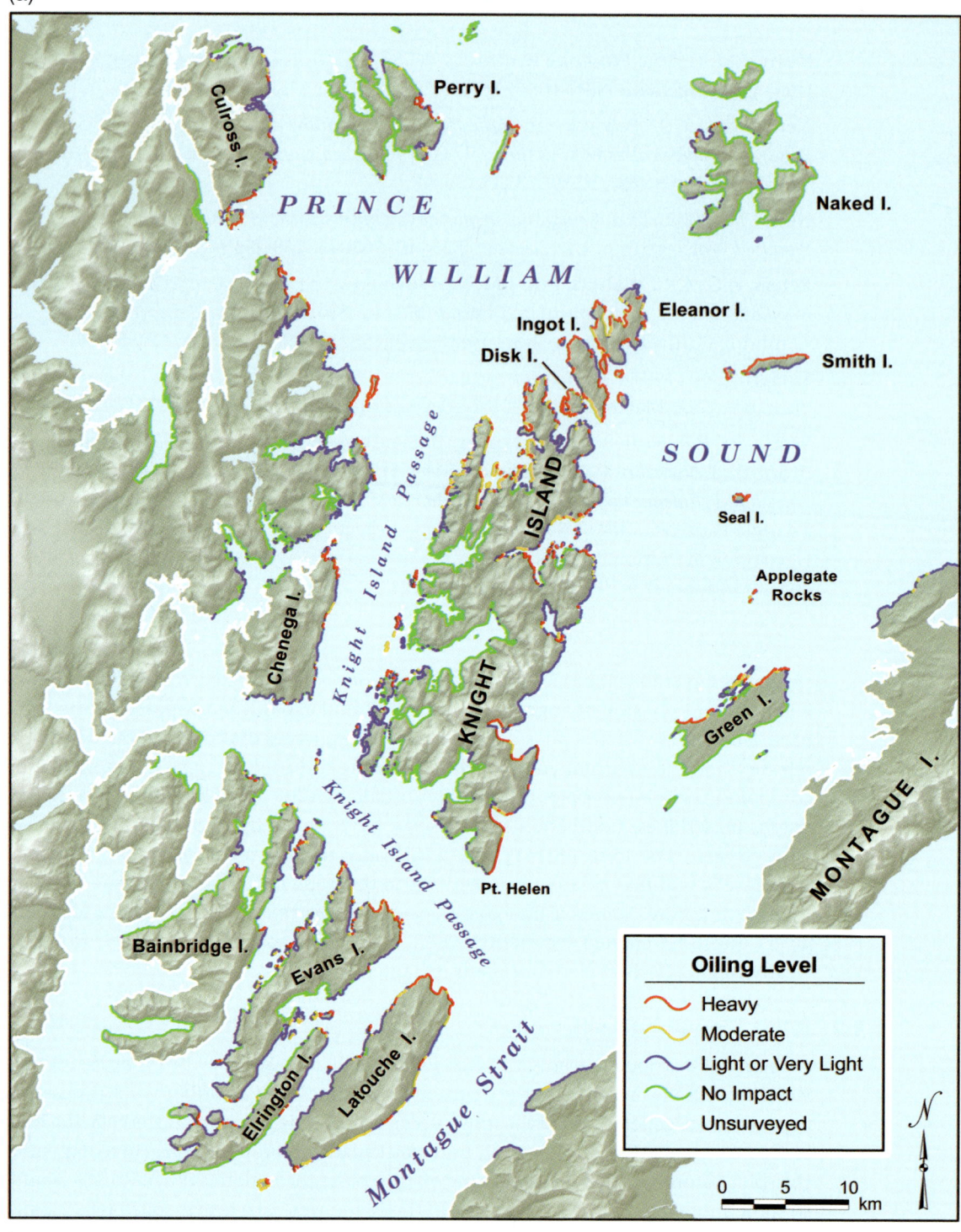

Figure 1.3 Levels of shoreline oiling recorded by Shoreline Cleanup Assessment Technique (SCAT) surveys during spring and summer, 1989. (a) Aerial surveys initially defined the coasts affected by the oil and ground SCAT surveys (Chapter 4) mapped the exact locations and concentrations of oil on the shorelines and the shore-zone character. The resulting information was used to determine the overall level of effort that would be required during the 1989 shoreline cleanup operations and to select appropriate treatment techniques. (b) Detailed ground mapping enabled cleanup activities to focus on locations with the highest concentrations of oil in the heavy and moderate oiling categories (defined in Chapter 4). This summary map for part of northern Knight Island illustrates that the distribution of oil was highly variable and that some sections of shoreline, such as Lower Herring Bay, were not oiled.

(b)

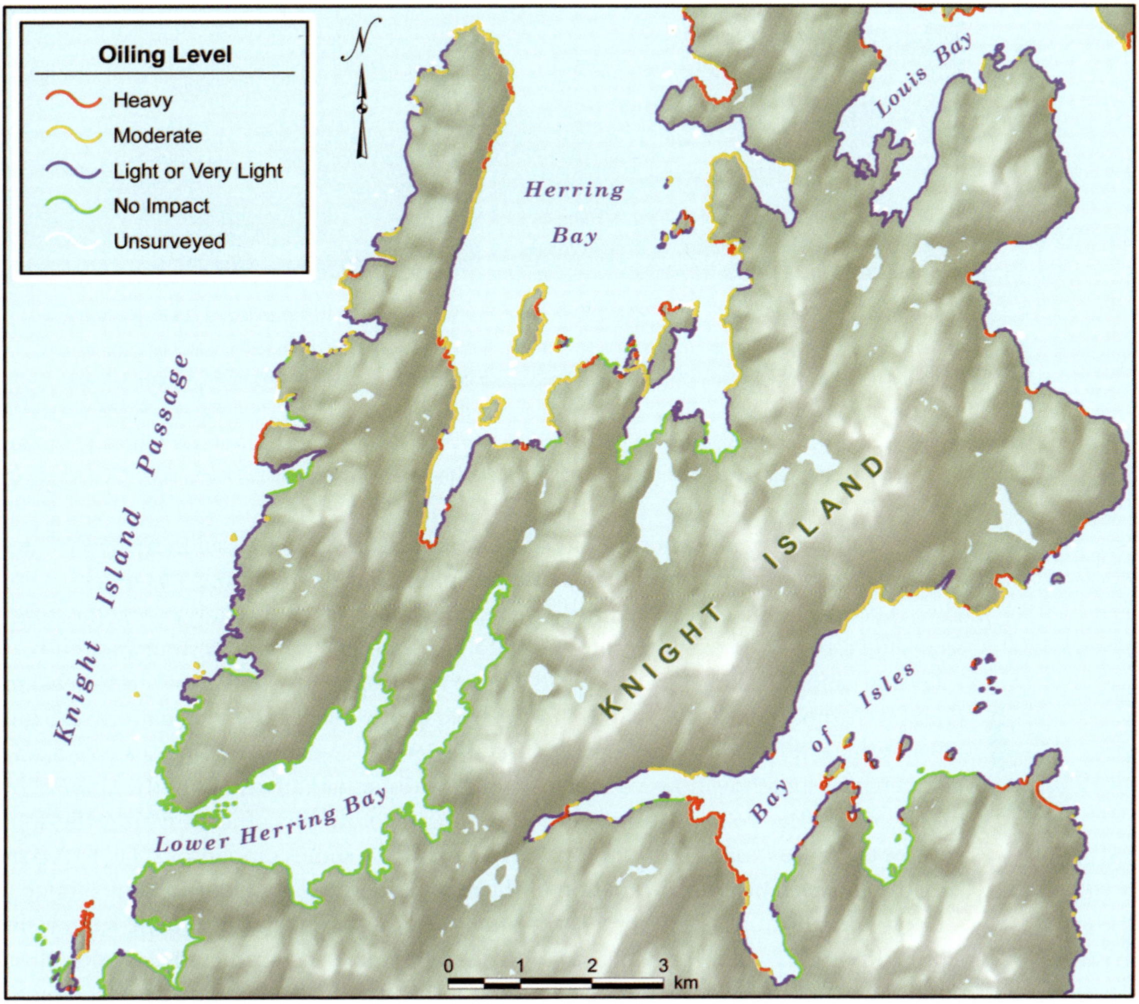

Figure 1.3 (*cont.*)

Free oil on the surface of open water was recovered by containing the oil with mechanical barriers ("booms") and removing it with skimmers. A basic boom consists of tubular floats with weighted vertical panels hanging below the water's surface. Skimmers remove floating oil from the water using a variety of methods, including suction and belts of absorbent materials.

Another method that may be used to remove surface oil is to break oil slicks into droplets small enough to disperse into the water column. This is done using chemical dispersants. Dispersants were a major tool for dealing with the BP Deepwater Horizon/Macondo blowout in the Gulf of Mexico in 2010.[5] Dispersants were *not* used, however, in the *Exxon Valdez* oil spill. Although there were several small-scale tests of dispersants, regulatory approval was not given for their use for a combination of technical, logistical, and political reasons.

---

[5] For updated information about the Deepwater Horizon/Macondo blowout and spill, see www.EoEarth.org/oceanoil.

Figure 1.4 Left: Ground-level view of cold-water flushing operations at Smith Island on May 2, 1989. Header hoses were used to lift the oil and hand-held hoses were used to flush the released oil down to the water line for recovery with skimmers. Right: Aerial view of shoreline cleanup operations in the Point Helen area on September 3, 1989, using barge-mounted, hot-water (60°C) washing equipment and shore-based cold-water flushing systems. The goal was to remove as much oil from the shorelines as possible to give the natural recovery process a head start. Chapter 11 provides a review of the biological effects of shoreline cleanup including these flushing techniques (Photos: Erich R. Gundlach).

A third method of oil removal from open water, which was also used in the Deepwater Horizon blowout but *not* in the *Exxon Valdez* spill, is burning. Burning was tested in the *Exxon Valdez* spill, but it was determined that winds and waves, particularly during the storm soon after the spill, mixed water into the oil to such a degree that the oil would not burn.

Overall, some 141 km (18%) of the 783 km of oiled shoreline were heavily oiled; 94 km (12%) were moderately oiled; 326 km (41%) were lightly oiled; and the remaining 223 km (29%) were only very lightly oiled (Neff *et al.*, 1995).[6] Even at the fine scale of a few kilometers of shoreline, however, there could be substantial differences in the amount of oil deposited (Fig. 1.3b). Consequently, simple categorizations of areas within the spill zone or beyond as "oiled" or "unoiled," while useful for some comparisons, fail to recognize the considerable spatial variation in the magnitude and extent of shoreline oiling.

Shoreline Cleanup Assessment Technique (SCAT) teams began shoreline surveys in April 1989. SCAT teams provided data on shoreline physical characteristics and oiling conditions, which were used to set priorities and methods for shoreline cleanup and to protect sensitive natural and cultural resources (Neff *et al.*, 1995). SCAT is discussed in detail in Chapters 4 and 5.

Four methods were used to clean oil from shorelines in the *Exxon Valdez* spill (www.valdezsciences.com):

1. Pooled oil, mousse patties, tar balls, and oiled fine-grain sediments were manually removed.
2. Cold-water washing was especially used on biologically productive, lower intertidal zones. It involved low-pressure pumping of ambient-temperature seawater through fire hoses to wash oil down the shore to the waterline, where it was collected by skimmers or sorbents (Figs. 1.4, 3.3, 4.3).
3. Seawater warmed to 27–60°C was used to mobilize weathered oil that adhered to rocks; once the oil was removed from the rocks, cold-water washing was used to push it to the waterline for removal (Figs. 1.4, 3.3, 4.3).

---

[6] These categorizations refer to the width of the surface band of oil on the shoreline, not the total amount of oil present.

4. Bioremediation increased the rate of oil biodegradation by natural hydrocarbon-eating bacteria by adding fertilizers to oiled areas (Chapter 8).

Cleanup activities in 1989 were suspended for safety reasons in mid-September as the winter storm season began. Between September 1989 and February 1990, 90% of the surface and subsurface oil on exposed shorelines and 70% of the oil on intermittently exposed and sheltered shorelines were removed by natural processes (Michel *et al.*, 1991). Shoreline cleanup resumed during the summers of 1990 and 1991 and the spring of 1992. The final response update, released by the 17th USCG District Commander in June 1992, concluded (Leschine *et al.*, 1993, p. 198):

> There is still some oil remaining on the shorelines impacted by the *EXXON VALDEZ* oil spill. The oil left is generally oil mousse, which is very high in water content, weathered and has lost most of its toxicity. This oil is primarily located in areas protected from the elements, behind rocks and boulders or is below the surface. Algae, mussels, periwinkles and other marine life are recolonizing in strength on these shorelines. Removal of the remaining weathered, generally benign oil, would require invasive cleanup measures which would disrupt the environmental recovery process that is well underway. The consensus and judgment of state and federal agencies involved in the spill response is that additional cleanup would cause unacceptable environmental harm. Accordingly, the FOSC has determined cleanup "complete" [quotation marks in original].

As a result of the intensive cleanup efforts and natural removal, much of the oiled shoreline appeared to be free of oil within a year or two of the spill (Fig. 1.5). Frequent

Green Island (1989–92)

Block Island (1989–92)

Figure 1.5 Photographs showing the disappearance of surface oil from heavily oiled shorelines on Green Island and Block Island in PWS, 1989–92 (Photos: Exxon Corporation).

field surveys and chemical analyses confirmed that concentrations of polycyclic aromatic hydrocarbons (PAH) from *Exxon Valdez* oil in the water column peaked within a week of the spill and declined rapidly to background levels at most locations by June 1989 and at all locations by the spring of 1990 (Wolfe *et al.*, 1994; Neff and Stubblefield, 1995; Chapter 3). Oil remaining on or in beach sediments declined substantially between 1989 and 1992. By 1993, oiled shoreline had decreased to 14 km, 92% of which was only very lightly oiled (Wolfe *et al.*, 1994; Neff *et al.*, 1995; Chapter 6).

In 1993, the State of Alaska surveyed shores in PWS originally oiled by the spill, finding subsurface oil residues (SSOR) in scattered locations along about 7 km of shoreline and surface oil residues along 4.8 km of shoreline (Gibeaut and Piper, 1998). This pattern was consistent with 1991–92 joint surveys conducted by Exxon and representatives of government and landowners, including the National Oceanic and Atmospheric Administration (NOAA), US National Park Service, US Forest Service, US Fish and Wildlife Service, Alaska Department of Environmental Conservation, Alaska Department of Fish and Game, Alaska Department of Natural Resources, Alaska Native groups (Chugach, Chenaga, Kodiak, Port Graham), and others.

Sediment chemistry and toxicity tests showed that the total PAH (TPAH) concentrations and toxicity of beach sediments in the spill zone decreased rapidly following the oil spill, as the lighter and more toxic fractions weathered away. Toxicity,[7] as measured by US Environmental Protection Agency (EPA) protocols, was restricted to upper intertidal sediments at about 10% of heavily oiled boulder/cobble/gravel shores in 1990–91 (Boehm *et al.*, 1995). By 1999, the median sediment TPAH value was well below the EPA's toxicity threshold (Page *et al.*, 2002).

A 2001 survey, which focused primarily on locations that had been heavily oiled in 1989 and still contained substantial oil residues in 1990–93, mapped 0.78 hectares of SSOR (Short *et al.*, 2004). Extrapolating from those data, Short *et al.* (2002, 2004) estimated that 7.8 hectares of shoreline in PWS were still contaminated with SSOR in 2001. Based on data from the 2001 shoreline survey, Michel *et al.* (2006, p. 1) concluded that "surface oil primarily occurs as highly weathered residues that pose little continuing ecological risk." Most of the remaining SSOR was highly weathered and confined to widely scattered patches in the middle and upper intertidal zones, where it was sequestered under a surface armor of boulders and cobbles (Boehm *et al.*, 2008; Taylor and Reimer, 2008). This location made it largely inaccessible to oil-degrading microbes and animals that live or forage in the intertidal zone (i.e., not bioavailable; Neff *et al.*, 2011; Chapters 6, 7).

The long-term persistence of small, sequestered remnants from major oil spills in the intertidal zone at specific shoreline locations has been well documented for many major marine oil spills (Vandermeulen and Gordon, 1976; Owens *et al.*, 1987; Vandermeulen and Singh, 1994; Chapter 6). In all cases, the shorelines have physical and geomorphological features that contribute to the sequestration and persistence of SSOR. These features are discussed in more detail in Chapters 5, 6, and 7, but the important point is

---

[7] Toxicity is the inherent property of a chemical or mixture of chemicals, such as petroleum, that produces harmful effects in an organism after exposure to a specific concentration for a specific duration. In environmental toxicology, the "dose" of the chemical(s) is usually expressed as the concentration and duration in the external medium (air, water, sediment) to which the organism is exposed or in the tissues of the organism.

that the factors that promoted the persistence of isolated spill remnants also substantially reduced their bioavailability (Bressler and Gray, 2003; Neff *et al.*, 2006).

1.3.3 Other sources of oil

When assessing the effects of an oil spill on an ecosystem, other sources of hydrocarbons in the environment must also be considered. These nonspill hydrocarbons may not be very important in the immediate aftermath of a spill of the magnitude of the *Exxon Valdez* spill, but they could confound the results of later studies of potential chronic effects, when the presence of spilled oil is much reduced or eliminated.

The *Exxon Valdez* oil spill was only one of several sources of petroleum hydrocarbons identified in PWS intertidal and subtidal sediments following the spill (Kvenvolden *et al.*, 1993; Page *et al.*, 1995, 1999; Chapter 6). There are natural and anthropogenic sources of petroleum hydrocarbons, forming a petroleum-related (petrogenic) background throughout PWS seafloor sediments. For example, the suspended sediments carried by the Alaska Coastal Current into PWS contain hydrocarbons derived from active oil seeps and eroding petroleum-source rocks along the northeastern GOA coastline (Page *et al.*, 1996).

There is also a long history of human and industrial activities in PWS. Some of these activities have used fossil fuels, contributing additional hydrocarbons to the environment (Lethcoe and Lethcoe, 2001; Wooley, 2002). In particular, the chronic release of fuel and combustion residues from boats and the many historical industrial sites in PWS constitute a major source of PAH in some areas (Kvenvolden *et al.*, 1995; Bence *et al.*, 1996; Page *et al.*, 1996; Boehm *et al.*, 2003), as do the fuel-storage tanks ruptured in the 1964 earthquake (Fig. 1.6) (Kvenvolden *et al.*, 1993).

Figure 1.6 Three of the many PWS locations where chronic inputs of fuel and combustion residues from past industrial activity contribute to locally high PAH levels in the nearshore marine environment. (a) Leaking fuel oil tanks from cannery ruins above the beach at Port Ashton, Sawmill Bay, Evans Island (2001). (b) Boiler and fuel oil tank, Mummy Bay, Knight Island (2001). (c) Fish processing plant ruins, Thumb Bay, Knight Island (2001) (Photos: David S. Page).

1.4 Documenting exposure pathways

Organisms, populations, or ecosystems can suffer detrimental effects from spilled oil either directly, through exposure to the oil or its weathered residues, or indirectly, through deleterious effects on prey populations or habitat suitability. The greatest concerns are usually about direct exposure, since the effects are often immediate and measurable. As noted in Box 1.2, when a potential effect of an oil spill (e.g., increased mortality, reduced abundance, physiological stress) has been documented, it is critical

Box 1.2 Exposure and exposure pathways

The studies and findings described in this book rely heavily on determining both exposure to spilled oil and the means (pathways) by which oil or its components reach the habitats or organisms of concern. Two types of exposure assessment are described in the natural resource damage assessment (NRDA) regulations under the Comprehensive Environmental Response, Compensation, and Liability Act (CERCLA) (US Department of the Interior, 1986, 1987) and the Oil Pollution Act of 1990 (OPA 90) (Hugenin et al., 1996): Type A and Type B. Type A assessments involve the use of computer models to determine damages that result from the discharge of oil into the environment, whereas Type B assessments employ laboratory and field studies to accomplish this aim. The Trustees used the Type B assessment guidance (US Department of the Interior, 1986) (*Exxon Valdez* Oil Spill Trustee Council, 1990), which states that:

- "Exposed to" or "exposure of" means that all or part of a natural resource is, or has been, in physical contact with oil or with media containing oil.
- "Pathway" means the route or medium through which oil or a hazardous substance is or was transported from the source of the discharge or release to the injured resource.

In the Type B injury determination phase, the Trustees determined whether an injury to one or more natural resources had occurred. They also determined, based on the exposure pathway, exposure assessment, and the nature of the injury, whether the injury resulted from the discharge of oil.

The Trustees and Exxon used the concepts and framework of an ecological risk assessment, later codified by EPA (US Environmental Protection Agency, 1992), as the basis for the design of an oil-spill NRDA. Because the oil-spill ecological risk assessment included a problem formulation that incorporated pathway assessment, an exposure assessment, and an injury assessment to evaluate the causal links between the disturbance (the spill or spill-response activities) and harmful effect (injury) to natural resources, the rigor of the NRDA was strengthened.

Characterizing exposure quantitatively is difficult, so exposure assessments performed according to CERCLA and OPA 90 guidance frequently have relied heavily on models that link spill trajectory, oil fates, and biological effects to estimate damages from a spill to natural resources (French, 1998). Because models necessarily make assumptions about exposure pathways and use indirect measures of exposure, it is important to conduct scientifically rigorous, quantitative field assessments of exposure pathways and actual exposure to validate the model results, particularly as time passes after the spill. Visible oil on the sea surface and shore usually disappears relatively rapidly after a spill and the oil weathers to demonstrably less harmful residues, rendering modeling predictions less certain and empirical data even more important. In the absence of direct measurements, it becomes more difficult for models to differentiate between chronic injuries resulting from exposure to oil versus natural environmental change.

The postspill temporal and spatial weathering patterns of petroleum in the environment are complex. Identifying exposure pathways requires comparing the distribution and weathering of oil with historical and seasonal distributions of the biota of concern in the spill path. It also requires considering any natural-history aspects of the organisms that may increase or decrease the likelihood of exposure. This information should be collected along the spill path for the main environmental compartments (air, water, shoreline and subtidal sediments, and tissues of species of concern and their prey) at different times and analyzed for chemical components to determine how concentrations change in absolute amounts and relative to each other. Models examining the effects and fate of oil should be used to aid field spill assessments.

The exposure assessment should rely on direct measurements of spill residues, particularly the more toxic oil fractions, in all environmental compartments identified in the pathway assessment. These measurements should be repeated periodically to document the temporal changes in distributions, concentrations, compositions, and physical forms of spill residues (Chapters 4, 6; Neff *et al.*, 2011). Indirect measures of exposure, such as exposure biomarkers in tissues of species, can be used to confirm the results of the direct measurements of exposure, but they should not be used as the sole evidence of exposure (Chapter 9).

Exxon Valdez Oil Spill Trustee Council (1990). *The 1990 State/Federal Natural Resource Damage Assessment and Restoration Plan for the Exxon Valdez Oil Spill*. Anchorage, AK, USA: *Exxon Valdez* Oil Spill Trustee Council; Volumes I, II. [http://www.evostc.state.ak.us/Universal/Documents/Publications/Workplans/1990Workplan.pdf]

French, D.P. (1998). Evolution of oil trajectory, fate and impact assessment models. In *Proceedings: Oil Spill 98 Conference, July 29–31, 1998, Wessex Institute of Technology, Ashurst Lodge, Ashurst, Southampton, United Kingdom*. Southampton, UK: Wessex Institute of Technology. [http://www.asascience.com/about/publications/pdf/1998/wessex98.pdf]

Hugenin, M.T., D.H. Haury, J.C. Weiss, D. Heldon, C.-A. Manen, E. Reinharz, and J. Michel (1996). *Injury Assessment. Guidance Document for Natural Resource Damage Assessment under the Oil Pollution Act of 1990*. Silver Spring, MD, USA: National Oceanic and Atmospheric Administration, Damage Assessment and Restoration Program. [http://www.darrp.noaa.gov/library/pdf/iad.pdf]

Neff, J.M., D.S. Page, and P.D. Boehm (2011). Potential for exposure of sea otters and harlequin ducks in the Northern Knight Island area of Prince William Sound, Alaska, to shoreline oil residues 20 years after the *Exxon Valdez* oil spill. *Environmental Toxicology and Chemistry* **30**(3): 659–762.

US Department of the Interior (1986). Natural resource damage assessments. Final rule. *Federal Register* **51**(148): 27674–27753.

US Department of the Interior (1987). Natural resource damage assessments. Final rule. *Federal Register* **52**(54): 9042–9100.

US Environmental Protection Agency (1992). *Framework for Ecological Risk Assessment*. Washington DC, USA: US Environmental Protection Agency, Risk Assessment Forum; EPA/630/R-92/001. [http://rais.ornl.gov/documents/FRMWRK_ERA.PDF]

that an exposure pathway also be identified and measured. This is accomplished through exposure assessments that connect the spilled oil to a potential injury to biological resources. Exposure pathways may include: direct contact with oil or oiled sediments; consumption of oiled prey; contact with oil dissolved in the water column; or contact with air-borne, volatile components of oil. Determinations of exposure pathways figure importantly in Chapters 11–15 and are incorporated into an overall ecological risk-assessment framework in Chapter 16.

1.5 The context: regulations, definitions, and litigation

Studies following the *Exxon Valdez* oil spill drew heavily from well-established scientific principles and experience in assessing other oil spills over several decades. They were designed and undertaken to establish a factual scientific foundation for understanding short- and long-term fates and effects of the spilled oil and to assess recovery from those effects. But they were also conducted in a context of laws and regulations that dictate the process for oil-spill damage assessments, as well as the accompanying litigation over responsibilities and payments for damages

To gain a better understanding of the short- and long-term effects of the oil spill, both the federal and Alaska governments (the *Exxon Valdez* Oil Spill Trustee Council; hereafter "Trustees"[8]) and Exxon recruited teams of experts in key disciplinary areas. Initially, both parties focused on documenting the distribution, condition, and toxicity of oil in the water and on the shorelines. By the summer and fall of 1989, both parties began to turn their attention to longer-term issues, although they continued to cooperate on cleanup-related studies and surveys through 1992.

The Trustee studies were conducted using the framework of a natural resource damage assessment (NRDA) (derived from the Comprehensive Environmental Response, Compensation, and Liability Act (CERCLA) and associated regulations) focusing on injuries to biological resources and the restoration and recovery of these resources. The Trustees published their first restoration plan in August 1989 (*Exxon Valdez* Oil Spill Trustee Council, 1989). As the amount and toxicity of oil in the environment rapidly diminished, Exxon's focus also shifted to evaluating long-term injuries and recovery.

The NRDA regulations under CERCLA[9] and the Oil Pollution Act of 1990 (OPA 90) – which was not enacted until after the 1989 *Exxon Valdez* spill and consequently did not "govern" the spill or the response to it – define recovery in terms of the return of an injured resource to baseline services, where baseline services reflect the condition that would have been expected had the incident not occurred (Hugenin *et al.*, 1996; Box 1.3). This definition leaves considerable room for interpretation. What is "baseline"? How does one determine what a resource's condition would have been absent the spill? Differences in interpretations led to disagreements about the status of salmon,

---

[8] The Trustee Council was formed following the *Exxon Valdez* spill to oversee the assessment of natural-resource damages and the restoration of the damaged ecosystem. The Council consists of representatives of the Alaska Departments of Fish and Game, Environmental Conservation, and Law and the US Departments of Agriculture (Forest Service), Interior, and Commerce (National Oceanic and Atmospheric Administration). See http://www.evostc.state.ak.us.
[9] US Code of Federal Regulations 43 CFR §§11.14(e), 11.72.

Box 1.3 Defining terms

Determining how or whether an oil spill had short- or long-term environmental consequences requires clear, operational definitions of terms. For example, terms such as *impact*, *effect*, and *injury* conjure up different images and can influence how hypotheses are framed, studies are designed, statistical analyses are conducted, data are interpreted, or what conclusions are reached. Clarity in how terms are used can help to avoid confusion and disagreements. We illustrate how we use these terms in this book in Figure 1.3.1.

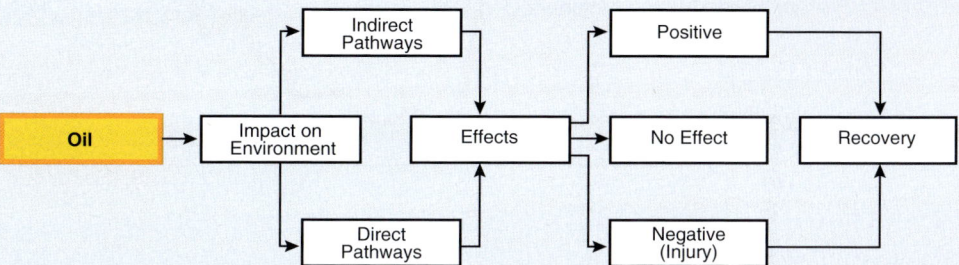

Figure 1.3.1 Sequence between the release of oil into the environment and potential consequences and recovery, to illustrate how the terms "impact," "effect," "injury," and "recovery" are used in this book.

Impact, effect, and injury

"Impact" and "effect" have often been used interchangeably to describe the consequences of an oil spill. We use *impact* as a verb, to refer to a physical contact between spilled oil and the environment, as when oil is deposited on the water, shoreline, or organisms. This contact is the first stage leading to (causing) consequences of the spill. We describe these consequences as *effects* of the spill. For there to be effects on biological resources, there must be an identifiable pathway. The pathway may be direct, through exposure of organisms to the oil or its constituents (Box 1.2), or indirect, as, for example, through disruption of the abundance or availability of prey for foragers resulting from the spill. Because indirect pathways may be subtle and several steps removed from direct contact with oil or its constituents, they are more often inferred than empirically documented. Most assessments of spill-related effects therefore focus on direct exposure. Increasingly, computer models of biological effects rather than field studies are used as part of the natural resource damage assessment (NRDA) process to determine indirect effects (French-McCay, 2009).

Effects may be positive, negative, or neutral (i.e., no effects). The usual focus is on negative effects – "injuries" – rather than positive effects. As defined by the Comprehensive Environmental Response, Compensation, and Liability Act (CERCLA) and NRDA regulations, "injury means a measurable adverse change, either long- or short-term, in the chemical or physical quality or the viability of a natural resource resulting either directly or indirectly from exposure to a discharge of oil or release of a hazardous substance, or exposure to a product of reactions resulting from the discharge of oil or release of a hazardous substance".[1] A negative effect or injury is recognized by a sudden departure from the previous state of the system (Fig. 1.3.2). For an injury to be caused by an oil spill:

- There must be a direct exposure pathway (Box 1.2) or a plausible and documented pathway for indirect effects;
- There must be measurable effects on organisms, populations, or ecosystems; and
- These effects must meet or exceed some threshold of what is considered harmful.

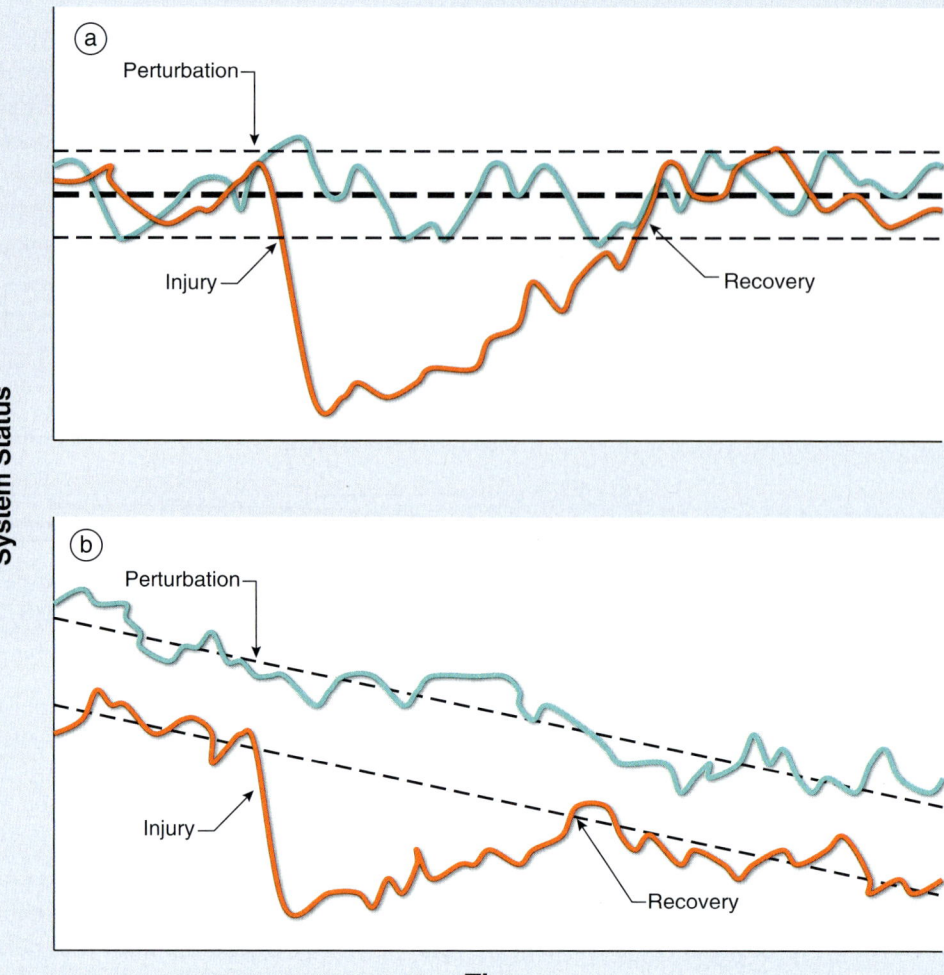

Figure 1.3.2 Hypothetical illustrations of injury and recovery. In panel a, the system is in steady-state equilibrium. The blue line depicts the dynamics of a system in the absence of a perturbation (or in a suitable reference area), with the dashed lines indicating the long-term mean and a window of variation (e.g., ± 95% confidence interval). The red line shows the deviation from the "normal" system state following a disturbance and its subsequent return to the window of natural variation (i.e., recovery). In panel b, populations in reference (blue line) and impacted (red line) areas differ and are trending down over time (i.e., dynamic equilibrium). Recovery occurs when the two areas follow parallel trajectories over time, even though the system status may continue to differ between the areas. See Chapter 10 for a discussion of different assumptions about the equilibrium status of a system.

Large deposits of oil on shorelines or oiled seabirds are unambiguous evidence of impact and, in most cases, effects. Many effects may be less obvious, however, and determining whether they are adverse and a consequence of the oil spill or something else requires rigorous scientific study. The presence of oil in the environment does not, by itself, indicate injury; the linkages in Figure 1.3.1 may be broken at any point, so there may be no pathways, no exposure, or no effects on resources even though oil impacted the environment. Likewise, simply because an environmental change coincides with the spill does not necessarily indicate that there were spill effects. The issue is not so much when (or whether) oil completely disappears from an affected area, but when (or if) it no longer poses a risk of injury, based on its concentration, bioavailability, and toxicity.

Recovery

Once an effect (usually an injury) has been determined, attention shifts to documenting recovery. NRDA regulations under the Oil Pollution Act of 1990 (OPA 90) define recovery as "the return of injured natural resources and services to baseline"[2] (Fig. 1.3.2a), where "baseline" is "the condition of the natural resources and services that would have existed had the incident not occurred."[3] The CERCLA NRDA (US Department of the Interior, 1986) guidance focuses on the recovery period, which means either "the longest length of time required to return the services of the injured resource to their baseline condition, or a lesser period of time selected by the authorized official and documented in the Assessment Plan." Determining whether natural-resource services have been effectively restored does not require that the recovered ecosystem or other resource necessarily be identical to the one lost, only that all important and measurable attributes of the injured resource have returned to some designated baseline condition.

This guidance leaves considerable latitude for interpretation. One might assume that if the source of stress (e.g., oil, hydrocarbons, other chemicals) is removed or diminished, the resource will return to its prespill condition. However, in PWS (and nearly all ecosystems), environmental conditions and ecological processes change over multiple temporal and spatial scales in response to climatic cycles, natural disturbances, or other biological changes in the ecosystem. Using prespill information to specify a fixed target for recovery is inappropriate because, due to natural changes, the prespill conditions cannot be realized. Moreover, information on prespill conditions is not often available.

Therefore, it is often more appropriate to measure "recovery" by evaluating "baseline" in terms of the difference between an impacted area and an appropriate reference location. If the impacted and reference areas differ intrinsically, independent of their oiling history, then recovery might be assessed by comparing trends in the areas over time (Fig. 1.3.2b). The disappearance of the difference over time then indicates that the effects of the incident no longer exist: the resource has recovered. This does not mean, for example, that no PAH remain in the environment, but only that PAH concentrations in affected areas or resources cannot be distinguished from background or baseline concentrations, however determined.

Given these considerations, we define "recovery" operationally, as the disappearance over time of a previously documented adverse or deleterious change in an attribute of a system that was caused by the accident or environmental disruption, rather than by some other factor(s).

Selecting reference areas carefully is essential. Habitat characteristics and population dynamics of focal species may differ between the spill and unoiled reference areas. Conditions may vary over time in different ways in different areas. Differences in scale can confound comparisons: the loss or addition of a few individuals to a small population in a localized area will have a far greater effect on overall counts and population trends than the loss or addition of the same number of individuals to a larger population surveyed over a much broader area. Unless they are accounted for, such differences in time, space, or scale may compromise comparisons and statistical analyses.

Notes

[1] US Code of Federal Regulations 43 CFR §§ 11.14(v); http://ecfr.gpoaccess.gov/cgi/t/text/text-idx?c=ecfr&tpl=/ecfrbrowse/Title43/43cfr11_main_02.tpl.

[2] US Code of Federal Regulations 15 CFR § 990.30.

[3] US Code of Federal Regulations 15 CFR § 990.30; the CERCLA regulation included the additional phrase "taking into account both natural processes and those that are the result of human activities"; US Code of Federal Regulations 43 CFR §§ 11.14(e) and (gg), 11.72.

French-McCay, D. (2009). State-of-the-art and research needs for oil spill impact assessment modeling. In *Proceedings of the Thirty-Second Arctic and Marine Oilspill Program (AMOP) Technical Seminar, June 9–11, 2009, Vancouver, British Columbia, Canada*. Ottawa, ON, Canada: Environment Canada; pp. 601–641.

US Department of the Interior (1986). Natural resource damage assessments. Final rule. *Federal Register* **51**(148): 27674–27753.

---

marine birds, sea otters, and other organisms (Chapters 11–15). We discuss these issues and how we use the terms *impact*, *effect*, *injury*, and *recovery* in Box 1.3.

As the specter of protracted and contentious litigation emerged, the studies sponsored by the Trustees and by Exxon continued to investigate the potential for continued oil exposure, chronic effects, and recovery for many years. These studies were driven in part by the civil lawsuit between the United States and the State of Alaska and Exxon (which was settled in 1991), the Reopener provision of the settlement,[10] and litigation over damages (e.g., disruption of commercial salmon and herring fisheries, effects of oiling on land values) with private plaintiffs (for which some Trustee- and Exxon-contracted scientists acted as expert witnesses) (see Box 1.4). Given their positions in the litigation, the Trustee studies tended to investigate ongoing damages to biological resources, whereas those conducted by Exxon scientists focused more on ascertaining the time-course of recovery.

The scientific studies of the *Exxon Valdez* spill and its consequences were driven less by the search for knowledge for knowledge's sake that characterizes science in the public mind, but more by the need to support decisions about where to deploy cleanup crews, identify which resources merited attention, determine how restoration funds should be allocated, and inform legal arguments. The differences in the positions of the Trustees and Exxon in the legal proceedings did not necessarily diminish the scientific process (good science was done by both groups), but they did influence how questions were posed, how hypotheses were framed and tested (Chapter 10), and how interpretations and conclusions were drawn.

1.6 Conclusions

The *Exxon Valdez* oil spill spurred a proliferation of remarkable scientific studies that, in turn, prompted discussions about methodologies, the importance of alternative explanations, the confounding effects of environmental variation in time and space, and the value of a multidisciplinary approach (see Chapters 10 and 17). Much was learned along

---

[10] http://www.evostc.state.ak.us/pdf/settlement/agreement_consent_decree093091.pdf; see Box 1.4.

Box 1.4 The legal context

In the wake of the *Exxon Valdez* oil spill, a tsunami of civil and criminal litigation washed through federal and Alaskan state courts. Lawsuits were filed, some within days of the spill, asserting thousands of claims on behalf of fishers, boat owners, fish processors, land owners, local businesses, Alaska Native corporations, municipalities, the State of Alaska, and the Federal government. There were also criminal charges brought by the Federal government alleging violations of the Clean Water Act, Refuse Act, Migratory Bird Treaty Act, Ports and Waterways Safety Act, and Dangerous Cargo Act.

In general, the litigation can be separated into three categories: federal criminal cases; Alaskan and federal civil cases concerning compensation of private (i.e., nongovernmental) plaintiffs for various damage claims; and Alaskan and federal civil cases concerning claims of damages to the environment. Of the three categories, the studies discussed in this book are relevant principally to the governmental civil cases for environmental damages and, to a much lesser extent, the private civil cases for compensatory damages; they are not germane to the criminal cases.

Exxon and the federal government settled the criminal cases in October 1991.[1] Exxon paid $100 million restitution for environmental injuries and a fine of $150 million. Half of the restitution payment went to the State of Alaska, the other half to the federal government. From the fine, $13 million went to the national Victims of Crime Fund, $12 million went to the North American Wetlands Conservation Fund, and $125 million was forgiven Exxon because it took responsibility for the spill cleanup.

The private lawsuits largely concerned compensation for claims of economic (or business) losses due to: lost commercial fishing harvests because salmon and herring fisheries were closed by the state of Alaska; lower fish prices paid to fishers by processors; reductions in values of fishing boats and fishing permits; and lowered real estate values because of the spill. There were also claims of damage to archaeological resources. To the extent that the testimony of scientists was required concerning alleged biological injuries to salmon and herring populations or archaeological harm, both Exxon and *Exxon Valdez* Oil Spill Trustee Council (hereafter "Trustees") scientists served as expert witnesses for these lawsuits. A number of these private lawsuits were dismissed by the courts or settled out of court. Most of the remaining cases were consolidated into two trials conducted in Alaska State and Federal District Courts in Anchorage, Alaska, during May–September 1994. (Several other trials were conducted years later when appellate courts reinstated previously dismissed cases.) In 1994, the two juries awarded plaintiffs a total of $299.7 million in compensatory damages. After the jury awards were adjusted for compensation already paid to the plaintiffs before the trials[2] and for interest due the plaintiffs, the 1994 plaintiffs received $57.6 million in compensation as a result of the two civil lawsuits.

The most publicized result of the 1994 federal trial, however, was the jury's award of $5 billion in punitive damages to the plaintiffs. Punitive damages are intended to punish and deter the defendant (Exxon) from similar conduct in the future rather than to compensate plaintiffs for damages suffered (which is the function of compensatory damages). In late 2006, the jury's punitive award was reduced by the Federal District Court for Alaska to $2.5 billion in compliance with a ruling from the Ninth Circuit

Court of Appeals. The issue was finally resolved by the US Supreme Court in 2008, which limited punitive damages to $507.5 million.

Of greatest relevance to the scientific investigations considered in this book are the governmental lawsuits for environmental damages. In 1989, suits were filed by the State of Alaska and the US Government for the expenses incurred by these entities in responding to the spill and to enforce state and federal environmental laws prohibiting oil spills and harm to environmental resources. These claims were brought under the provisions of the Alaska Oil Spill Act, Clean Water Act, Trans Alaska Pipeline Act, and other federal and state statutes, as well as maritime and common law. By taking these actions, the governments sought compensation on behalf of the general public for all environmental damages resulting from the spill. It is these environmental damages – their existence, cause, nature, degree, and longevity – that led to over 20 years of scientific studies by Exxon and the Trustees.

On September 20, 1991, the governments settled with Exxon in a consent decree that resolved the civil claims for recovery of damages to natural resources resulting from the oil spill.[3] Under the settlement, Exxon agreed to pay $900 million to the governments over a 10-year period. Of this, $213.1 million was to reimburse the governments for their cleanup costs, and $686.9 million was to compensate for natural resource damages and for damage-assessment and restoration costs stemming from the spill. These funds are administered by the Trustees. From 1992 through 2010, $246 million was spent for Trustee-sponsored scientific research, 86% of it on surveys and monitoring to determine the extent and duration of damages to natural resources. Another $375.4 million was spent on land acquisition, under the premise that protecting habitat helps prevent additional injury to species due to intrusive development or loss of habitat. Some of the funds ($172.6 million) were invested in a Trust Fund to support future restoration expenses.[4]

The consent decree also contained a Reopener for Unknown Injury[5] (the "Reopener"). Since the Reopener provides the legal context for both Trustee and Exxon scientific studies, its full text follows:

> 17. Notwithstanding any other provision of this Agreement, between September 1, 2002, and September 1, 2006, Exxon shall pay to the Governments such additional sums as are required for the performance of restoration projects in Prince William Sound and other areas affected by the Oil Spill to restore one or more populations, habitats, or species which, as a result of the Oil Spill, have suffered a substantial loss or substantial decline in the areas affected by the Oil Spill provided, however, that for a restoration project to qualify for payment under this paragraph the project must meet the following requirements: (a) the cost of a restoration project must not be grossly disproportionate to the magnitude of the benefits anticipated from the remediation; and (b) the injury to the affected population, habitat, or species could not reasonably have been known nor could it reasonably have been anticipated by any Trustee from any information in the possession of or reasonably available to any Trustee on the Effective Date.
>
> 18. The amount to be paid by Exxon for the restoration projects referred to in Paragraph 17 shall not exceed $100,000,000.
>
> 19. The Governments shall file with Exxon, 90 days before demanding any payment pursuant to paragraph 17, detailed plans for all such restoration projects,

together with a statement of all amounts they claim should be paid under Paragraph 17 and all information upon which they relied in the preparation of the restoration plan and the accompanying cost statement.

On August 21, 2006, the US Department of Justice and the State of Alaska's Department of Law sent Exxon a Demand of Cost of Restoration under Reopener for Unknown Injuries[6] for $92.2 million. Exxon disputes the validity of that request based on the consent decree's language and requirements and on the findings of the scientific studies discussed in this volume. As of this writing, the status of the Demand has not been finally resolved.

The Reopener had the effect of directing much of the scientific study since the consent decree toward determining the likelihood, occurrence, or extent of potential unanticipated injuries to populations, habitats, or species. Exxon and the Trustees (through the settlement funds) separately sponsored hundreds of scientific studies of the spill and its effects. The two groups of scientists often conducted research on the same topics, in small part to address concerns arising in the private plaintiffs' litigation, but largely to address the governments' concerns over perceived or anticipated environmental damage. Much of the Trustee-sponsored research following the settlement concentrated on documenting continuing harm from the spill, particularly whether or how lingering subsurface oil residues (SSOR) might affect individual organisms, particularly fish, seaducks, and sea otters. This emphasis, in turn, led several Exxon-supported studies to investigate and model these potential linkages, as detailed in Chapters 12–16. The context of the litigation certainly influenced the topics studied and, in part, the timing of some of the scientific studies. The adversarial nature of that litigation may have affected the tone of the publicity arising from the studies, but ultimately the give and take of opinions about scientific results and their interpretations was no different from the healthy discourse of scientific analyses that is central to the scientific process. In the end, the litigation and the adversarial pressures it provided reinforced the rigor of the scientific work by the Trustees and Exxon.

Notes

[1] http://www.evostc.state.ak.us/facts/settlement.cfm.
[2] Compensation was already paid to the plaintiffs from a settlement with Alyeska Pipeline Service Company (which operates the Trans Alaska Pipeline and the Valdez tanker terminal), the Trans Alaska Pipeline Liability Fund (a statutory fund to compensate for oil spill damages related to the pipeline), and by Exxon (from its voluntary claims program).
[3] http://www.evostc.state.ak.us/facts/settlement.cfm.
[4] http://www.evostc.state.ak.us/facts/restorationplan.cfm.
[5] Paragraphs 17–19 (pages 18–19) of http://www.evostc.state.ak.us/pdf/settlement/agreement_consent_decree093091.pdf.
[6] http://www.evostc.state.ak.us/Files.cfm?doc=/Store/Event_Documents/Reopener_Demand_Letter.pdf&.

the way. The following chapters do more than review the results of the scientific studies conducted by the many parties; they convey a sense of how and why the science was done the way it was, and they distill out the important lessons that can help to bring good science to bear on future spills or environmental disturbances.

REFERENCES

Adams, J., P. Lavin, and A. Turrini (2002). *Prince William Sound Biological Hot Spots*. Anchorage, AK, USA: National Wildlife Federation.

Alaska Department of Fish and Game (2012). *Catalog of Waters Important for the Spawning, Rearing or Migration of Anadromous Fishes and its Associated Atlas*. Anchorage, AK, USA: Alaska Department of Fish and Game, Division of Sport Fish. [http://www.adfg.alaska.gov/cfanc/sfpublic/SARR/AWC/]

Anderson, P.J. and J.F. Piatt (1999). Trophic reorganization in the Gulf of Alaska following ocean climate regime shift. *Marine Ecology Progress Series* **189**: 117–123.

Bence, A.E., K.A. Kvenvolden, and M.C. Kennicutt, II (1996). Organic geochemistry applied to environmental assessments of Prince William Sound, Alaska, after the *Exxon Valdez* oil spill: A review. *Organic Geochemistry* **24**(1): 7–42.

Blanchard, A.L., H.M. Feder, and M.K. Hoberg (2010). Temporal variability of benthic communities in an Alaskan glacial fjord, 1971–2007. *Marine Environmental Research* **69**(1): 95–107.

Boehm, P.D., D.S. Page, E.S. Gilfillan, W.A. Stubblefield, and E.J. Harner (1995). Shoreline Ecology Program for Prince William Sound, Alaska, following the *Exxon Valdez* oil spill: Part 2 – Chemistry and toxicology. In *Exxon Valdez Oil Spill: Fate and Effects in Alaskan Waters*. P.G. Wells, J.N. Butler, and J.S. Hughes, eds. Philadelphia, PA, USA: American Society for Testing and Materials; ASTM Special Technical Publication 1219; ISBN-10: 0803118961; pp. 347–397.

Boehm, P.D., J.M. Neff, J.S. Brown, D.S. Page, W.A. Burns, A.W. Maki, and A.E. Bence (2003). The chemical baseline as a key to defining continuing injury and recovery of Prince William Sound. In *Proceedings of the 2003 International Oil Spill Conference (Prevention, Preparedness, Response and Restoration – Perspectives for a Cleaner Environment), April 6–11, 2003, Vancouver, British Columbia, Canada*. Washington DC, USA: American Petroleum Institute; API Publication I4730B (paper), I4730A (CD); 275–283.

Boehm, P.D., D.S. Page, J.S. Brown, J.M. Neff, J.R. Bragg, and R.M. Atlas (2008). Distribution and weathering of crude oil residues on shorelines 18 years after the *Exxon Valdez* spill. *Environmental Science & Technology* **42**(24): 9210–9216.

Brady, J., K. Schultz, E. Simpson, E. Biggs, S. Sharr, and K. Robertson (1990). *Prince William Sound Area Annual Finfish Management Report 1988*. Cordova, AK, USA: Alaska Department of Fish and Game; Regional Information Report 2C90–02. [http://www.adfg.alaska.gov/FedAidPDFs/RIR.2C.1990.02.pdf]

Bressler, D.C. and M.R. Gray (2003). Transport and reaction processes in bioremediation of organic contaminants. 1. Review of bacterial degradation and transport. *International Journal of Chemical Reactor Engineering* **1**(1): 1–16; published online June 19, 2003.

Coll, S. (2012). *Private Empire: ExxonMobil and American Power*. New York, NY, USA: Penguin Group (USA); ISBN-10: 1594203350; ISBN-13: 9781594203350.

Connell, J.H. (1961). The influence of interspecific competition and other factors on the distribution of the barnacle *Chthamalus stellatus*. *Ecology* **42**(4): 710–722.

Davidson, A. (1990). *In the Wake of the Exxon Valdez. The Devastating Impact of the Alaska Oil Spill*. San Francisco, CA, USA: Sierra Club Books; ISBN-10: 0871566141.

de Laguna, F. (1956). *Chugach Prehistory: The Archaeology of Prince William Sound*. Seattle, WA, USA: University of Washington Press; University of Washington Publications in Anthropology; Volume 13.

Exxon Valdez Oil Spill Trustee Council (1989). *The State/Federal Natural Resource Damage Assessment Plan for the Exxon Valdez Oil Spill*. Anchorage, AK, USA: *Exxon Valdez* Oil Spill Trustee Council. [http://www.evostc.state.ak.us/Universal/Documents/Publications/Workplans/1989Workplan.pdf]

Finney, B.P., I. Gregory-Eaves, M.S.V. Douglas, and J.P. Smol (2002). Fisheries productivity in the northeastern Pacific Ocean over the past 2,200 years. *Nature* **416**(6882): 729–733.

Fitzhugh, W.W. and V. Chaussonnet (1994). *Anthropology of the North Pacific Rim*. Washington DC, USA: Smithsonian Institution Press; ISBN-10: 1560982020; ISBN-13: 9781560982029.

Francis, R.C. and S.R. Hare (1994). Decadal-scale regime shifts in the large marine ecosystems of the North-east Pacific: A case for historical science. *Fisheries Oceanography* **3**(4): 279–291.

Gaichas, S., G. Skaret, J. Falk-Petersen, J.S. Link, W. Overholtz, B.A. Megrey, H. Gjødsaeter, W.T. Stockhausen, A. Dommasnes, K.D. Friedland, and K. Aydin (2009). A comparison of community and trophic structure in five marine ecosystems based on energy budgets and system metrics. *Progress in Oceanography* **81**(1–4): 47–62.

Gibeaut, J.C. and E. Piper (1998). *1993 Shoreline Oiling Assessment of the Exxon Valdez Oil Spill*. Juneau, AK, USA: Alaska Department of Environmental Conservation, Office of Restoration and Damage Assessment; *Exxon Valdez* Oil Spill Restoration Project 93038 Final Report. [http://www.evostc.state.ak.us/Files.cfm?doc=/Store/FinalReports/1993-93038-Final.pdf&]

Hare, S.R. and N.J. Mantua (2000). Empirical evidence for North Pacific regime shifts in 1977 and 1989. *Progress in Oceanography* **47**(2–3): 103–145.

Harrison, O.R. (1991). An overview of the *Exxon Valdez* oil spill. In *Proceedings of the 1991 International Oil Spill Conference (Prevention, Behavior, Control, Cleanup), March 4–7, 1991, San Diego, California*. Washington DC, USA: American Petroleum Institute Technical Publication 4529; pp. 313–319.

Haven, S.B. (1971). Effects of land-level changes on intertidal invertebrates, with discussion of post earthquake ecological succession. In *The Great Alaska Earthquake of 1964: Biology*. Washington DC, USA: National Academy of Science, National Academy Press; NAS Publication 1604; pp. 82–126.

Hugenin, M.T., D.H. Haury, J.C. Weiss, D. Heldon, C.-A. Manen, E. Reinharz, and J. Michel (1996). *Injury Assessment. Guidance Document for Natural Resource Damage Assessment under the Oil Pollution Act of 1990*. Silver Spring, MD, USA: National Oceanic and Atmospheric Administration, Damage Assessment and Restoration Program. [http://www.darrp.noaa.gov/library/pdf/iad.pdf]

Keeble, J. (1991). *Out of the Channel. The Exxon Valdez Oil Spill in Prince William Sound*. New York, NY, USA: HarperCollins; ISBN-10: 091005553X (paperback); ISBN-10: 0910055548 (hardback).

Kvenvolden, K.A., P.R. Carlson, C.N. Threlkeld, and A. Warden (1993). Possible connection between two Alaskan catastrophes occurring 25 yr apart (1964 and 1989). *Geology* **21**(9): 813–816.

Kvenvolden, K.A., F.D. Hostettler, P.R. Carlson, J.B. Rapp, C.N. Threlkeld, and A. Warden (1995). Ubiquitous tarballs with a California-source signature on the shorelines of Prince William Sound, Alaska. *Environmental Science & Technology* **29**(10): 2684–2694.

Leacock, E. (2005). *The Exxon Valdez Oil Spill*. New York, NY, USA: Facts On File; ISBN-10: 0816057540.

Lebedoff, D. (1997). *Cleaning Up. The Story Behind the Biggest Legal Bonanza of Our Time*. New York, NY, USA: Simon & Schuster, The Free Press; ISBN-10: 0684837064.

Leschine, T.M., J. McGee, R. Gaunt, A. van Emmerik, D.M. McGuire, R. Travis, and R. McCready (1993). T/V Exxon Valdez *Oil Spill: Federal On Scene Coordinator's Report*. Washington DC, USA: United States Department of Transportation, United States Coast Guard; Report DOT-SRP-94–1; National Technical Information Service Order Number PB94–121845 (Volume 1) and PB-121852 (Volume 2).

Lethcoe, J. and N. Lethcoe (2001). *History of Prince William Sound, Alaska (Revised 2nd Edition)*. Valdez, AK, USA: Prince William Sound Books; ISBN-10: 1877900125; 1877900125; ISBN-13: 9781877900129.

Litzow, M.A. (2006). Climate regime shifts and community reorganization in the Gulf of Alaska: how do recent shifts compare with 1976/1977? *ICES Journal of Marine Science: Journal du Conseil* **63**(8): 1386–1396.

Losey, R.J. (2005). Earthquakes and tsunami as elements of environmental disturbance on the northwest coast of North America. *Journal of Anthropological Archaeology* **24**(2): 101–116.

Loughlin, T.R., ed. (1994). *Marine Mammals and the Exxon Valdez*. San Diego, CA, USA: Academic Press; ISBN-10: 0124561608.

Mantua, N.J. and S.R. Hare (2002). The Pacific decadal oscillation. *Journal of Oceanography* **58**(1): 35–44.

Mantua, N.J., S.R. Hare, Y. Zhang, J.M. Wallace, and R.C. Francis (1997). A Pacific interdecadal climate oscillation with impacts on salmon production. *Bulletin of the American Meteorological Society* **78**(6): 1069–1079.

Michel, J., M.O. Hayes, W.J. Sexton, J.C. Gibeaut, and C. Henry (1991). Trends in natural removal of the Exxon Valdez oil spill in Prince William Sound from September 1989 to May 1990. In *Proceedings of the 1991 International Oil Spill Conference (Prevention, Behavior, Control, Cleanup), March 4–7, 1991, San Diego, California*. Washington DC, USA: American Petroleum Institute Technical Publication 4529; pp. 181–187.

Michel, J., Z. Nixon, and L. Cotsapas (2006). *Evaluation of Oil Remediation Technologies for Lingering Oil from the Exxon Valdez Oil Spill in Prince William Sound, Alaska*. Juneau, AK, USA: National Oceanic and Atmospheric Administration, National Marine Fisheries Service; *Exxon Valdez* Oil Spill Restoration Project 050778 Final Report. [http://www.evostc.state.ak.us/Files.cfm?doc=/Store/FinalReports/2005–050778-Final.pdf&]

Mundy, P.R., ed. (2005). *The Gulf of Alaska: Biology and Oceanography*. Fairbanks, AK, USA: Alaska Sea Grant College Program, University of Alaska; Publication AK-SG–05–01; ISBN-10: 156612090x.

Neff, J.M., A.E. Bence, K.R. Parker, D.S. Page, J.S. Brown, and P.D. Boehm (2006). Bioavailability of PAH from buried shoreline oil residues thirteen years after the *Exxon Valdez* oil spill: a multispecies assessment. *Environmental Toxicology and Chemistry* **25**(4): 947–961.

Neff, J.M., E.H. Owens, S.W. Stoker, and D.M. McCormick (1995). Shoreline oiling conditions in Prince William Sound following the *Exxon Valdez* oil spill. In Exxon Valdez *Oil Spill: Fate and Effects in Alaskan Waters*. P.G. Wells, J.N. Butler, and J.S. Hughes, eds. Philadelphia, PA, USA: American Society for Testing and Materials; ASTM Special Technical Publication 1219; ISBN-10: 0803118961; pp. 12–346.

Neff, J.M., D.S. Page, and P.D. Boehm (2011). Exposure of sea otters and harlequin ducks in Prince William Sound, Alaska, to shoreline oil residues 20 years after the *Exxon Valdez* oil spill. *Environmental Toxicology and Chemistry* **30**(3): 659–672.

Neff, J.M. and W.A. Stubblefield (1995). Chemical and toxicological evaluation of water quality following the *Exxon Valdez* oil spill. In Exxon Valdez *Oil Spill: Fate and Effects in Alaskan Waters*. P.G. Wells, J.N. Butler, and J.S. Hughes, eds. Philadelphia, PA, USA: American Society for Testing and Materials; ASTM Special Technical Publication 1219; ISBN-10: 0803118961; pp. 141–177.

Ott, R. (2005). *Sound Truth and Corporate Myth$: The Legacy of the* Exxon Valdez *Oil Spill*. Cordova, AK, USA: Dragonfly Sisters Press; ISBN-10: 0964522667.

Ott, R. (2008). *Not One Drop: Betrayal and Courage in the Wake of the* Exxon Valdez *Oil Spill*. White River Junction, VT, USA: Chelsea Green Publishing; ISBN-10: 1933392584; ISBN-13: 9781933392585.

Overland, J.E., S. Rodionov, S. Minobe, and N. Bond (2008). North Pacific regime shifts: definitions, issues and recent transitions. *Progress in Oceanography* **77**(2–3): 92–102.

Owen, B.M., D.A. Argue, H.W. Furchgott-Roth, G.J. Hurdle, and G. Mosteller (1995). *The Economics of a Disaster: The* Exxon Valdez *Oil Spill*. Westport, CT, USA: Quorum Books; ISBN-10: 0899309879.

Owens, E.H., W. Robson, and B. Humphrey (1987). Observations from a site visit to the *Metula* spill 12 years after the incident. *Spill Technology Newsletter (Environment Canada)* **12**(3): 83–96.

Page, D.S., P.D. Boehm, G.S. Douglas, and A.E. Bence (1995). Identification of hydrocarbon sources in the benthic sediments of Prince William Sound and the Gulf of Alaska following the *Exxon Valdez* oil spill. In Exxon Valdez *Oil Spill: Fate and Effects in Alaskan Waters*. P.G. Wells, J.N. Butler, and J.S. Hughes, eds. Philadelphia, PA, USA: American Society for Testing and Materials; ASTM Special Technical Publication 1219; ISBN-10: 0803118961; pp. 41–83.

Page, D.S., P.D. Boehm, G.S. Douglas, A.E. Bence, W.A. Burns, and P.J. Mankiewicz (1996). The natural petroleum hydrocarbon background in subtidal sediments of Prince William Sound, Alaska, USA. *Environmental Toxicology and Chemistry* **15**(8): 1266–1281.

Page, D.S., P.D. Boehm, G.S. Douglas, A.E. Bence, W.A. Burns, and P.J. Mankiewicz (1999). Pyrogenic polycyclic aromatic hydrocarbons in sediments record past human activity: a case study in Prince William Sound, Alaska. *Marine Pollution Bulletin* **38**(4): 247–260.

Page, D.S., P.D. Boehm, W.A. Stubblefield, K.R. Parker, E.S. Gilfillan, J.M. Neff, and A.W. Maki (2002). Hydrocarbon composition and toxicity of sediments following the *Exxon Valdez* oil spill in Prince William Sound, Alaska, USA. *Environmental Toxicology and Chemistry* **21**(7): 1438–1450.

Paine, R.T. (1966). Food web complexity and species diversity. *American Naturalist* **100**(910): 65–75.

Peterson, W.T. and F.B. Schwing (2003). A new climate regime in northeast Pacific ecosystems. *Geophysical Research Letters* **30**(17): 1896–1899; DOI: 10.1029/2003GL017528.

Plafker, G. (1965). Tectonic deformation associated with the 1964 Alaska Earthquake. *Science* **148**(3678): 1675–1687.

Plafker, G. (1969). *Tectonics of the March 27, 1964, Alaska Earthquake.* Washington DC, USA: US Geological Survey Professional Paper 543-I; plate 2, scale 1:500,000. [*Plate 2*: http://www.dggs.alaska.gov/webpubs/usgs/p/oversized/p0543ipt02.PDF] [*Paper*: http://www.dggs.alaska.gov/webpubs/usgs/p/text/p0543i.PDF]

Reimnitz, E. (1966). *Late Quaternary History and Sedimentation of the Copper River Delta and Vicinity, Alaska.* San Diego, CA, USA: University of California; Ph.D. Dissertation; UMI (ProQuest) Publication Order No. 6614473. [http://disexpress.umi.com/dxweb]

Reimnitz, E. and N.F. Marshall (1965). Effects of the Alaska earthquake and tsunami on recent deltaic sediments. *Journal of Geophysical Research* **70**(10): 2363–2376.

Rice, S.D., R.B. Spies, D.A. Wolfe, and B.A. Wright, eds (1996). *Proceedings of the* Exxon Valdez *Oil Spill Symposium.* Bethesda, MD, USA: American Fisheries Society; Symposium 18; ISBN-10: 0913235954; ISSN: 08922284.

Royer, T.C. and C.E. Grosch (2006). Ocean warming and freshening in the northern Gulf of Alaska. *Geophysical Research Letters* **33**: L16605; doi:10.1029/2006GL026767.

Royer, T.C., C.E. Grosch, and L.A. Mysak (2001). Interdecadal variability of Northeast Pacific coastal freshwater and its implications on biological productivity. *Progress in Oceanography* **49**(1–4): 95–111.

Royer, T.C., J.A. Vermersch, T.J. Weingartner, H.J. Neibauer, and R.D. Nuench (1990). Ocean circulation influencing the *Exxon Valdez* oil spill. *Oceanography* **3**(2): 3–10.

Shanks, A.L. and W.G. Wright (1986). Adding teeth to wave action: The destructive effects of wave-borne rocks on intertidal organisms. *Oecologia* **69**(3): 420–428.

Sharma, G.D. (1979). *The Alaskan Shelf: Hydrographic, Sedimentary, and Geochemical Environment.* New York, NY, USA: Springer-Verlag; ISBN-10: 0387903976; ISBN-13: 9780387903972.

Short, J.W., M.R. Lindeberg, P.M. Harris, J. Maselko, J.J. Pella, and S.D. Rice (2004). Estimate of oil persisting on the beaches of Prince William Sound 12 years after the *Exxon Valdez* oil spill. *Environmental Science & Technology* **38**(1): 19–25.

Short, J.W., M.R. Lindeberg, P.M. Harris, J. Maselko, and S.D. Rice (2002). Vertical oil distribution within the intertidal zone 12 years after the *Exxon Valdez* oil spill in Prince William Sound Alaska. In *Proceedings of the Twenty-Fifth Arctic and Marine Oilspill Program (AMOP) Technical Seminar, June 11–13, 2002, Calgary, Alberta, Canada.* Ottawa, ON, Canada: Environment Canada; pp. 57–72.

Smith, C. (1992). *Media and Apocalypse: News Coverage of the Yellowstone Forest Fires,* Exxon Valdez *Oil Spill, and Loma Prieta Earthquake.* Westport, CT, USA: Greenwood Press; Contributions to the Study of Mass Media and Communications, Number 36; ISBN-10: 0313277257; ISSN: 07324456.

Spies, R.B., ed. (2007). *Long-Term Ecological Change in the Northern Gulf of Alaska*. Amsterdam, The Netherlands: Elsevier; ISBN10: 0444529608; ISBN-13: 9780444529602.

Springer, A.M. (2007). Seabirds in the Gulf of Alaska. In *Long-Term Ecological Change in the Northern Gulf of Alaska*. R.B. Spies, ed. Amsterdam, the Netherlands: Elsevier; ISBN10: 0444529608; ISBN-13: 9780444529602; pp. 311–335.

Taylor, E. and D. Reimer (2008). Oil persistence on beaches in Prince William Sound – A review of SCAT surveys conducted from 1989 to 2002. *Marine Pollution Bulletin* 56(3): 458–474.

Trenberth, K.E. and T.J. Hoar (1996). The 1990–1995 El Niño-Southern Oscillation event: longest on record. *Geophysical Research Letters* 23(1): 57–60; DOI:10.1029/95GL03602.

Trenberth, K.E. and T.J. Hoar (1997). El Niño and climate change. *Geophysical Research Letters* 24(23): 3057–3060; DOI:10.1029/97GL03092.

Trites, A.W., V.B. Deecke, E.J. Gregr, J.K.B. Ford, and P.F. Olesiuk (2006). Killer whales, whaling, and sequential megafaunal collapse in the North Pacific: a comparative analysis of the dynamics of marine mammals in Alaska and British Columbia following commercial whaling. *Marine Mammal Science* 23(4): 751–765.

Vandermeulen, J.H. and D.C. Gordon (1976). Reentry of 5-year oil stranded Bunker C fuel oil from a low-energy beach into water, sediments, and biota of Chedabucto Bay, Nova Scotia. In *Pollution Symposium – 13th Pacific Science Congress (Mankind's Future in the Pacific): Sublethal Effects of Pollution on Aquatic Organisms, August 18–30, 1975, Vancouver, British Columbia. Journal of the Fisheries Research Board of Canada* 33(9): 2002–2010.

Vandermeulen, J.H. and J.G. Singh (1994). *Arrow* oil spill, 1970–90: Persistence of 20-yr weathered Bunker C fuel oil. *Canadian Journal of Fisheries and Aquatic Sciences* 51(4): 845–855.

Wells, P.G., J.N. Butler, and J.S. Hughes, eds (1995). Exxon Valdez *Oil Spill: Fate and Effects in Alaskan Waters*. Philadelphia, PA, USA: American Society for Testing and Materials; ASTM Special Technical Publication 1219; ISBN-10: 0803118961.

Wheelwright, J. (1994). *Degrees of Disaster. Prince William Sound: How Nature Reels and Rebounds*. New York, NY, USA: Simon & Schuster; ISBN-10: 0671702416.

Wiens, J.A., R.H. Day, S.M. Murphy, and M.A. Fraker (2010). Assessing cause-effect relationships in environmental accidents: Harlequin ducks and the *Exxon Valdez* oil spill. In *Current Ornithology*. C.F. Thompson, ed. New York, NY, USA: Springer; Volume 17; ISBN-13: 9781441964205; pp. 131–189.

Wilson, B.W. and A. Torum (1972). Effects of the tsunamis: an engineering study. In *The Great Alaska Earthquake of 1964: Oceanography and Coastal Engineering*. Washington DC, USA: National Academy of Science, National Academy Press; NAS Publication 1604; pp. 361–523.

Wilson, F.H. and C.P. Hults (2008). *Preliminary Integrated Geologic Map Databases for the United States: Digital Data for the Reconnaissance Geologic Map for Prince William Sound and the Kenai Peninsula, Alaska*. Washington DC, USA: US Geological Survey; Open-File Report 2008–1002. [pubs.usgs.gov/of/2008/1002]

Wolfe, D.A., M.J. Hameedi, J.A. Galt, G. Watabayashi, J. Short, C. O'Claire, S. Rice, J. Michel, J.R. Payne, J. Braddock, S. Hanna, and D. Sale (1994). The fate of the oil spilled from the *Exxon Valdez*. *Environmental Science & Technology* **28**(13): A560–A568.

Wooley, C.B. (2002). The myth of the "pristine environment": past human impact on the Gulf of Alaska coast. *Spill Science and Technology Bulletin* **7**(1–2): 89–104.

Yarborough, L.F. (1997). *Site Specific Archaeological Restoration at SEW-440 and SEW-488*. Anchorage, AK, USA: US Department of Agriculture and US Forest Service, Chugach National Forest; *Exxon Valdez* Oil Spill Restoration Project 95007B Final Report. [http://www.evostc.state.ak.us/Files.cfm?doc=/Store/FinalReports/1995–95007B-Final.pdf&]

Yarborough, M.R. and L.F. Yarborough (1996). *Uqciuvit: A Multicomponent Site in Northwestern Prince William Sound, Alaska*. Anchorage, AK, USA: US Department of Agriculture, Forest Service, Chugach National Forest; Final Site Report; Contract No. 53-01 14-7-00132.

Yarborough, M.R. and L.F. Yarborough (1998). Prehistoric maritime adaptations of Prince William Sound and the Pacific coast of the Kenai Peninsula. *Arctic Anthropology* **35**(1): 132–145.

Yatsu, A., K.Y. Aydin, J.R. King, G.A. McFarlane, S. Chiba, K. Tadokoro, M. Kaeriyama, and Y. Watanabe (2008). Elucidating dynamic responses of North Pacific fish populations to climatic forcing: influence of life-history strategy. *Progress in Oceanography* **77**(2–3): 252–268.

CHAPTER TWO

The phases of an oil spill and scientific studies of spill effects

Paul D. Boehm, Erich R. Gundlach, and David S. Page

2.1 Introduction and overview

Following a marine oil spill, it is important to know where the oil goes, how it changes chemically, how long the oil persists in various environmental compartments such as water or sediments, and what biological resources are affected. As an oil spill progresses over time, the behavior of the oil and the impacted areas and levels of risk to people and biological resources such as fish and wildlife change. Scientific studies provide the most benefit to cleanup efforts and the protection of people and biological resources in the area when they are coordinated and focused on the most pressing questions based on the phase of the oil spill. Over several decades, previous marine oil spills have shown a consistent pattern; understanding this pattern can help predict where the oil will go, how it changes chemically, where it will persist, and what living things are likely to be affected. Similarly, this predictability, coupled with specific observations at each spill, can help to provide a framework for designing and conducting studies that can address key questions at critical junctures in the evolution of the spill.

Water and air are the first environmental media affected during the early phase of any marine spill. Animal and plant life (or "biological resources") can be affected immediately – as can humans involved in spill cleanup. The initial exposures to the chemicals in petroleum and the resulting effects can be acute, but short-lived. This is because once the spilled oil is no longer moving on or in the water, concentrations of harmful chemicals decrease rapidly owing to dilution, dispersion, and degradation (collectively known as "weathering"). Likewise, the evaporation of the volatile hydrocarbon components of fuels or crude oil immediately following a spill first increases, then decreases. By contrast, the effects on shoreline biological resources from oil that reaches land may persist. The last area potentially to be affected is bottom sediments, where oil can be transported before or after it reaches land (Chapter 4).

Oil in the Environment: Legacies and Lessons of the Exxon Valdez *Oil Spill*, ed. J. A. Wiens. Published by Cambridge University Press. © Cambridge University Press 2013.

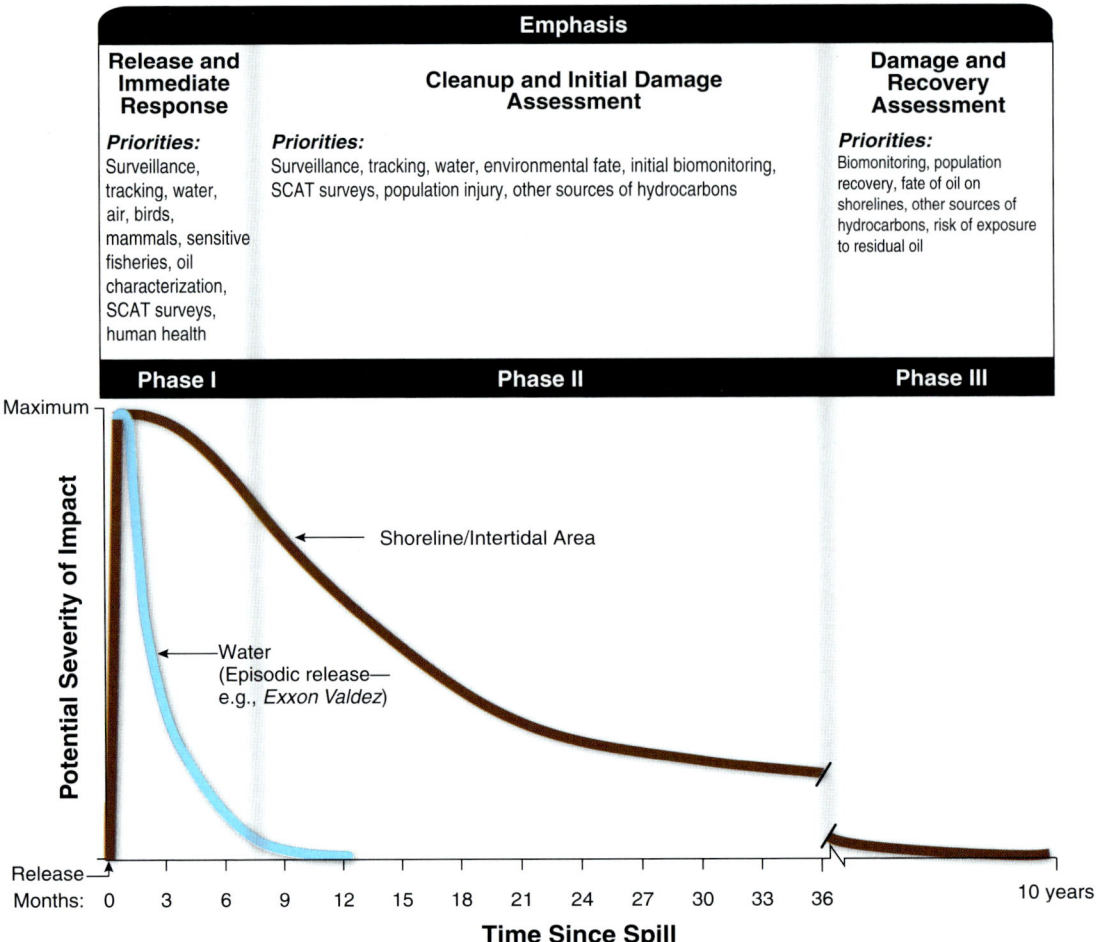

Figure 2.1 General phases of investigation of an oil spill. Scientific studies are implemented and evolve with the movement and fate of the oil. The timing of phases and studies will vary according to the specific spill scenario and events. Studies may begin in one phase and extend into another.

Many strategic and scientific questions are addressed during a spill, including:

- What are the exposure pathways from the spilled oil to biological resources?
- What concentrations of petroleum chemicals are found in the water and in shoreline sediments?
- To what concentrations are biological resources exposed? Do these exposure levels exceed toxicological thresholds?
- What specific biota have been affected and how severely?
- What habitats have been contaminated and to what extent?
- How quickly and to what extent does the spilled oil weather?
- How quickly do the concentrations and related toxicity decline, either because of cleanup or natural processes?
- By what measures should injury and recovery be assessed?

To begin answering these questions – whether assessing a tanker accident, a pipeline rupture, or the prolonged release of oil from an ocean platform – scientists need to

Table 2.1 The characteristics of the five major oil groups and their behaviors (modified from Boehm and Page, 2007).

Group I: Very Light (jet fuels, gasoline)
- Highly volatile and flammable (evaporates quickly)
- High toxicity from soluble compounds
- Toxicity causes localized severe effects on water column and intertidal resources
- Resource recovery rate determines effect duration
- Do not form emulsions in water
- Removal methodologies are limited for safety and efficacy reasons

Group II: Light (diesel, No. 2 fuel oil, high API gravity oils, light crude oils)
- Up to 50% of Group II crudes will remain as a residue
- Moderate toxicity from soluble compounds
- May cause long-term intertidal contamination
- Has potential for subtidal impacts (dissolution, mixing, sorption to suspended sediments)
- Chemical dispersal not usually warranted
- Can form emulsions in water

Group III: Medium (most crude oils)
- About 50% will evaporate within 24 hours
- May cause long-term intertidal effects and severely affect birds and fur-bearing mammals
- Chemical dispersion may be possible for a period of time dependent on location
- Can form emulsions in water
- Quick cleanup is most effective

Group IV: Heavy (heavy crude oils, Bunker C)
- Limited evaporation or dissolution
- Significantly impacts intertidal areas and birds and fur-bearing mammals (coating and ingestion)
- Long-term sediment contamination possible
- Weathers very slowly
- Chemical dispersion seldom effective
- Can form emulsions in water
- Shoreline cleanup is difficult

Group V: Very Heavy (low API gravity oils, No. 6 fuel oil, asphalt)
- Can be denser than water; may float, remain neutrally buoyant, have floating and sinking fractions, or may sink
- No evaporation when submerged
- Very viscous to semisolid
- Effects from smothering, not direct toxicity
- Weathers very slowly
- Can form emulsions in water under appropriate conditions
- Subsurface cleanup methodologies are limited

know where the oil is going, how the oil is changing physically and chemically, and what areas and resources may be impacted and in need of protection. In general, early scientific investigations aim to link the distribution across space and time of potentially harmful oil components with exposure to these components (Gentile et al., 1993). These links are examined through a series of focused oil-spill investigations that generally follow the phases of spill development over time.

The scale of the *Exxon Valdez* spill required substantial scientific studies over many years to inform both cleanup and the assessment of short- and long-term environmental effects, including those related to natural resource damage assessment (NRDA) and civil litigation of damages. A general progression of exposure and impact can be depicted in simplified form (Fig. 2.1). The details and exact timing of these phases are highly dependent on the spill-release scenario, the characteristics of the oil (Table 2.1), and the characteristics of the receiving environment and its biological resources. Predicting the transport and behavior of spilled oil is important because it frames the sequence of scientific studies (Fig. 2.1). With the *Exxon Valdez* spill, extensive studies of the water column and its resources gave way over time to a focus on the nearshore and shoreline resources.

2.2 The phases of an oil spill

Based on the driving priorities and roughly defined time periods, major oil spills can be divided into three overlapping phases (Boehm and Page, 2007; Fig. 2.1). Such was the case with the *Exxon Valdez* spill. In this chapter we present an overview of Exxon- and government-supported studies by phase, providing a framework for prioritization and timing following a marine oil spill.

Several factors combine to define the scientific questions addressed during an oil spill. These factors include the circumstances of the spill, details of the transport and fate of the oil within the various environmental compartments, issues relating to spill response and cleanup, and the exposure and risk to humans and to natural resources. The phases relate to where the oil moves and its rate of removal by active cleanup and natural processes. Although these phases are relevant to every oil spill, the timing and emphasis will shift depending on the particular features of the event. Scientists must make time-critical decisions about collecting data very early on, often in response to priorities set by response managers and other authorities. After the initial fate of the oil is determined, scientists can plan for later-phase investigations. The priorities and investigation components of most oil spills are summarized in Table 2.2.

For the *Exxon Valdez* spill, priorities were first based on the movement of oil on and in the water prior to shoreline impact. Later, studies focused on determining shoreline impact locations and the rates of removal (by weathering or cleanup) of oil from the shorelines (Short *et al.*, 2007). The three study phases were (Boehm and Page, 2007):

Phase 1 – release and immediate response (days to several months). This phase occurs when the spill is in progress and oil is on the water and continues until the oil is mostly on the shorelines. Immediate concerns are stopping the release, tracking the oil, protecting human health and fisheries, and identifying potentially affected natural resources. Surveying and measuring the extent of the offshore and shoreline "footprints" of the oil are central to the studies in this phase. Distribution information informs study designs and site selection for detailed ecological and injury assessments in Phases 2 and 3.

Phase 2 – cleanup (weeks to several years). As crews clean oil from shorelines, scientists center on continuing to track the distribution of oil on the shores (Harrison, 1991) and ensuring the responders' safety. Measurements center on quantifying the amounts of oil on and buried within shorelines, measuring temporal changes in the distribution and chemistry of oil, and assessing initial exposures and injury to natural resources.

Phase 3 – recovery (1 to 10 years or more). The focus is on assessing injury and monitoring recovery of affected species, as well as assessing the overall ecological recovery of the affected area. Weathered, visible oil may remain on the surface of shorelines, and some oil may persist as subsurface residues (Chapters 4, 6, and 7).

In each phase, scientists develop studies of the habitats affected and biological resources of greatest concern. Priorities for sampling (e.g., air, water, sediment, tissues) and types of analyses (visual observations versus screening-level analyses versus detailed analyses) are properly sequenced and merged to answer evolving questions and concerns (Table 2.2).

Table 2.2 Overview of oil-spill investigation phases

PHASE 1			
Description	Resource priorities	Measurement and sampling priorities	Data/information
Release and Immediate Response: Spill in progress; hours to 3 months; oil on water surface; some stranded but remobilized by waves and tides.	Human exposure; diving birds; marine mammals; reptiles; surface feeding fish; anadramous fish; fish in lakes and streams; human health issues.	Aerial overflights; shoreline surveys (air and ground level – SCAT[a]); air sample monitoring; collection of offshore and nearshore water associated with surface oil; source oil samples from release point; weathered oil samples; nearshore and stream-mouth water samples.	Distribution maps of oil on water surface and shorelines; BTEX[b] concentrations in air; PAH concentrations in water; characterization of released oil; PAH fingerprints of spill oil; background/reference site samples. Oil characterization and weathering. Observations of effects on surface dwelling/feeding wildlife. Data used to assess environmental fate, determine risks to fisheries, measure acute exposures to oil, and to design future studies.

PHASE 2			
Description	Resource priorities	Measurement and sampling priorities	Data/information
Cleanup Period: Weeks to several years; natural recovery and active shoreline cleanup.	Exposed receptors in Phase 1 (above) that may be still exposed to oil: intertidal and nearshore subtidal invertebrates and fish; shoreline-feeding/nesting birds and shoreline-feeding mammals.	Aerial overflights; shoreline surveys (air and ground level – SCAT); nearshore and stream-mouth water samples; subtidal or nearshore lake/river bottom sediments; intertidal or shoreline invertebrates; mussels for monitoring; worms, clams, etc., as prey for foraging birds and mammals; fish and mammals for biomarkers; sampling of fish eggs and water from spawning streams; location and sampling of nonspill reference sites including human use sites; subtidal/nearshore bottom sediments. Shoreline studies begin. Emphasis on biota serving as prey for foragers and on mussels as biomonitors of bioavailable PAH from remaining surface and buried residues; sampling of associated sediments; sampling of reference sites (biota and sediments); comprehensive determination of and sampling of other hydrocarbon sources (baseline).	Distribution maps of oil on water surface and shoreline; PAH concentrations in all samples; oil and PAH source verification (fingerprinting); data used to determine cleanup endpoints. Determination of bioavailability; sampling and analysis of biota and risk to wildlife.

PHASE 3			
Description	Resource priorities	Measurement and sampling priorities	Data/information
Recovery Period: Natural recovery; some weathered and visible oil residues may remain; risk is lowered owing to reduced bioavailability; some deeply buried oil (in shorelines) or inert asphaltic residues may remain.	All from Phase 2; remaining receptors at risk from residual oil.	Detailed scientific studies of injury to specific resources; population-level studies of birds and mammals; shoreline surveys and expansion of offshore sea-bottom studies, if warranted; emphasis on bioavailability for any remaining oil residues; sampling of associated sediments. Possible deployment of semipermeable membrane devices in areas devoid of bivalves; also sediments from shoreline and/or nearshore subtidal in support of sediment toxicity studies.	Surveys to assess the form, location, and extent of remaining spill residues; potential for exposure of biota; fish and wildlife population assessments. Includes chemical measurements (e.g., PAH) in all samples; PAH source verification (fingerprinting); includes bioavailability measurements. Data used to assess remaining risks and to refine estimates of actual return to baseline (reference site) conditions.

Source: Adapted from Boehm and Page (2007).
[a] Shoreline Cleanup Assessment Technique; see Chapter 4.
[b] Benzene, toluene, ethylbenzene, and xylenes; see Chapter 1, Box 1.1.

2.3 Studies in Phase 1: release and immediate response

Depending on the release scenario, this is typically the shortest phase, lasting only a few months. It requires rapid mobilization and a substantial number of critical response activities and measurements. This phase represents the greatest period of risk to surface-dwelling and diving birds, as well as marine mammals. Tracking the movement of oil and its distribution in the water column is critical to determining the risk to people, natural resources, commercial fisheries, and subsistence foods.

In Phase 1, oil moves on and in the water. Oil components in the water column occur as dissolved chemicals and as oil droplets or particulates (Chapter 3). Weathering rapidly disperses oil on the water surface and dilutes it within the water column. Countermeasures – such as open-water skimming, chemical dispersant application, or *in situ* burning – may be used.

One of the more important early efforts in the *Exxon Valdez* spill was the development of a rigorous water-sample-collection and chemical-analysis program. Thousands of water samples were collected and tracked (Chapter 3). The need for high-quality measurements of petroleum components (especially the polycyclic aromatic hydrocarbons [PAH]) at low detection limits drove quality-control studies and the selection and qualification of laboratories.

2.3.1 Issues addressed

The number of issues and the types of data collected in Phase 1 involve the movement, fate, and distribution of the oil on and in the water, the development and prioritization of cleanup efforts, and the need to identify potentially affected natural resources and people. These issues and the types of data needed in Phase 1 are usually the most complex. Efforts target the following questions:

- Are people at risk or potentially at risk in the spill area?
- What are the characteristics of the oil (Table 2.1)?
- Where is the oil going; how is its composition changing; and how quickly is it dissipating?
- Are fish or wildlife resources at risk in the spill area?
- What are the initial and maximum exposures of biological resources to oil and its chemical components?
- What are the response priorities and environmental tradeoffs of cleanup or other countermeasures?
- What sites are appropriate to select as unoiled reference locations?

Answering these questions requires extensive review of existing baseline information. It also requires rapid data collection and sample analysis, especially targeting data that may change or be quickly lost, adding to the chaos of Phase 1.

2.3.2 Data collection

The initial movements of oil on the surface and in the water column are related to the chemical and physical characteristics of the oil. To characterize the oil, large volumes (tens of liters) of representative samples of the oil should be collected quickly from the point of release. This oil should then be fully characterized, both physically and chemically.

Multiple oil samples are required, as oil properties may vary during release. The samples should be documented under chain-of-custody and appropriately stored for initial analyses and possible later use as reference material for laboratory studies. Additional surface-oil samples should also be collected throughout Phase 1 to measure temporal changes of the oil as it weathers. In offshore spills, oil stranded on shorelines will be substantially altered from that spilled. Similarly, oil that reaches the surface from a sea-bottom blowout will differ from the oil at the wellhead (Brown et al., 2011). Documenting changes in the oil over time is also important in river spills, where weathering occurs as oil is transported downstream.

Water measurements receive considerable attention during Phase 1, as these data are especially important for exposure assessment (Chapter 3) and are the quickest to change or be lost (Robilliard et al., 1997). Initial water samples document oil movement and quantify resource exposures.

Identifying biological resources that might be affected by spilled oil is an important component of Phase 1. Although resources in the general area may be known from prespill baseline data, surveys to confirm their abundances are needed. Aerial surveys often identify areas in which marine mammal and bird densities are associated with oiling. Environmental Sensitivity Index (ESI) maps, such as those generated by the National Oceanic and Atmospheric Administration (NOAA) and others, can help identify biological, economic, and cultural resources potentially at risk and the consequences of any responses. Spill contingency plans identify these resources, their locations, and response strategies for protection. These initial maps are crucial to the design and conduct of Phase 2 and Phase 3 damage-assessment efforts.

In Phase 1 of the *Exxon Valdez* spill, the large release of medium-weight crude oil over a short period (some 10 hours) on the surface combined with a major storm several days later spread oil out over a large area of western Prince William Sound (PWS). This area has important marine mammal, fisheries, and avian biological resources (see Harrison, 1991; Chapter 1); pink salmon (*Oncorhynchus gorbuscha*), Pacific herring (*Clupea pallasii*), marine-oriented birds, and sea otters (*Enhydra lutris*) were later identified as the most important species affected (see Chapters 12–15).

Aerial surveillance (Kelso and Kendziorek, 1991; Chapter 4) and water-column monitoring studies (Neff and Stubblefield, 1995; Short and Harris, 1996a, b; Boehm *et al.*, 2007; Chapter 3) began immediately. A major driver of scientific studies was the potential acute, but short-lived, exposure of water-column organisms (e.g., salmon eggs and alevins, herring eggs and larvae; Chapters 12 and 13), especially in critical habitat areas. Hence, the major focus was on collecting and analyzing surface and near-surface water samples. Seabird and mammal mortality and initial injuries were also assessed (Chapters 14 and 15), and extensive tracking and mapping of oil on the shorelines (Chapter 4) began.

As Phase 1 progressed from oil on the water to oil on the shorelines, the general objectives of the investigation remained the same: to track the oil, provide information to assist response efforts, and assess affected resources – particularly those associated with the water column. Systematic mapping of shoreline oil was vital (Owens and Sergy, 2000; Chapter 4). To quantify the fate of the oil, where the oil went and what happened to it (i.e., mass-balance calculations) were determined (Wolfe *et al.*, 1994).

During Phase 1, studies also focused on human health impacts, including the safety of commercial and subsistence fishing (Brown *et al.*, 1996; Yender *et al.*, 2002), cleanup worker exposure, and possible public exposure. Spill-trajectory maps and targeted data helped to identify potentially impacted fisheries and determine if and when closure and re-opening notices should be issued (Yender *et al.*, 2002). Experiences from the *Exxon Valdez* oil spill have served to shape more recent oil-spill seafood safety investigations (Field *et al.*, 1999).

2.4 Studies in Phase 2: cleanup

Phase 2 of an oil spill begins when the oil's initial transport pathways and fate have been largely established and the location of oil on shorelines and in sediments is known. During this phase, oil has settled on shorelines and is being actively removed by cleanup workers and weathering. In practice, Phase 1 and Phase 2 priorities naturally overlap, as they did in the *Exxon Valdez* spill: oil continued to move within PWS from tidal processes and active shoreline cleanup, so water monitoring continued through the summer of 1989.

Studies in Phase 2 involve monitoring shoreline cleanup and collecting initial data for natural resource damage assessments. Oil tracking and measurement that occur in Phase 2 are focused on the evolution of oil on the shoreline and the oil's changing chemistry, toxicology, rate of weathering, and its loss. Natural resource damage assessment becomes more targeted as the geographic scope of the impact zone narrows.

Phase 2 of the *Exxon Valdez* spill lasted until cleanup was completed in June 1992, 3 years after the release. Short *et al.* (2007) estimated initial annual rates of removal of buried shoreline oil during 1989–91 (see Chapters 4 and 6).

2.4.1 Issues addressed

In Phase 2, scientific studies begin determining whether, where, and to what extent oil has been transported to shorelines and offshore sediments, and whether natural resources in these habitats have suffered injuries (Table 2.2). Questions include:

- What are the amounts of oil and the concentrations of its chemical constituents on the shorelines, in subtidal sediments, and in key biota and prey species?
- What is the rate of oil loss from natural processes and active cleanup?
- Are biological, cultural, and other resources exposed to the spill? If so, which ones? What special precautions must be taken in cleanup to safeguard them?
- To what extent have key resources been injured, and which ones warrant study?
- What are the appropriate designs of scientific studies?

In Phase 2 of the *Exxon Valdez* spill, continuing, extensive shoreline surveys produced oiling maps (Chapters 1 and 4). These maps helped to identify locations for additional cleanup and study based on proximity to sensitive natural resources and/or human use. In general, and in the *Exxon Valdez* spill in particular, the Phase 2 studies provide data for damage and injury assessments in Phase 3.

Additionally, Phase 2 investigations begin to focus on oil that has penetrated shoreline sediments and has become isolated from the normal weathering processes and effectively sequestered (Boehm *et al.*, 2008; Page *et al.*, 2008; Li and Boufadel, 2010; Pope *et al.*, 2011; Chapters 4, 6, and 7). Scientific investigations in Phase 2 begin to focus on locating and surveying these deposits, monitoring their behavior over time, and assessing the possible exposure of wildlife and their prey species to these residues (Neff *et al.*, 2006). These sequestered deposits can persist much longer than the majority of shoreline oil, which does not become isolated (Hayes *et al.*, 1991; Michel and Hayes, 1993; Owens *et al.*, 2008; Page *et al.*, 2008; Li and Boufadel, 2010; Pope *et al.*, 2011).

2.4.2 Data collection

Sampling in this phase focuses on both the intertidal and subtidal zones. Understanding the evolution of oiled shorelines helps to determine the potential exposures of resident biota (especially during Phase 3) and supports decision-making on cleanup plans.

Exposure assessments become very important in Phase 2. Through these studies, scientists understand the potential injury to biota through both direct and indirect uptake of chemicals. Exposure assessments rely on chemical analyses of sediments and "sentinel species" for quantified, direct estimates of exposure. In the *Exxon Valdez* spill, analysis of the chemical body burdens of ubiquitous populations of blue mussels (*Mytilus trossulus*) was important for studies of the intertidal environment in Phases 2 and 3 (Shigenaka and Henry, 1995; Boehm *et al.*, 1996, 2004; Short and Babcock, 1996). Beginning a few months after the spill, and continuing through 2004 (e.g., Boehm *et al.*, 2007), water-quality estimates began to rely heavily on mussel PAH chemistry as water-column PAH levels rapidly decreased to levels at or below their analytical detection limits (Neff and Burns, 1996).

In Phase 2 of the *Exxon Valdez* spill, scientists quantified the initial effects of the spill on the resources in the intertidal zone and the risk posed by remaining oil. These effects and risks were quantified using spatially arrayed sediment sampling and companion biota measurements, based on the extent of shoreline oiling determined in Phase 1 and continuing into Phase 2 (Chapters 4 and 6). These investigations addressed the bioavailability of and risk from oil residues (Neff et al., 2006). The analyses of intertidal and subtidal prey species provided valuable data for the wildlife-exposure studies implemented in Phase 3.

An important objective of Phase 2 is to determine other nonspill hydrocarbon sources that are part of the prespill background. These may include petrogenic sources (fossil fuels), pyrogenic sources (combustion products) that are widespread in the environment (Neff, 1979; Brown et al., 1980), and biogenic sources (natural organic matter). High-quality chemical analyses are needed for these determinations. The compositions of PAH and particular geochemical biomarkers need to be compared with those from other local and regional sources. Sediment and biological samples with elevated PAH concentrations should be chemically "fingerprinted" to confirm that the hydrocarbons found are from the oil spill (Boehm et al., 1997; Bence et al., 2006; see Chapter 1, Box 1.1). Fingerprinting identifies the source and geographical extent of exposure, which are critical to considerations of recovery in Phase 3.

It is important to develop the capability to differentiate other hydrocarbon sources that are part of the prespill background early in a spill investigation, and it becomes critical when other hydrocarbon sources have a significant input relative to the spilled oil. In the *Exxon Valdez* studies, examination of the chemical background began during Phase 2 (Kvenvolden et al., 1993; Page et al., 1995a, b; Boehm et al., 2001) and continued into Phase 3 (Bence et al., 2006). An important finding late in Phase 2 was the presence of a pervasive, nonspill, natural petroleum hydrocarbon background in subtidal sediments (Page et al., 1995a; Boehm et al., 2001; Chapter 6). This finding addressed concerns that spilled oil was distributed throughout the Sound, and it redirected research to define the prespill background from *all* sources.

Data collection and a data-management system for all components of the spill response and damage assessment should be formally documented in Phase 2, as realities suggest that a single system is seldom fully achieved during Phase 1. Well-developed, scientifically driven study plans cannot be properly implemented unless accurate and validated data and related information are available (in an accessible and usable form) to make informed decisions.

2.5 Studies in Phase 3: recovery

Phase 3 generally begins when active cleanup of the shorelines winds down, although many Phase 2 studies may continue. In Phase 3, studies focusing on natural resource injury and recovery assessment use the information generated in Phases 1 and 2 to focus, refine, and enhance further scientific studies. The Phase 3 period may develop into the longest phase of scientific investigation of the oil spill, potentially lasting more than a decade (Fig. 2.1).

2.5.1 Issues addressed

The major issues addressed in Phase 3 are determining the injury to and recovery of injured biological resources and assessing the potential harm from remaining oil residues. While tracking and surveying of oil in the environment continue, several phase-specific questions guide the investigations:

- Where is the remaining oil? Was any overlooked in the survey programs undertaken in Phase 2?
- At what rate is the oil disappearing naturally?
- When does exposure to oil residues decline to levels that no longer pose a risk to biota?
- Have biological populations been affected?
- When do biological components of the system evidence recovery?

2.5.2 Data collection

Studies in Phase 3 consist of well-designed, statistically robust studies that focus on identifying the locations and chemical character of any remaining oil, assessing the bioavailability of those residues to resources of concern, assessing the risk to shoreline biota and wildlife that feed on shoreline prey, and determining population-level injury to and recovery of particular species of interest.

Phase 3 investigations of the *Exxon Valdez* spill initially focused on a broad array of species. Over time, a particular emphasis developed on harlequin ducks (*Histrionicus histrionicus*) (Esler *et al.*, 2002; Chapter 14) and sea otters (Bodkin *et al.*, 2002; Chapter 15). This emphasis was in response to concerns that these species could potentially be exposed to oil residues buried at some specific shoreline locations (Neff *et al.*, 2011; see Chapter 16).

Extensive post-2000 surveys of remaining oil deposits initiated by Short and co-workers (Short *et al.*, 2004) and carried forward in detail by others (e.g., Boehm *et al.*, 2008; Page *et al.*, 2008) helped to define the areas and specific locations where oil and its potential risk to biota remained. Phase 3 also included monitoring of residues in prey and sentinel species (Neff *et al.*, 2006; Payne *et al.*, 2008).

Studies in Phase 3 are most successful when the results can be applied to the area as a whole (or to a well-defined subset of the area) and projected forward in time. Often, a modified sediment-triad-type approach is designed, incorporating Phase 1 and 2 study results. This approach was used after the *Exxon Valdez* oil spill (Page *et al.*, 1995b; Chapter 11), where sediment chemistry, infaunal community composition, mussel studies/chemical analyses, and toxicity-test results were collected in combination. Specific prey species were also sampled at the same sites (Neff *et al.*, 2006). This coordinated effort provided a comprehensive view of the exposure environment and the potential for injury to biological resources.

2.6 Lessons learned

Experience in conducting scientific studies of oil spills over more than four decades has demonstrated the value of designing and sequencing investigations in relation to the general phases of spill evolution described here. The lessons learned from the spill response and damage assessment for the *Exxon Valdez* spill have reinforced and built upon those from earlier investigations. Collectively, these investigations provide a framework for studying the effects of future oil spills. This framework is a phased approach to addressing the scientific objectives of supporting oil-spill response efforts and assessing natural-resource damages. Here are the principal lessons we have learned:

- As an oil spill progresses through the release and immediate response, cleanup, and recovery phases, the affected areas and biological resources change, as do the questions that need to be scientifically addressed.

- The design and implementation of scientific studies are guided by the questions posed during each phase. Using the results of these studies to design appropriate follow-up studies is crucial. This ensures that decisions about cleanup or subsequent activities are based on good science, and that resource injury and recovery assessments are rigorous and scientifically sound.
- Some studies need be launched immediately, others at later stages. The immediate focus should be on valuable data that will be lost if not collected very early on (Phase 1). Spill tracking and offshore and shoreline surveys provide critical input to the design and implementation of subsequent studies (Phases 2 and 3).
- Conducting investigations in overlapping steps promotes the effective use of data, as information collected in one phase forms the foundation for study design in subsequent phases.
- To generate scientifically defensible data in a timely manner, sample collection and supporting analytical chemistry programs – with rigorous quality control – need to be designed early in Phase 1. Observations, samples, and data collected in the early phase of a spill will invariably be important to later studies and analyses of spill effects; consequently, such material must be managed and archived until all spill issues are resolved.
- Historical data about the spill-affected area and its resources can help in designing appropriate studies and understanding the effects of the spill. Samples and data from outside the spill-affected area can aid in developing an understanding of baseline conditions and the context of spill effects.

REFERENCES

Bence, A.E., D.S. Page, and P.D. Boehm (2006). Advances in forensic techniques for petroleum hydrocarbons: The *Exxon Valdez* experience. In *Oil Spill Environmental Forensics: Fingerprinting and Source Identification*. Z. Wang and S.A. Stout, eds. Amsterdam, The Netherlands: Elsevier Academic Press; ISBN-10: 0123695236; ISBN-13: 0123695236; pp. 449–487.

Bodkin, J.L., B.E. Ballachey, T.A. Dean, A.K. Fukuyama, S.C. Jewett, L. McDonald, D.H. Monson, C.E. O'Clair, and G.R. VanBlaricom (2002). Sea otter population status and the process of recovery from the 1989 *Exxon Valdez* oil spill. *Marine Ecology Progress Series* **241**: 237–253.

Boehm, P.D., G.S. Douglas, W.A. Burns, P.J. Mankiewicz, D.S. Page, and A.E. Bence (1997). Application of petroleum hydrocarbon chemical fingerprinting and allocation techniques after the *Exxon Valdez* oil spill. *Marine Pollution Bulletin* **34**(8): 599–613.

Boehm, P.D., P.J. Mankiewicz, R. Hartung, J.M. Neff, D.S. Page, E.S. Gilfillan, J.E. O'Reilly, and K. Parker (1996). Characterization of mussel beds with residual oil and the risk to foraging wildlife 4 years after the *Exxon Valdez* oil spill. *Environmental Toxicology and Chemistry* **15**(8): 1289–1303.

Boehm, P.D., J.M. Neff, and D.S. Page (2007). Assessment of polycyclic aromatic hydrocarbon exposure in the waters of Prince William Sound after the *Exxon Valdez* oil spill: 1989–2005. *Marine Pollution Bulletin* **54**(3): 339–367.

Boehm, P.D. and D.S. Page (2007). Exposure elements in oil spill risk and natural resource damage assessments: A review. *Human and Ecological Risk Assessment* **13**(2): 418–448.

Boehm, P.D., D.S. Page, J.S. Brown, J.M. Neff, J.R. Bragg, and R.M. Atlas (2008). Distribution and weathering of crude oil residues on shorelines 18 years after the *Exxon Valdez* spill. *Environmental Science & Technology* **42**(24): 210–216.

Boehm, P.D., D.S. Page, J.S. Brown, J.M. Neff, and W.A. Burns (2004). Polycyclic aromatic hydrocarbon levels in mussels from Prince William Sound, Alaska, USA, document the return to baseline conditions. *Environmental Toxicology and Chemistry* **23**(12): 2916–2929.

Boehm P.D., D.S. Page, W.A. Burns, A.E. Bence, P.J. Mankiewicz, and J.S. Brown (2001). Resolving the origin of the petrogenic hydrocarbon background in Prince William Sound, Alaska. *Environmental Science & Technology* **35**(3): 471–479.

Brown, D.W., D.G. Burrows, C.A. Sloan, R.W. Pearce, S.M. Pierce, J.L. Bolton, S.C. Brassell, and G. Eglinton (1980). Environmental chemistry: an interdisciplinary subject. Natural and pollutant organic compounds in contemporary aquatic environments. In *Analytical Techniques in Environmental Chemistry: 1st International Congress Proceedings*. J. Albaiges, ed. Oxford, UK: Pergamon Press; Pergamon Series on Environmental Science; ISBN-10: 0080238092; ISBN-13: 9780080238098; pp. 1–22.

Brown, D.W., D.G. Burrows, C.A. Sloan, R.W. Pearce, S.M. Pierce, J.L. Bolton, K.L. Tilbury, K.L. Dana, S.-L. Chan, and U. Varanasi (1996). Survey of Alaskan subsistence invertebrate seafoods collected in 1989–1991 to determine exposure to oil spilled from the *Exxon Valdez*. In *Proceedings of the* Exxon Valdez *Oil Spill Symposium*. S.D. Rice, R.B. Spies, D.A. Wolfe, and B.A. Wright, eds. Bethesda, MD, USA: American Fisheries Society; Symposium 18; ISBN-10: 0913235954; ISSN: 08922284; pp. 844–855.

Brown, J.S., D. Beckmann, L. Bruce, L. Cook, and S. Mudge (2011). PAH depletion ratios document the rapid weathering and attenuation of PAHs in oil samples collected after the Deepwater Horizon. In *Proceedings of the 2011 International Oil Spill Conference (Promoting the Science of Spill Response), May 24–26, 2011, Portland, Oregon, USA*. Washington DC, USA: American Petroleum Institute.

Esler, D., T.D. Bowman, K.A. Trust, B.E. Ballachey, T.A. Dean, S.C. Jewett, and C.E. O'Clair (2002). Harlequin duck population recovery following the *Exxon Valdez* oil spill: Progress, process and constraints. *Marine Ecology Progress Series* **241**: 271–286.

Field, L.J., J.A. Fall, T.S. Nighswander, N. Peacock, and U. Varanasi, eds (1999). *Evaluating and Communicating Subsistence Seafood Safety in a Cross-Cultural Context: Lessons Learned from the Exxon Valdez Oil Spill*. Pensacola, FL, USA: Society for Environmental Toxicology and Chemistry; ISBN-13: 9781880611296.

Gentile, J.H, M.A. Harwell, W. van der Schalie, S.B. Norton, and D.J. Rodier (1993). Ecological risk assessment: a scientific perspective. *Journal of Hazardous Materials* **35**(2): 241–253.

Harrison, O.R. (1991). An overview of the *Exxon Valdez* oil spill. In *Proceedings of the 1991 International Oil Spill Conference (Prevention, Behavior, Control, Cleanup), March 4–7, 1991, San Diego, California*. Washington DC, USA: American Petroleum Institute Technical Publication 4529; pp. 313–319.

Hayes, M.O., J. Michel, and D.C. Noe (1991). Factors controlling initial deposition and long-term fate of spilled oil on gravel beaches. In *Proceedings of the 1991 International Oil Spill Conference (Prevention, Behavior, Control, Cleanup), March 4–7, 1991, San Diego, California*. Washington DC, USA: American Petroleum Institute Technical Publication 4529; pp. 453–460.

Kelso, D.D. and M. Kendziorek (1991). Alaska's response to the *Exxon Valdez* oil spill. *Environmental Science & Technology* **25**(1): 16–23.

Kvenvolden, K.A., F.D. Hostettler, J.B. Rapp, and P.R. Carlson (1993). Hydrocarbons in oil residues on beaches of islands of Prince William Sound, Alaska. *Marine Pollution Bulletin* **26**(1): 24–29.

Li, H. and M.C. Boufadel (2010). Long-term persistence of oil from the *Exxon Valdez* spill in two-layer beaches. *Nature Geoscience* **3**(2): 96–99.

Michel, J. and M.O. Hayes (1993). Persistence and weathering of *Exxon Valdez* oil in the intertidal zone – 3.5 years later. In *Proceedings of the 1993 International Oil Spill Conference (Prevention, Preparedness, Response), March 29–April 1, 1993, Tampa, Florida*. Washington DC, USA: American Petroleum Institute Publication 4580; pp. 279–286.

Neff, J.M. (1979). *Polycyclic Aromatic Hydrocarbons in the Aquatic Environment: Sources, Fates, and Biological Effects*. Barking, Essex, UK: Applied Science Publishers, Ltd.; ISBN-10: 0853348324.

Neff, J.M., A.E. Bence, K.R. Parker, D.S. Page, J.S. Brown, and P.D. Boehm (2006). Bioavailability of polycyclic aromatic hydrocarbons from buried shoreline oil residues thirteen years after the *Exxon Valdez* oil spill: a multispecies assessment. *Environmental Toxicology and Chemistry* **25**(4): 947–961.

Neff, J.M. and W.A. Burns (1996). Estimation of polycyclic aromatic hydrocarbon concentrations in the water column based on tissue residues in mussels and salmon: an equilibrium partitioning approach. *Environmental Toxicology and Chemistry* **15**(12): 2240–2253.

Neff, J.M., D.S. Page, and P.D. Boehm (2011). Exposure of sea otters and harlequin ducks in Prince William Sound, Alaska, USA, to shoreline oil residues 20 years after the *Exxon Valdez* oil spill. *Environmental Toxicology and Chemistry* **30**(3): 659–672.

Neff, J.M. and W.A. Stubblefield (1995). Chemical and toxicological evaluation of water quality following the *Exxon Valdez* oil spill. In Exxon Valdez *Oil Spill: Fate and Effects in Alaskan Waters*. P.G. Wells, J.N. Butler, and J.S. Hughes, eds. Philadelphia, PA, USA: American Society for Testing and Materials; ASTM Special Technical Publication 1219; ISBN-10: 0803118961; pp. 141–177.

Owens, E.H. and G.A. Sergy (2000). *The SCAT Manual: A Field Guide to the Documentation and Description of Oiled Shorelines (Second Edition)*. Edmonton, AB, Canada: Environment Canada.

Owens, E.H., E. Taylor, and B. Humphrey (2008). The persistence and character of stranded oil on coarse-sediment beaches. *Marine Pollution Bulletin* **56**(1): 14–26.

Page, D.S., P.D. Boehm, G.S. Douglas, and A.E. Bence (1995a). Identification of hydrocarbon sources in the benthic sediments of Prince William Sound and the Gulf of Alaska following the *Exxon Valdez* oil spill. In Exxon Valdez *Oil Spill: Fate and Effects in Alaskan Waters*. P.G. Wells, J.N. Butler, and J.S. Hughes, eds. Philadelphia, PA, USA:

American Society for Testing and Materials; ASTM Special Technical Publication 1219; ISBN-10: 0803118961; pp. 41–83.

Page, D.S., P.D. Boehm, and J.M. Neff (2008). Shoreline type and subsurface oil persistence in the *Exxon Valdez* spill zone of Prince William Sound, Alaska. In *Proceedings of the Thirty-First Arctic and Marine Oilspill Program (AMOP) Technical Seminar, June 3–5, 2008, Calgary, Alberta. Canada*. Ottawa, ON, Canada: Environment Canada; pp. 545–563.

Page, D.S., E.S. Gilfillan, P.D. Boehm, and E.J. Harner (1995b). Shoreline Ecology Program for Prince William Sound, Alaska, following the *Exxon Valdez* oil spill: Part I – Study design and methods. In Exxon Valdez *Oil Spill: Fate and Effects in Alaskan Waters*. P.G. Wells, J.N. Butler, and J.S. Hughes, eds. Philadelphia, PA, USA: American Society for Testing and Materials; ASTM Special Technical Publication 1219; ISBN-10: 0803118961; pp. 263–295.

Payne, J.R., W.B. Driskell, J.W. Short, and M.L. Larsen (2008). Long term monitoring for oil in the *Exxon Valdez* spill region. *Marine Pollution Bulletin* **56**(12): 2067–2081.

Pope, G.A., K.D. Gordon, and J.R. Bragg (2011). Fundamental reservoir engineering principles explain lenses of shoreline oil residue twenty years after the *Exxon Valdez* oil spill. In *Proceedings of the Society of Petroleum Engineers' Americas E&P Health, Safety, Security, and Environmental Conference, March 21–23, 2011, Houston, Texas*. Houston, TX, USA: Society for Petroleum Engineers; SPE Paper 141809.

Robilliard, G.A., P.D. Boehm, and M.J. Amman (1997). Ephemeral data collection guidance manual, with emphasis on oil spill NRDAs. In *Proceedings of the 1997 International Oil Spill Conference (Improving Environmental Protection – Progress, Challenges, Responsibilities) April 7–10, 1997, Fort Lauderdale, Florida, USA*. Washington DC, USA: American Petroleum Institute Special Technical Publication 4651; pp. 1029–1030.

Shigenaka, G. and C.B. Henry, Jr., (1995). Use of mussels and semipermeable membrane devices to assess bioavailability of residual polynuclear aromatic hydrocarbons three years after the *Exxon Valdez* oil spill. In Exxon Valdez *Oil Spill: Fate and Effects in Alaskan Waters*. P.G. Wells, J.N. Butler, and J.S. Hughes, eds. Philadelphia, PA, USA: American Society for Testing and Materials; ASTM Special Technical Publication 1219; ISBN-10: 0803118961; pp. 239–260.

Short, J.W. and M.M. Babcock (1996). Prespill and postspill concentrations of hydrocarbons in mussels and sediments in Prince William Sound. In *Proceedings of the* Exxon Valdez *Oil Spill Symposium*. S.D. Rice, R.B. Spies, D.A. Wolfe, and B.A. Wright, eds. Bethesda, MD, USA: American Fisheries Society; Symposium 18; ISBN-10: 0913235954; ISSN: 08922284; pp. 149–166.

Short, J.W. and P.M. Harris (1996a). Chemical sampling and analysis of petroleum hydrocarbons in near-surface seawater in Prince William Sound after the *Exxon Valdez* oil spill. In *Proceedings of the* Exxon Valdez *Oil Spill Symposium*. S.D. Rice, R.B. Spies, D.A. Wolfe, and B.A. Wright, eds. Bethesda, MD, USA: American Fisheries Society; Symposium 18; ISBN-10: 0913235954; ISSN: 08922284; pp. 17–28.

Short, J.W. and P.M. Harris (1996b). Petroleum hydrocarbons in caged mussels deployed in Prince William Sound after the *Exxon Valdez* oil spill. In *Proceedings of the* Exxon Valdez *Oil Spill Symposium*. S.D. Rice, R.B. Spies, D.A. Wolfe, and B.A. Wright, eds. Bethesda, MD, USA: American Fisheries Society; Symposium 18; ISBN-10: 0913235954; ISSN: 08922284; pp. 29–39.

Short, J.W., G.V. Irvine, D.H. Mann, J.M. Maselko, J.J. Pella, M.R. Lindeberg, J.M. Payne, W.B. Driskell, and S.D. Rice (2007). Slightly weathered *Exxon Valdez* oil persists in Gulf of Alaska beach sediments after 16 years. *Environmental Science & Technology* **41**(4): 1245–1250.

Short J.W., M.R. Lindeberg, P.M. Harris, J.M. Maselko, J.J. Pella, and S.D. Rice (2004). Estimate of oil persisting on the beaches of Prince William Sound 12 years after the *Exxon Valdez* oil spill. *Environmental Science & Technology* **38**(1): 19–25.

Wolfe, D.A., M.J. Hameedi, J.A. Galt, G. Watabayashi, J. Short, C. O'Clair, S. Rice, J. Michel, J.R. Payne, J. Braddock, S. Hanna, and D. Sale (1994). Fate of the oil spilled from the *Exxon Valdez*. *Environmental Science & Technology* **28**(13): 561A–568A.

Yender, R., J. Michel, and C. Lord (2002). *Managing Seafood Safety after an Oil Spill*. Seattle, WA, USA: National Oceanic and Atmospheric Administration, Hazardous Materials Response Division, Office of Response and Restoration. [http://docs.lib.noaa.gov/noaa_documents/NOS/ORR/963_seafood2.pdf]

PART II

OIL IN THE ENVIRONMENT

INTRODUCTION

When oil is spilled into a marine environment, it immediately begins to undergo changes in its form and constituents as it is moved by wind, waves, and currents to other places. If the spill occurs close to land, some of the oil will be deposited on shorelines. Over time, much of the deposited oil is removed by cleanup efforts, bioremediation, or natural processes. Some of the oil may end up beneath the shoreline surface, particularly in locations sheltered from natural weathering. Understanding what happens to spilled oil and the forces affecting its fate is an essential prerequisite to assessing its potential effects on valued natural and cultural resources. That is the focus of the chapters in this section.

In Chapter 3, Paul Boehm, Jerry Neff, and David Page describe the physical and chemical factors that affect oil in water and how these factors came into play in the *Exxon Valdez* spill. As the oil changes and undergoes weathering over time, it is essential to sample the composition of the oil. This requires careful attention to sampling design. In the *Exxon Valdez* spill, intensive sampling of the water column showed that perhaps one quarter of the spilled oil evaporated from the water's surface within a few days, and concentrations of polycyclic aromatic hydrocarbons (PAH) had returned to background levels within a few months.

When oil strikes a shoreline, it is deposited unevenly. In order to marshal effective cleanup efforts and anticipate where natural or cultural resources may be at greatest risk, it is important to identify the location, form, and quantity of the stranded oil. This was the focus of the Shoreline Cleanup Assessment Technique (SCAT) surveys conducted immediately following the *Exxon Valdez* spill. In Chapter 4, Edward Owens and Douglas Reimer describe the development and implementation of SCAT surveys and their effectiveness in directing the massive cleanup activities after the spill. Chris Wooley and James Haggarty follow in Chapter 5 by illustrating the use of SCAT surveys to identify and record cultural sites that could be at risk from the spill or the cleanup activities. These surveys also resulted in the discovery of previously unknown cultural sites, more than doubling the number of known shoreline archaeological sites in Prince William Sound and the northern Gulf of Alaska.

The potential effects of oil on the shorelines were of greatest concern following the *Exxon Valdez* spill, so knowing what happened to that oil was vital. In Chapter 6, David Page, Paul Boehm, John Brown, Erich Gundlach, and Jerry Neff follow the fate of the oil that came ashore, describing where it was deposited, why some locations were oiled more than others, and how oil disappeared over the years (and why, in a few isolated locations, it persisted). The level of shoreline oiling decreased dramatically over the 2 years following the spill owing to cleanup and natural weathering, particularly severe winter storms, and it continued to diminish over the next 15 years. Importantly, no subsurface oil residues (SSOR) were subsequently found in locations that had not already been identified during the shoreline surveys after the spill.

Intensive sampling of shorelines following the *Exxon Valdez* spill showed that a small amount of SSOR did remain in a few locations many years later, fueling debates about possible exposure to organisms that live or feed on the shoreline. Gary Pope, Kimberly Gordon, and James Bragg address these issues in Chapter 7 using a combination of conceptual and mathematical models of hydrogeological dynamics derived from reservoir engineering. The models, in combination with direct measurements of residue levels and toxicities, explain the apparent paradox of how oil residues can persist as isolated subsurface patches, yet pose little exposure risk to biota.

One of the natural processes contributing to the diminishment of shoreline oil following a spill is biodegradation by bacteria that feed on naturally occurring hydrocarbons. In Chapter 8, Ronald Atlas and James Bragg discuss efforts to enhance this process by the use of fertilizers to promote bacterial growth ("bioremediation") on shorelines of Prince William Sound. After extensive testing in the field and the laboratory (conducted in cooperation with the US Environmental Protection Agency), bioremediation was used on several hundred shoreline sites that had been identified by the initial SCAT surveys. The rates of natural biodegradation of hydrocarbons were substantially accelerated by the judicious addition of fertilizer, especially in highly porous shorelines where nutrients and oxygenated seawater could reach the surface and subsurface oil residues. Using bioremediation years after the spill to help remove remaining pockets of SSOR, however, was shown to be both unnecessary (because exposure risk was negligible) and not feasible (because fluid flow through the substrate would be inadequate to carry the nutrients to the bacteria).

CHAPTER THREE

Oil in the water column

Paul D. Boehm, Jerry M. Neff, and David S. Page

3.1 Introduction

When crude oil or petroleum products are released during a marine oil spill, organisms living in the water or feeding at the surface are the first to be affected. Oil on water or mixed into the water column may injure aquatic species of all types. Understanding the potential for injury to organisms from exposure to oil requires fully studying physical and chemical effects and quickly communicating the results. The risks to the public from the consumption of fish or other species normally harvested from the water can also be serious. A comprehensive water-assessment program provides quantitative data to address multiple concerns.

The *Exxon Valdez* oil spill was, until recently, the most comprehensively sampled oil spill in history and remains the most exhaustively studied oil spill. In fact, the thoroughness of the data – and the disappearance of most oil slicks and sheens by the end of the summer of 1989 – enabled all commercial fisheries to be reopened in 1990, much earlier than had been anticipated. Techniques and protocols established during the *Exxon Valdez* spill have been used in subsequent spills, most notably in the 2010 Deepwater Horizon oil spill in the Gulf of Mexico.

The nature and the extent of oil-spill effects on organisms, populations, and ecosystems (i.e., biological resources) vary widely. Biological resources using the water surface may be injured by exposure to surface slicks, and those living in the water column may be injured by exposure to dissolved and dispersed oil in the water column. Risk of injury is directly related to oil's physical properties, chemical composition, and concentrations, and to its physical and chemical transformations as it spreads on or through the water column (National Research Council, 1985; Wolfe *et al.*, 1994; Chapter 1, Box 1.1; Chapter 2, Table 2.1). How oil transforms (or "weathers") determines its fate, persistence, and toxicity (National Research Council, 1985; Neff, 1990). Weathering varies with the release scenario – whether at the surface, as in the *Exxon Valdez* tanker oil spill in 1989 (Fig. 3.1a), or subsurface, as in the Deepwater Horizon well blowout in 2010 (Fig. 3.1b). Where the spill occurred and environmental conditions during and after (e.g., time of year, weather) also affect the fate and transport of oil, including any direct effects on the nature and extent of shoreline impacts (Chapters 6, 7, and 11). All of these factors influence the types of

Oil in the Environment: Legacies and Lessons of the Exxon Valdez *Oil Spill*, ed. J. A. Wiens. Published by Cambridge University Press. © Cambridge University Press 2013.

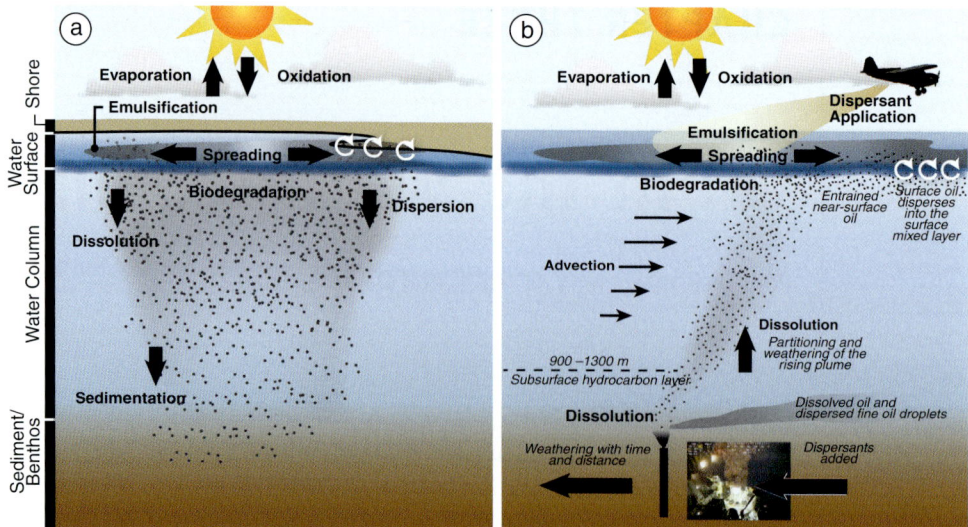

Figure 3.1 Processes acting on spilled oil under two release scenarios. (a) Surface release. (b) Subsurface release as in a seafloor oil well blowout or pipeline break.

scientific investigations needed, the locations investigated, and the methods required to track and quantify the oil and its chemical constituents.

The exposure of organisms to oil in the water column is ephemeral (Robilliard et al., 1997), but it can last longer near the shoreline, where tides, waves, and shoreline cleanup can remobilize beached oil and disperse it back into the water column. Because of its transient nature, the spread and transport of surface slicks should be monitored, usually from aerial surveillance, and boat-based water sampling should begin as soon after the release as possible and be repeated frequently. Water samples should be collected based on a systematic sampling design, particularly in the spill zone and in areas important to biological resources (Neff and Stubblefield, 1995). In parallel, sampling should also be done at the perimeter of the spill area and in reference areas outside of it. This helps to define the extent of impact and the normal, background levels of hydrocarbons from hydrocarbon sources other than the spill. The resulting data can then form a basis for designing scientific studies throughout the oil-spill phases (Chapter 2) to address five general questions:

1. Is there an exposure pathway from the release to specific biological resources at the water surface and/or in the water column?
2. What are the levels and durations of exposure of surface water, pelagic, and nearshore/intertidal resources to oil components in the water?
3. What is the geographic extent and water volume in which organisms are exposed to spill hydrocarbons?
4. What are the forms (dissolved or particulates/droplets) and fates of key toxic oil constituents (e.g., polycyclic aromatic hydrocarbons [PAH]) in the water column?
5. What are the typical background concentrations of PAH in the water column adjacent to the spill area?

In this chapter, we discuss scientific approaches for detecting and monitoring the transport and weathering of oil components in the water column and the exposure of biological resources under various release scenarios. We then go into greater detail, using the *Exxon Valdez* oil spill as a case study.

3.2 Overview of oil in the water column

3.2.1 Petroleum: chemicals, behavior, and key processes

Crude oil and refined petroleum products are complex mixtures of thousands of organic chemicals of fossil and recent biogenic origin and small amounts of inorganic salts and metals. Most crude and refined oils contain the same classes of compounds but differ in the relative amounts depending on the geologic age of the crude oil and the refining processes (National Research Council, 1985; Neff, 1990). Hydrocarbons represent 50% to nearly 100% of the total composition (see Chapter 1, Box 1.1). The constituents usually monitored during oil spills include:

- Total petroleum hydrocarbons (TPH), a useful measure of the total oil concentration in the water. TPH is measured by gas chromatography (GC) with flame ionization detection (FID) (Page *et al.*, 1995; Reddy and Quinn, 1999; Boehm *et al.*, 2007).
- Volatile aromatic hydrocarbons – namely the monoaromatic benzene, toluene, ethylbenzene, and xylenes (BTEX). Although toxic and frequently the most abundant aromatic hydrocarbons in crude and light fuel oils, they rarely persist in the water column long enough to injure biological resources (Neff *et al.*, 2000). BTEX is measured by GC with low-resolution mass spectrometry (MS).
- Polycyclic aromatic hydrocarbons (PAH), both unsubstituted (or "parent") and alkyl-substituted compounds (Sauer and Boehm, 1991). A typical crude oil contains less than 1–3% total PAH (TPAH) (Neff, 2002), although concentrations can be higher (Requejo *et al.*, 1996). TPAH is typically dominated by the naphthalene compounds and is measured by GC/MS using selected ion monitoring (SIM). The Alaskan North Slope crude oil released from the *Exxon Valdez* has about 1.3% TPAH, of which 96% are 2- and 3-ring PAH (Bence *et al.*, 2007). The low molecular weight, 2- and 3-ring alkyl-PAH probably contribute most to the aquatic toxicity of crude oil, but are not the sole toxicants, particularly in weathered oil (Barron *et al.*, 1999; Neff *et al.*, 2000, 2005). Therefore, it is important to monitor concentrations of individual parent and alkyl 2- through 6-ring PAH in the water column inside and near the spill zone because they identify spill residues and quantify the exposure of biological resources.

The type of oil spilled, including its chemical characteristics, influences its behavior in the water column and how scientists decide to investigate it (Fig. 3.2) (Chapter 2, Table 2.1). In general, two primary classes of oil are involved in oil spills: refined products and crude oils. Refined products include gasoline, middle distillates (e.g., diesel and No. 2 fuel oil), and heavy distillates and residual oils (e.g., bunker and No. 6 fuel oils). Within each of these primary classes, the character of the oil is determined by the types of hydrocarbons present – specifically, the number of carbons in any hydrocarbon, which, in turn, affects the solubility, volatility, and persistence of the oils (Fig. 3.2). Crude oils are classified into five groups according to specific gravity (API Gravity) and viscosity (National Research Council, 1999). Lighter crudes, with a higher fraction of low molecular weight hydrocarbons, are more volatile, more water soluble, and less persistent in the environment than are heavier crudes (see Chapter 2, Table 2.1).

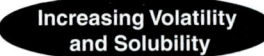

Figure 3.2 The type of oil spilled influences its behavior in the water column. Crude and fuel oil groups are based on API Gravity (API) and specific gravity (SG). Group I: API > 45, SG < 0.8 g/cm^3; Group II: API = 35–45, SG = 0.8–0.85 g/cm^3; Group III: API = 17.5–35, SG = 0.85–0.95 g/cm^3; Group IV: API = 10–17.5, SG = 0.95–1.0 g/cm^3; Group V: API < 10, SG > 1.0 (modified from International Tanker Owners Pollution Federation, 2011), compared to the SG of fresh and seawater at 4°C of 1.00 and 1.02 g/cm^3, respectively.

3.2.2 Release scenarios and ephemeral data

The collection of data on the distribution, physical properties, and chemistry of surface oil and oil constituents in the water column is central to scientific investigations in the early phase of an oil spill (Chapter 2). In most spills from vessels, where crude or fuel oil is released on the sea surface (e.g., *Exxon Valdez*, *North Cape*, *Selendang Ayu*, *Prestige*), the concentrations of hydrocarbons in the water column to which organisms are exposed and the resulting biological effects are likely to be short-lived owing to the low persistence of surface oil slicks and dissolved and dispersed oil constituents as they are mixed into the subsurface water by waves. At the other end of the release spectrum are prolonged oil releases over many months that occur through oil-well blowouts in subsurface waters of varying depths, such as the Ixtoc I (Jernelöv and Lindén, 1981) and Deepwater Horizon (Camilli *et al.*, 2010).

Spills can also involve oil fires, in which the generation of combustion products can influence a spill's behavior and thereby investigation priorities and strategies (e.g., *Haven*, *Mega Borg*, *Burmah Agate*). Oil spills may involve light crude oils or fuel oil with abundant water-soluble constituents, as in the *Braer* and *North Cape* spills, respectively (Kingston, 1999; Reddy and Quinn, 1999), to very heavy fuels with few water-soluble constituents, as in the *M/V Prestige* spill (Diéz *et al.*, 2007). The release scenario, as well as the application of chemical dispersants at or below the water surface, affects the distribution, composition, and persistence of oil constituents in the water column and directly affects the investigation priorities and scientific strategies employed in studying the dynamics of the oil and its potential impacts.

Historically, most oil spills in the marine environment occur over a very short duration, so there may be limited opportunity to collect sufficient field samples to quantify exposure and document injury. Consequently, several models have been developed to compensate for the lack of empirical data. These models have become increasingly prominent in recent years and are commonly applied to most oil spills. A widely used exposure assessment–injury assessment model is the *Spill Impact Model*

Application Package (SIMAP) (French-McCay, 2004). Other models have also been developed for injury assessment, but they all use a similar approach that includes a hydrodynamic model coupled with chemical-fate and transport models to determine the most probable chemical concentrations within an exposure area. Important outputs of models, such as water-column concentrations over time and space, should be validated with empirical data. In this regard, models rely heavily on empirical field data for model calibration and then for model validation, in which the model outputs are compared to field data. Thus, water sampling is very important even where models will play a role in injury assessment.

3.2.3 Key processes and environmental factors

Petroleum constituents on the water's surface can evaporate and be photodegraded, processes that alter oil before it is mixed into the water (Fig. 3.1a). Whether released at the surface or into the subsurface (Fig. 3.1b), petroleum constituents can undergo processes that partition oil components into different physical forms (i.e., into dissolved, droplet, and particulate phases) and alter the form and composition of the oil through processes collectively called weathering. Most of these processes occur in every marine oil spill, but which ones dominate depend on the physical and chemical properties of the oil and the environment in which the oil was spilled. Response actions can also affect the fate of the spilled oil. In numerous incidents, chemical dispersants have been applied to break up surface oil slicks, preventing intact oil slicks from impacting the shorelines. Chemical dispersion initially increases the concentration of petroleum in the water column and facilitates droplet formation, dissolution, entrainment, and rapid biodegradation of oil constituents.

The environment in which an oil spill occurs, including the geographic location, depth of oil release, time of year, and weather conditions, influences the dominant fate and transport processes affecting the oil and its effects on water-column biological resources. Oil spilled in arctic environments remains more viscous and evaporates more slowly than oil spilled in tropical environments. Those processes affecting the physical and chemical properties of spilled oil also govern whether it is appropriate to apply dispersants (National Research Council, 2005). Where chemical dispersants are used, water-column measurements provide key information on dispersant effectiveness and the concentrations of petroleum constituents.

3.2.4 Sampling strategies: the four-dimensional approach

When an oil spill occurs, it is critical to respond quickly to contain the release, track the movement of the oil, and document the initial exposure conditions. These are the major priorities in the first phase of a spill (the Release and Immediate Response Phase; Chapter 2), which begins immediately after the release and continues for weeks to months until surface slicks disappear. During this phase, the oil is being released and/or is still moving on or in the water as it partitions, dilutes, degrades, and is distributed more broadly. Water-column sampling strategies – where to sample, at what depths to sample, and at what frequency to obtain water samples – must be developed within days of the release. The sampling strategy should be based on oil type and behavior, response strategies (i.e., possible use of chemical dispersants or *in situ* burning), surface oil-trajectory modeling and aerial surveillance, a consideration of ocean currents, and other factors.

The overall management of a sampling and analytical program by dedicated scientists is essential to its success. Water-quality monitoring programs after oil spills generate many samples, and laboratory capacity to analyze them is often limited. To ensure that data meet quality objectives, sample processing should be carefully managed, lest unqualified labs be allowed to learn on the job. Analytical data must be reviewed both by experienced auditors and environmental chemists. Data need to be managed in a relational database that can provide rapid access to geospatial water-column chemistry results.

A water-quality monitoring program requires a four-dimensional (4D) effort to sample at different directions and distances from the release, at multiple depths below the water surface, and at different times after the release (length + width + depth + time = 4D). Maximum exposures (i.e., oil concentrations directly under surface slicks or within contaminated water below the surface) need to be characterized as part of a deliberately biased sampling strategy. It is equally important to implement an unbiased sampling strategy to characterize water-column concentrations in the spill zone and define the limits of exposures in relation to background levels from reference areas.

Specific considerations for a sampling strategy include:

- Identifying sampling locations to characterize the area and depth of contamination and ecologically relevant parcels of water (e.g., spawning areas).
- Determining sampling depths based on oil release characteristics and biological resources at risk.
- Measuring how the oil is partitioned in the water column, which is important for determining the toxicological relevance of chemical measurements.
- Determining the frequency and duration of sampling.
- Adjusting sampling locations and intensity as the spatial spread and chemical makeup of the oil change over time.

The focus of sampling will change as the oil moves, and the questions being asked will shift from offshore to nearshore and shoreline concerns.

3.2.4.1 Selection of sampling locations

Generating data that accurately represent the various areas and depths of the spill zone requires the careful selection of sampling locations and replication. During Phase 1, the most obvious locations to sample are those around and under the surface oil slick. This area is considered the spill zone and includes the area with the highest concentrations of TPAH in the water column. Concentration gradients should be measured with respect to water depths, distances from the release point, and locations within the spill zone. Samples should also be collected in areas outside the spill zone to define baseline or reference conditions. Biologically important areas such as fish spawning areas, sensitive nearshore habitats (e.g., sea grass and marsh habitats), known nursery and migratory habitats for important commercial and subsistence species, and other sensitive water bodies should be given a high priority. In all spills, spill-trajectory information coupled to information from Environmental Sensitivity Index (ESI) maps generated by the National Oceanic and Atmospheric Administration (NOAA) (Petersen et al., 2002) should be used to identify appropriate areas to sample.

Determining the depths at which samples will be collected should be based on a careful consideration of the combination of the release scenario, the oil type, dispersant

Figure 3.3 Oil on nearshore water during *Exxon Valdez* shoreline cleanup, June 30, 1989. (Photo: Erich R. Gundlach).

usage, and the biologically relevant depth zones. For surface or near-surface releases, including areas where dispersant application occurred, the highest PAH concentrations will usually be within the top few meters of the water column, with lower concentrations below the top 10 m. In the *Exxon Valdez* spill, sampling was concentrated in the top 30 m of the water column. When oil is released at depth, as with a subsurface well blowout or the sinking of an oil tanker, it can form subsurface plumes, with elevated concentrations of both dissolved and droplet chemical constituents. Sample collections following subsurface releases should profile the entire water column to capture the distribution of the oil and to identify and track any subsurface plumes that may develop. Sample locations should be selected to detect these anomalies, if present, and track them as they move, disperse, and degrade.

As the investigations shift from the immediate response to the shoreline cleanup phase, Phase 2 (Chapter 2), sampling shifts to locations near oiled shorelines where oil can be resuspended into the water column during shoreline cleanup (Fig. 3.3).

Once cleanup ends, chemical concentrations in water samples are expected to be low, so especially sensitive analytical methods must be used to detect these low levels of PAH and other compounds. While sampling will lessen, there is still great value in repetitive sampling in offshore habitats of the key commercial and subsistence fisheries, even after scientists are satisfied that background levels have been restored. This helps verify, for public safety and confidence, that no oil remains in the water column.

3.2.4.2 Sampling baseline locations

Spilled oil is not the only source of PAH in the environment (Chapters 1, 5, and 6; Wooley, 2002). Therefore, water-column sampling should include reference areas unimpacted by the spilled oil in order to identify other sources and baseline concentrations of PAH in and around the spill zone. These other sources are part of the chemical background or baseline conditions. They may include natural hydrocarbon seeps and inputs from current and historic human activities that contribute hydrocarbons to the water column within and adjacent to the spill zone (Page et al., 1995; Wooley, 2002; Boehm et al., 2003). The cleanup activities themselves and ongoing commercial activities (e.g., boat traffic) also contribute background hydrocarbons. Urban runoff from storm drains can also be a major source of PAH in marine waters near population centers, where most small spills occur (National Research Council, 1985). A rigorous natural resource damage assessment (NRDA) requires that locations to collect water-column data be selected to include measures of TPAH concentrations and the composition of PAH that characterize background conditions absent the spill.

3.2.5 Sampling and data collection methods

Because PAH are among the most toxic components in oil, their detection and quantification are the principal targets of water sampling after a spill. Different sampling approaches can help to address different questions and can be used in combination to inform the overall sampling strategy. Some field methodologies, such as *in situ* fluorometry, in which the presence and types of aromatic hydrocarbons are measured (e.g., Boehm and Fiest, 1982), can provide guidance on where to sample more intensively or broaden the spatial coverage of data collection. Other sampling methods capture the exposure conditions at a single point in time and space, while passive (e.g., semipermeable membrane devices [SPMDs]) and biological (e.g., caged mussels) sampling can integrate water PAH concentrations over time at a given location.

The instruments and methods used after a spill are specific to the information required, but the most common methods include some form of oceanographic characterization along with the direct collection of discrete water samples.

3.2.5.1 Oceanographic characterization

For most marine spills, chemical sampling should be guided by oceanographic factors, such as water-column structure and water currents. Knowledge of water movements is especially important for subsurface spills, particularly at deeper offshore locations. Water-column characterization usually involves deploying conductivity–temperature–depth (CTD) instrument packages, which measure physical characteristics and structure. CTD is important for selecting sampling depths and for providing inputs to oil fate and transport modeling.

Additional sensors are typically deployed with the CTD, field spectrofluorometers and dissolved oxygen sensors being the most important. Fluorometric methods have been used to quantify PAH in discrete water samples aboard ships (Boehm and Fiest, 1982) and as part of towed systems (American Petroleum Institute, 1998; Robertson, 2001). Field fluorometry is generally the method of choice for monitoring oil concentrations in water when chemical dispersants are applied (National Research Council, 2005). Following the Deepwater Horizon blowout, fluorometry was used extensively because of the depth of the oil released, the depth of the water column impacted, and the use of chemical dispersants (Camilli et al., 2010).

3.2.5.2 Direct water sampling

The analysis of appropriately collected water samples provides the most comprehensive quantitative assessment of water-column chemistry after a spill. Fluorometry was not used in the *Exxon Valdez* spill; instead, PAH concentrations were assessed based on discrete water and caged-mussel samples and detailed analytical methods (Neff and Stubblefield, 1995; Short and Harris, 1996b; Boehm *et al.*, 2007). Collecting and analyzing an appropriately large number of water samples – with consideration of spatial distribution, depth, and time using standard protocols – documents the concentrations and fates of oil in 4D. Several methods can be used to collect these water samples, depending on the depth of collection (Robertson, 2001). Sampling protocols and laboratory capacity must be in place to keep analysis within regulatory-mandated maximum hold limits and to generate robust analytical data.

Ensuring that no contamination artifacts are introduced to the water sample is of primary importance. Contamination can come from the surface oil slick, exhaust, diesel, or many other sources. The ability to chemically identify (fingerprint) these sources should be part of the analytical protocol for water investigations (Boehm *et al.*, 1997; see Chapter 1, Box 1.1).

The most important component of water-column exposure assessments is chemical data on unfiltered water samples. The chemistry of these water samples can be used to differentiate between dissolved and dispersed (droplet) forms of oil. Alternatively, samples may be filtered to separate dissolved from particulate/droplet forms of oil (e.g., Payne and Driskell, 2003). Filtration separates the operationally defined dissolved fraction (i.e., material that passes through a filter) from particulate/droplet forms of oil in the water column so that they can be quantified separately. The dissolved fraction is the most relevant to toxicity assessments because the organic chemicals in this phase are the most bioavailable and most likely to cause toxicity.

3.2.5.3 Passive samplers

In later phases of a spill, when chemical concentrations are expected to be low and/or a time-averaged sample of the water column is preferred, passive samplers can be used to collect dissolved hydrocarbons from the water. SPMDs and caged or indigenous mussels are commonly used to passively sample dissolved, lipid-soluble organic chemicals in water (e.g., Boehm *et al.*, 2004, 2005; Carls *et al.*, 2004). Both SPMDs and mussels can provide an estimate of the average biologically available chemical concentration (e.g., TPAH) in the water column over time. If deployed correctly and not in direct contact with oil (e.g., free oil or dispersed droplets), SPMD can provide data on water-column PAH concentrations over days to months (Boehm *et al.*, 2005).

The tissues of mussels and some other animals can be used to calculate the biologically available fraction of PAH in the water column (Neff and Burns, 1996). In the United States and elsewhere, monitoring programs collect time-series data on chemical concentrations in coastal mussels, which can provide baseline information for a natural resource damage assessment (Lauenstein and Daskalakis, 1998). In contrast to water samples, which provide snapshots of TPAH in the water at a given time and place, PAH residues in mussel tissues represent an estimate of the time-integrated exposure concentration in the water column. Short and Harris (1996b) used caged mussels to track spill hydrocarbons at depths to 25 m in the months following the *Exxon Valdez* spill.

3.2.6 Analysis for hydrocarbons

TPH, BTEX, and PAH form the basis for most chemical measurements in the water column after an oil spill. Geochemical biomarkers, however, such as steranes and triterpanes (Stout and Wang, 2007; Chapter 8, Box 8.1), are essential for source identification and fingerprinting of oil residues, especially in sediments and oil samples. Chemical indicators of dispersants (National Oceanic and Atmospheric Administration, 2010) are also important to incorporate when appropriate. Because concentrations of PAH in the water column are typically low, particularly as time passes, the detection capacity needs to be sufficiently robust: 1–10 parts per trillion for individual PAH (Sauer and Boehm, 1991; Douglas et al., 2004).

Although measurements of the TPAH concentration in the water column are important, it is also essential to record the detailed composition of the extended suite of PAH in water samples, inclusive of the 2- to 6-ringed parent and alkylated PAH. It is not sufficient to measure only the 16 nonalkylated EPA priority-pollutant PAH (Sauer and Boehm, 1991). The alkylated PAH are the most abundant PAH compounds in crude and refined oils, and both are toxicologically important and essential for identification of PAH sources in sediments, biota, and water (Bence et al., 2007; Stout and Wang, 2007). A complete data set provides a strong scientific basis for linking PAH from an oil spill to injury of biological resources.

3.3 The *Exxon Valdez* oil spill

Following the *Exxon Valdez* spill, of the 11 million gallons (40 million liters) of oil spilled into Prince William Sound (PWS):

- 20–30% evaporated in the first few days;
- ~40% washed ashore in PWS; and
- 7–11% washed ashore along the coast of the western Gulf of Alaska (GOA) (Wolfe et al., 1994).

Much of the remaining oil (19–33%) dispersed or dissolved from surface slicks or sheens into the upper water column in the weeks after the spill (Wolfe et al., 1994; Neff and Stubblefield, 1995). By the end of April 1989, a month after the spill, large oil slicks were rare (Galt et al., 1991; Taft et al., 1995), although localized oil sheens near shores being cleaned were common at heavily oiled locations until the fall of 1989.

3.3.1 Water sampling programs

PWS is home to wildlife and to important commercial, subsistence, and recreational fisheries. Because of immediate concerns that oil could injure these biological resources, within 6 days of the spill Exxon and government agencies began a massive water-sampling program throughout the spill zone. Sampling continued into 1990 (Fig. 3.4; Boehm et al., 2007). The undertaking was all the more remarkable because of the spill area's remoteness. The sampling strategy quickly focused on offshore surface waters (upper 30 m) and on nearshore areas and bays where Environmental Sensitivity Index maps identified important habitats. The strategy had both an area-wide facet, to quantify and delineate water-column impacts in general, and a resource-specific facet, to target nearshore fisheries (e.g., herring spawning habitats) and protected species (e.g., sea otters, *Enhydra lutris*, seabirds) with repeated sampling during the spring and summer of 1989 and 1990 and with shorter sampling surveys near oiled shores in 2002 and 2005.

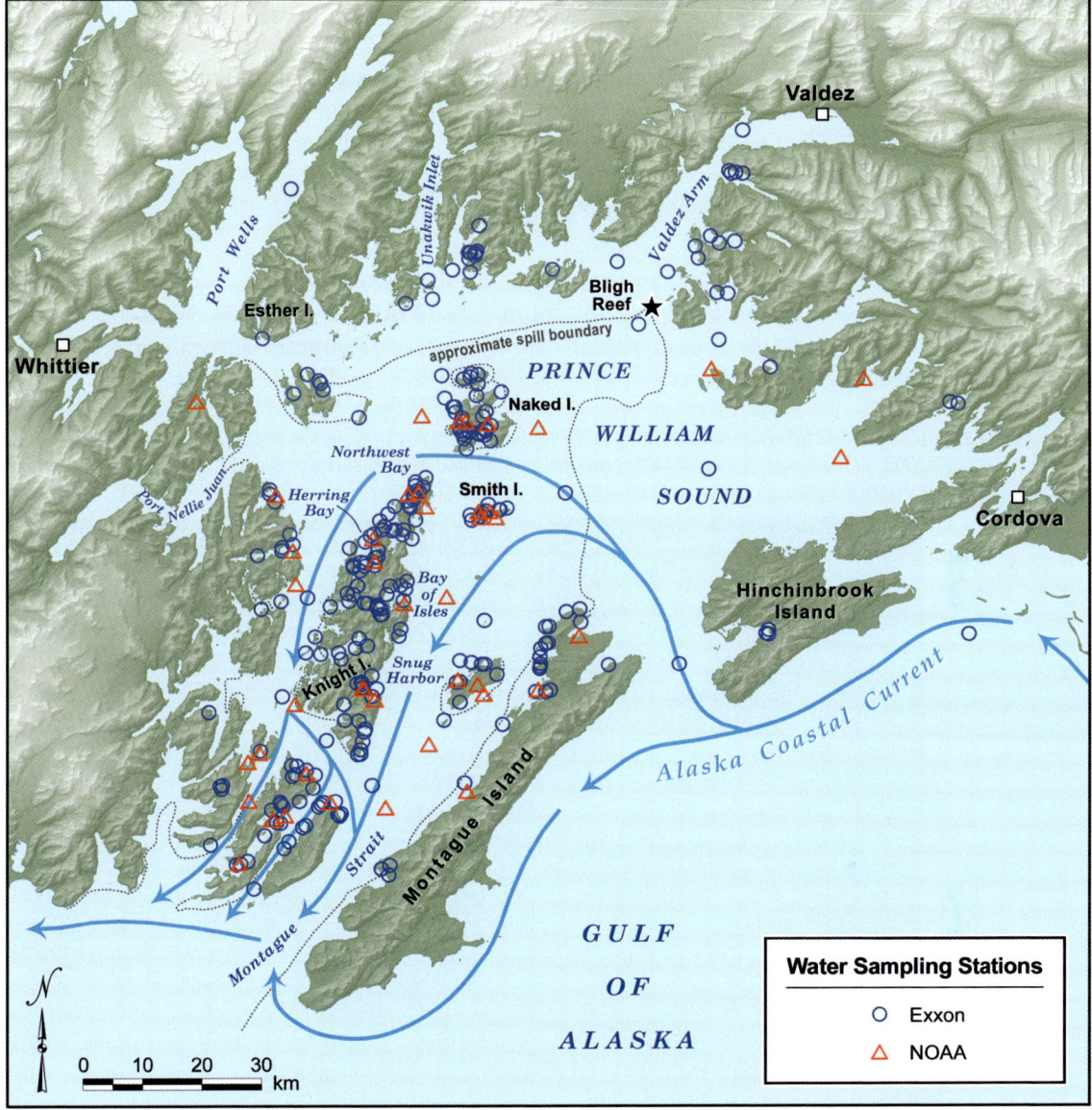

Figure 3.4 A comprehensive water-column sampling program was undertaken during the first year of the spill. The Prince William Sound map shows main features, main current flow, and all 1989 and 1990 water-sampling locations for Exxon- and NOAA-supported studies. (From Boehm et al., 2007, with permission from Elsevier)

The sampling program's objectives were to document the distribution and composition of petroleum hydrocarbons in the water column and to determine whether petroleum hydrocarbons posed a risk to biological resources (Neff and Stubblefield, 1995; Short and Harris, 1996a, b; Payne et al., 2005). Between 1989 and 2005, the Exxon and NOAA programs collected more than 5000 water samples for TPH analysis from 491 spill zone and reference locations in PWS (Boehm et al., 2007). Roughly half of these samples were analyzed for PAH by NOAA and half by Exxon. Water analyses were supplemented by indirect estimates of water-column PAH concentrations using indigenous (Neff and Burns, 1996) and caged mussels (Short and Harris, 1996b).

3.3.2 Data for water samples, 1989–2005

3.3.2.1 Data sources

Water-chemistry data sources include the Exxon chemistry database and data for the NOAA studies recorded in the *Exxon Valdez* Trustee Hydrocarbon Database (Boehm et al., 2007; *Exxon Valdez* Oil Spill Trustee Council, 2009; www.valdezsciences.com). The PAH data reported here are for near-surface and subsurface waters (most 0.5–10 m deep) from PWS nearshore and offshore stations in the spill zone and at nonspill-zone or reference locations.

3.3.2.2 Data presentation and discussion

The TPAH data are summarized in Table 3.1 and Figure 3.5. These values are calculated as the sum of the concentrations of individual PAH analytes measured in the samples, less the concentration of parent naphthalene, a ubiquitous laboratory contaminant (Boehm et al., 2007).

Table 3.1 Measured TPAH concentrations (µg/L) in PWS water samples collected by NOAA and Exxon in 1989, 1990, 2002, and 2005 spill path (oiled), historical human and industrial activity, and unoiled (nonspill path) reference sites.

Year	Statistic	Measured Water TPAH (µg/L)		
		Oiled	Reference Human Activity	Reference Unoiled
Exxon-Supported Studies				
1989	No. of samples	1091		302
	Arithmetic mean ± SD	0.31 ± 2.0		0.05 ± 0.1
	Median	0.04		0.03
	Range	ND–41.7		ND–1.28
1990	No. of samples	215		70
	Arithmetic mean ± SD	0.014 ± 0.02		0.05 ± 0.16
	Median	0.01		0.01
	Range	ND–0.145		ND–0.94
2005	No. of samples	41	26	3
	Arithmetic mean ± SD	0.002 ± 0.003	0.02 ± 0.03	0.001 ± 0.0001
	Median	0.001	0.012	0.001
	Range	0.001–0.02	0.001–0.12	0.001–0.001
NOAA Studies				
1989	No. of samples	182		51
	Arithmetic mean ± SD	0.97 ± 1.42		0.45 ± 0.68
	Median	0.49		0.29
	Range	ND–14.8		0.02–4.58
2002	No. of samples	4		3
	Arithmetic mean ± SD	0.09 ± 0.06		0.004 ± 0.002
	Median	0.08		0.005
	Range	0.02–0.17		0.002–0.006

ND = Not detected (< 0.001 µg/L).
All values are given as TPAH minus parent naphthalene, a ubiquitous laboratory contaminant (see Boehm et al., 2007).
Source: From Boehm et al. (2007); *Exxon Valdez* Oil Spill Trustee Council (2009).

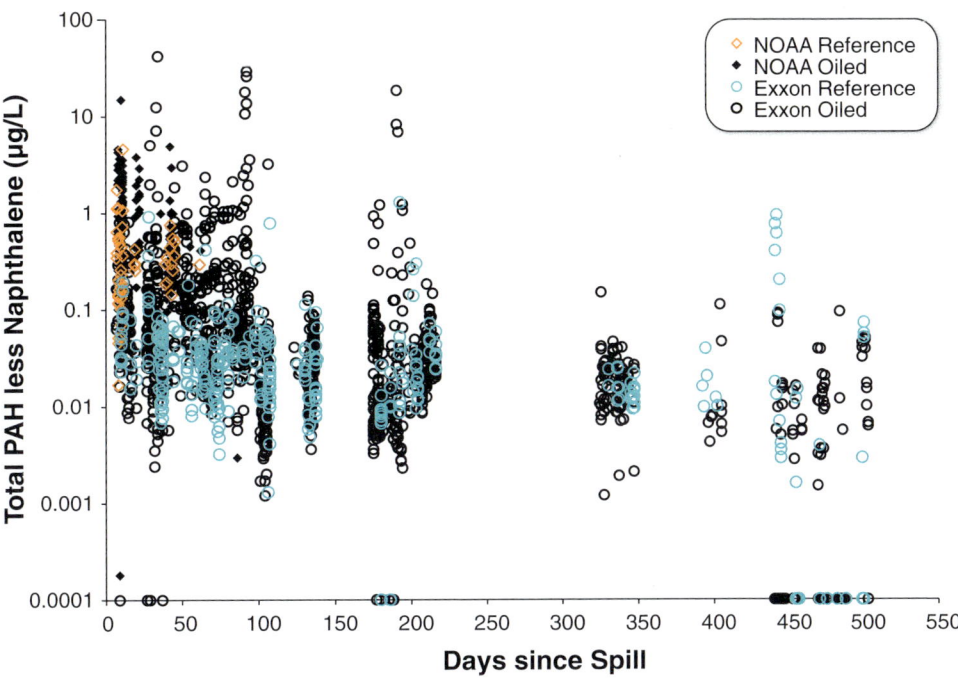

Figure 3.5 Measured TPAH concentrations (μg/L) in NOAA and Exxon water samples collected in 1989 and 1990. (From Boehm et al., 2007, with permission from Elsevier)

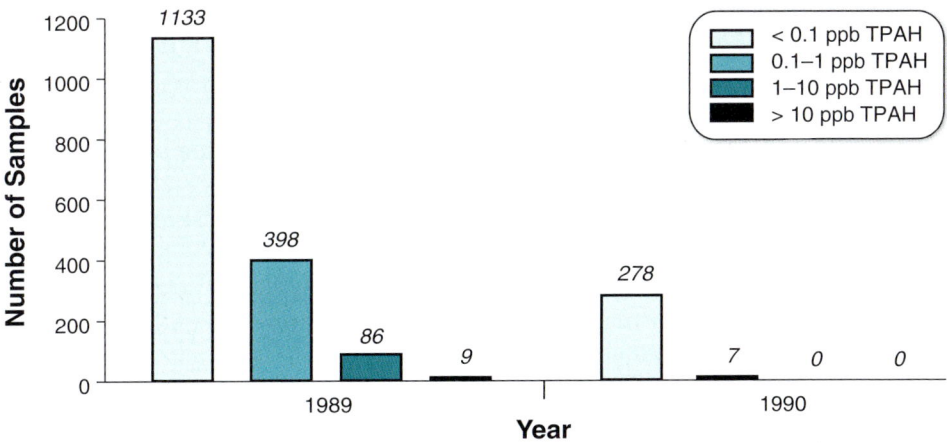

Figure 3.6 Summary, by concentration range, of measured TPAH concentrations (μg/L) for NOAA and Exxon water samples collected in PWS in 1989 and 1990.

No water-column PAH measurements were made during the first week after the spill. Wolfe et al. (1994) estimated that approximately 23% of the oil released was dispersed into the near-surface water column over an area of $10 \times 10^8 \, m^2$ during a storm that occurred 3 days after the spill, producing maximum theoretical concentrations of TPAH in the top 10 m of the water column of ~12 μg/L. This estimated peak TPAH concentration preceded the start of herring spawning in the spill zone by 12 days (see Chapter 13).

Concentrations of TPAH in the upper 30 m of the spill zone's water column were measured from March 31 through April 1989. TPAH ranged from < 0.01 to 41.6 μg/L, with most samples containing < 1 μg/L (Fig. 3.5). The high spatial and temporal

Table 3.2 TPAH concentrations (μg/L), estimated from PAH residues in NOAA caged mussels, from oiled and unoiled (reference) shores in PWS in 1989, 1990, and 1991.

Year	Statistic	Estimated Water TPAH (ug/L)	
		Oiled	Reference
1989	No. of samples	141	52
	Arithmetic mean ± SD	0.1 ± 0.19	0.04 ± 0.03
	Median	0.03	0.02
	Range	0.003–1.0	0.002–0.17
1990	No. of samples	66	15
	Arithmetic mean ± SD	0.01 ± 0.01	0.010 ± 0.01
	Median	0.01	0.009
	Range	0.0003–0.06	0.0003–0.03
1991	No. of samples	26	3
	Arithmetic mean ± SD	0.02 ± 0.01	0.01 ± 0.01
	Median	0.01	0.005
	Range	0.003–0.04	0.005–0.02

Source: From Boehm et al. (2007).

variability in 1989 was due to oil moving around PWS and becoming periodically remobilized from shorelines.

Of the entire 1989 NOAA and Exxon data set of 1626 water samples, 9 samples contained TPAH concentrations greater than 10 μg/L (Fig. 3.6), which is the State of Alaska's water-quality standard (Alaska Department of Environmental Conservation, 2003). Eighty water samples contained concentrations of 1–10 μg/L. Samples from 1989 in which TPAH exceeded 10 ppb were rare, and none over 1 μg/L was found in 1990 (Fig. 3.6). The marked decrease in TPAH concentration after April through midsummer 1989 corresponded to the decrease in surface oil slicks and sheens, which had become rare by midsummer 1989 (Taft et al., 1995).

The highest concentration of TPAH (41.7 μg/L) was observed in an arm of Northwest Bay, Eleanor Island (see Map 3, p. vii), within 100 m of the shoreline during the period when oil containment booms were in place and shorelines were being actively cleaned in 1989. These water samples contained oil droplets, as evidenced by the chemical composition in the sample resembling whole-oil PAH.

3.3.2.3 Estimated water TPAH concentrations from mussel-tissue data

Estimated TPAH concentrations in water derived from mussel-tissue PAH measured in the NOAA caged-mussel program, as calculated by the regression method of Neff and Burns (1996), are presented in Table 3.2 and Figure 3.7. Figure 3.7a shows that bioavailable PAH decreased from a maximum of 1.0 μg/L in 1989 (median value = 0.032 μg/L) to background range levels of less than 0.100 μg/L in 1990 and 1991, when the NOAA caged-mussel program ended. Results from caged mussels deployed at herring spawning sites (Fig. 3.7b) showed that ambient water TPAH were in the background range below 0.100 μg/L for 1989–91. This demonstrates the low level of exposure during times of herring spawning and hatching of herring larvae (Boehm

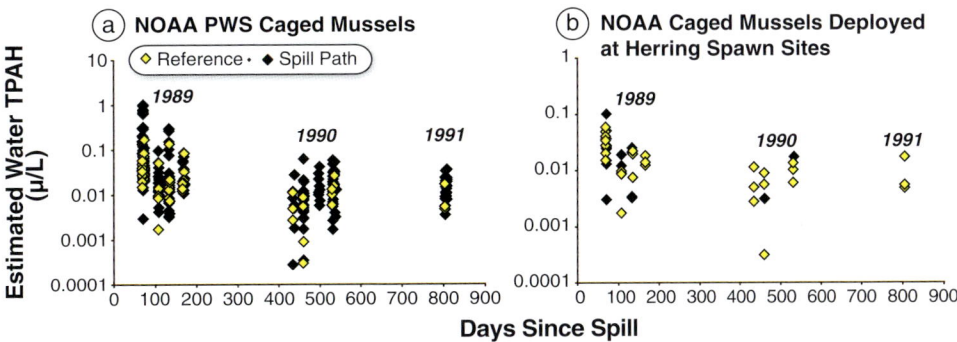

Figure 3.7 TPAH concentrations (µg/L) in offshore water in 1989–91, estimated from PAH residues in NOAA caged mussels (see Boehm et al., 2007). (a) Estimated TPAH concentrations in offshore water at all spill path and unoiled reference locations in 1989–91. (b) Estimated TPAH concentrations in offshore water off spill-path and in unoiled, reference herring spawning shores in 1989–91. (From Boehm et al., 2007, with permission from Elsevier)

et al., 2007). The data shown in Figure 3.7b illustrate the utility of direct water measurements coupled with water TPAH calculated from mussel-tissue PAH concentrations as a means of evaluating the exposure of biota to dissolved spill-oil fractions.

3.3.2.4 Baseline water TPAH

Measured and estimated background TPAH concentrations in water from unoiled and historical industrial sites in PWS (Table 3.1) ranged from ND (i.e., < 0.001) to 4.58 µg/L (water samples) and ND to 0.17 µg/L (NOAA caged-mussel samples) during 1989–91 (Table 3.2). TPAH concentrations derived from intertidal mussel-tissue PAH at unoiled reference sites ranged from ND to 0.25 µg/L during 1992–2004 (Boehm et al., 2007). By 1994, the range of TPAH concentrations (as derived from tissue PAH concentrations in intertidal mussels) was the same in the spill zone as at reference sites (Boehm et al., 2007).

The results of the NOAA and Exxon water-sampling programs present a consistent picture of the changes in ambient-water PAH concentrations over time, from elevated levels during the month after the spill to background levels by 1990–91. TPAH concentrations provide an effective means to assess acute and long-term exposure of and ecological risk to marine organisms. The water-quality data (Table 3.1 and Figs. 3.5–3.7) show that TPAH concentrations in the spill zone were not high enough and elevated concentrations did not persist long enough to cause lethal or sublethal effects to early life stages of fish (Boehm et al., 2007). These data indicate that PAH related to oil residues had returned to concentrations found in reference areas within 12 months after the *Exxon Valdez* spill (Fig. 3.5) – most even sooner.

3.3.2.5 Value of the water-sampling program

Not only did the comprehensive water-sampling program have scientific and injury-assessment value, but it was also highly effective in communicating the extent of oil exposure and risk experienced by fish and other species. The 1989 results were communicated to scientific, governmental, and public audiences in January 1990, as soon as the analyses were completed and the data tabulated. The data alleviated concerns about water-body contamination. In fact, there were never any substantive water-column-risk issues raised following the release of the water-quality data. In that regard, it is important to make the measurement program more expansive than what might be required by science alone; the many "zero" concentrations had the effect of easing public concern, based on real data.

3.4 Comparison with the Deepwater Horizon oil spill

The 1989 *Exxon Valdez* spill was, until the 2010 Deepwater Horizon spill, the most sampled and studied spill in terms of the number of detailed, low-level, PAH data points obtained in the water column.[1] Although the spill scenarios were very different, they shared a high priority for collecting water-column-chemistry data early on. Samples were collected during the release at both offshore and nearshore locations, and continued until samples showed background levels of chemicals. Discrete water samples were obtained at multiple depths and rigorous analytical chemical techniques were applied to obtain PAH concentration and compositional data.

Several major differences in sampling emphasis were driven by the spill scenarios. While *Exxon Valdez* sampling focused on the top 30 m of the water column, Deepwater Horizon sampling covered the entire water column, from sea floor (about 1500 m) to the surface (Boehm et al., 2011; Reddy et al., 2011). The use of chemical dispersants during Deepwater Horizon,[2] both at the deep point of release and on the water's surface, necessitated additional detailed sampling at multiple depths: oceanographic and fluorometry sensors were used extensively, and other chemicals related to chemical dispersants were measured in addition to PAH.

In both spills, the concentration of PAH and other chemicals decreased very rapidly after the releases ended (Boehm et al., 2007, 2011).

3.5 Lessons learned

The *Exxon Valdez* oil spill remains the largest tanker oil spill in United States' history. The spill response and data-collection efforts were unprecedented in their scope and volume, and the data generated have, for more than 20 years, provided insights into the fate and transport of oil in the water. Important lessons were learned from the water-column investigations of the *Exxon Valdez* oil spill that have influenced the response and sampling efforts in subsequent spills.

- The need for urgency in collecting short-lived water-quality data emphasizes the importance of preparedness and proper sampling strategies. When a spill occurs, it is essential to mobilize immediately and to develop sampling plans geared to the spill scenario at hand. Otherwise, the opportunity to collect short-lived chemical-concentration and organism-exposure data (i.e., "ephemeral" data) may be lost.
- Sampling design relies on spill tracking and surveillance. Sampling designs need to consider where the oil is moving to obtain a comprehensive data set that includes many samples across length, width, depth, and time. Sampling stations should provide sufficient geographic coverage to support conclusions about the state of the spill zone

---

[1] For updated information about the Deepwater Horizon/Macondo blowout and spill, see www.EoEarth.org/oceanoil.

[2] Dispersants were not used in the *Exxon Valdez* spill.

as a whole. Adherence to these principles will produce a dataset capable of defining the exposure conditions experienced by at-risk biological resources.

- Background levels of petroleum constituents are present in all aquatic systems. Many sources of petroleum hydrocarbon and PAH exist, even in remote coastal and offshore areas. PAH from fossil fuels, industrial and natural combustion sources, and biosynthesis are ubiquitous in the marine environment and must be measured as part of the oil-spill assessment.

- Analytical data need to be of high specificity, sensitivity, precision, and accuracy from the outset. Accurate characterizations of spilled and background hydrocarbons, exposure assessments, and, ultimately, injury assessments require high-quality analytical data. Detection limits must be low – single parts per trillion (ng/L) – for individual parent and alkylated PAH in the complex mixture in environmental samples.

- A spill chemistry program needs to be tightly directed and managed. The challenge of large numbers of samples and limited high-quality laboratory capacity needs to be carefully managed by a dedicated scientist who understands environmental data. Rigorous quality-control standards need to be implemented and enforced.

- Indigenous bivalves, such as mussels, should be used to supplement direct water measurements. Mussels, oysters, and other filter-feeding bivalves naturally accumulate organic chemicals from the water column. These sentinel species should be used to estimate water-column concentrations of PAH when water concentrations are too low to measure directly. They are preferable to SPMDs, which can be contaminated by direct contact with undissolved oil.

- PAH concentrations in water rapidly return to background levels. Within 3–6 months after a surface spill, water-column PAH concentrations usually return to background levels, even at the most contaminated areas. Although oil at concentrations high enough to cause injury can occur in specific locations immediately after a spill, the area of such concentrations and effects rapidly diminishes. Natural dispersion and biodegradation account for these rapid decreases.

- Water sampling and analysis are critically important elements of scientific assessments of an oil spill. A rigorous water-chemistry effort conducted at major oil spills as part of an overall exposure and injury assessment is one of the most important investigative elements following the spill. With the safety of commercial and subsistence seafood of great concern and millions of dollars' worth of commercial fisheries at risk, water programs produce results that are valuable to scientists, regulators, and the public.

REFERENCES

Alaska Department of Environmental Conservation (2003). *Water Quality Standards: As Amended through June 26, 2003*. Juneau, AK, USA: Alaska Department of Environmental Conservation, Division of Water; 18 AAC 70. [http://dec.alaska.gov/water/wqsar/wqs/pdfs/70mas.pdf]

American Petroleum Institute (1998). *Element 2: Water Column and Source Sampling, Guidelines for the Scientific Study of Oil Spill Effects*. Washington DC, USA: American Petroleum Institute.

Barron, M.G., T. Podrabsky, S. Ogle, and R.W. Ricker (1999). Are aromatic hydrocarbons the primary determinant of petroleum toxicity to aquatic organisms? *Aquatic Toxicology* **46**(3–4): 253–268.

Bence, A.E., D.S. Page, and P.D. Boehm (2007). Advances in forensic techniques for petroleum hydrocarbons: the *Exxon Valdez* experience. In *Oil Spill Environmental Forensics: Fingerprinting and Source Identification*. Z. Wang and S.A. Stout, eds. San Diego, CA, USA: Academic Press; ISBN-10: 0123695236; ISBN-13: 9780123695239; pp. 449–487.

Boehm, P.D., J.S. Brown, D.S. Page, W.A. Burns, J.M. Neff, A.W. Maki, and A.E. Bence (2003). The chemical baseline as a key to defining continuing injury and recovery of Prince William Sound. In *Proceedings of the 2003 International Oil Spill Conference (Prevention, Preparedness, Response and Restoration – Perspectives for a Cleaner Environment), April 6–11, 2003, Vancouver, British Columbia, Canada*. Washington DC, USA: American Petroleum Institute; pp. 275–283.

Boehm, P.D., L.L. Cook, and K.J. Murray (2011). Aromatic hydrocarbon concentrations in seawater: Deepwater Horizon oil spill. In *Proceedings of the 2011 International Oil Spill Conference (Promoting the Science of Spill Response), May 24–26, 2011, Portland, Oregon, USA*. Washington DC, USA: American Petroleum Institute.

Boehm, P.D., G.S. Douglas, W.A. Burns, P.J. Mankiewicz, D.S. Page, and A.E. Bence (1997). Application of petroleum hydrocarbon chemical fingerprinting and allocation techniques after the *Exxon Valdez* oil spill. *Marine Pollution Bulletin* **34**(8): 599–613.

Boehm, P.D. and D.L. Fiest (1982). Subsurface distributions of petroleum from an offshore well blowout. The IXTOC I blowout, Bay of Campeche. *Environmental Science & Technology* **16**(2): 67–74.

Boehm, P.D., J.M. Neff, and D.S. Page (2007). Assessment of polycyclic aromatic hydrocarbon exposure in the waters of Prince William Sound after the *Exxon Valdez* oil spill: 1989–2005. *Marine Pollution Bulletin* **54**(3): 339–356.

Boehm, P.D., D.S. Page, J.S. Brown, J.M. Neff, and A.E. Bence (2005). Comparison of mussels and semi-permeable membrane devices as intertidal monitors of polycyclic aromatic hydrocarbons at oil spill sites. *Marine Pollution Bulletin* **50**(7): 740–750.

Boehm, P.D., D.S. Page, J.S. Brown, J.M. Neff, and W.A. Burns (2004). Polycyclic aromatic hydrocarbon levels in mussels from Prince William Sound, Alaska, USA, document the return to baseline conditions. *Environmental Toxicology and Chemistry* **23**(12): 2916–2929.

Camilli, R., C.M. Reddy, D.R. Yoerger, B.A.S. Van Mooy, M.V. Jakuba, J.C. Kinsey, C.P. McIntyre, S.P. Sylva, and J.V. Maloney (2010). Tracking hydrocarbon plume transport and biodegradation at Deepwater Horizon. *Science* **330**(6001): 201–204.

Carls, M.G., L.G. Holland, J.W. Short, R.A. Heintz, and S.D. Rice (2004). Monitoring polynuclear aromatic hydrocarbons in aqueous environments with passive low-density polyethylene membrane devices. *Environmental Toxicology and Chemistry* **23**(6): 1416–1424.

Diéz, S., E. Jover, J.M. Bayona, and J. Albaigés (2007). *Prestige* oil spill. III. Fate of a heavy oil in the marine environment. *Environmental Science & Technology* **41**(9): 3075–3082.

Douglas, G.S., W.A. Burns, A.E. Bence, D.S. Page, and P. Boehm (2004). Optimizing detection limits for the analysis of petroleum hydrocarbons in complex environmental samples. *Environmental Science & Technology* **38**(14): 3958–3964.

***Exxon Valdez* Oil Spill Trustee Council** (2009). Exxon Valdez *Trustee Council Hydrocarbon Database*. Anchorage, AK, USA: *Exxon Valdez* Oil Spill Trustee Council; Project 090290. [http://www.afsc.noaa.gov/ABL/Habitat/ablhab_exxonvaldez_hydrocarbon_database.htm]

French-McCay, D.P. (2004). Oil spill impact modeling: Development and validation. *Environmental Toxicology and Chemistry* **23**(10): 2441–2456.

Galt, J.A., W.J. Lehr, and D.L. Payton (1991). Fate and transport of the *Exxon Valdez* oil spill. *Environmental Science & Technology* **25**(2): 202–209.

International Tanker Owners Pollution Federation (2011). *ITOPF Handbook 2011/ 2012*. London, UK: International Tanker Owners Pollution Federation. [http://www.itopf.com/news-and-events/documents/itopfhandbook2011.pdf]

Jernelöv, A. and O. Lindén (1981). Ixtoc I: a case study of the world's largest oil spill. *Ambio* **10**(6): 299–306.

Kingston, P. (1999). Recovery of the marine environment following the *Braer* spill, Scotland. In *Proceedings of the 1999 International Oil Spill Conference (Beyond 2000 – Balancing Perspective), March 8–11, 1999, Seattle, Washington*. Washington DC, USA: American Petroleum Institute; Special Technical Publication 4686B; pp. 103–109.

Lauenstein, G.G. and K.D. Daskalakis (1998). US long-term coastal contaminant temporal trends determined from mollusk monitoring programs, 1965–1993. *Marine Pollution Bulletin* **37**(1–2): 6–13.

National Oceanic and Atmospheric Administration (2010). *Analytical Quality Assurance Plan: Mississippi Canyon 252 (Deepwater Horizon) Natural Resource Damage Assessment, Version 2.2*. Washington DC, USA: National Oceanic and Atmospheric Administration.

National Research Council (1985). *Oil in the Sea: Inputs, Fates and Effects*. Washington DC, USA: National Research Council; National Academy Press; ISBN-10: 0309034795.

National Research Council (1999). *Spills of Non-floating Oils: Risk and Response*. Washington DC, USA: National Research Council; National Academies Press; ISBN-10: 0309065909.

National Research Council (2005). *Oil Spill Dispersants: Efficacy and Effects*. Washington DC, USA: National Research Council, National Academies Press; ISBN-10: 030909562X.

Neff, J.M. (1990). Composition and fate of petroleum and spill-treating agents in the marine environment. In *Sea Mammals and Oil: Confronting the Risks*. J.R. Geraci and D.J. St. Aubin, eds. San Diego, CA, USA: Academic Press; ISBN-13: 9780122806001; pp. 1–33.

Neff, J.M. (2002). *Bioaccumulation in Marine Organisms: Effect of Contaminants from Oil Well Produced Water*. Amsterdam, The Netherlands: Elsevier Science; ISBN-10: 0080437168.

Neff, J.M. and W.A. Burns (1996). Estimation of polycyclic aromatic hydrocarbon concentrations in the water column based on tissue residues in mussels and salmon: an equilibrium partitioning approach. *Environmental Toxicology and Chemistry* **15**(12): 2240–2253.

Neff, J.M., S. Ostazeski, W. Gardiner, and I. Stejskal (2000). Effects of weathering on the toxicity of three offshore Australian crude oils and a diesel fuel to marine animals. *Environmental Toxicology and Chemistry* **19**(7): 1809–1821.

Neff, J.M., S.A. Stout, and D.G. Gunster (2005). Ecological risk assessment of polycyclic aromatic hydrocarbons in sediments: Identifying sources and toxicity. *Integrated Environmental Assessment and Management* **1**(1): 22–33.

Neff, J.M. and W.A. Stubblefield (1995). Chemical and toxicological evaluation of water quality following the *Exxon Valdez* oil spill. In Exxon Valdez *Oil Spill: Fate and Effects in Alaskan Waters*. P.G. Wells, J.N. Butler, and J.S. Hughes, eds. Philadelphia, PA, USA: American Society for Testing and Materials; ASTM Special Technical Publication 1219; ISBN-10: 0803118961; pp. 141–177.

Page, D.S., P.D. Boehm, G.S. Douglas, and A.E. Bence (1995). Identification of hydrocarbon sources in the benthic sediments of Prince William Sound and the Gulf of Alaska following the *Exxon Valdez* oil spill. In Exxon Valdez *Oil Spill: Fate and Effects in Alaskan Waters*. P.G. Wells, J.N. Butler, and J.S. Hughes, eds. Philadelphia, PA, USA: American Society for Testing and Materials; ASTM Special Technical Publication 1219; ISBN-10: 0803118961; pp. 41–83.

Payne, J.R. and W.B. Driskell (2003). The importance of distinguishing dissolved versus oil-droplet phases in assessing the fate, transport, and toxic effects of marine oil pollution. In *Proceedings of the 2003 International Oil Spill Conference (Prevention, Preparedness, Response and Restoration: Perspectives for a Cleaner Environment), April 6–11, 2003, Vancouver, British Columbia, Canada*. Washington DC, USA: American Petroleum Institute; Special Technical Publication I 4730B; pp. 771–778.

Payne, J.R., W.B. Driskell, M.R. Lindeberg, W. Fournier, M.L. Larsen, J.W. Short, S.D. Rice, and D. Janka (2005). Dissolved- and particulate-phase hydrocarbons in interstitial water from Prince William Sound intertidal beaches containing buried oil thirteen years after the *Exxon Valdez* oil spill. In *Proceedings of the 2005 International Oil Spill Conference (Prevention, Preparedness, Response and Restoration – Raising Global Standards), May 15–19, 2005, Miami, Florida, USA*. Washington DC, USA: American Petroleum Institute; pp. 83–88.

Petersen, J., J. Michel, S. Zengel, M. White, C. Lord, and C. Plank (2002). *Environmental Sensitivity Index Guidelines. Version 3.0*. Seattle, WA, USA: National Oceanographic and Atmospheric Administration, Ocean Service, Hazardous Materials Response Division; NOAA Technical Memorandum NOS OR&R 11.

Reddy, C.M., J.S. Arey, J.S. Seewald, S.P. Sylva, K.L. Lemkau, R.K. Nelson, C.A. Carmichael, C.P. McIntyre, J. Fenwick, G.T. Ventura, B.A.S. Van Mooy, and R. Camilli (2011). Composition and fate of gas and oil released to the water column during the Deepwater Horizon oil spill. *Proceedings of the National Academy of Sciences of the United States of America*. DOI /10.1073/pnas.1101242108.

Reddy, C.M. and J.G. Quinn (1999). GC-MS analysis of total petroleum hydrocarbons and polycyclic aromatic hydrocarbons in seawater samples after the *North Cape* oil spill. *Marine Pollution Bulletin* **38**(2): 126–165.

Requejo, A.G., R. Sassen, T. McDonald, G. Denoux, M.C. Kennicutt, II, and J.M. Brooks (1996). Polynuclear aromatic hydrocarbons (PAH) as indicators of the source and maturity of marine crude oils. *Organic Geochemistry* **24**(10–11): 1017–1033.

Robertson, S.B. (2001). Guidelines and methods for determining oil spill effects. In *Proceedings of the 2001 International Oil Spill Conference (Global Strategies for Prevention, Preparedness, Response, and Restoration), March 26–29, 2001, Tampa, Florida*. Washington DC, USA: American Petroleum Institute; Special Technical Publication 14710; pp. 1545–1548.

Robilliard, G.A., P.D. Boehm, and M.J. Amman (1997). Ephemeral data collection guidance manual, with emphasis on oil spill NRDAs. In *Proceedings of the 1997 International Oil Spill Conference (Improving Environmental Protection – Progress, Challenges, Responsibilities), April 7–10, 1997, Fort Lauderdale, Florida, USA*. Washington DC, USA: American Petroleum Institute; Special Technical Publication 4651; pp. 1029–1030.

Sauer, T. and P. Boehm (1991). The use of defensible analytical chemical measurements for oil spill natural resource damage assessment. In *Proceedings of the 1991 International Oil Spill Conference (Prevention, Behavior, Control, Cleanup), March 4–7, 1991, San Diego, California*. Washington DC, USA: American Petroleum Institute; Technical Publication 4529; pp. 363–369.

Short, J.W. and P.M. Harris (1996a). Chemical sampling and analysis of petroleum hydrocarbons in near-surface seawater of Prince William Sound after the *Exxon Valdez* oil spill. In *Proceedings of the* Exxon Valdez *Oil Spill Symposium*. S.D. Rice, R.B. Spies, D.A. Wolfe, and B.A. Wright, eds. Bethesda, MD, USA: American Fisheries Society; Symposium 18; ISBN-10: 0913235954; ISSN: 08922284; pp. 17–28.

Short, J.W. and P.M. Harris (1996b). Petroleum hydrocarbons in caged mussels deployed in Prince William Sound after the *Exxon Valdez* oil spill. In *Proceedings of the* Exxon Valdez *Oil Spill Symposium*. S.D. Rice, R.B. Spies, D.A. Wolfe, and B.A. Wright, eds. Bethesda, MD, USA: American Fisheries Society; Symposium 18; ISBN-10: 0913235954; ISSN: 08922284; pp. 29–39.

Stout, S.A. and Z. Wang (2007). Chemical fingerprinting of spilled or discharged petroleum: methods and factors affecting petroleum fingerprints in the environment. In *Oil Spill Environmental Forensics: Fingerprinting and Source Identification*. Z. Wang and S.A. Stout, eds. San Diego, CA, USA: Academic Press; ISBN-10: 0123695236; pp. 1–54.

Taft, D.G., D.E. Egging, and H.A. Kuhn (1995). Sheen surveillance: An experimental monitoring program subsequent to the 1989 *Exxon Valdez* shoreline cleanup. In Exxon Valdez *Oil Spill: Fate and Effects in Alaskan Waters*. P.G. Wells, J.N. Butler, and J.S. Hughes, eds. Philadelphia, PA, USA: American Society for Testing and Materials; ASTM Special Technical Publication 1219; ISBN-10: 0803118961; pp. 215–238.

Wolfe, D.A., M.J. Hameedi, J.A. Galt, G. Watabayashi, J. Short, C. O'Clair, S. Rice, J. Michel, J.R. Payne, J. Braddock, S. Hanna, and D. Sale (1994). The fate of the oil spilled from the *Exxon Valdez*. *Environmental Science & Technology* **28**(13): 561A–568A.

Wooley, C.B. (2002). The myth of the "pristine environment": past human impact on the Gulf of Alaska coast. *Spill Science and Technology Bulletin* **7**(1–2): 89–104.

CHAPTER FOUR

Surveying oil on the shoreline

Edward H. Owens and P. Douglas Reimer

4.1 Introduction

Responding rapidly to oil that has reached shorelines is critical for minimizing risk to people and a host of other organisms, including many that have limited mobility. The coastal zone and tidal shorelines are among the most productive ecosystems and are sensitive spawning habitats for many marine animals. They are also traditional commercial and subsistence food sources and are recreation and tourist destinations. For spill response efforts to be effectively prioritized and targeted over a large area, it is necessary to determine where oil has stranded and where resources or activities are most at risk. The challenge is greater when a spill occurs in a remote area, as was the case with the *Exxon Valdez* oil spill.

In this chapter, we describe how the Shoreline Cleanup Assessment Technique (SCAT) process was created in 1989 to meet this challenge. We show how responses were mobilized, how shorelines were surveyed, and how guidelines and recommendations to deal with oil on the shorelines were generated and implemented. We conclude with lessons learned that may help streamline and focus responses to other oil spills or environmental accidents.

Information generated by the SCAT process supported planning and cleanup decisions that were the foundation for the 1989 shoreline response operation at both the strategic and tactical levels (Owens and Teal, 1990a). The multiyear shoreline surveys provided data for subsequent treatment activities and documented the changes in oiling conditions resulting from the combined effects of cleanup and natural weathering processes (Neff *et al.*, 1995). The SCAT process was the cornerstone of the *Exxon Valdez* response in Alaska from 1989 through 1993 and again 20 years later during the Deepwater Horizon response in the Gulf of Mexico, not to mention countless small and large response operations on coasts, lakes, and rivers worldwide. In each case, SCAT data are the basis for the development of shoreline treatment recommendations and provide a detailed record of changes in shoreline oiling conditions due to cleanup and natural cleaning.

Oil in the Environment: Legacies and Lessons of the Exxon Valdez *Oil Spill*, ed. J. A. Wiens. Published by Cambridge University Press. © Cambridge University Press 2013.

4.2 Background and survey objectives

Oil began to wash up on the shores of Prince William Sound (PWS) within a few days of the *Exxon Valdez* oil spill on March 24, 1989, and cleanup crews responded by recovering mobile oil on the beaches around Naked Island and the northern parts of the Knight Island group (Fig. 1.3a and Map 3, p. vii). From the beginning, it was evident that shoreline cleanup would be a major element of the response operation. An environmental scientist was assigned to be the Shoreline Cleanup Technical Advisor in Exxon's spill-management team and to work with the Operations Division (Table 4.1), continuing in that role for the next 4 years.

The shorelines of southcentral Alaska are a complex mix of bedrock outcrops, sands, pebbles, cobbles, and boulders (Chapter 1). Mapping this varied shoreline required a systematic protocol that could identify the primary characteristics and include the range of substrate types that might exist within a stretch of shoreline. Many sections of coast are a combination of sands and coarse sediments resting on underlying bedrock platforms, with scattered boulders over the top. In PWS, the primary shore types are exposed or sheltered bedrock/cobble (72%) and mixed pebble/cobble or boulder/cobble/pebble (27%) (Owens, 1991a, b; Chapter 6).

The overall objective of the SCAT program during the 1989 field season was to support Exxon's spill-management team as it established cleanup priorities and selected appropriate cleanup tactics. Exxon created and managed the SCAT program, which was staffed by Woodward-Clyde Consultants. The United States Coast Guard (USCG) report from the Federal On Scene Coordinator (FOSC) (Leschine *et al.*, 1993, p. 135) notes the importance of SCAT, stating that "at the foundation of the shoreline cleanup organizational system were the shoreline assessment teams, SCAT, organized by Exxon to conduct detailed shoreline surveys."

Table 4.1 Exxon's Spill Management Team.

General Manager	
DEPARTMENTS/DIVISIONS	**OPERATIONS DIVISION**
Safety	Operations Manager
Operations	....
• Shoreline Operations	PWS Shoreline Operations Manager
• Free Oil Operations	• Operations Support
Salvage	• Operations Services Coordinator
Disposal	• Spill Specialist
Operations Support (logistics)	• Task Force Coordinator – Task Forces, 1, 2, and 3
Surveillance and Tracking	• Task Force Coordinator – Task Forces 4, 5, and 6
Oil Spill Chemical Advisor	Outside PWS Valdez Coordinator
Ecology	• Seward Coordinator
Shoreline Cleanup Technical Advisor	• Homer Coordinator
	• Kodiak Coordinator
	Safety Coordinator

Shoreline oiling information was collected in the SCAT program by three, sequenced field-survey activities:

1. **Aerial Surveys** collected information on shoreline oiling, including the location and extent of the oil coverage and the oil's character. Reconnaissance flights over PWS were followed by systematic aerial-videotape shoreline surveys. The reconnaissance flights provided a perspective on the size of the affected area and on the scale of the cleanup response that would be required. Mapping based on the systematic video surveys identified sections of oiled shoreline for more detailed ground surveys. The ground survey teams walked the shorelines to document the exact distribution and character of the stranded oil to generate treatment recommendations. A geographical information system (GIS) was created to store and manage the field data (Box 4.1).
2. **Ground Surveys** provided shoreline oiling information, along with cleanup recommendations, to Exxon's spill-management team (Table 4.1) and all response organizations. SCAT information formed the basis of Exxon's May 1, 1989, Shoreline Restoration Plan, which was approved by the USCG (Leschine *et al.*, 1993). Exxon used the field teams' data and cleanup recommendations to develop site-specific plans for oiled shorelines that were submitted for review by the Interagency Shoreline Cleanup Committee (ISCC) (Teal, 1990, 1991; Neff *et al.*, 1995).
3. **Inspections** evaluated oiling conditions after cleanup was completed. The SCAT teams conducted inspection surveys to ensure that the treatment actions had achieved the desired objectives or to recommend additional treatment where that was appropriate.

The SCAT teams were staffed by independent scientists from the business and academic communities, selected for their experience and expertise. Each team had a geologist, a biologist, and a cultural resources specialist (Neff *et al.*, 1995). The geologist focused on the shore-zone materials, the shoreline character, and the oiling conditions. The biologist documented the plant and animal communities and identified any ecological constraints that might apply to cleanup activities so additional injuries could be minimized. The cultural resources specialist provided expertise on the many documented and undocumented sites that had been used by people in historic and prehistoric times and identified constraints to prevent potential damage to these resources during cleanup (Chapter 5).

Box 4.1 The Geographical Information System (GIS): 1989

The GIS database used in 1989 was run on an IBM 386 computer with a 25-MHz processor, 12 megabytes of memory, and 350 megabytes of hard drive storage. To put this into perspective with today's technology, a standard smart phone has more processing power and considerably more storage capacity than the entire system used to maintain and process all the SCAT data in 1989.

4.3 The SCAT process

The SCAT process has a sequence of steps to document oiling conditions, support operations, and inspect the shorelines after cleaning.

4.3.1 Step 1: Detection and documentation of shoreline oiling in 1989

4.3.1.1 Aerial reconnaissance and videotape mapping (April)

An aerial shoreline-videotape mapping survey covered over 8000 km of coastline to locate and characterize oiled shorelines. Helicopters typically flew at slow speeds (often less than 50 knots) and low altitudes (under 100 m) with the rear door open or removed in order to record high-quality imagery. (Fixed-wing aircraft were not suitable, as they could not follow the complex shoreline configurations that typify this coastal region.) The aerial reconnaissance and mapping of shorelines was based out of Valdez and was separate from the many other aerial missions taking place at the time, which focused on surveillance and tracking of oil on the water. In the first month after the spill, the aerial shoreline-survey teams focused on defining the scale of the problem within PWS, but by the end of April, it became evident that it was necessary to extend the survey program into the Gulf of Alaska (GOA). Aerial surveys in the GOA continued until mid-June, eventually covering almost 4000 km of shoreline. For logistical reasons, a separate SCAT program was created, based in Kodiak, to survey the coasts outside of PWS in the Kenai-Kodiak (GOA) region.

At the same time, the Alaska Department of Environmental Conservation (ADEC) was flying aerial surveys to record shoreline oiling directly on maps without the use of the videotape record. Exxon and ADEC aerial-survey data were later compared and found to be in general agreement.

The primary source of data used in the aerial shoreline mapping process is the audio commentary on the tape, rather than the visual backdrop of the video. It is critical that the aerial observer be able to recognize and document shoreline sediment size. Aerial survey data can be used to generate maps at a range of scales that can be an order of magnitude more detailed than published charts and maps, which is often necessary for segment mapping and planning cleanup operations. Categorizing the types of sediment and relationships to surrounding bedrock provides a direct understanding of the potential for oil penetration, persistence, and remobilization – information that is crucial to the development of treatment priorities and recommendations. In addition, biologically sensitive habitats can be

Table 4.2 "Degree of oiling" category definitions and the lengths of PWS shorelines meeting the definitions.

	Definitions	Lengths
A. 1989		
Heavy	> 6 m wide	141 km
Moderate	3–6 m wide	94 km
Light	< 3 m wide	326 km
Very light	< 10% oil cover	223 km
No oil	No oil observed	667 km
B. 1990		
Wide	> 6 m wide and > 50%	21 km
Medium	> 6 m and 10–50%	
or	3–6 m and > 10%	46 km
Narrow	< 3 m wide and > 10%	80 km
Very light	< 10% oil cover	323 km
No oil	No oil observed	689 km

identified and mapped as part of this first phase so that ecological considerations and constraints can be incorporated into the selection of appropriate shoreline cleanup tactics.

A second key element is that the observer be able to recognize oil from the air. This is relatively straightforward where there are high concentrations of stranded oil but becomes increasingly more difficult as the oil cover decreases. In addition, observers must distinguish between the many dark-colored shoreline materials that can be mistaken for stranded oil: lichen, algae, mussels, heavy mineral sands, and black volcanic bedrock. An aerial reconnaissance identifies areas where "heavy" and "moderate" concentrations (Table 4.2) of oil are stranded. Lower concentrations often are not visible from the air, depending on the type of shoreline and substrates, and require subsequent ground surveys to accurately document the distribution and character of the oil.

A third important element of SCAT surveys, whether they are aerial or ground, is to document shore character and oil conditions on a checklist or form that can be entered into a database and linked to individually mapped sections of shoreline. Systematic shoreline-mapping protocols had been developed prior to the *Exxon Valdez* spill as part of extensive surveys in British Columbia and the Atlantic provinces of Canada (Owens, 1983; Howes et al., 1995). Checklists had also been used for several years as part of shoreline-response training programs by Environment Canada, so that scientists knew in 1989 the types of information that would be required to make decisions regarding shoreline cleanup strategies and tactics (e.g., Owens, 1979, 1987, 1990). The SCAT process was a success primarily because these systematic shoreline-mapping protocols and oil-documentation checklists were integrated in a practical manner based on easily understood terminology.

An essential step in the SCAT shoreline mapping process is the division of the shoreline into planning and operational work units, called "segments." Segment boundaries

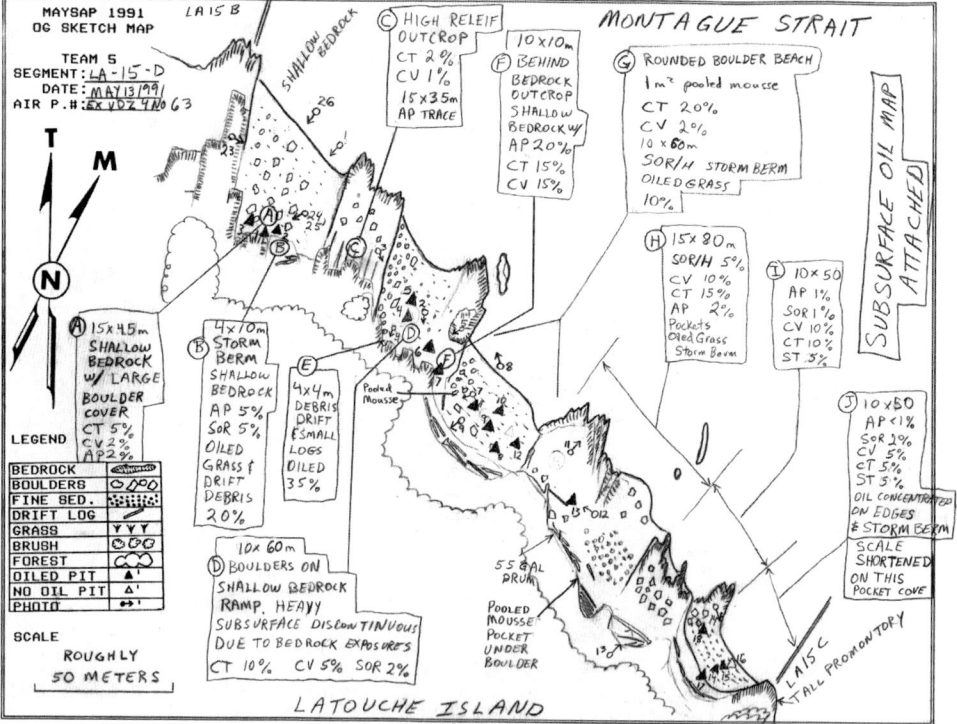

Figure 4.1 Example of a SCAT sketch map.

were established on the basis of prominent geological features (such as a headland), changes in shoreline or substrate type, a change in oiling conditions, or the boundary of an operations area. Each segment had relatively homogenous shoreline character and oiling conditions. As the oiling conditions changed in 1990 and subsequent years, many segments were broken down into "subdivisions" to further define cleanup areas.

Individual segments were assigned a unique alphanumeric identifier based on a geographic area (e.g., LA-15-D was subdivision D of segment 15 on Latouche Island; Figs. 4.1 and 4.2). This convention enabled all response parties to collect data for the same sections of shoreline and compare among each other and over time as oiling conditions changed.

The segments were the building blocks of the shoreline response planning process, and cleanup priorities and site-specific tactics were linked to each segment. As a result of the 1989 SCAT program, the value of prespill segmentation and shoreline mapping was recognized and many coasts of North America have been segmented and mapped based on aerial-video surveys (e.g., Owens et al., 2003).

4.3.1.2 Detailed ground SCAT (April–September)

The second phase of Step 1 of the SCAT process involved detailed ground-mapping surveys that further described a shoreline's character, documented the distribution and character of stranded oil, identified potential ecological or archeological/cultural resource constraints on cleanup, and identified possible safety concerns. This level of detail enabled Exxon's planners and the ISCC decision-makers to generate site-specific instructions for Operations to treat the oil while minimizing negative effects due to cleanup itself.

The first systematic ground surveys were conducted on April 13. The GIS database (Box 4.1) was set up the following day, enabling field observations to be easily and quickly accessible to decision-makers and planners. Four teams worked in PWS out of Valdez, and four worked in the GOA out of Seward and Kodiak; overall, more than 5500 km were ground-surveyed in 1989 (Owens and Teal, 1990a). A total of 550 shoreline segments covering 1450 km were defined in PWS. Each segment was subdivided into across-shore zones, perpendicular to the water, based on changes in shoreline or substrate type or oiling conditions, which were linked directly to tidal elevations. Standard forms and terminologies were developed to provide a consistent data set that could be used by all of the field teams to describe the shoreline character and the oiling conditions for each segment. A sketch was drawn for each segment to provide a context for the information recorded on the form: examples of a sketch map and a Shoreline Oiling Summary (SOS) form are shown in Figures 4.1 and 4.2. Information generated for individual segments was managed through the GIS database, reviewed by Exxon's planners, and cleanup recommendations were submitted to the ISCC (Teal, 1990, 1991).

The 1989 surveys were successful, even though they were conducted initially very much as a seat-of-the-pants operation (Box 4.2), largely because everyone on the team was experienced in shoreline mapping and most of the team leaders had prior experience from other oil spills. These experienced teams were able to adjust and modify data collection in the very early stages to create a consistent procedure that was then used throughout the operating area.

ADEC carried out ground-shoreline surveys in parallel with Exxon in 1989. Their independent results recorded only very slight differences from the Exxon SCAT data on lengths of oiled shoreline. In the spring of 1990, the ISCC introduced a single, multidisciplinary team approach for integrating surveys that included federal and state government representatives. This integration of responders with local authorities and

MAYSAP SHORELINE OILING SUMMARY

TEAM NO. 5

PAGE ____ OF ____

SEGMENT LA 15

SUBDIVISION D

DATE MAY, 13 /91

OG ____ BIO ____

ADEC ____ LANDMANAGER ____ for ____

EXXON ____ USCG/NOAA ____

TIME 07:00 to 09:30 TIDE LEVEL -2.5 ft. to 1.0 ft. ENERGY LEVEL: [✓] H [ ] M [ ] L

SURVEYED FROM: [✓] FOOT [ ] BOAT [ ] HELO WEATHER: [✓] SUN [✓] CLOUDS [ ] FOG [ ] RAIN [ ] SNOW

TOTAL LENGTH SHORELINE SURVEYED: 450 m NEAR SHORE SHEEN: [ ] BR [ ] RB [✓] SL [ ] NONE

EST. OIL CATEGORY LENGTH: W 10 m M 365 m N — m VL 35 m NO 20 m US 20 m

LOC	SURFACE OIL CHARACTER										SURFACE SEDIMENT	SHORE SLOPE	AREA		ZONE				
	AP	MS	TB	SOR	CV	CT	ST	FL	DB	NO	TYPE	VHML	WIDTH m	LENGTH m	S	UI	MI	LI	NOTES
A	S				S	S					RB	H	15	45		✓			Shallow Bedrock
B	S			S					P		BR Logs	H	4	10	✓				Oiled Grass & Drift Debris
C	T				S	S					BR	VH	15	35	✓	✓			Mostly Bedrock
D				S	S	S					BR	H	10	60	✓	✓			Boulders on Shallow Bdrk
E									P		Logs Debris	H	4	4	✓				Oiled Spruce Needles & Drift
F	P				P	P					BR	H	10	10		✓			Protected by Bedrock Outcrop
G		T		S	S	P					B	M	10	60	✓	✓			Rounded Boulder Beach
H	S	T		S	S	P					RB	H	15	80	✓	✓			Shallow Bedrock
I	S			S	S	S	S				RB	H	10	50		✓			Shallow Bedrock
J	T			S	S	S	S				BC	M	10	50	✓	✓			Boulder Beach

DISTRIBUTION: C = 91–100%; B = 51–90%; P = 11–50%; S = 1–10%; T = <1%

SLOPE: V = VERTICAL; H = HIGH ANGLE; M = MEDIUM ANGLE; L = LOW ANGLE PHOTO ROLL # MAYSAP- 5 - 13 / 14 FRAMES 22-26 / 1-14

PIT NO.	PIT DEPTH (cm)	SUBSURFACE OIL CHARACTER							OILED ZONE	CLEAN BELOW	H2O LEVEL	SHEEN COLOR	PIT ZONE				SURFACE-SUBSURFACE	
		OP	HOR	MOR	LOR	OF	TR	NO	cm-cm	Y/N	(cm)	BRSN	S	UI	MI	LI	SEDIMENTS	NOTES
1	15			✓					0-15	Y	10	BR		✓			B-BPR	Bedrock Pocket
2	10	✓							0-10	Y	—	—		✓			B-CPR	Bedrock Pocket
3	10		✓						0-10	Y	5	BR		✓			B-CPR	Bedrock Pocket
4	20	✓							5-15		15	BR		✓			B-CPR	Bedrock Pocket
				✓					15-20	Y		R		✓				
5	20		✓						8-15	Y	10	BR		✓			B-CPR	Bedrock Pocket
				✓					15-20	Y		BR		✓				
6	10	✓							0-10	Y	9	B		✓			B-CPR	Bedrock Pocket
7	15		✓						0-15	Y	10	B		✓			B-CPR	Bedrock Pocket

SHEEN COLOR: B = BROWN; R = RAINBOW; S = SILVER; N = NONE

OG COMMENTS: AREA (A) CT, CV, AP ON PROTECTED BEDROCK HOLLOWS & BACK SIDES OF BOULDERS. AREA (B) OILED GRASS & VEGETATION IN STORM BERM. PITS 1, 2 & 3 IN BEDROCK POCKETS WHERE DIGGING WAS POSSIBLE. HEAVY OILING SUBSURFACE ~5×10m BUT NOT CONTINUOUS DUE TO SURFACE BEDROCK EXPOSURES.

Figure 4.2 Example of a Shoreline Oiling Summary (SOS) form.

inhabitants had been the key element of the *Nestucca* oil-spill surveys earlier in 1989 in British Columbia (Owens, 1990) and was introduced for the *Exxon Valdez* response through the ISCC process and the interagency SCAT teams.

Although different in scope, within a few weeks the aerial and ground surveys were coordinated and integrated. As shoreline oiling data from the daily aerial surveys were processed, ground teams were deployed to the more heavily oiled shorelines. This interaction was particularly important early in the response to best

MAYSAP SHORELINE OILING SUMMARY (cont.)

PAGE ____ OF ____
TEAM NO. ____
SEGMENT LA-15   SUBDIVISION D   DATE MAY, 13 / 91

PIT NO.	PIT DEPTH (cm)	SUBSURFACE OIL CHARACTER						OILED ZONE cm-cm	CLEAN BELOW Y/N	H2O LEVEL (cm)	SHEEN COLOR				PIT ZONE			SURFACE-SUBSURFACE SEDIMENTS	NOTES
		OP	HOR	MOR	LOR	OF	TR	NO				B	R	S	N	S	UI	MI	LI		
8	40		✓						20-40	N	—				—		✓			BC-BC	Boulders at Bottom
9	35	✓							15-30	Y	—				—		✓			BC-BC	" "
10	35						✓		?- ?		20			S				✓		BC-BCP	Free Silver Sheen
11	35						✓		?- ?		25			S		✓				BC-BC	Silver Sheen
12	40		✓						20-40	N	25	B				✓				BC-BCP	
13	20	✓							0-20	N	—				—		✓			BC-BCP	SOR/H on Surface
14	35		✓						5-15	Y	—				—		✓			BC-CPB	
15	30		✓						5-20	Y	—				—		✓			BC-CPB	
16	35						✓		-										✓	BC-CPB	
17	40	✓							10-20		—				—					BC-BPG	
				✓					20-35	Y						✓					
18	25	✓							15-25	N	10	B R				✓				BC-BPG	Big Boulders under

SHEEN COLOR: B = BROWN; R = RAINBOW; S = SILVER; N = NONE

OG COMMENTS:
Pit 13 BETWEEN LARGE BOULDERS. LIMITED PIT DIGGING OPPORTUNITIES

SUBSURFACE OIL EXTENDS THE FULL WIDTH OF THE BEACH AT ⓙ
ROUGHLY 20m x 50m   SOR/H IN STORM BERM

Figure 4.2 (cont.)

expedite the positioning of ground SCAT teams as well as the deployment of operational resources and cleanup efforts.

In September 1989, the "Marry Map" study compared data generated by the aerial-videotape survey with data from ground SCAT surveys. The results were remarkably close. Areas categorized as heavily, moderately, and lightly oiled as mapped by aerial surveys (574 km for PWS and 1270 km for GOA) were confirmed by the ground teams (561 km for PWS (Table 4.2) and 1171 km for GOA).

Box 4.2 Things were very different in those days …

The 1989 surveys were exceptional for several reasons, primarily because the process enabled shoreline-treatment activities to be directed in a considered manner to minimize any potential additional effects from the treatment activities themselves. These surveys were carried out with few of the supporting tools that we now take for granted. There was no Geographical Positioning System (GPS), so all navigation was carried out using maps and charts. These maps and charts were woefully inaccurate in many areas because they had not been revised following the 1964 earthquake, which lifted the shoreline as much as 10 m in some areas of PWS. Some islands became headlands and inlets became lakes, whereas where the shoreline dropped, some headlands became islands and low-lying coastal areas were inundated. Along sections of the Kenai and Alaska Peninsulas, fjord glaciers had substantially retreated, exposing shorelines that were indicated on maps as being covered with ice. There were no recent vertical aerial photographs and no accurate digitized shoreline maps to help with navigation or defining locations. There were no "on-line" satellite images that we take for granted today (e.g., Google Earth). For the aerial over-flights, all flight-path and location information was hand-drawn onto enlarged photocopies of pre-1964 topographic maps.

Initially, the SCAT team had one laptop computer (see Box 4.1). Data were managed with word processing DOS software, as spreadsheets had not been developed at that time. All photographs had to be processed as prints or slides, so that sometimes it was 10 days or longer after a survey before the teams could look at the photographs for a particular site. With no Internet and no e-mails, field teams in remote areas or onboard vessels had to send in daily forms and reports by fax. A bank of five or six fax machines in one corner of the SCAT room in the Exxon Command Post in Valdez lay virtually silent all day until about 4 o'clock in the afternoon, when one by one they woke and in unison began to spew reams of paper into collection boxes.

Communications were difficult because the mountainous terrain in the region limited the use of radios until line-of-sight relay towers were installed in the summer of 1989. Cell phones did not exist, and we had only a few International Maritime Satellite phones that had to be shared between the teams.

4.3.2 Step 2: SCAT support to shoreline cleanup in 1989

The SCAT data were the critical drivers for information generation and the shoreline-cleanup decision process. The SCAT teams submitted their documentation and recommendations to Exxon, which was a nonvoting member of the ISCC but who generated shoreline assessment summaries and cleanup work orders for the ISCC to review.

The ISCC was initially chaired by the Exxon Shoreline Cleanup Technical Advisor, and later by the National Oceanic and Atmospheric Administration (NOAA). It included representatives from all federal and state agencies, as well as interested local community groups (e.g., Chugach Alaska Corporation, Cordova District Fishermen United, the Prince William Sound Conservation Alliance). The daily ISCC meetings were an important forum for all agencies and interested parties to contribute to the discussion and provide direction for the operational activities, and they approved the shoreline cleanup work orders. Eventually, there were separate ISCCs in Valdez (for PWS), Seward, Homer, and Kodiak.

The primary shoreline treatment methods used in 1989 were a combination of flooding, washing, and flushing (Figs. 1.4, 3.3, and 4.3; www.valdezsciences.com). This choice was

Figure 4.3 Flooding, washing, and flushing on northern Smith Island, May 1989. (Photos: Edward H. Owens)

driven in part by the recognition that the removal of shoreline sediments could likely result in shoreline retreat, as there are few contemporary sources to replace any coarse material that would be removed. At first, ambient-temperature seawater was pumped through header hoses, remobilizing the stranded oil for collection by floating skimmers. As the oil weathered, washing and flushing with higher pressures were needed to release the oil. By midsummer, the seawater was heated to around 40–60 °C (Chapter 1, Fig. 1.4).

Field trials tested the effectiveness of other tactics, such as chemical cleaners and a water-injection system, but none proved suitable or acceptable. Bioremediation offered the potential to accelerate natural biodegradation and was applied to 119 km of shoreline in PWS (Chapter 8). A few other site-specific tactics were approved, such as the use of treated peat as a sorbent on a seal pupping haulout at Applegate Rocks, but the majority of the oiled segments were treated by flooding, washing, and flushing.

By the end of 1989, 426 km in PWS had been treated and the USCG had determined that they required no further cleanup. The USCG also approved no further cleanup for an additional 1048 km that had been treated in the GOA (Leschine et al., 1993).

The fact that the present-day oil spill response plans for PWS employ the same shoreline treatment tactics that played such a prominent role in 1989 testifies to their effectiveness and to the legacy of lessons learned from the *Exxon Valdez* spill response.

4.3.3 Step 3: Postcleanup shoreline inspections and monitoring in winter 1989/90

Summer shoreline-treatment operations ended on September 15, as scheduled in the May 1 Shoreline Restoration Plan. This preplanned cut-off was a safety measure based on the anticipated onset of winter wind and wave conditions in PWS and the GOA. An Exxon aerial-videotape survey between September 30 and October 28 resurveyed segments to document the shoreline conditions after the treatment program for the year had been completed (Table 4.2). Also in September, ADEC conducted a Post Treatment Assessment Survey, called "Beachwalk." The results of these surveys formed the basis for planning the spring 1990 SCAT field program. A Spring Oiling Prediction Map for PWS was completed in January 1990 based on primary substrate type (i.e., bedrock/pebble/cobble beach, etc.), 1989 oiling conditions, and wave exposure. The objective of this study was to generate an estimate of the possible scale and magnitude of the 1990 cleanup program. The mapping process identified segments where oil would most likely persist and which, therefore, would require additional cleanup.

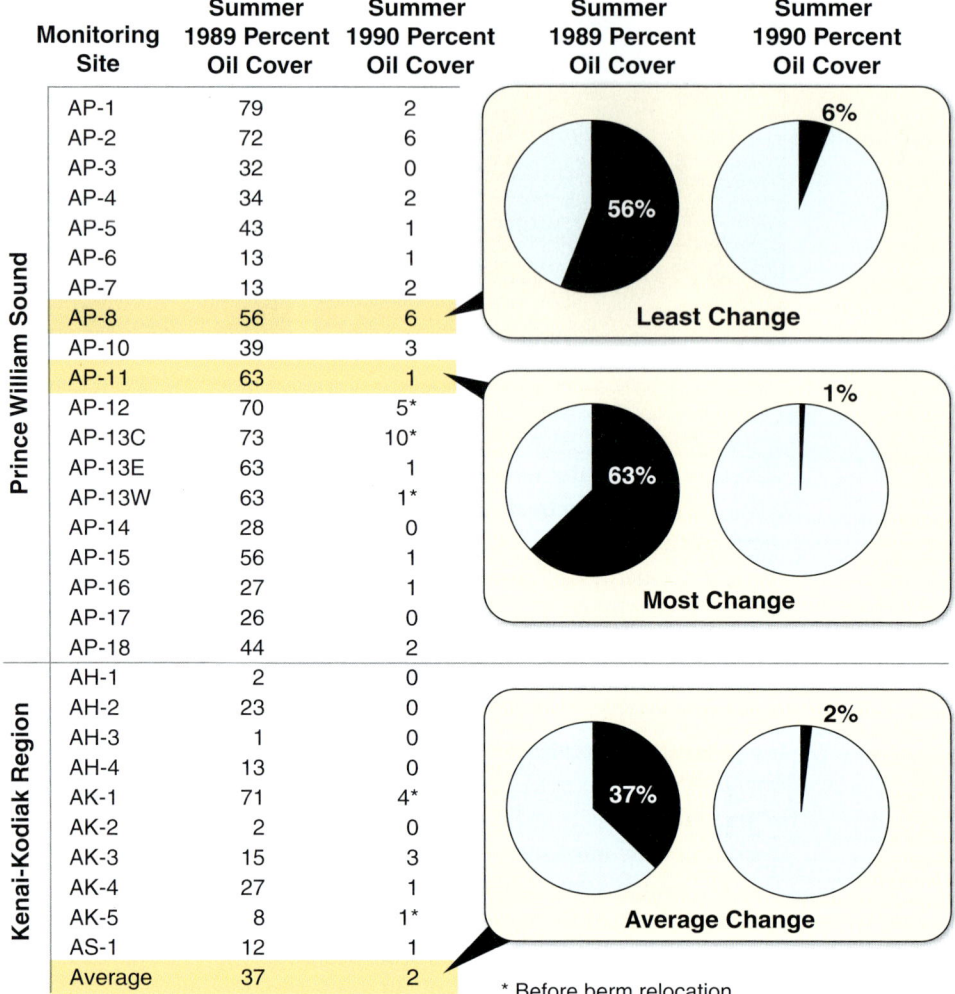

Figure 4.4 Summary of changes in surface oil cover at detailed monitoring sites in PWS and GOA, summer 1989 and summer 1990.

Shoreline surveys continued through the winter with a monitoring program that was now based out of Anchorage. This program was different from SCAT. Whereas the previous aerial and ground surveys had focused on the distribution and character of the stranded oil, the winter monitoring program focused on documenting trends for oil fate and persistence over time (Fig. 4.4).

Monitoring locations were selected to represent the range of shore types and oil conditions. In May 1989, Exxon's SCAT teams had established six shoreline monitoring sites in PWS; by September 1989, this program was expanded to 18 sites in PWS and 10 in the GOA (Jahns et al., 1991). These 28 sites were selected for detailed documentation (the "A Program") and were monitored monthly. The A Program included multiple shoreline types, oil-cover transects, subsurface pits, biological transects, and sample collection (Figs. 4.5 and 4.6) (Neff et al., 1995).

A second, concurrent program involved 36 reconnaissance sites (the "B Program") where single across-shore profiles were monitored monthly. In addition, five time-lapse cameras were set up in PWS to record wave conditions and shoreline changes

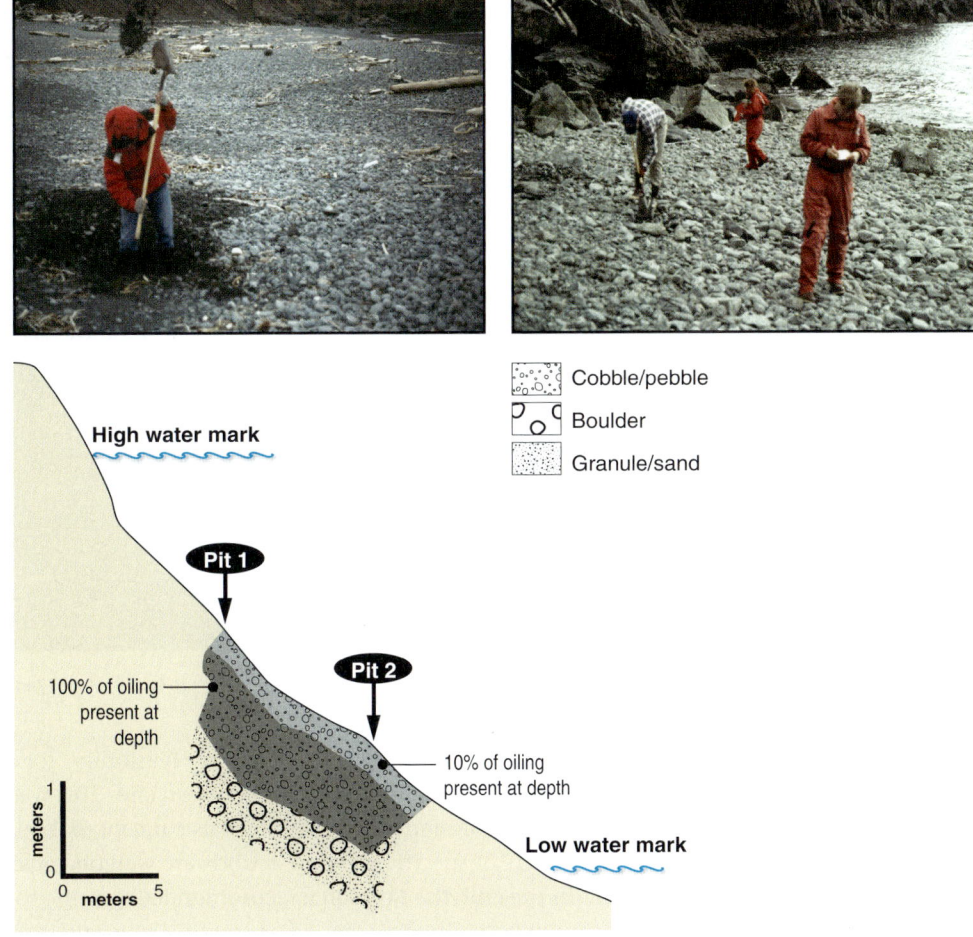

Figure 4.5 Digging pits on mixed-sediment (sand–pebble–cobble) beaches and an example of a subsurface oil sketch. (Photos: Edward H. Owens)

continuously during November 1989–April 1990 (Owens and Teal, 1990b). Other sites in the area were monitored by federal and state agencies primarily based on single across-shore profiles, as opposed to Exxon's program of mapping and characterizing oil at each location in addition to the beach profile data.

One significant observation during the winter monitoring surveys was that stranded oil concentrations in the intertidal zone continued to decrease, even in very sheltered areas with limited wave action. Exxon researchers sought to explain this observation in the laboratory using oiled pebbles and cobbles and identified the process by which clays remove oil from sediments (Bragg and Yang, 1995). This process, known as "mineral-oil flocculation" (or "clay-oil flocculation" at the time), was associated with the presence of very fine-grained sediments (glacial flour) in the waters and on the shorelines of PWS (Chapter 6, Box 6.1). Numerous subsequent field and laboratory studies have confirmed that fine-grained sediments, whether clays or other materials, can play a critical role in the natural cleaning of oiled shorelines anywhere in the world and in fresh- and salt-water environments (Lee *et al.*, 2001, 2002; Owens and Lee, 2003; Sun and Zheng, 2009).

During the winter, the ISCC discussed potential cleanup tactics for subsurface oil on the cobble and boulder shorelines of Point Helen on southern Knight Island. A special

Figure 4.6 Beach profiling at a monitoring site, Sleepy Bay, Latouche Island, May 9, 1989. (Photo: Edward H. Owens)

scientific task force was assembled to review the feasibility for constructing a "rockwasher" to remove, clean, and replace sediments on site. This was one of the first applications of the Net Environmental Benefit Analysis concept (Baker, 1999), in which potential negative impacts of specific cleanup actions are weighed against the benefits of removing the oil. In the end, the task force recommended that the rockwasher would cause more harm than would allowing the oil to weather naturally.

4.4 Shoreline surveys 1990 and later

In January and February 1990, Exxon organized a Fast Assessment Shoreline Survey Team (FASST), in which six interagency teams and an Exxon Operations representative surveyed 190 km of shoreline where concentrations of "wide band oil" were believed to remain (Table 4.2).

From April through June 1990, a Spring Shoreline Assessment Team (SSAT) survey involved 20 teams with 5–6 persons each that were deployed in PWS (16) and GOA (4). SSAT surveyed 1875 km of shoreline, and 598 of the 1035 subdivisions surveyed were recommended for cleanup. The remaining 437 subdivisions had "no treatment required" recommendations. The SCAT forms and terminology were modified slightly from those used in 1989 as the oiling conditions had changed (Neff *et al.*, 1995). The summary surface-oiling categories became wide, medium, narrow, and very light (Table 4.2); these would later be changed back to the original heavy, moderate, light, and very light categories, with the inclusion of oiling distribution and thickness in the categorization. An interagency Technical Advisory Group (TAG), which replaced but performed the same function as the ISCC, reviewed data generated from SSAT (Teal, 1991). The oil character in 1990 had changed owing to weathering, so the cleanup program shifted

Figure 4.7 (a) Sediment relocation (Ushagat Island), and (b) mixing (Latouche Island). (Photos: Edward H. Owens)

away from predominantly flooding, washing, and flushing that had been appropriate in 1989 to bioremediation (Chapter 8), sediment relocation, and mixing (tilling).

In 1990, mechanical equipment was used at 30 locations where oil on mixed pebble/cobble beaches had been stranded above the normal high tide to move these oiled sediments to the intertidal zone (Fig. 4.7a). This oil had been stranded during spring high tides, and relocating the sediments down into the intertidal zone accelerated natural sediment cleaning (Owens *et al.*, 1991). Similarly, rakes and tines were used on beaches to expose oiled sediments to waves (Fig. 4.7b). In some locations, clean surface sediments were side cast and subsurface oiled sediments were spread over the intertidal zone where they could be reworked by wave action.

A third survey, the August Shoreline Assessment Program (ASAP), was carried out over the summer. The objectives of this survey were to determine whether additional cleanup was required in 1990 and to prioritize shorelines for spring 1991 surveys. ASAP surveyed 822 sites and found that 99.3 km of shoreline were still oiled: 60.6 km in PWS and the remainder in the GOA. Most (61%) were very lightly oiled and only 6.1 km (6%) were found to have heavy oiling. As a result of ASAP, 521 subdivisions were recommended for survey in the 1991 May Shoreline Assessment Program (MAYSAP), including 163 first-priority locations (Chapter 6).

MAYSAP involved three separate field programs between April and June. The survey covered 429 subdivisions in PWS and 148 subdivisions in the GOA, from which the TAG generated treatment recommendations for 20 shoreline subdivisions (www.valdezsciences.com).

From 1991 onward, the emphasis of shoreline surveys shifted away from supporting cleanup to focus on the fate and effects of stranded oil (Chapter 6). These surveys were conducted separately by Exxon and the *Exxon Valdez* Oil Spill Trustee Council (Trustee) agencies (see e.g., Hayes and Michel, 1999; Gillfillan *et al.*, 2001; Short *et al.*, 2004; Taylor and Reimer, 2005; Irvine *et al.*, 2006).

Cleanup in 1992 was based on a Final Shoreline Assessment Program (FINSAP) in PWS and the Kenai Peninsula. FINSAP was boat-based, small in scale, and combined shoreline assessment with actual cleanup. The team visited 82 sites in May and June, of which 63 required treatment.

On June 10, 1992, the USCG issued their final pollution report, declaring the *T/V Exxon Valdez* cleanup operation closed (Leschine *et al.*, 1993). However, Exxon

Table 4.3 Shoreline Assessment Surveys 1989–1993.

1989 April–June	Aerial videotape and mapping (Exxon)	~14 000 km
1989 April–May	Aerial mapping (ADEC)	
1989 April–September	SCAT (Exxon)	4 teams and 1611 km in PWS 4 teams and 3922 km in GOA
1989 September–October	"Walkathon" (ADEC)	
1989 September–October	Aerial videotape and mapping (Exxon)	
1990 February–March	FASST (interagency)	6 teams 170 km in PWS 20 km in GOA
1990 March–June	SSAT (interagency)	20 teams 16 teams and 1043 km in PWS 4 teams and 832 km in GOA
1990 July–August	ASAP (interagency)	3 teams in PWS 2 teams in GOA
1991 April–June	MAYSAP (interagency)	3 phases 6 teams: 5 teams and 429 subdivisions in PWS and 1 team and 148 subdivisions in GOA
1992	FINSAP (interagency)	2 teams combining SCAT and operations; 76 locations in PWS and 6 along the Kenai Peninsula
1993 June–July	PostSAP (Exxon)	1 team; 97 locations in PWS and 34 in GOA

continued to monitor selected locations and conducted a Post Shoreline Assessment Program (POSTSAP) in June and July 1993, surveying 97 locations in PWS and 34 in GOA (Table 4.3). One study of note, Exxon's Detailed Shoreline Assessment Program in July 1993 on Squirrel Island, involved surveys along 15 transects (four stations on each) and dug 36 pits to determine the location and character of any subsurface oil.

There is general agreement that by 1992 about 800 metric tons of the estimated 20 000 metric tons of stranded oil remained (2%; Wolfe et al., 1994). By 2001, about 55.6 metric tons of oil (0.14–0.24% of the amount originally stranded) remained sequestered in the subsurface of a few coarse-sediment beaches (Short et al., 2004; Chapter 7). By 2002, surface oil covered less than 200 m^2 of the approximately 110 000 m^2 previously documented to have oil (Taylor and Reimer, 2005).

4.5 The legacy: SCAT in 2011

The shoreline survey and mapping procedures created in 1989, and the concept of SCAT teams supporting the decision-making process and shoreline cleanup, have stood the test of time. SCAT has evolved and been applied to river, freshwater, and Arctic environments and to low-concentration tar-ball oiling conditions (Owens and Sergy, 2000, 2004). The SCAT program was again the critical driver for information generation and treatment recommendations on the Deepwater Horizon shoreline response in 2010–12 in the Gulf of Mexico (Santner et al., 2011). The SCAT concept has been adopted with only minor modifications for use in Canada, the European Union, France,

New Zealand, the Mediterranean, the United Kingdom, and the United States (e.g., Cramer *et al.*, 1991; Jacques *et al.*, 1998; Kerambrun, 2006; Maritime and Coastguard Agency, 2007; International Maritime Organization and United Nations Environment Programme, 2009). SCAT terms and definitions have been translated into French, Russian, and Spanish.

The fundamental elements of the SCAT concept – standard terms and definitions, standard shoreline oiling data-collection forms, and interagency survey teams – constitute the nucleus of every program. In the United States, the SCAT process has become an integral part of the Incident Command System (ICS) and has become more formalized as part of many oil-spill response or contingency plans.

Designing a SCAT program involves generating semiquantitative data on the location of stranded oil, oil concentrations and characteristics, and environmental sensitivities and constraints. When done properly, this process supports decision-makers and cleanup strategies. SCAT data, however, may not be adequate for quantitative monitoring of the long-term persistence of oil, at least in cases such as the *Exxon Valdez*.

One key advance has been the development of Shoreline Treatment Recommendation (STR) forms, by which SCAT field teams summarize shoreline and oiling character, develop recommendations for managers and planners, and identify operational constraints for each segment. STR forms can act as work orders used to direct shoreline cleanup operations. In the ICS for oil spills, STRs are standard for assigning activities to cleanup teams or task forces.

SCAT teams continue to perform an inspection role once cleanup has been completed. Similar to a form developed following the *Exxon Valdez* spill, a Shoreline Inspection Report (SIR) form is now used to document that cleanup has been sufficient. The SIR is signed by the responsible party and federal and state/provincial representatives, bringing operational closure to sections or segments of shoreline.

4.6 Lessons learned

- The only oil that remains more than two decades after the spill is at locations where oil was initially documented by the 1989 SCAT surveys.
- SCAT identified where oil was located and this information was used to prioritize cleanup. This critical function has been adopted in basically the same form on many coastal and river spills since the *Exxon Valdez*. However, not every element of the survey may be needed for every spill.
- Aerial surveys worked very well following the *Exxon Valdez*. Differences in operational conditions, shoreline types, and the character and distribution of the oil determine which survey techniques are best suited to a particular event.
- The use of standard forms, terms, and definitions is a cornerstone of the SCAT process.
- Segmentation of the shoreline is crucial to database generation and data management. Segmentation has also become a key design element of prespill shoreline mapping (e.g., Owens *et al.*, 2003).
- Surveys to map oil conditions must document where sections of shoreline have been inspected but are not oiled. This category of "No oil observed" is as important in many ways as are the oiling categories.

- When the same team leaders resurveyed the area over the several years of the SCAT program, there was a high level of continuity and consistency to the observations and documentation process.
- The generation of a consensus on oiling conditions and treatment recommendations, as embodied in the interagency SCAT teams and the ISCC and TAG committees, is a sound basis for a cooperative response operation.
- The SCAT process best focuses on operational support, rather than on the fate and persistence of oil or on the effects of stranded oil on living organisms. SCAT only considers the persistence of oil as part of the cleanup decision process, rather than as part of a longer term scientific monitoring effort.
- Shoreline oiling data and information that are generated by a SCAT program are a major contribution to assessment studies, even though its focus is operational support for cleanup.

In Memoriam

The systematic data-based mapping system that is the foundation of the SCAT process and of all prespill shoreline mapping is a legacy of our good friend and mentor, Don Howes (1948–2011). We dedicate this story of the initial evolution of the SCAT process to Don. He instigated systematic shore-zone mapping from aerial video surveys in 1979 in British Columbia: a system that has been applied throughout that province, Canada, and worldwide with little change. Don was a SCAT Team Lead on the *Exxon Valdez* (1989–93) and on the Deepwater Horizon (2010–11) responses. Throughout his life, Don gave without seeking credit. All of us who map shorelines and who use the SCAT process worldwide today owe Don. We who followed in his footsteps simply expanded his original concept. He played ice hockey and ran SCAT surveys to the end.

REFERENCES

Baker, J.M. (1999). Ecological effectiveness of oil spill countermeasures: how clean is clean? *Pure and Applied Chemistry* **71**(1): 135–151.

Bragg, J.R. and S.H. Yang (1995). Clay-oil flocculation and its role in natural cleansing in Prince William Sound following the *Exxon Valdez* oil spill. In Exxon Valdez *Oil Spill: Fate and Effects in Alaskan Waters*. P.G. Wells, J.N. Butler, and J.S. Hughes, eds. Philadelphia, PA, USA: American Society for Testing and Materials; ASTM Special Technical Publication 1219; ISBN-10: 0803118961; pp. 178–214.

Cramer, M.A., E.H. Owens, D.E. Howes, and G.A. Sergy (1991). Spill response manuals for the coasts of British Columbia. In *Proceedings of the Fourteenth Arctic and Marine Oilspill Program (AMOP) Technical Seminar, June 12–14, 1991, Vancouver, British Columbia, Canada*. Ottawa, ON, Canada: Environment Canada; pp. 551–577.

Gillfillan, E.S., D.S. Page, K.R. Parker, J.M. Neff, and P.D. Boehm (2001). A 10-year study of shoreline conditions in the *Exxon Valdez* spill zone, Prince William Sound, Alaska. In *Proceedings of the 2001 International Oil Spill Conference (Prevention, Behavior, Control, Cleanup), March 26–29, 2001, Tampa, Florida*. Washington DC, USA: American Petroleum Institute; Technical Publication 14710; pp. 559–567.

Hayes, M.O. and J. Michel (1999). Factors determining the long-term persistence of *Exxon Valdez* oil in gravel beaches. *Marine Pollution Bulletin* **38**(2): 92–101.

Howes, D.E., J. Harper, and E. Owens (1995). *British Columbia Physical Shore-Zone Mapping System*. Victoria, BC, Canada: British Columbia Ministry of Environment, Resource Inventory Committee, Coastal Task Force. [http://www.ilmb.gov.bc.ca/risc/pubs/coastal/pysshore/index.htm]

International Maritime Organization and United Nations Environment Programme (2009). *Regional Information System: Part D: Operational Guides and Technical Documents, Section 13: Mediterranean Guidelines on Oiled Shoreline Assessment*. Valletta, Malta: International Maritime Organization (IMO) and United Nations Environment Programme (UNEP), Regional Marine Pollution Emergency Response Centre for the Mediterranean Sea (REMPEC); Project ME/XM/6030–08–11.[http://www.rempec.org/admin/store/wyswiglmg/file/Information%20resources/Guidelines/RIS%20D13/EN/RIS%20D13-Mediterranean%20Guidelines%20on%20Oiled%20Shoreline%20Assessment%20(EN).pdf]

Irvine, G.V., D.H. Mann, and J.W. Short (2006). Persistence of 10-year old *Exxon Valdez* oil on Gulf of Alaska beaches: The importance of boulder-armouring. *Marine Pollution Bulletin* **52**(9): 1011–1022.

Jacques, T.G., A.J. O'Sullivan, and E. Donnay (1998). *Polscale: A Guide, Reference System and Scale for Quantifying and Assessing Coastal Pollution and Clean-up Operations in Oil-polluted Coastal Zones*. Brussels, Belgium: European Commission, Office for Official Publications of the European Communities; May 29, 1998; ISBN-10: 9282818152; ISBN-13: 9789282818152.

Jahns, H.O., J.R. Bragg, L.C. Dash, and E.H. Owens (1991). Natural cleaning of shorelines following the *Exxon Valdez* spill. In *Proceedings of the 1991 International Oil Spill Conference (Prevention, Behavior, Control, Cleanup), March 4–7, 1991, San Diego, California*. Washington DC, USA: American Petroleum Institute; Technical Publication 4529; pp. 167–176.

Kerambrun, L. (2006). *Surveying Sites Polluted by Oil – An Operational Guide for Conducting an Assessment*. Brest, France: Cedre (Centre de documentation, de recherche et d'expérimentations sur les pollutions accidentelles des eaux). [http://www.cedre.fr/en/publication/operational-guidesurveying/surveying.php]

Lee, K., P. Stoffyn-Egli, and E.H. Owens (2001). Natural dispersion of oil in a freshwater ecosystem: Desaguadero Pipeline Spill, Bolivia. In *Proceedings of the 2001 International Oil Spill Conference (Global strategies for prevention, preparedness, response, and restoration), March 26–29, 2001, Tampa, Florida*. Washington DC, USA: American Petroleum Institute; Publication 14710; pp. 1445–1448.

Lee, K., P. Stoffyn-Egli, and E.H. Owens (2002). The OSSA II pipeline oil spill: natural mitigation of a riverine oil spill by oil-mineral aggregate formation. *Spill Science & Technology Bulletin* **7**(3–4): 149–154.

Leschine, T.M., J. McGee, R. Gaunt, A. van Emmerik, D.M. McGuire, R. Travis, and R. McCready (1993). *T/V Exxon Valdez Oil Spill: Federal On Scene Coordinator's Report*. Washington DC, USA: United States Department of Transportation, United States Coast Guard; Report DOT-SRP-94–1; National Technical Information Service Order Number PB94–121845 (Volume 1).

Maritime and Coastguard Agency (2007). *The UK SCAT Manual (Shoreline Cleanup Assessment Technique): A Field Guide to the Documentation of Oiled Shorelines in the UK*. J. Moore, ed. Southampton, UK: Maritime & Coastguard Agency (MCA), Counter Pollution and Response; Corp 119. [http://www.dft.gov.uk/mca/corp119ext.pdf]

Neff, J.M., E.H. Owens, S.W. Stoker, and D.M. McCormick (1995). Shoreline oiling conditions in Prince William Sound following the *Exxon Valdez* oil spill. In Exxon Valdez *Oil Spill: Fate and Effects in Alaskan Waters*. P.G. Wells, J.N. Butler, and J.S. Hughes, eds. Philadelphia, PA, USA: American Society for Testing and Materials; ASTM Special Technical Publication 1219; ISBN-10: 0803118961; pp. 312–346.

Owens, E.H. (1979). *Prince Edward Island: Coastal Environments and the Cleanup of Oil Spills*. Ottawa, ON, Canada: Environment Canada, Environmental Impact Control Directorate; Economic and Technical Review Report EPS 3-EC-79-5.

Owens, E.H. (1983). The application of videotape recording (VTR) techniques for coastal studies. *Shore & Beach* **51**(1): 29–33.

Owens, E.H. (1987). Estimating and quantifying oil contamination on the shoreline. *Marine Pollution Bulletin* **18**(3): 110–118.

Owens, E.H. (1990). Suggested improvements to oil spill response planning following the *Nestucca* and *Exxon Valdez* incidents. In *Proceedings of the Thirteenth Arctic and Marine Oilspill Program (AMOP) Technical Seminar, June 6–8, 1990, Edmonton, Alberta, Canada*. Ottawa, ON, Canada: Environment Canada; pp. 439–450.

Owens, E.H. (1991a). Shoreline conditions following the *Exxon Valdez* oil spill as of fall 1990. In *Proceedings of the Fourteenth Arctic and Marine Oilspill Program (AMOP) Technical Seminar, June 12–14, 1991, Vancouver, British Columbia, Canada*. Ottawa, ON, Canada: Environment Canada; pp. 579–606.

Owens, E.H. (1991b). *Changes in Shoreline Oiling Conditions 1-½ Years after the 1989 Prince William Sound Spill*. Seattle, WA, USA: Woodward-Clyde; unpublished report. [www.valdezsciences.com]

Owens, E.H. and K. Lee (2003). Interaction of oil and mineral fines on shorelines: Review and assessment. *Marine Pollution Bulletin* **47**(9–112): 397–405.

Owens E.H., P.D. Reimer, A. Lamarche, S.O. Marchant, and D.K. O'Brien (2003). Pre-spill shoreline mapping in Prince William Sound, Alaska. In *Proceedings of the Twenty-Sixth Arctic and Marine Oilspill Program (AMOP) Technical Seminar, June 10–12, 2003, Victoria, British Columbia, Canada*. Ottawa, ON, Canada: Environment Canada; pp. 233–251.

Owens, E.H. and G.A. Sergy (2000). *The SCAT Manual: A Field Guide to the Documentation and Description of Oiled Shorelines (Second Edition)*. Edmonton, AB, Canada: Environment Canada.

Owens, E.H. and G.A. Sergy (2004). *The Arctic SCAT Manual: A Field Guide to the Documentation of Oiled Shorelines in Arctic Regions*. Edmonton, AB, Canada: Environment Canada.

Owens, E.H. and A.R. Teal (1990a). Shoreline cleanup following the *Exxon Valdez* oil spill – Field data collection within the SCAT program. In *Proceedings of the Thirteenth Arctic and Marine Oilspill Program (AMOP) Technical Seminar, June 6–8, 1990, Edmonton, Alberta, Canada*. Ottawa, ON, Canada: Environment Canada; pp. 411–421.

Owens, E.H. and A.R. Teal (1990b). A brief overview and initial results from the winter shoreline monitoring program following the *Exxon Valdez* incident. In *Proceedings of the Thirteenth Arctic and Marine Oilspill Program (AMOP) Technical Seminar, June 6–8, 1990, Edmonton, Alberta, Canada*. Ottawa, ON, Canada: Environment Canada; pp. 451–470.

Owens, E.H., A.R. Teal, and P.R. Haase (1991). Berm relocation during the 1990 shoreline cleanup program following the *Exxon Valdez* oil spill. In *Proceedings of the Fourteenth Arctic and Marine Oilspill Program (AMOP) Technical Seminar, June 12–14, 1991, Vancouver, British Columbia, Canada*. Ottawa, ON, Canada: Environment Canada; pp. 607–630.

Santner, R., M. Cocklan-Vendl, B. Stong, J. Michel, E.H. Owens, and E. Taylor (2011). Deepwater Horizon MC252-Macondo Shoreline Cleanup Assessment Technique (SCAT) Program. In *Proceedings of the 2011 International Oil Spill Conference (Promoting the Science of Spill Response), May 24–26, 2011, Portland, Oregon, USA*. Washington DC, USA: American Petroleum Institute; Paper No. 2011-270.

Short, J.W., M.R. Lindeberg, P.M. Harris, J.M. Maselko, J.J. Pella, and S.D. Rice (2004). Estimate of oil persisting on the beaches of Prince William Sound 12 years after the *Exxon Valdez* oil spill. *Environmental Science & Technology* **38**(1): 19–25.

Sun, J. and X. Zheng (2009). A review of oil-suspended particulate matter aggregation: a natural process of cleansing spilled oil in the aquatic environment. *Journal of Environmental Monitoring* **11**(10): 1801–1809.

Taylor, E. and P.D. Reimer (2005). SCAT surveys of Prince William Sound beaches – 1989 to 2002. In *Proceedings of the 2005 International Oil Spill Conference (Prevention, Preparedness, Response, and Restoration – Raising Global Standards), May 15–19, 2005, Miami Beach, Florida, USA*. Washington DC, USA: American Petroleum Institute; Special Technical Publication I 4781B; pp. 801–806.

Teal, A.R. (1990). Shoreline cleanup following the *Exxon Valdez* oil spill – The decision process for shoreline cleanup. In *Proceedings of the Thirteenth Arctic and Marine Oilspill Program (AMOP) Technical Seminar, June 6–8, 1990, Edmonton, Alberta, Canada*. Ottawa, ON, Canada: Environment Canada; pp. 423–429.

Teal, A.R. (1991). Shoreline cleanup: reconnaissance, evaluation and planning following the *Exxon Valdez* oil spill. In *Proceedings of the 1991 International Oil Spill Conference (Prevention, Behavior, Control, Cleanup), March 4–7, 1991, San Diego, California*. Washington DC, USA: American Petroleum Institute; Special Technical Publication 4529; pp. 149–152.

Wolfe, D.A., M.J. Hameedi, J.A. Galt, G. Watabayashi, J. Short, C. O'Clair, S. Rice, J. Michel, J.R. Payne, J. Braddock, S. Hanna, and D. Sale (1994). The fate of the oil spilled from the *Exxon Valdez*. *Environmental Science & Technology* **28**(13): 561A–568A.

CHAPTER FIVE

Ancient sites and emergency response: cultural resource protection

Chris B. Wooley and James C. Haggarty

5.1 Introduction

For over 4000 years, humans have lived, fished, hunted, and developed industries in Prince William Sound (PWS). When *Exxon Valdez* oil spilled in 1989, hundreds of archaeological sites and artifacts were potentially at risk. Many were at least superficially known, but most were not. Archaeologists and other cultural resource experts had to move quickly to identify vulnerable sites and work with others engaged in the response to safeguard Alaska's heritage from oil and incidental harm during cleanup. Exxon created the *Exxon Valdez* Cultural Resource Program (hereafter the "Program") to address potential threats to archaeological sites and to comply with state and federal laws.

The 28 archaeologists contracted to Exxon in 1989 coordinated with archaeologists from governmental agencies and Alaska Native organizations to assess the integrity of sites, identify oil concerns, and protect sites from impacts during shoreline cleanup. The archaeological contractors and their agency counterparts were knowledgeable in all areas of cultural-resource management, laws, and the region's pre- and postcontact history.[1]

Two aspects of past human history in the spill area were relevant to shoreline cleanup. First, there were numerous known archaeological sites in the spill area that might be impacted by the spill and many other sites that had not yet been documented. Second, various postcontact human activities had contributed oil to area shorelines before the spill, confounding attempts to isolate and document the effects of the *Exxon Valdez* spill. This chapter deals only with the first issue (see also Chapter 1). For information related to the second concern, see Page *et al.* (2002), Wooley (2002), and Chapter 6.

---

[1] "Precontact" refers to the time before Europeans first had contact with Native Alaskans in 1741. The "postcontact period" refers to 1741 to the present.

Oil in the Environment: Legacies and Lessons of the Exxon Valdez *Oil Spill*, ed. J. A. Wiens. Published by Cambridge University Press. © Cambridge University Press 2013.

The Program was extraordinary in terms of the size of the survey area and the emergency nature of the response, which necessitated immediate access to agency data and representatives in order to make informed decisions. There was little background information on how to undertake such a project; only one example of "oil spill archaeology" was referenced in the literature, and this dealt with a 1978 spill on the Oregon–Washington border that only involved archaeological monitoring of heavy equipment used during cleanup (Benson, 1987). Additionally, the coastal area affected by the *Exxon Valdez* spill had never been the focus of systematic archeological surveys.

The Program was developed and implemented with input from the State Historical Preservation Officer, state and federal governmental agencies, and Alaska Native organizations. All partners helped the Program achieve its fundamental goals of site identification and protection during the 4 years of shoreline cleanup. In the process, the Program collected a substantial amount of new information regarding the history of the Alutiiq region (PWS, Kenai Peninsula, Kodiak Island, and the Gulf of Alaska [GOA]).

The archaeological information collected and synthesized by the Program proved invaluable to state and federal land managers with cultural-resource site-management responsibilities. In addition, the new procedures developed during the spill response – such as treatment constraints, training of response personnel, and instituting cultural resource policies and procedures – have been applied during subsequent spill responses. Many of the policies and procedures were formally adopted in 2002 as the *Alaska Implementation Guidelines for Federal On-Scene Coordinators for the Protection of Historic Properties during Emergency Response under the National Oil and Hazardous Substance Pollution Contingency Plan* (Alaska Regional Response Team, 2002).

The Program provided site-protection strategies that could be implemented without unduly affecting cleanup. After years of additional site monitoring, agency damage assessments would conclude that minimal impacts to sites had occurred from oiling, cleanup, or other spill-related activities (Reger *et al.*, 2000).

5.2 The Exxon Cultural Resource Program

The Cultural Resource Program developed out of the Shoreline Cleanup Assessment Technique (SCAT) surveys (Chapter 4). SCAT teams surveyed affected shorelines to provide accurate descriptions of shoreline conditions. The data enabled informed decisions regarding shoreline treatment priorities and cleanup operations, including the protection of cultural resources. The general design included:

- Extensive shoreline surveys by archaeologists to identify and document cultural sites.
- Survey data synthesized with existing site data to document sensitivities for consideration in developing cleanup plans.
- Archaeological protection strategies based on site-specific sensitivities (i.e., constraints on cleanup techniques).
- Educational programs to ensure that all cleanup workers were aware of laws and procedures related to the protection of cultural resources.
- Special responses for issues such as artifact finds and vandalism.

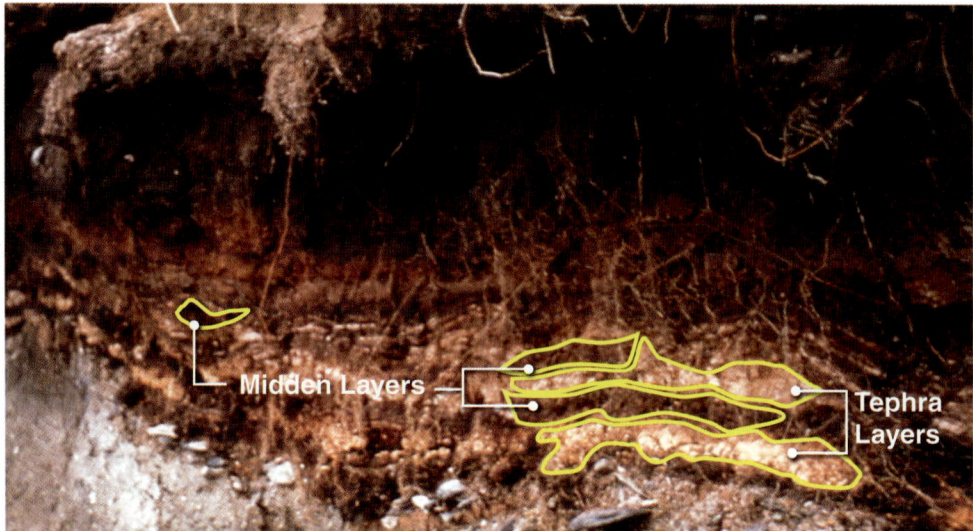

Figure 5.1 Eroding coastal midden, west coast Kodiak Island. The middens are the brown layers in the mixed deposit on top of the uniformly gray and somewhat rocky base, which is most likely old beach deposit. The tephra layers are the whitish layers interspersed between the dark brown midden layers. Many prehistoric sites like this were affected by shoreline subsidence and subsequent erosion associated with the 1964 earthquake.
(Photo: Chris B. Wooley)

Such a large-scale archeological protection project under emergency conditions had never been attempted before. Given the region's rich human history, there was a high probability of discovering new sites, ranging from a single stone tool on the beach; to large (100 m × 30 m) coastal middens[2] (Fig. 5.1) with semisubterranean house depressions; to abandoned mines, canneries, and World War II harbor-defense installations (see Chapter 1, Fig. 1.6).

5.2.1 A cooperative approach

Throughout the program, Exxon consulted with the Alaska State Historic Preservation Officer, federal agencies responsible for cultural resources, Alaska Natives and corporations, and private landowners. Alaska Native tribes and villages – including Tatitlek, Chenega Bay, Port Graham, and Nanwalek (formerly English Bay and before that Alexandrovsk) – were represented by a cultural resource specialist from the Chugach Alaska Corporation.

This level of formal consultation among affected parties is triggered any time the federal government begins an "undertaking" – such as leading the response to a major oil spill – with the potential to affect significant cultural resources. The National Historic Preservation Act (NHPA), Section 106, governs this process. NHPA Section 106 does not provide a detailed, step-by-step approach to protecting cultural resources

---

[2] Refuse deposits.

but instead describes a general process with three separate stages: identification, evaluation, and (if necessary) mitigation.

Cultural resource specialists from all affected parties were very involved at every step in the program. In 1989, cleanup plans first had to be approved by the Alaska State Historic Preservation Officer. They were then reviewed by the Interagency Shoreline Cleanup Committee (ISCC) composed of up to 24 agencies, Alaska Native groups, and interested parties. The ISCC passed the plan to the Federal On Scene Coordinator (FOSC), the US Coast Guard officer in charge of the overall cleanup, for final approval.

In 1990, to streamline the process, a Technical Advisory Group (TAG) replaced the ISCC. A Cultural TAG (CTAG), composed of eight agencies and Alaska Native groups, addressed archaeological issues. CTAG implemented an array of protections for cultural resources – including site avoidance, site inspection, and site monitoring – that still allowed cleanup activities to proceed. The State Historic Preservation Officer and then the FOSC exercised final approvals of cleanup plans. CTAG continued to operate on a much reduced scale during 1991 and 1992 (when cleanup ended) as the cleanup effort was reduced in scope and knowledge of the area increased.

5.2.2 Cultural resource site data before 1989

At the time of the *Exxon Valdez* oil spill, cultural resource information was scattered and out-of-date. Many of the previously recorded archaeological sites on or near shorelines had not been revisited by an archaeologist since Frederica de Laguna did her pioneering work in the 1930s (de Laguna, 1956).

The need for compiled and accessible data was recognized immediately. Although some information was available in 1989, primarily through the Alaska Office of History and Archaeology, the data were incomplete. Extensive file research was necessary to collect, collate, update, and use these records effectively during the response. Federal agencies, Alaska Native organizations, and other landowners held additional site information that also needed to be compiled and merged with state data.

5. 3 Methods

SCAT survey data were used to evaluate the potential effects of proposed cleanup and to modify shoreline-segment cleanup plans to prevent damage to sensitive resources. A shoreline segment (1989) or subdivision (1990 onward) was defined as a continuous length of shoreline with similar geological characteristics that could be addressed by a single cleanup plan. For each of the 1500 shoreline segments, there was a field report with proposed cleanup plans, biological and nonconfidential archaeological data, proposed cultural resource constraints, and other guidelines to be followed during cleanup of the segment (Teal, 1991). A segment might contain a single cultural site, multiple cultural sites, or no sites at all.

Standard archaeological field methods were used consistently during the 4-year field program, including systematic shoreline surveys; site testing with soil probes and test units; site documentation; and collection of diagnostic surface artifacts, samples for ^{14}C radiocarbon dating, and tephra samples (fragmental fallout material from volcanic eruptions). In the field, archaeologists used notebooks, standardized recording forms, sketch maps, 35-mm film, and videotape to document sites and their exact locations on the landscape (this work was done before GPS became widely available). Standardized recording forms were developed for confidential, site-specific data and program tasks (field surveys, site inspections, site monitoring, and site incidents).

Subsurface testing for the presence of cultural deposits was conducted only where such deposits were suspected. Tests were conducted using either 2.5-cm diameter Oakfield soil samplers or 20-cm^2 shovel tests to the depth necessary to determine the presence of cultural material. There was no mechanical screening of sediments, but shell, bone, tephra, samples for ^{14}C dating, and artifacts were collected when encountered. No further subsurface testing was conducted once a site was identified as cultural in origin.

Site-specific descriptions, maps, photos, etc., were filed in a museum-style archive separate from the publicly available SCAT files. This process allowed the Program to maintain confidentiality of site locations, yet still protect sites during cleanup. All landowners received copies of all files pertaining to sites on their lands. The Alaska Office of History and Archaeology received copies of all project maps, field notes, shoreline segment files, and other data. This helped the office maintain a complete record of site information for the Alaska Heritage Resources Survey, a restricted, statewide database of Alaska's cultural resource sites (Alaska Office of History and Archaeology, 2011).

5.3.1 Studies of ^{14}C dating contamination by crude oil

When crude oil contaminates organic materials, including artifacts, it can create incorrect ^{14}C dating results. Because of their extreme age, crude oils are deficient in ^{14}C. When present in samples, they can make samples appear older than they really are. Several government-sponsored studies addressed this problem.

Mifflin and Associates, Inc. (1991) carried out laboratory studies of the effect of fresh and weathered Prudhoe Bay crude oil from the *Exxon Valdez* spill on ^{14}C dating of wood, peat, and charcoal samples taken from outside the spill area. The studies confirmed that both fresh and weathered oils made samples appear older than they really were, but that cleaning could correct the error. Mifflin and Associates also showed that weathered crude oil penetrates organic material poorly, even in a high-speed centrifuge. This suggests that the potential for oil contamination of organic archaeological materials decreased with time as oil weathered. Nonetheless, investigators need to be cautious with ^{14}C dating when there is a possibility of oil contamination.

Prompted by Mifflin and Associates, Reger *et al*. (1992) investigated 13 intertidal sites in the spill area to determine whether oiling had any effect on ^{14}C dating of organic artifacts. Four of the 13 sites had artifacts that could be dated by means other than ^{14}C. Only two of the sites contained sediments that tested positive for traces of *Exxon Valdez* oil. The study concluded that ^{14}C dating was still reliable for artifacts at all four sites. The ^{14}C dates obtained were well within the range of dates established by comparison with artifact assemblages from nonoiled sites in the region.

5.3.2 Cultural resource constraints

Constraints to protect cultural sites during cleanup included site avoidance, access restrictions, and stipulations that treatment take place only in the presence of an archaeological monitor. In conjunction with a blanket prohibition on unauthorized access to areas upland of shorelines (which remained in effect throughout the cleanup), site avoidance was the primary constraint. During 1989, when cleanup was scheduled for areas that had not been fully assessed by SCAT personnel, the State Historic Preservation Officer requested additional archaeological monitoring. The most restrictive constraint, the need to have an archaeologist on site during cleanup, was included in about 6% of the site-specific cleanup plans during the 4-year cleanup.

At each site, program archaeologists recorded conditions before and after cleanup. In some cases, they personally cleaned areas near scattered artifacts in the intertidal zone to ensure that no artifacts were damaged or removed. This approach minimized impacts to sites and conveyed the importance of site protection to treatment crews and agency representatives.

There was always the possibility that additional subsurface sites that were not visible could be present, but careful inspection and monitoring of cleanup areas where such deposits were possible ensured that they were protected.

5.3.3 Training and educational programs

Program archaeologists met regularly with cleanup supervisors before sensitive shoreline segments were cleaned, providing guidance for site protection, strict adherence to subdivision constraints, and treatment monitoring when appropriate.

A concerted effort was made to instill in cleanup-crew members an awareness that they had a legal responsibility to avoid collecting artifacts and to report any artifacts they might encounter to Program staff. This policy was implemented by presentations and briefing handouts. A training video was produced in 1989 and refined in subsequent years to ensure that all cleanup personnel were fully aware of the need to protect cultural resources. A simple written cultural resource policy and a video briefing have become standard for subsequent spill responses in Alaska.

A particular concern of training was the prevention of site vandalism. Postcleanup assessments and *Exxon Valdez* Oil Spill Trustee damage assessments generally concluded that some vandalism occurred in the early days, when cleanup crews had not been as well briefed as they were later in the cleanup.

Throughout the 1970s and 1980s, cultural resource managers had been attempting to prevent illicit collecting and selling of artifacts by keeping site data confidential and through public education stressing the historical value of undisturbed sites. In an effort to strengthen this goal, Exxon provided staff time and funding to support and develop the first Alaska Archaeology Week programs in 1990 and 1991 – a cooperative project with the National Park Service, Minerals Management Service, and the Anchorage Museum of History and Art. This ongoing annual event (now Alaska Archaeology Month) increases public awareness of the importance of cultural resource preservation.

5.4 Field survey results

Program archaeologists surveyed over 5000 km of PWS and GOA shorelines during the 1989–92 field seasons. All shorelines with the potential to be affected by oil and subsequent cleanup were surveyed at a reconnaissance level in 1989 and at a more intensive level in 1990–92. Hundreds of treatment areas were also inspected and monitored. Project activities in 1991 and 1992 focused primarily on training cleanup personnel and monitoring cleanup operations.

Information about 200 previously known sites near oiled shorelines was updated. The Program located, documented, and reported 326 new sites to the Alaska Heritage

Table 5.1 The 1989 oiling conditions near cultural resource sites on shorelines potentially affected by the *Exxon Valdez* oil spill.

1989 Shoreline Oiling Category[e]	Number and location of sites on potentially affected shorelines[a]								
	Upland only[b]		Upland/ intertidal[c]		Intertidal only[d]		All sites		
	PWS	GOA	PWS	GOA	PWS	GOA	PWS	GOA	Total
No oil	20	107	6	93	10	13	36	213	249
Very light	6	76	7	94	0	9	13	179	192
Light	9	10	5	8	2	1	16	19	35
Moderate	11	5	3	7	1	0	15	12	27
Heavy	51	18	10	14	10	3	71	35	106
Total near oil	77	109	25	123	13	13	115	245	360
Total near oil	186		148		26		360		
Total no oil	127		99		23		249		
Total sites	313		247		49		609		609

[a] Highest level of initial oiling within 200 m of site location.
[b] Sites entirely in upland areas.
[c] Sites with both intertidal and upland portions.
[d] Sites entirely in intertidal areas.
[e] Based on 1989 SCAT initial oiling maps (Owens, 1991).

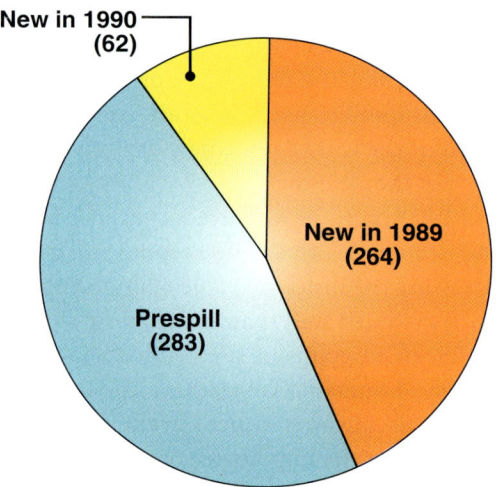

Figure 5.2 Number of cultural resource sites identified in the spill area as of January 1991 (total = 609).

Resource Survey (Fig. 5.2), increasing the number of known sites from 283 to 609. New cultural resource data were gathered from 526 of the 609 cultural resource sites, including data from two intertidal sites containing intact subsurface cultural material. Only 83 of the 609 known sites in the area affected by the spill were not inspected by Program archaeologists, and none of these was near shorelines that involved cleanup.

More sites were recorded on potentially affected shorelines in the GOA than in PWS (Table 5.1). The sites in the GOA were evenly distributed among locations near unoiled shorelines and oiled shorelines, whereas less than one-third of the sites in PWS

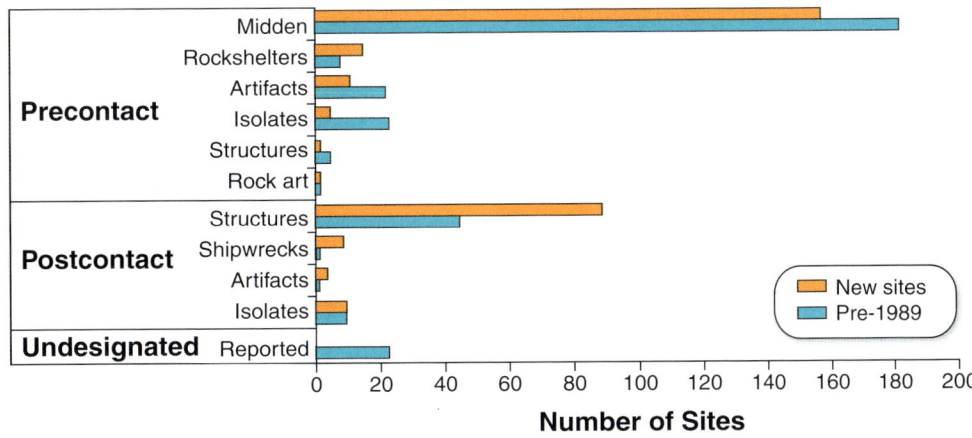

Figure 5.3 Types of cultural resource sites identified in the spill area as of January 1991.

were near unoiled shorelines. Of the sites near oiled shorelines, roughly 60% in PWS were near heavily oiled shorelines; in the GOA, only 14% of the sites were near heavily oiled shorelines and almost 75% were near very lightly oiled shorelines, no doubt reflecting the overall patterns of shoreline oiling in the GOA versus PWS. Of the sites in PWS, most were in upland locations; only 13 sites were found entirely within the intertidal zone.

Of the 609 sites identified, about 70% were categorized as primarily precontact (Fig. 5.3). The remaining 30% were classified as primarily postcontact, although many sites included both precontact and postcontact components. The geographic distribution of new sites is approximately the same as the distribution of previously known sites, illustrating the success of field surveys in identifying a full range of site types.

Nonconfidential project reports describing the goals, methods, implementation, and results of the program were written each year and reviewed by permitting agencies and Alaska Native organizations prior to publication (www.valdezsciences.com; Mobley et al., 1990; Betts et al., 1991; Haggarty et al., 1991; Haggarty and Wooley, 1994). These and other reports (Mobley and Haggarty, 1989; Haggarty and Wooley, 1990; Erlandson et al., 1992; Mobley and Eldridge, 1992; Moss, 1992; Moss and Erlandson, 1992; Wooley et al., 1992; Wooley and Haggarty, 1993, 1995) provide a much better understanding of the culture and history of the Alutiiq area than existed prior to the spill.

5.5 Effects on cultural resources from oil, erosion, and vandalism

Only very minor damage to cultural resources from the spill and cleanup occurred. Oil caused short-term, cosmetic injuries to some intertidal sites and isolated artifacts. Some artifacts were collected and cleaned, while others were left to be cleaned by tidal waters.

During SCAT surveys, archaeologists identified many previously unrecorded sites that had been vandalized, indicating that prespill vandalism was a serious problem. During the 4-year field program, only two incidents of vandalism were directly related to spill cleanup, one of which was successfully prosecuted by federal authorities. The Program's archaeologists carefully monitored sensitive locations, followed up all incidents fully, and reported them to pertinent landowners and, when appropriate, to law-enforcement agencies (Mobley *et al.*, 1990; Haggarty *et al.*, 1991).

In separate surveys of more than 20 archaeological sites most likely to have sustained damage by direct oiling, there was no evidence that direct oiling caused injury (Reger *et al.*, 1992; Dekin *et al.*, 1993). The most widely observable injury was from coastal erosion in areas of tectonic activity, rather than from the spill (Dekin *et al.*, 1993). There was also no evidence that the spill oiled any upland deposits, although some upland sites do have widespread low-level hydrocarbon contamination from other sources (Dekin *et al.*, 1993).

An Archaeological Index Site Monitoring program concluded that, while vandalism continued at oil-spill sites where erosion had exposed artifacts to view, it was a small problem overall. Oil effects also appeared to be negligible. Presenting a contrary position, Jesperson and Griffin (1992) reviewed program and governmental agency archaeological documentation for evidence of site injury and concluded that 19 of the 609 sites in the spill area showed "substantive" injury due to direct oiling or cleanup. An additional 16 sites showed "circumstantial" evidence of injury. The findings of Jesperson and Griffin were not supported by the Cultural Resource Program's documentation, nor were they supported by later field surveys that found no effects of oil or cleanup at four of the 19 sites Jesperson and Griffin had reported as sustaining substantive injury (Dekin *et al.*, 1993).

5.6 Archeology in PWS

The Program made a unique and lasting contribution to the archaeology of the Alutiiq region by identifying large village sites, site remnants, and rare intertidal sites containing organic artifacts and subtidal deposits. These sites yielded data that have greatly increased our understanding of Alutiiq cultural developments and the geological processes that affect them. Dates for two sites extended the known age of human habitation in Prince William Sound to earlier than 4000 years ago (Haggarty *et al.*, 1991). Also, the Program dated selected tephra layers from sites on the outer Kenai Peninsula coast to establish an initial chronological sequence for the area.

Two archaeological sites discovered in the Knight Island area during the initial 1989 SCAT surveys were subsequently investigated by the US Forest Service and found to have been occupied intermittently over 2000–3000 years (Yarborough, 1997). Site analyses resulted in valuable information on prehistoric technology, subsistence, seasonality, and site use.

Archaeologists documented well-preserved organic material (plant fibers, wood, bark, bone, etc.) at a previously known site on Afognak Island, and identified two new intertidal sites in PWS in 1989–90. Further investigations provided rare glimpses into past woodworking and fiber-processing technologies, including spruce-root grommet weaving and cedar carving.

The two PWS intertidal sites are rare wet sites. One site is on a raised relic peat deposit, a landform that subsided rapidly many hundreds of years ago until it was uplifted during the 1964 earthquake. Artifacts – including a stone lamp, a wooden splitting wedge, shaft fragments, wood chips, sharpened stakes, and large timbers with chopping marks – preserve information about prehistoric Chugach use of organic materials. This is important because organic materials almost always decompose in dry sites with acidic soils. At this site, however, they were preserved because of an ancient tectonic event that created unique preservation conditions. There was little oiling in the area, so no response activities took place near the site.

The second, smaller wet site was located in an area that de Laguna (1956) had investigated in the 1930s and believed had eroded away. Program archaeologists returned to the site during a very low tide and identified an extensive area of eroding intertidal peat that contained carved wooden tools, some with fiber grommets. Chugach Alaska Corporation and Office of History and Archaeology archaeologists (Reger et al., 1992) later investigated the site, collected artifacts, and, through ^{14}C dating, determined the deposits to be nearly 2000 years old.

Additional archaeological investigations in the region have been inspired by the Program. Mobley and Eldridge (1992) evaluated the use of forest products in the PWS area. Moss and Erlandson (1992) analyzed the distribution and age of defensive sites in the North Pacific, including forts from the Alutiiq region found during Program surveys. Moss (1992) discussed the regional implications of new site data generated by the Program. Survey results published in project reports (Mobley et al., 1990; Haggarty et al., 1991) and initial paleodemographic studies (Erlandson et al., 1992) provided a foundation for subsequent investigations in the Kodiak area (the Alutiiq Museum's Community Archaeology Projects), the Outer Kenai Peninsula (National Park Service and Smithsonian Institution's Kenai Fjords project), and in PWS (US Forest Service investigations) that further refined many questions raised by the abundant precontact Alutiiq region cultural resources.

In the Kodiak archipelago, several sites documented during the 1989 SCAT surveys also produced significant scientific results during subsequent investigations. In 1993, Afognak Native Corporation archaeologists returned to a Malina Creek site and determined that the site had been wet in the past but was now drying out. Their excavations salvaged important data that would otherwise have been lost over time. Basketry, game pieces, and other organic artifacts were preserved, and the excavations shed light on cultural transformations that later became manifested in elaborate Koniag traditions. A Horseshoe Cove site on Kodiak Island, located by SCAT archaeologists in 1989, had the remains of prehistoric houses and multiple layers of midden deposits, including extensive fish, shellfish, bird, and marine mammal remains. In 2004, the Alutiiq Museum excavated the site, revealing three distinct periods of precontact occupation over 4000 years and yielding other insights into the fishing technologies and food-storage capacities of the resident fishermen during the Early Kachemak and Koniag traditions (Saltonstall and Steffian, 2005).

Erlandson et al. (1992) used data collected during the spill response on the age, size, structure, and environmental context of 1295 coastal archaeological sites to examine variation in the spatial and temporal distribution of coastal sites in PWS, the outer Kenai Peninsula, the Kodiak Archipelago, and the southeast Alaska Peninsula. This was an initial, broad assessment of the general effects of environmental variation and demographic trends on pre- and postcontact settlement patterns in the Alutiiq region. The relationships between the distribution of archaeological sites and subsistence resources were examined in detail for a section of the Pacific coast of the Alaska Peninsula, and factors contributing to the differential age, density, and distribution of sites found in various parts of the region were identified.

Subsequent refinements of this big-picture approach by other archaeologists have further clarified the prehistory of the region.

5.7 Contributions to postcontact history in PWS

Cultural Resource Program staff documented numerous postcontact sites containing valuable information on the Russian fur trade, fox farming, mineral prospecting and mining, and World War II-era military defense sites. Some of these past activities are continuing sources of non-*Exxon Valdez* petroleum hydrocarbons to area shorelines (Page *et al.*, 2002; Wooley, 2002; Chapter 6). The complexity of the Alutiiq region's postcontact history is reflected in the overall range and diversity of these sites.

Fox and mink fur farming in the early twentieth century had a lasting impact on the human demography and ecology of the region. Sites recorded during Program surveys on Green Island, Storey Island, Seal Island, and Eleanor Island (see Map 3, p. vii) help document the history and associated ecological impacts of this relatively small but important industry. For example, a pigeon guillemot (*Cepphus columba*) restoration project in PWS has been hampered by predation on nests and adults by mink descended, in part, from fur-farm stock (Irons and Roby, 2011; Chapter 14).

Commercial fishing for Pacific salmon (*Oncorhynchus* spp.) and Pacific herring (*Clupea pallasii*) has also had a lasting effect on the human history of the region. Omar Humphrey started the first commercial operation in the region in 1889 (Wooley and Merrell, 2004), and Tarleton Bean, a US Commission of Fish and Fisheries officer, documented the status and health of the booming fishery in 1891 (Bean, 1891). The fishing industry evolved through the twentieth century. SCAT surveys or subsequent site visits documented several historic fishing-related sites, including the Sawmill Bay Cannery, Port Ashton, Port Benney, the San Juan Cannery, McClure Bay, and Port Nellie Juan.

Two especially significant historical artifacts found during SCAT surveys in 1989 were recorded and protected. A Program archaeologist identified the remains of a nearly complete kayak frame (Mobley *et al.*, 1990), likely constructed and abandoned during a brief resurgence of kayak building in PWS in the 1930s. The kayak was safeguarded during cleanup. Then, in 1991, Bureau of Indian Affairs archaeologists came back to excavate the kayak frame, and it is currently being preserved for future study.

A second rare historical artifact, a brass bell from a US Lighthouse Establishment buoy, was collected in 1989 by program archaeologists from an intertidal site in PWS. The bell is currently on loan to the Valdez Museum, where it is part of a display on the navigational history of PWS.

Investigations related to oil deposition in PWS before the *Exxon Valdez* spill focused on postcontact commercial and industrial sites (Page *et al.*, 1999). Many of these sites were related to early copper mining and prospecting companies, including the Kennecott Corporation, Knight's Island Consolidated Copper Company, and the Hubbard-Elliot Company. Mining legacy sites – including the Latouche Island town site (Fig. 5.4), the Beatson-Bonanza Copper Mine, and the Drier Bay Mine – were found to contain not only information on the history of mining in the area, but also petroleum products refined from Monterrey Formation (California) crude oil.

Figure 5.4 (a) Beatson copper mine facilities and town site at Latouche Island, Prince William Sound, 1917. (Photo: G.C. Martin, US Geological Survey) (b) Intertidal remains of the Beatson Mine site on Latouche Island circa 1995. The remains of the mine facility and town site were demolished in the late 1970s. Note the metal-rich tailings and drainage washing into PWS. (From Gray and Sanzolone, 1996).

5.8 Contributions to current postspill investigations

Academic and cultural resource management projects continue to benefit from the Program and CTAG – as have responders to oil spills and other emergencies. Site documentation has enhanced the ability of academic researchers to revisit questions of cultural origins and sequence in the region. The synergy developed among diverse researchers improved professional collaboration and site analysis. It has also enabled agency and Alaska Native cultural resource managers to better manage sites in their jurisdictions.

A direct result of SCAT was an improved understanding of the human history of PWS shorelines. The geomorphological data on coastal processes (Chaney, 1997; Mann *et al.*,

1998) enabled archaeologists to refine their understanding of where sites are located and the effect of past geological events, such as rapid seismic activity and erosion. Stone lamps, stone axes, hammerstones, and other heavy artifacts commonly found on shorelines in the spill area appeared when upland site deposits almost completely eroded. Shoreline surveys on the Sitkalidak Archaeological Survey Project between 1993 and 1996 used SCAT data to identify one of the oldest sites in the Kodiak Archipelago – right next to an area that a 1989 SCAT team noted as having high potential for additional upland sites (Fitzhugh, 2003).

The now widespread recognition of the need for systematic and accessible compilations of archaeological site data is a direct result of lessons learned during the *Exxon Valdez* spill. Alaska spill-response groups – including the Alaska Department of Environmental Conservation, Cook Inlet Spill Prevention Response, Inc., and the Alyeska Ship Escort/Response Vessel System – have established confidential, digital cultural resource data. This avoids their having to collect, compile, and organize these data during an emergency. Archaeological data management has also been addressed by British Columbia (Dickins *et al.*, 1990; Howes *et al.*, 1993, 1999).

The Archaeological Index Site Monitoring program enabled cultural resource managers to spend time in the field over many years collecting information on site conditions. This program would not have been possible without funding by the *Exxon Valdez* Oil Spill Trustee Council and support from the State of Alaska, US Fish and Wildlife Service, National Park Service, and Chugach National Forest (Reger *et al.*, 2000).

In 1992, the Alaska Regional Response Team established a Cultural Resources Working Group to develop guidelines for On Scene Coordinators to use during spill or hazardous substance release responses. An ad hoc committee finalized a *Programmatic Agreement on Protection of Historic Properties* in 1997. The agreement called for a historic properties specialists group to provide expertise during a response. In 2002, the *Alaska Implementation Guidelines* were developed. They were modeled on response strategies, information for emergency response personnel, the historic properties specialist checklist, and other policies and procedures developed by the Cultural Resource Program (Alaska Regional Response Team, 2002).

The same basic template for an emergency cultural resource response developed during the *Exxon Valdez* response was implemented during subsequent spills in Alaska – including the *Kuroshima* spill (1997, Unalaska Island), the Trans Alaska Pipeline System bullet hole spill (2001, near Livengood), and the *Selendang Ayu* spill (2006, Unalaska Island) – resulting in the identification and protection of many important cultural resource sites. Subsequent spill responses have benefited directly from the effective procedures for incorporating archaeologists into the SCAT model, which were established during the *Exxon Valdez* response and are now commonplace. Other spills have also followed the Program's model for keeping sensitive site data confidential and developing cleanup constraints.

In 1993, approximately 350 artifacts and samples collected during the project, along with the associated documentation, were transferred to the University of Alaska Museum of the North in Fairbanks. Project photos and videos, site survey records, field notes, and other archaeological files were turned over to the University of Alaska–Fairbanks' Elmer E. Rasmuson Library's Alaska and Polar Regions Collections.

Overall, the Cultural Resource Program laid a strong foundation for the identification and protection of cultural resource sites during an emergency response, while dramatically increasing academic and public awareness of the region's rich cultural heritage and the need to work cooperatively with landowners and other interested parties during the response.

5.9 Lessons learned

In the absence of any pre-existing model to use as a guide, Exxon's Cultural Resource Program managed to design and implement a large-scale project under emergency conditions in a remote coastal environment. They did so in full compliance with existing state and federal heritage conservation laws and regulations. The lessons learned, both large and small, were applied consistently over the duration of the 4-year Program to strengthen the Program's ability to identify and protect sensitive cultural resource sites during all stages of shoreline cleanup. The overall success of the Program in accomplishing its goals can best be measured by the wholesale adoption of many of the practices by subsequent projects, both in Alaska and elsewhere.

- Incorporation of cultural resource concerns as an integral part of SCAT surveys and the inclusion of archaeologists or trained cultural resource specialists on SCAT teams was the basis for successfully protecting cultural resources from damage.
- Centralized, accessible cultural resource site data enable coordination among the many different groups involved in the response to a major incident. Being proactive about compiling this information before an event is the best way to protect an area's cultural resources from unintended injuries associated with responses and/or vandalism.
- The need to protect cultural sites from vandalism by keeping site information confidential and the need for cleanup workers to know what needs to be protected are not mutually exclusive. Simple file controls, dissemination of information on a cleanup need-to-know basis, and careful monitoring can maintain necessary confidentiality while not hampering cleanup work.
- Consultative strategies, as required by laws and regulations, need to be established as early as possible in a spill response.
- Direct communication among all interested and affected parties is critical. The *Exxon Valdez* CTAG, for example, was an efficient and meaningful communication forum.
- All cleanup personnel must be trained in their legal responsibilities for site identification and protection. Cultural resource policies should be integrated into daily cleanup-worker safety briefings.
- The cleanup command structure's understanding of cultural resource policy is vital. A simple, written, one-page formal Cultural Resource Policy agreed to by the Federal On Scene Coordinator, the State On Scene Coordinator, and the Responsible Party (Exxon in this case) was effective in accomplishing this goal.
- Standardized cultural resource constraints incorporated into site-specific cleanup plans made site-protection actions both practical and effective.

REFERENCES

Alaska Office of History and Archaeology (2011). *Alaska Heritage Resource Survey*. Anchorage, AK, USA: Alaska Department of Natural Resources, Office of History and Archaeology. [http://dnr.alaska.gov/parks/oha/ahrs/ahrs.htm]

Alaska Regional Response Team (2002). *Alaska Implementation Guidelines for Federal On-Scene Coordinators for the Programmatic Agreement on Protection of Historic Properties during Emergency Response under the National Oil and Hazardous Substances Pollution Contingency Plan*. Anchorage, AK, USA: US Department of the Interior, Office of Environmental Policy and Compliance. [http://www.dec.state.ak.us/spar/perp/plans/uc/Annex%20M%20(Jan%2010).pdf]

Bean, T.H. (1891). Report on the salmon and salmon rivers of Alaska. *Bulletin of the US Fish Commission* **9**: 165–208.

Benson, C.L. (1987). Oil spill archaeology. *American Society for Conservation Archaeology Newsletter* **5**(5):16.

Betts, R.C., **C.B. Wooley**, **C.M. Mobley**, **J.C. Haggarty, and A. Crowell** (1991). *Site Protection and Oil Spill Treatment at SEL-188: An Archaeological Site in Kenai Fjords National Park, Alaska*. Anchorage, AK, USA: Exxon Shipping Company and Exxon Company, USA. [www.valdezsciences.com]

Chaney, G. (1997). Observations concerning Holocene tectonism and associated shoreline changes in Prince William Sound. Alaska. In *Site Specific Archaeological Restoration at SEW-440 and SEW-488*, L.F. Yarborough, ed. Anchorage, AK, USA: US Department of Agriculture, Forest Service, Chugach National Forest; *Exxon Valdez* Oil Spill Restoration Project 95007B Final Report, Appendix B, pp. 137–168. [http://www.evostc.state.ak.us/Files.cfm?doc=/Store/FinalReports/1995-95007B-Final.pdf&]

de Laguna, F. (1956). *Chugach Prehistory: The Archaeology of Prince William Sound*. Seattle, WA, USA: University of Washington Press; University of Washington Publications in Anthropology; Volume 13.

Dekin, A.A., **M.S. Cassell, J.I. Ebert, E. Camilli, J.M. Kerley, M.R. Yarborough, P.A. Stahl, and B.L. Turcy** (1993). Exxon Valdez *Oil Spill Archaeological Damage Assessment*. Juneau, AK, USA: US Department of Agriculture, Forest Service; Final Report; Contract No.53-0109-1-00325.

Dickins, D., **H. Rueggeberg, M. Poulin, I. Bjerkelund, J.C. Haggarty, L. Solsberg, J. Harper, A. Godon, P.D. Reimer, J. Booth, and K. Neary** (1990). *Oil Spill Response Atlas for the Southwest Coast of Vancouver Island*. Victoria, BC, Canada: British Columbia Ministry of Environment, Environmental Emergencies and Coastal Protection Branch; prepared by DF Dickins Associates, Ltd., Vancouver, BC, Canada.

Erlandson, J., **A. Crowell, C. Wooley, and J. Haggarty** (1992). Spatial and temporal patterns in Alutiiq paleodemography. *Arctic Anthropology* **29**(2): 42–62.

Fitzhugh, J.B. (2003). *The Evolution of Complex Hunter-Gatherers: Archaeological Evidence from the North Pacific*. New York, NY, USA: Kluwer Academic-Plenum Publishers; in the series *Interdisciplinary Contributions to Archaeology*, M. Jochim, ed. ISBN 9780306477539, ISBN 9780306478536.

Gray, J.E. and R.F. Sanzolone (1996). *Environmental Studies of Mineral Deposits in Alaska*. Denver, CO, USA: US Geological Survey, Information Services, US Geological Survey Bulletin 2156. [http://pubs.usgs.gov/bul/b2156/b2156.pdf]

Haggarty, J.C. and C. Wooley (1990). The 1990 *Exxon Valdez* Cultural Resource Program. *Alaska Anthropological Association Newsletter* 15(2).

Haggarty, J.C. and C. Wooley (1994). *Final Report of the Exxon Cultural Resource Program*. Anchorage, AK, USA: Exxon Company, USA. [www.valdezsciences.com]

Haggarty, J.C., C.B. Wooley, J.M. Erlandson, and A. Crowell (1991). *The 1990 Exxon Cultural Resource Program: Site Protection and Maritime Cultural Ecology in Prince William Sound and the Gulf of Alaska*. Anchorage, AK, USA: Exxon Shipping Company and Exxon Company, USA. [www.valdezsciences.com]

Howes, D., P. Wainwright, J. Haggarty, J. Harper, R. Frith, M. Morris, M. DeMarchi, K. Neary, C. Ogborne, and D. Reimer (1999). *Coastal Resource and Oil Spill Response Atlas for the West Coast of Vancouver Island*. Victoria, BC, Canada: British Columbia Ministry of Environment.

Howes, D., P. Wainwright, J. Haggarty, J. Harper, E. Owens, D. Reimer, K. Summers, J. Cooper, L. Berg, and R. Baird (1993). *Coastal Resources Oil Spill Response Atlas Southern Strait of Georgia*. Victoria, BC, Canada: British Columbia Ministry of Environment.

Irons, D. and D. Roby (2011). *Pigeon Guillemot Restoration Research in Prince William Sound, Alaska*. Anchorage, AK, USA: *Exxon Valdez* Oil Spill Trustee Council, Restoration Project 11100853. [http://www.evostc.state.ak.us/Projects/ProjectInfo.cfm?project_id=2190]

Jesperson, M. and K. Griffin (1992). *An Evaluation of Archaeological Injury Documentation: Exxon Valdez Oil Spill*. Anchorage, AK, USA: Alaska Department of Natural Resources, Office of History and Archaeology; US Department of the Interior, US National Park Service; Prepared at the direction of the Comprehensive Environmental Response, Compensation, and Liability Act Archaeological Steering Committee.

Mann, D.H., A.L. Crowell, T.D. Hamilton, and B.P. Finney (1998). Holocene geologic and climatic history around the Gulf of Alaska. *Arctic Anthropology* 35(1): 112–131.

Mifflin and Associates, Inc. (1991). Exxon Valdez *Oil Spill Damage Assessment Contamination of Archaeological Materials, Chugach National Forest: Radiocarbon Experiments and Related Analyses*. Juneau, AK, USA: US Department of Agriculture, US Forest Service; Final Report on Contract 53–0109–1–00305.

Mobley, C.M. and J.C. Haggarty (1989). The *Exxon Valdez* Cultural Resource Program. In *Proceedings of the British Columbia Oil Spill Prevention Workshop, September 14–15, 1989, Vancouver, British Columbia*. P.H. LeBlond, ed. Vancouver, BC, Canada: Department of Oceanography, University of British Columbia, Manuscript Report No. 52.

Mobley, C.M., J.C. Haggarty, C.J. Utermohle, M. Eldridge, R.E. Reanier, A. Crowell, B.A. Ream, D.R. Yesner, J.M. Erlandson, and P.E. Buck (1990). *The 1989 Exxon Valdez Cultural Resource Program*. Anchorage, AK, USA: Exxon Company, USA. [www.valdezsciences.com]

Mobley, C.M. and M. Eldridge (1992). Culturally modified trees in the Pacific northwest. *Arctic Anthropology* 29(2): 91–110.

Moss, M.L. (1992). Relationships between maritime cultures of southern Alaska: rethinking culture area boundaries. *Arctic Anthropology* **29**(2): 5–17.

Moss, M.L. and J.M. Erlandson (1992) Forts, refuge rocks, and defensive sites: The antiquity of warfare along the North Pacific Coast of North America. *Arctic Anthropology* **29**(2): 73–90.

Owens, E.H. (1991). *Changes in Shoreline Oiling Conditions 1-½ Years after the 1989 Prince William Sound Spill*. Seattle, WA, USA: Woodward-Clyde; unpublished report.[www.valdezsciences.com]

Page, D.S., A.E. Bence, W.A. Burns, P.D. Boehm, J.S. Brown, and G.S. Douglas (2002) A holistic approach to hydrocarbon source allocation in the subtidal sediments of Prince William Sound, Alaska, embayments. *Environmental Forensics* **3**(3–4): 331–340.

Page, D.S., P.D. Boehm, G.S. Douglas, A.E. Bence, W.A. Burns, and P.J. Mankiewicz (1999). Pyrogenic polycyclic aromatic hydrocarbons in sediments record past human activity: a case study in Prince William Sound, Alaska. *Marine Pollution Bulletin* **38**(4): 247–260.

Reger, D.R., J.D. McMahan, and C.E. Holmes (1992). *Effect of Crude Oil Contamination on Some Archaeological Sites in the Gulf of Alaska, 1991 Investigations*. Anchorage, AK, USA: Alaska Department of Natural Resources, Office of History and Archaeology; *Exxon Valdez* Oil Spill State/Federal Natural Resource Damage Assessment Archaeology Study Number 1 Final Report. [http://www.evostc.state.ak.us/Files.cfm?doc=/Store/FinalReports/1992-ARC1-Final.pdf&]

Reger, D.R., D. Corbett, A. Steffian, P. Saltonstall, T. Birkedal, and L.F. Yarborough (2000). *Archaeological Index Site Monitoring: Final Report*. Anchorage, AK, USA: Alaska Department of Natural Resources, Office of History and Archaeology; *Exxon Valdez* Oil Spill Restoration Project 99007A Final Report. [http://www.evostc.state.ak.us/Files.cfm?doc=/Store/FinalReports/1999–99007A-Final.pdf&]

Saltonstall, P.G. and A.F. Steffian (2005). *The Archaeology of Horseshoe Cove: Excavations at KOD-415, Uganik Island, Kodiak Archipelago, Alaska*. Anchorage, AK, USA: US Department of the Interior, Bureau of Indian Affairs, Alaska Region; Occasional Papers in Alaskan Field Archaeology No. 1.

Teal, A.R. (1991). Shoreline cleanup: reconnaissance, evaluation, and planning following the *Valdez* oil spill. In *Proceedings of the 1991 International Oil Spill Conference (Prevention, Behavior, Control, Cleanup), March 4–7, 1991, San Diego, California*. Washington DC, USA: American Petroleum Institute; Technical Publication 4529; pp. 149–152.

Wooley, C.B. (2002). The myth of the "pristine environment": past human impact on the Gulf of Alaska coast. *Spill Science and Technology Bulletin* **7**(1–2): 89–104.

Wooley, C.B. and J.C. Haggarty (1993). The hidden history of Chugach Bay. *Alaska Geographic* **20**(1): 12–17.

Wooley, C.B. and J.C. Haggarty (1995). Archaeological site protection: an integral component of the *Exxon Valdez* shoreline cleanup. In Exxon Valdez *Oil Spill: Fate and Effects in Alaskan Waters*. P.G. Wells, J.N. Butler, and J.S. Hughes, eds; Philadelphia,

PA, USA: American Society for Testing and Materials; ASTM Special Technical Publication 1219; pp. 933–949.

Wooley, C.B., J.C. Haggarty, and J.M. Erlandson (1992). Exxon's Cultural Resource Program: process and results of site protection in Prince William Sound and the Gulf of Alaska. Presented at the 57th Annual Meeting of the Society for American Archaeology, April 8–12, 1992, Pittsburgh, Pennsylvania.

Wooley, C.B. and B. Merrell (2004). Captain Omar J. Humphrey, letter from Alaska from the Pacific Steam Whaling Station, Prince William Sound, Alaska 1889. Presented at the Alaska Historical Society Annual Meeting, September 15–18, 2004, Anchorage, Alaska.

Yarborough, L.F. (1997). *Site Specific Archaeological Restoration at SEW-440 and SEW-488*. Anchorage, AK, USA: US Department of Agriculture, US Forest Service, Chugach National Forest; *Exxon Valdez* Oil Spill Restoration Project 95007B Final Report. [http://www.evostc.state.ak.us/Files.cfm?doc=/Store/FinalReports/1995–95007B-Final.pdf&]

CHAPTER SIX

Fate of oil on shorelines

David S. Page, Paul D. Boehm, John S. Brown, Erich R. Gundlach, and Jerry M. Neff

6.1 Introduction

Most oil tanker accidents occur near land. So when a marine oil spill occurs, it is usually not long before the spilled oil reaches shorelines. The shoreline is where the potential for harm to the environment and biological resources is the greatest, and where media attention and public concerns usually focus. Therefore, it is essential to determine the distribution, amount, composition, and fate of spilled oil on shorelines. This information forms the foundation for management decisions about cleanup during the early phases of the spill, assessments of long-term exposure and injury to biological resources, and long-term restoration strategies after the initial cleanup.

In this chapter, we consider the fate of shoreline oil following the *Exxon Valdez* oil spill, beginning with oil coming ashore in Prince William Sound (PWS) in 1989. This chapter picks up where Chapter 3 left off, describing where the oil was deposited, why some locations were oiled more than others, and how oil disappeared over time and why, in a few isolated locations, it persisted.

The basic scientific principles that govern the fate of oil on shorelines and the importance of shoreline type in predicting oil persistence were well understood prior to 1989, based on experience from previous oil spills. The "surprises" voiced by some after the *Exxon Valdez* spill and other major spills were not due to the emergence of new scientific principles; rather, because major oil spills occur infrequently and in widely scattered locations, local responders often are unaware of the scientific principles that control the fate of oil on shorelines. Wheels are often reinvented.

Studies over more than two decades following the *Exxon Valdez* spill show that the fate of *Exxon Valdez* oil on the shorelines of PWS and the western Gulf of Alaska (GOA) was determined primarily by three factors:

1. The geomorphology, geography, weather, and wave exposure of the spill site, which controlled where and in what amounts the shoreline was oiled;
2. The physical and chemical properties of the oil, which affect "weathering" – the dispersion, transformation, and loss of spill residues over time; and
3. The influence of cleanup and natural oil loss on the distribution and level of oil on the shorelines over time.

Oil in the Environment: Legacies and Lessons of the Exxon Valdez *Oil Spill*, ed. J. A. Wiens. Published by Cambridge University Press. © Cambridge University Press 2013.

This chapter is organized around the key issues in understanding the processes and conditions influencing the fate of oil on shorelines, starting with an initial review of general principles and the role of shoreline characteristics. Although we focus on the *Exxon Valdez* experience, key elements of the approach and the methodology for determining the fate of shoreline oil in this incident are common to all oil spills.

6.2 Overview of the fate of oil on shorelines

6.2.1 Understanding the fate of shoreline oil from past spills

Long-term monitoring at the sites of major spills before 1989 (National Research Council, 1985) – including *Arrow*, *Metula*, and *Amoco Cadiz* (Fig. 6.1) – helped scientists understand the relationship between shoreline type, oil penetration, and persistence (e.g., Gundlach and Hayes, 1978a, b; Gundlach et al., 1978, 1983; National Research Council, 1985; Hayes and Michel, 1999; Owens et al., 2008). In each case, physical processes and geomorphological features of the shoreline contributed to oil sequestration and long-term persistence (see Owens et al., 2008). Each place where oil residues persisted was in a location removed from active physical and (bio)chemical processes. Were this not the case, these spill remnants would have disappeared rapidly.

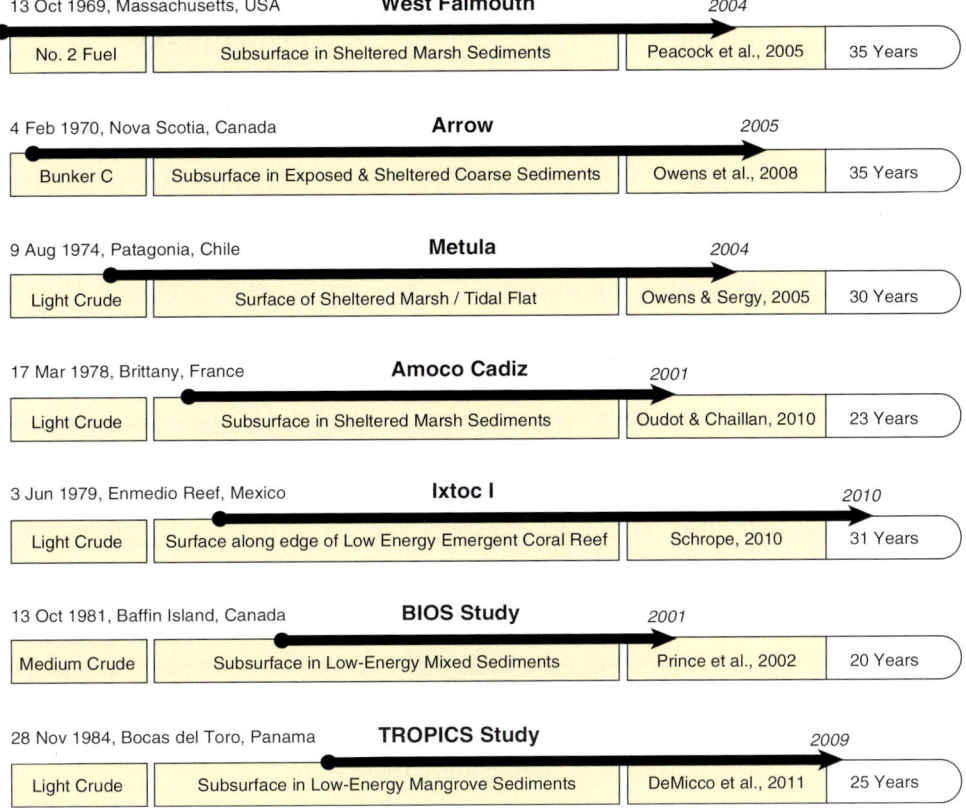

Figure 6.1 Examples of long-term studies of the persistence of surface and subsurface oil following marine oil spills before the *Exxon Valdez* spill.

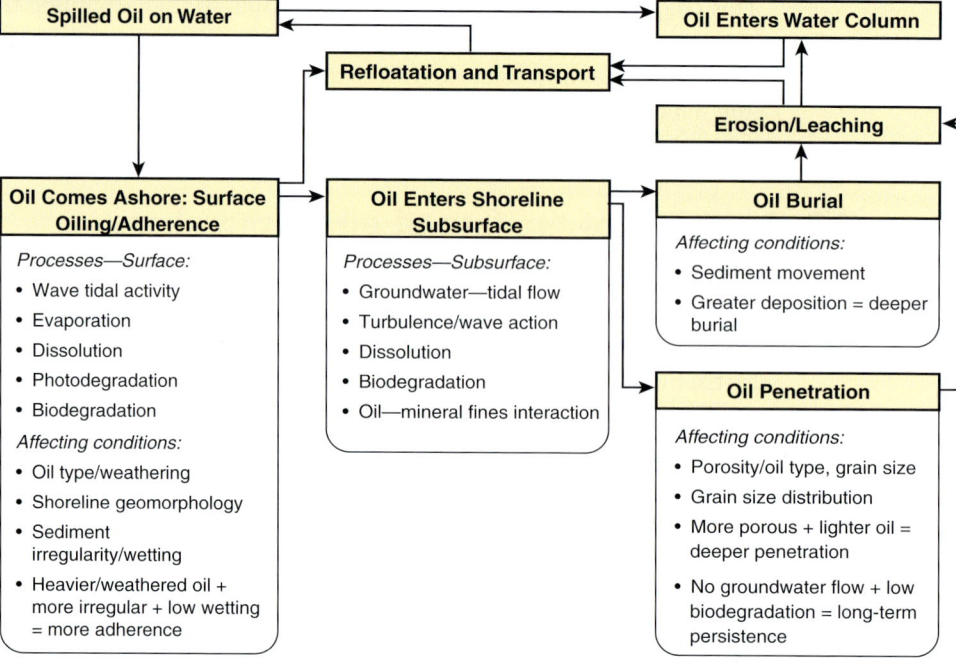

Figure 6.2 Major processes affecting the fate of oil on shorelines.

6.2.2 Shoreline oil surface deposits and penetration

The location and fate of oil on a shoreline are determined by the spill trajectory, the porosity and physical nature of a given shoreline location (Fig. 6.2), and the characteristics of the oil spilled. As oil washes ashore, waves and high tides move it to the highest levels of the shoreline. Under very heavy deposition, oil will coat the entire surface of an exposed intertidal zone as the tide level drops. On incoming tides that follow, much of the oil deposited in the active intertidal zone will be refloated and transported in ever-diminishing quantities to nearby shorelines. Onshore winds and the configuration of the shoreline contribute to the retention of oil in specific locations. Oil along the upper intertidal zone is often too high to be refloated, accounting for the high-tide oil band commonly observed at oil spills.

As oil weathers, it becomes more viscous and adheres more firmly to solid surfaces, remaining as a surface deposit. Following both the *Amoco Cadiz* spill and the *Exxon Valdez* spill, this occurred after about 1 month (Wolfe et al., 1994). Alternatively, on sand/gravel shorelines, oil may penetrate or become buried in subsurface sediments – called subsurface oil residues (SSOR). For the *Exxon Valdez* spill, SSOR was defined as deeper than 5 cm in the sediments below any armouring (Neff et al., 1995). Oil penetration varies with oil properties, the grain size of sediments, and the availability of open pores between grains of sediment (porosity) and their ability to allow water to pass through (permeability). Pores filled with water and/or fine-grained sediments inhibit oil penetration.

In most spills, SSOR form faster and deeper in the middle of the upper intertidal zone, where the beach is more porous, than in the lower shore (Fig. 6.2). If the pore space is fully open, as with coarse-grained sediments, oil will flush out after several tides. If the pore space is filled with fine-grained material, however, the incoming tide may not create enough water pressure to force out the oil, sometimes leading to long-term persistence or sequestration (Chapter 7). In PWS, most beaches in the low-tide zone had little or no penetration because they have tightly packed, fine-grained sediments or water-filled pore spaces. Rather, oil deposited in the low-tide zone was usually refloated on the incoming tide and either dispersed offshore or redeposited higher in the intertidal zone.

Oil may also become buried as sediments erode, shift, and accumulate as part of the natural beach cycle and other coastal processes, as observed in the *Amoco Cadiz* spill (Gundlach and Hayes, 1978b). This sometimes happened following the *Exxon Valdez* spill: repeated surveys in 1989–90 (Gundlach *et al.*, 1991) and 1989–97 (Hayes and Michel, 1998) noted oil burial and penetration related to observed changes in beach morphology at specific locations.

6.2.3 Factors determining the fate of oil on shorelines

6.2.3.1 Chemical factors

The amount and form of shoreline oil change during weathering, as more volatile components evaporate and less volatile residues on shore dissolve, disperse, biodegrade, and photo-oxidize (Chapter 8; Wolfe *et al.*, 1994). Oil weathering begins with the rapid evaporation or dissolution of more volatile and water-soluble components, particularly from slicks on the water surface. Biodegradation is slower, first affecting the linear alkanes and then branched alkanes (see Chapter 1, Box 1.1). For polycyclic aromatic hydrocarbons (PAH), biodegradation decreases with increasing ring numbers and degrees of alkylation, causing the absolute concentrations of all PAH to decrease during weathering. Because weathering causes a net decrease in the concentrations of both the total and the most toxic PAH, the toxicity of oil decreases as it weathers (Chapter 11). This was confirmed by the decreasing toxicity of intertidal sediments from spill sites in PWS between 1990 and 1993 (Page *et al.*, 2002).

6.2.3.2 Shoreline types in PWS and characteristics affecting oil persistence

In 1978, Gundlach and Hayes developed a shoreline Environmental Sensitivity Index (ESI). The ESI ranks shoreline types on a scale of 1 to 10 (10 being most sensitive) based on predicted sensitivity to disturbance from oil spills and cleanup operations (Table 6.1). The ESI was first applied to areas adjacent to and including PWS several years before the *Exxon Valdez* spill (Hayes, 1980). The ESI helps to categorize the types of shorelines in PWS (Table 6.1, Fig. 6.3) and identify five of the most important physical characteristics in understanding where and why oil from the *Exxon Valdez* spill persisted:

1. *Large boulders and bedrock outcrops.* Stranded oil can pool around the base or in the "wave shadow" of large boulders and bedrock outcrops, usually at the top of sheltered shores. There it forms a hard asphaltic surface layer over time (Taylor and Reimer, 2008). The boulders and the asphaltic skin isolate the oil in the interior of the deposit, limiting its exposure to air, water, and hydrocarbon-degrading microbes. Some of these asphaltic surface deposits do weather to particles that crumble, which can be removed by storms or waves. But more sheltered deposits can persist as asphaltic mats more than 5–10-cm thick. Following the *Exxon Valdez* spill, this type of asphaltic surface deposit persisted at widely scattered locations in PWS (Taylor and Reimer, 2008) and on the Alaska Peninsula (Irvine *et al.*, 2006).
2. *Surface boulder/cobble/gravel armor layer.* A surface layer of boulder and cobble-sized material (known as "armor") can protect SSOR sequestered in subsurface low-porosity, finer-grained sediments against waves and physical disturbance (Fig. 6.4).

Table 6.1 Shoreline Environmental Sensitivity Index (ESI) ranking for PWS shorelines, categories used, and percentage of oiled shoreline in each category in Exxon-supported shoreline studies (Chapter 11).

ESI #[a]	ESI Shoreline Types in PWS[b]	Categories for Exxon Shoreline Study[c] (% of Oiled PWS Shoreline)
1	Exposed rocky shore	Exposed bedrock/rubble (~17%)
2	Exposed wave-cut platform	
3	Fine-grained sand beach	Not considered (not present)
4	Coarse-grained sand beach	
5	Mixed sand and gravel beach	Mixed pebble/gravel (~3.7%)
6	Gravel beach	Boulder/cobble/gravel (~23%)
7	Exposed tidal flat	Not considered (< 0.1%)
8	Sheltered rocky shore	Sheltered bedrock/rubble (~56%)
9	Sheltered tidal flat	Marsh/tidal flat (< 0.1%)
10	Marshes	

[a]ESI #1 = least sensitive to oil spills based on persistence and potential for environmental injury, #10 = most sensitive.
[b]National Oceanic and Atmospheric Administration, 2007
[c]Chapter 11

Figure 6.3 Examples of the four major shoreline types in the *Exxon Valdez* oil-spill zone in PWS. (a) Exposed bedrock/rubble shore (16.8% of the spill zone): northeastern Smith Island, SM005B, 1990. (b) Sheltered bedrock/rubble shore (56.6% of the spill zone): Herring Bay KN117A, 2007. (c) Boulder/cobble/gravel shore (22.9% of the spill zone): northern Eleanor Island, EL107A, 2008 (note the wave-battered logs in the foreground). (d) Mixed pebble/gravel shore (3.7% of the spill zone): southeastern Eleanor Island, EL010A, 1996. (Photos: David S. Page).

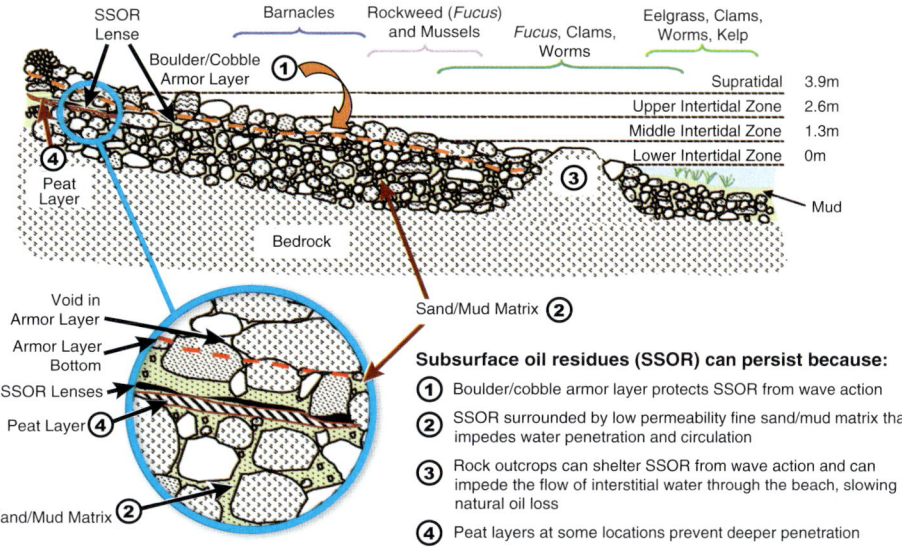

Figure 6.4 Cross-sectional representation of an exposed boulder/cobble/gravel shoreline showing the typical location of SSOR and describing those factors that promote persistence. Natural oil loss is more rapid at shorelines with unimpeded water flow between sediment grains. (Adapted from Taylor and Reimer, 2008).

SSOR persistence in armor-layer beaches is similar to that on exposed beaches, except that the lower wave energy further slows natural loss. Hayes *et al.* (2010) describe this process, with particular emphasis on the spill zone in PWS.

3. *Headlands and other geomorphic features.* Shoreline areas behind headlands, offshore ledges, and other features are sheltered from erosional wave energy, enabling SSOR persistence. Bedrock outcrops and subsurface bedrock structures on gravel beaches can impede tide- and wave-driven water flow through pore spaces on the beach, slowing natural oil loss (Chapter 7).

4. *Low-permeability subsurface sediments.* Even if a beach is sufficiently porous to enable oil penetration, oil still needs subsurface low-permeability, fine-grained sediment to persist. Low permeability inhibits water flow between sediment grains and further slows oil removal by inhibiting the delivery of oxygen and nutrients to support biodegradation (see Chapters 7 and 8; Hayes *et al.*, 2010). This set of conditions is particularly relevant to PWS because the 1964 earthquake lifted shorelines 1–3 m in most of the spill zone (see Chapter 1, Fig. 1.2). The uplift from the earthquake moved nearshore deposits of subtidal silt and clay up into the intertidal zone below a coarse-sediment veneer.

5. *Organic carbon-rich sediments.* Oil can be absorbed and persist in anoxic, peat-sediment salt marshes (Peacock *et al.*, 2005; Owens and Sergy, 2005). Exposed peat sediment and shallow, subsurface peat may absorb oil, where SSOR degrades very slowly because of the high organic-carbon content and anoxic character of the peat. If waterlogged, however, subsurface peat layers also prevent deeper penetration of oil (Fig. 6.4).

6.3 The *Exxon Valdez* experience

As with all marine oil spills, immediately after the *Exxon Valdez* oil spill it was critical to develop a clear record of the locations of oiled shorelines and, equally important, of locations of shorelines that were *not* oiled. The *Exxon Valdez* released most of the oil into northeastern PWS within 5 hours of its grounding (Wolfe et al., 1994). During the following week, the movements and fates of floating oil were influenced primarily by the weather and the net counterclockwise surface currents that flow northward through Hinchinbrook Entrance into north and central PWS and then southward, mostly along the eastern shore of Knight Island, exiting through Montague Strait (Galt et al., 1991; see Map 1, p. v). The surface currents and a major storm during the third day of the spill drove the floating oil to the southwest, causing much of it to strand in heavy concentrations on exposed north- and east-facing shorelines.

Because of storm-generated mixing, the oil lost a large fraction of the more volatile and water-soluble hydrocarbons during this period. The storm also provided high wave energy that caused floating oil to form a water-in-oil emulsion ("mousse") containing ~70% water, increasing the volume and viscosity of the oil mass. By day 38 of the spill, ~41% of the total oil spilled had stranded on PWS shorelines and ~13% on GOA shorelines (Wolfe et al., 1994). After exiting PWS, floating, emulsified oil further degraded and dispersed in the large reaches of the GOA. As a result, the shorelines of the Kodiak Island group and the Kenai and Alaska peninsulas were more lightly and sporadically oiled relative to PWS, and mostly with this more-weathered mousse. These areas had a limited number of reports of persistent oil (e.g., Irvine et al., 2006) and received much less scientific attention, so we focus here on the long-term fate of oil on PWS shorelines.

6.3.1 Characterizing the spill zone: baseline determination

A common mistake is to assume that the spill itself is the only source of hydrocarbons in the spill zone. The presence of nonspill petroleum and natural hydrocarbon sources has been a factor in other oil spills (Page et al., 1979, 1988). The identification and quantification of pre-existing hydrocarbon sources in a spill area is fundamental to establishing baseline conditions for any postspill impact assessment.

In the case of the *Exxon Valdez* oil spill, there were two major sources of nonspill hydrocarbons in PWS:

1. *Natural, petroleum-related hydrocarbons in subtidal sediments*. Page et al. (1995, 1996) identified two distinct petroleum sources in PWS subtidal sediments. The first had a high proportion of sulfur-containing PAH (dibenzothiophenes) and was from the *Exxon Valdez* spill and diesel fuel refined from Alaska North Slope crude oil. It was found only in very limited and specific nearshore locations. The second had a low proportion of sulfur-containing PAH and was from natural hydrocarbons associated with fine-grained sediments. These sediments are transported by the Alaska Coastal Current from petroleum-source rocks and oil seeps in eastward GOA into PWS, where they settle to the seafloor (Boehm et al., 1995; Page et al., 1995, 1996; O'Clair et al., 1996; Short et al., 2007b). Age-dating of sediment cores indicates that this process has been going on for at least the past 160 years, and probably for many thousands of years (Page et al., 1996). The magnitude is large: an estimated 1.4×10^7 metric tons of sediment, containing about 12 metric tons of PAH, are deposited in PWS annually (Page et al., 1997).
2. *Human-activity-related hydrocarbons in intertidal and shallow subtidal sediments*. PWS was an important mining and fish-processing center during the early to mid-twentieth century (Page et al., 1999, 2006; Wooley, 2002; Chapters 1 and 5). In addition to inputs from routine fossil-fuel use during their operation, many

mining and cannery facilities were destroyed during the 1964 earthquake and remaining fuel oil spilled from ruptured storage tanks. These facilities were in sheltered bays, and their ruins are visible today (Chapter 1, Fig. 1.6 and Chapter 5, Fig. 5.4). Kvenvolden *et al.* (1995) showed that surface tar deposits sampled at 61 locations throughout the northern and western parts of PWS were from a California petroleum source (Monterrey Formation), not the 1989 spill, and were most likely related to oil released during the 1964 earthquake. Human-activity-related baseline hydrocarbon sources are significant. For example, Page *et al.* (2006) mapped 36 000 m^2 of intertidal sediments with total PAH (TPAH) >2500 ng/g dry weight from nonspill petroleum and combustion-related sources at nine former mine and cannery sites.

Shoreline-monitoring programs for the *Exxon Valdez* spill fall primarily into two time periods: the cleanup/response period from 1989–92 and the period of natural oil loss after 1992.

6.3.2 The first three years (1989–92)

Important study elements during this period documented the changes in extent and character of spill residues, defined the level of exposure of sediments and biota to spill residues, characterized shoreline physical factors that can affect biological communities and oil persistence, and identified nonspill hydrocarbon sources.

Within 2 months of the *Exxon Valdez* spill, most of the stranded oil on the shoreline consisted of viscous, sticky mousse. In places, oil penetrated coarse sediments and formed SSOR (Galt *et al.*, 1991; Wolfe *et al.*, 1994). Shoreline oiling was very patchy, with large differences in the levels and distribution, even within a short length of shoreline (Neff *et al.*, 1995; Chapter 1, Fig 1.4 (right)). Most, however, was in the upper third to half of the intertidal zone; there were only a few instances of surface oil in the lower intertidal zone (Michel *et al.*, 1991).

Over the next 2 years, the level of shoreline oiling decreased dramatically owing to cleanup and natural weathering, particularly severe winter storms. Not surprisingly, later shoreline surveys (Neff *et al.*, 1995; Chapter 4) revealed that the locations on north- and east-facing shores that were most heavily oiled in 1989 also had the most SSOR remaining in 1991 and beyond.

6.3.2.1 Oil transport to subtidal sediments

In PWS, hydrocarbons did not accumulate in subtidal sediments except in shallow, subtidal areas near certain heavily oiled, sheltered shorelines (Payne *et al.*, 1991; Boehm *et al.*, 1995, 1998; Page *et al.*, 1995, 1996; O'Clair *et al.*, 1996). At those locations where this occurred, the transport of oil residues to nearshore subtidal sediments was caused primarily by intertidal sediment erosion during intensive shoreline cleanup, where oil associated with sediments washed offshore (Wolfe *et al.*, 1994; Erich R. Gundlach, 1989, personal observation).

Other conditions that can cause oil to be transported from the shoreline to subtidal areas were not factors for the *Exxon Valdez* spill. The association of oil with sediment particles – which decreases oil's buoyancy and allows it to sink – was not widespread because of limited intertidal, fine-grained sediment sources (Page *et al.*, 1995, 1996). There were some places that had small amounts of fine-grained sediment ("glacial flour") originating from glaciers emptying into PWS (Bragg and Yang, 1995). Although this glacial flour did associate with loosely aggregated particles of oil (mineral-oil flocs; Box 6.1), it was dispersed and biodegraded in the water column. Additionally, evaporation of lighter oil fractions can increase the oil density and promote sinking; in the *Exxon Valdez* case, however, evaporation did not increase the oil density enough for it to sink.

Box 6.1 Mineral particle-oil flocculation and biodegradation

In late 1989–early 1990, a significant laboratory observation was made at the same time that field observations indicated that upon further weathering, oil residues were somehow being transformed in a manner that enhanced their removal from shorelines even with only gentle water movement. Laboratory tests of bioremediation in columns packed with oiled sediment from PWS showed that the physical appearance of oil residue immersed in seawater changed and much of the residue ceased to stick to sediments (Bragg *et al.*, 1992). In the field, cleanup workers found that oil residue could be washed from hands and boots with cold water, rather than requiring kerosene or other cleaners. What had changed?

Within hours after being immersed in seawater, oil on sediment in laboratory experiments appeared no longer to adhere strongly to sediments but instead to exist as loosely aggregated, fuzzy droplets typical of a flocculated emulsion. These slightly buoyant droplets (flocs) tended to concentrate at the upper interstitial contact points in pores (Fig. 6.1.1). However, when water was drained at low tide, the oil again appeared as black and sticky as it was when originally spilled. Closer examination revealed that the modified oil had been transformed into a water-external emulsion (Fig. 6.1.1); a "drop" of oil actually consisted of thousands of loosely flocculated, individual micron-sized oil droplets, and the oil-in-water emulsion was stabilized by fine mineral particles. Analyses by electron microscope and X-ray diffraction showed that the mineral fines were mostly 1 μm or less in size and consisted of clays, quartz, and feldspars – typical of mineral particles in PWS glacial flour. Microbial oil degradation increased the concentrations of polar compounds in the residual oil, thereby helping the formation of mineral particle-oil flocs (Bragg *et al.*, 1992). Most of the flocs exhibited almost neutral buoyancy since they contained 60–80% water by volume.

 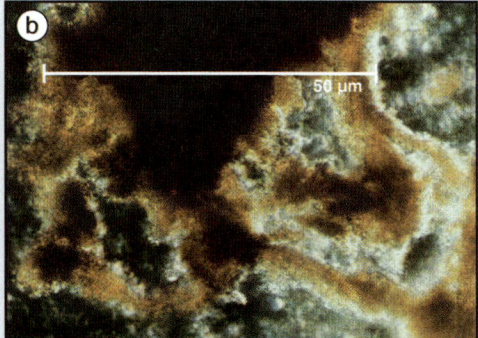

Figure 6.1.1 (a) Photo of flocculated oil residue on sediments in a microcosm. A fluffy, colloidal emulsion was observed within hours after the sediments were submerged under sea water. (b) Photomicrograph of flocculated oil (water external emulsion) shows outer edges that are covered with fine mineral particles of about 1 μm or less. (Photos: James R. Bragg)

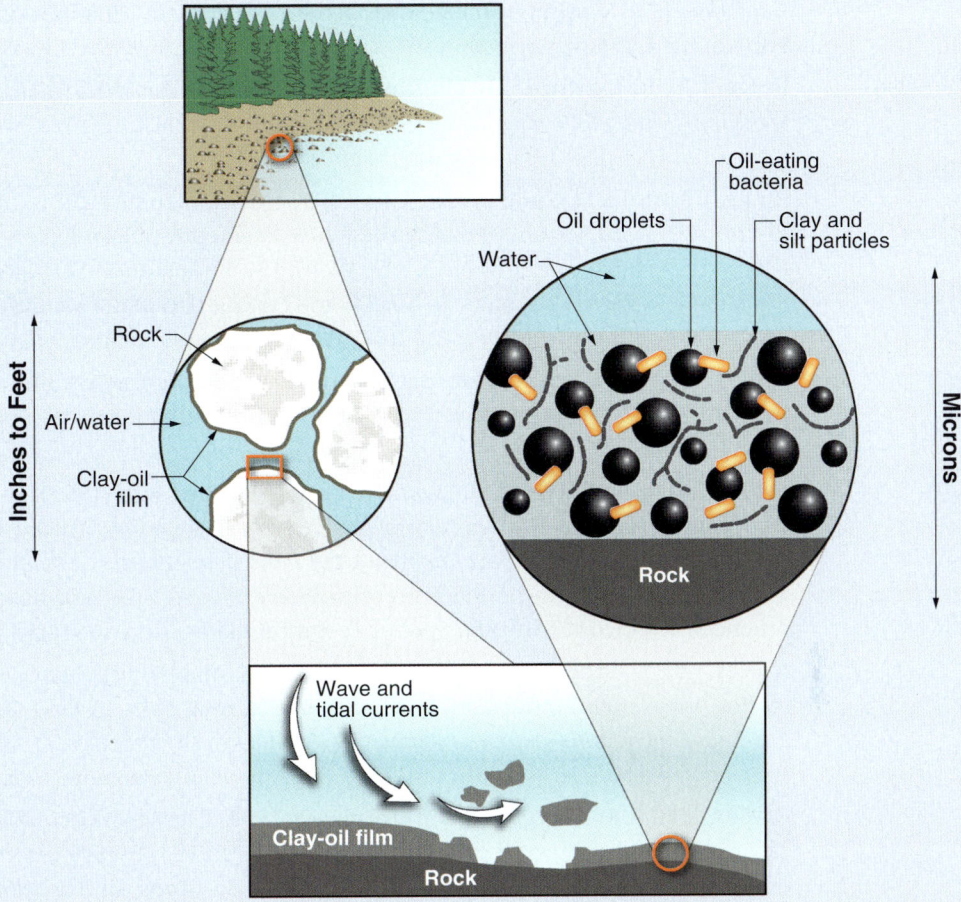

Figure 6.1.2 Biodegradation and clay/oil (or mineral-particle/oil) flocculation are synergistic natural cleansing processes that contribute to degrading and removing oil residues. As biodegradation forms more polars in the oil, flocculation is enhanced, which in turn increases the oil–water interfacial area and access of microbes to the oil. (From Bragg et al., 1992).

Wave-tank tests showed that even gentle waves efficiently removed virtually all of the transformed oil from sediments (Bragg and Yang, 1995). The increased surface area between the micron-sized oil droplets and water enhanced rates of biodegradation such that, once the neutrally buoyant flocs washed to sea, their hydrocarbons were almost certainly biodegraded rapidly (Weise et al., 1999; Warr et al., 2009) (Fig. 6.1.2).

Studies of other spills confirm that these floc structures and oil-removal process are not unique to PWS (Bragg and Owens, 1994, 1995; Owens et al., 1994; Lee et al., 1997; Lee, 2002). This oil-transport process continues to be of current research interest (Niu, 2011).

Bragg, J.R. and E.H. Owens (1994). Clay-oil flocculation as a natural cleansing process following oil spills. Part 1: Studies of shoreline sediments and residues from past spills. In *Proceedings Seventeenth Arctic and Marine Oilspill Program (AMOP) Technical Seminar, June 9–10, 1994, Vancouver, British Columbia, Canada*. Ottawa, ON, Canada: Environment Canada; pp. 1–23.

Bragg, J.R. and E.H. Owens (1995). Shoreline cleansing by interactions between oil and fine mineral particles. In *Proceedings of the 1995 International Oil Spill Conference (Achieving and Maintaining Preparedness), February 22–March 2, 1995, Long Beach,*

California, USA. Washington DC, USA: American Petroleum Institute Special Technical Publication 4620; pp. 219–227.

Bragg, J.R., R.C. Prince, J.B. Wilkinson, and R.M. Atlas (1992). *Bioremediation for Shoreline Cleanup following the 1989 Alaskan Oil Spill*. Houston, TX, USA: Exxon Company, USA. [http://www.valdezsciences.com]

Bragg, J.R. and S.H. Yang (1995). Clay-oil flocculation and its role in natural cleansing in Prince William Sound following the *Exxon Valdez* oil spill. In *Exxon Valdez Oil Spill: Fate and Effects in Alaskan Waters*. P.G. Wells, J.N. Butler, and J.S. Hughes, eds. Philadelphia, PA, USA: American Society for Testing and Materials; ASTM Special Technical Publication 1219; ISBN-10: 0803118961; pp. 178–214.

Lee, K. (2002). Oil–particle interactions in aquatic environments: Influence on the transport, fate, effect and remediation of oil spills. *Spill Science & Technology Bulletin* 8(1): 3–8.

Lee, K., T. Lunel, P. Wood, R. Swannell, and P. Stoffyn-Egli (1997). Shoreline cleanup by acceleration of clay-oil flocculation processes. In *Proceedings of the 1997 International Oil Spill Conference (Improving Environmental Protection – Progress, Challenges, Responsibilities), April 7–10, 1997, Fort Lauderdale, Florida, USA*. Washington DC, USA: American Petroleum Institute; Special Technical Publication 4651; pp. 235–249.

Niu, H. (2011). Modeling the long term fate of oil-mineral-aggregates (OMAs) in the marine environment and assessment of their potential risks. In *Proceedings of the 2011 International Oil Spill Conference (Promoting the Science of Spill Response), May 24–26, 2011, Portland, Oregon, USA*. Washington DC, USA: American Petroleum Institute.

Owens, E.H., J.R. Bragg, and B. Humphrey (1994). Clay-oil flocculation as a natural cleaning process following oil spills. Part 2: Implications of study results in understanding past spills and for future response decisions. In *Proceedings Seventeenth Arctic and Marine Oilspill Program (AMOP) Technical Seminar, June 9–10, 1994, Vancouver, British Columbia, Canada*. Ottawa, ON, Canada: Environment Canada; pp. 25–37.

Warr, L.N., J.N. Perdrial, M. Lett, A. Heinrich-Saimeron, and M. Khodja (2009). Clay mineral-enhanced bioremediation of marine oil pollution. *Applied Clay Science* **46**(4): 337–345.

Weise, A.M., C. Nalewawajko, and K. Lee (1999). Oil-mineral fine interactions facilitate oil biodegradation in seawater. *Environmental Technology* **20**(8): 811–824.

6.3.2.2 Fate of shoreline oil in PWS 1989–92

Studies of the fate of *Exxon Valdez* oil on the shoreline from 1989–92 occurred when shoreline oil concentrations were highest. The findings are consistent with previously known factors affecting shoreline oil persistence, as discussed above.

As previously noted, the heaviest oil concentrations were on the north- and east-facing shorelines of the spill zone in PWS. In September 1989, the deepest deposits of SSOR, with an average penetration depth of 50 cm, were found in the upper shore of heavily oiled, exposed boulder/cobble/gravel beaches (Owens, 1991). Later surveys confirmed that this beach type is the most likely to have persistent SSOR (Michel and Hayes, 1993). Sheltered tidal flats, with low sediment permeability, had little oil penetration (Boehm *et al.*, 2007).

Although rare, oil was absorbed into intertidal peat deposits at some sheltered shorelines. This happened at the surface and shallow subsurface of the middle- and upper-tidal zones where the anaerobic environment slowed natural oil biodegradation. One example is a peat bog in the Bay of Isles, Knight Island, where oil was absorbed into surface peat deposits in 1989 and persisted for more than a decade (Rice, 2002).

6.3.2.3 Shoreline surveys to determine oil fate: SCAT, NOAA, and NRDA

Shoreline Cleanup Assessment Technique (SCAT) surveys (Chapter 4) and surveys by the State of Alaska (Gundlach et al., 1991) began during the cleanup and response phase (Chapter 2). These surveys supported the cleanup by defining and characterizing the spill zone and providing key information on the distribution and form of remaining oil. They also became a foundation for later studies and a point of reference for evaluating long-term oil persistence.

Consistent with other oil spills, the surveys found an initial rapid oil loss (~75–90% per year) from the shoreline between 1989 and 1992 due to natural physical processes, biodegradation, and cleanup (Wolfe et al., 1994; Neff et al., 1995). The natural loss of surface oil during 1989–90 winter storms was ~90% at exposed shorelines and ~70% at intermittently exposed and sheltered shorelines (Michel et al., 1991). About 90% of SSOR in the upper 20 cm of sediments was removed by mixing of beach sediments during these severe winter storms, and ~40% was removed in the 25- to 45-cm layer.

Beginning in 1990, SSOR became more important as the amount of surface oil declined rapidly (Neff et al., 1995). Joint federal, state, and Exxon SCAT surveys in 1990, 1991, and 1992 (Chapter 4) specifically focused on mapping areas of SSOR and their degree of oiling by digging pits. Surveys visually determined the level of SSOR in pits according to the following categories (Neff et al., 1995): oil-filled pores; heavy, moderate, and light oil residues; oil film/trace; and no oil observed. Because the remaining oil rapidly disappeared from lightly oiled shorelines, the area covered by these surveys decreased from year to year. Because the National Oceanic and Atmospheric Administration (NOAA) determined that light oil residues and lower oiling levels "do not pose any significant environmental concern" (Kennedy, 1991), the 1991 and later surveys only reported areas with oil-filled pores and moderate and heavy oiling levels. This is consistent with NOAA sediment-quality guidelines (Long et al., 1998).

A principal goal of the SCAT surveys was to assess oil persistence over time. The 1991 May Shoreline Assessment Program (MAYSAP) (http://www.valdezsciences.com/maysap.cfm) provides a point of reference for post-1991 data. Sites were surveyed after 1990 only if surveys found moderate to heavy surface oil and the presence of SSOR in 1989–90. It is unlikely that the MAYSAP survey missed significant amounts of SSOR because of its scope, its basis in prior surveys, and the imperative of identifying sites requiring further cleanup. As expected from the original spill trajectory, most of the 434 sites surveyed during MAYSAP were on east- or north-facing shores. The MAYSAP surveys found that an estimated area of ~51 000 m^2 of moderate or heavy surface oil or oil-filled pores remained in PWS in 1991 (Neff et al., 1995).

In addition to the 1989–92 SCAT shoreline surveys, NOAA conducted periodic intertidal surveys at up to 18 oiled sites in PWS from 1989 to 1997 (Michel and Hayes, 1993; Hoff and Shigenaka, 1999). NOAA confirmed the strong correlation between shore geomorphology and oil persistence. For shorelines where oil had penetrated more than 50 cm in 1989, such as southeastern Knight Island and northern Smith Island, only the deepest layer of subsurface oil (>25 cm deep) remained by August 1992 (Michel and Hayes, 1993).

Lastly, several shoreline studies were conducted as part of the injury assessment phase of the natural resource damage assessment (NRDA). Two studies are particularly relevant because they included a systematic sampling and analysis of intertidal sediments and biota: the Exxon Shoreline Ecology Program (SEP) (Chapter 11) and the NOAA Hazardous Materials (HAZMAT) Biological Monitoring Study (BMS) (Chapter 11; Hoff and Shigenaka, 1999).

The Exxon SEP had two components: (1) a 1990 sampling of 64 randomly chosen sites over four oil levels (none, light, moderate, heavy) and four shoreline types (Fig. 6.3); and (2) sampling of 11 nonrandomly selected "worst-case," heavily oiled sites of special concern (Chapter 11). At all sites, sediment samples were collected at different tide levels and mussels (*Mytilus trossulus*) were collected from the middle intertidal zone for hydrocarbon analysis. For the randomly chosen sites, sediment TPAH concentrations at 30 of 48 oiled sites were indistinguishable from those at unoiled reference sites 15–17 months after the spill. Most of the sites with TPAH concentrations above background were in the middle- to upper-tidal zone at heavily oiled boulder/cobble/gravel sites. For the nonrandomly chosen sites, TPAH decreased steadily during the first 2 years after the spill (Boehm *et al.*, 1995), but were still one-half to one order of magnitude higher than the heavy oiled random-site category. Therefore, the nonrandomly selected sites truly represented worst-case situations. By 1991, persisting oil deposits were limited to a small number of locations where the initial oiling was very heavy or cleanup activities were limited. The SEP survey underscores the importance of randomly selecting survey sites; the worst-case locations did not represent the spill zone as a whole and could not enable one to generalize results.

The NOAA BMS (Hoff and Shigenaka, 1999) annually monitored the fate of residual oil and the recovery of intertidal life from 1989–97. The study covered nine nonrandomly chosen locations and three shoreline types: rocky, boulder/cobble, and mixed/soft (Chapter 11). Like the SEP, samples of sediments and mussels were collected for hydrocarbon analysis. Elevated PAH concentrations were measured in mussels in 1990 and 1991 at some heavily oiled sites, but by 1993 the average PAH concentration in mussels from heavily oiled sites was not significantly different from that at reference sites (Chapter 11).

6.3.3 Longer term fate of shoreline oil: 1993–2008

Over the next 15 years, shoreline studies focused on assessing how much area was still oiled as a function of shoreline type and location, the degree of oil weathering, and whether oil residues were at locations and in a form accessible to biota.

A key question on the longer term fate of the spill is whether biota could be exposed to and injured by spill remnants – the issue of "lingering effects." For a contaminant to cause toxic effects, it must be two things. The first is *bioaccessible*, meaning it is in a form and location where it can be released into the medium the organism inhabits. The second is *bioavailable*, meaning the contaminant can freely pass through an organism's cellular membranes and be absorbed into its tissues (Semple *et al.*, 2004). Understanding the bioavailability of toxins in intertidal oil residues (mainly PAH) is essential for a scientifically valid, quantitative exposure and injury assessment of biological resources (Semple *et al.*, 2004).

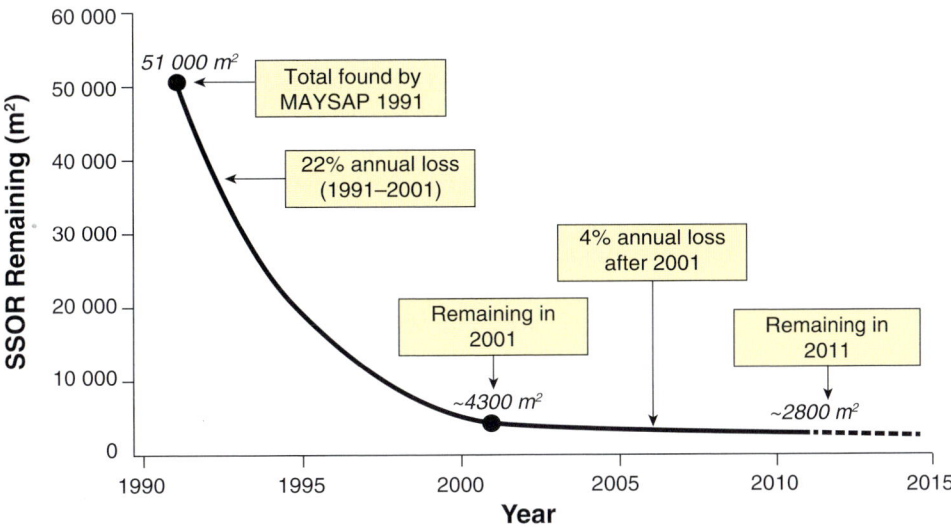

Figure 6.5 Natural SSOR loss in PWS is a biological and physical process that occurred at an average rate of ~22% per year between 1991 and 2001 and at ~4% per year after 2001 (Short *et al.*, 2004, 2007b). As SSOR weathers and becomes more inert, natural oil loss slows. Applying these average loss rates to the total area of ~51 000 m^2 of moderate and heavy SSOR mapped in PWS in 1991 yields a 2001 estimate of ~4300 m^2 and a 2011 estimate of ~2800 m^2. This places an upper limit on remaining SSOR.

6.3.3.1 Rate of natural oil loss in PWS

By the end of the 1992 cleanup season, cleanup and natural weathering had removed most surface oil residues (SOR) in PWS or the SOR had formed inert, asphaltic layers. SSOR, however, remained at specific locations mapped by MAYSAP and described by Michel and Hayes (1993). After the rapid decline in shoreline oil in the first 3 years, the average natural loss of less accessible SOR and SSOR slowed to ~22% per year from 1991 to 2001 (Short *et al.*, 2004). The average decline for the least accessible remnants was even slower, estimated at ~4% per year after 2001 (Short *et al.*, 2007a). Figure 6.5 shows natural SSOR loss from 1991 through 2011. The decreasing rates of natural oil loss observed in PWS are consistent with predictions from earlier coastal spills (e.g., Reed *et al.*, 1989).

6.3.3.2 Measuring the long-term fate of shoreline oil

The long-term fate and bioavailability of shoreline oil was assessed in two ways:

1. *Systematic hydrocarbon analysis of native mussels* (Boehm *et al.*, 2004; Carls *et al.*, 2004; Page *et al.*, 2005; Payne *et al.*, 2008). Mussels were sampled at the established survey sites (SEP and BMS) and at specific, heavily oiled mussel beds. Because mussels feed by filtering large volumes of water, thereby bioaccumulating any fat-soluble chemicals present, they are widely used as *in situ* biomonitors for chemical contaminants. The NOAA Status and Trends Mussel Watch Program is one example of the successful use of this approach (Lauenstein and Daskalakis, 1998). Time-series, mussel-tissue hydrocarbon data yielded valuable information about the natural loss of bioavailable PAH residues following the *Exxon Valdez* spill.

By 1999, the TPAH concentrations in mussels from worst-case oiled sites were similar to those from reference sites (Boehm et al., 2004; Carls et al., 2004; Page et al., 2005; Neff et al., 2006, 2011). Page et al. (2005) analyzed data for mussels collected from 11 sites that were heavily oiled in 1989. They used 1990–2002 time-series PAH concentrations to calculate an estimated 2.4-year half-life of TPAH in mussels. This translates into a mean loss of bioaccumulated TPAH of 25% per year, very close to the 22% decline for sediment reported by Short et al. (2004) from 1991 to 2001. This indicates that TPAH concentrations in mussels are in equilibrium with those in sediments where they live. However, TPAH concentrations in mussels from sites containing SSOR were always much lower than the TPAH concentrations in the sediments (Boehm et al., 1996; Neff et al., 2006). This was particularly the case after 2002, however, when the average TPAH concentrations in mussels and in sediments collected on 17 worst-case oiled shores in 2002 were 23.5 and 1570 ng/g dry weight, respectively. The large discrepancy indicates that PAH in SSOR have a very low bioaccessibility and bioavailability and, therefore, no longer present an exposure risk to animals (Chapter 16).

2. *Shoreline surveys with site-survey and pit-sediment sampling* (Short et al., 2002, 2004, 2006; Neff et al., 2006; Boehm et al., 2008; Page et al., 2008; Taylor and Reimer, 2008). NOAA, Research Planning, Inc. (RPI), and Exxon performed 11 shoreline surveys after 2000 to map the geographic and intertidal distribution and quantity of SSOR (Table 6.2). These surveys generated information on the form, location, and extent of remaining SSOR in PWS by digging pits at each site according to a systematic plan, recording the visual oiling level in each pit, and collecting sediment and biota samples for hydrocarbon analysis (e.g., Neff et al., 2006). The results of these 2001–08 surveys give a full, consistent picture of the locations and types of shorelines where persistent SSOR was found or was likely to be found. They also allowed direct comparison of site-survey data from different years (Chapter 8) to assess natural oil loss. We focused on SSOR (not SOR) because Michel et al. (2006) concluded that after 2000 "surface oil occurs primarily as highly weathered residues that pose little continuing ecological risk." (This was based on the 2001 NOAA survey and was confirmed by risk analyses; Chapter 16.) Results of the 2001 NOAA shoreline survey at 91 oiled sites are particularly useful for making comparisons between areas of SSOR mapped in 1991 and after 2000 because estimates of areal extent (i.e., m^2) of SSOR at different oiling levels are available on a site-by-site basis (Michel et al., 2006).

6.3.3.3 Natural oil loss after 1991

Surveys from 2001 to 2008 found limited and declining quantities of SSOR at a small number of locations, demonstrating that SSOR loss continues by weathering (Short et al., 2004, 2007a; Page et al., 2008; Chapter 8). For example, comparing 2002 survey results to 2007 at a northwestern Smith Island site shows considerable natural oil loss, most likely due to major winter storms in early 2007 (Page et al., 2008). Similar comparisons of pit samples between 2002 and 2007 at other sites indicate that, by 2007, SSOR deposits were more fragmented, more degraded, and had lower levels of oil (Boehm et al., 2008; Chapter 8). Although the rate is slower than during 1991–2001, biodegradation and dispersion do continue (Fig. 6.5).

Table 6.2 A summary of post-2000 shoreline surveys that mapped SSOR at spill sites in Prince William Sound. OF = oil film; LOR = light oil residues; MOR = moderate oil residues; HOR = heavy oil residues.

Year	# of sites	Study plan	# Pits Total	No oil	OF/ LOR	MOR/ HOR	Comments
NOAA Surveys							
2001	91[a,b,c]	Randomly selected beach sites, ≤ 100-m long, primarily from a pool of moderate and heavily oiled sites known from 1990–93 surveys. Random pits dug within a grid at 0.5-m tide-height increments from +1.8 m to +4.8 m tide height.[a,b]	4249[b]	3902	255	92	Total SSOR area at all levels = 7784 m^2, with ~70% in light categories.[b,c,d]
2003	29[e,f]	Similar to 2001 NOAA survey except survey focused on N. Knight Island area; included 9 sites that had been surveyed in 2001. Random pitting was within a grid at 1-m tide-height increments from −0.2 m to +4.8 m.[f]	1140	1080	35	25	Total SSOR area not reported. Few oiled pits in low-tide zone.[f]
2005	10[g]	Sites selected from randomly chosen sites surveyed in 2001 and found to have SSOR. Random pit within a grid at 1-m tide-height increments.	240	225	11	4	Report a total of 1260 m^2 of SSOR at all oiling levels.[g] No detailed data available.
Research Planning Inc. (RPI)							
2007	106[h,i]	106 shoreline sites selected using a SSOR encounter probability model.[i] Random pitting was within sampling cells at 1-m tide-height increments from +0.8 m to +3.8 m with field methods similar to NOAA 2001, 2003 surveys.	4526	4473	23	30	SSOR found at 11 of the 106 sites. Most of pits with HOR SSOR were from 3 sites.
2008	27[h,i]	Designed to identify locations with a model-predicted >90% probability of encountering MOR or HOR SSOR.[i] 27 sampling units established nonrandomly at 13 locations.	318	334	12	20	10 sampling units had no SSOR.
Exxon Surveys							
2002	39[j]	35 NOAA 2001 survey sites where SSOR was found and four where SSOR was not found were surveyed using 1991–92 SCAT protocols to identify SSOR deposits.	1182	815	232	135	SSOR found at 33 of the 39 sites, primarily in the middle- to upper-tide zone.
2002	24[k]	17 NOAA 2001 sites with SSOR; 5 reference sites; 2 historic industrial sites. Pitting done on a grid at 10-m intervals on a transect parallel to the water line at 3 tide elevations: +3, +2, and 0.0 m above mean low low-water.	710	–	–	–	SSOR residues were observed in upper- and middle-intertidal sediments at 16 sites. Pits not recorded for visual oiling level.
2004	28	28 NOAA 2001 and 2003 survey sites in the N. Knight Island area where SSOR was found were surveyed using 1991–1992 SCAT protocols to identify and characterize SSOR deposits.	151	131	12	8	SSOR was found at 9 sites.

Table 6.2 (cont.)

Year	# of sites	Study plan	# Pits Total	No oil	OF/ LOR	MOR/ HOR	Comments
2004	14	12 sites where SSOR was found in the NOAA 2001 and 2003 surveys in the N. Knight Island area and two reference sites using the 2002 grid-survey methodology.[k]	226	168	43	15	2 of the sites accounted for most of SSOR found.
2007	24[l]	Sites correspond to those with most of SSOR residues identified by prior NOAA surveys using the 2002 grid-survey methodology, but at four tide elevations: +3, +2, +1 and 0.0 m.[k]	746	534	162	40	Most of heavy SSOR at 3 sites. Deposits of SSOR more weathered and discontinuous compared with prior surveys.
2008	22	Sites correspond to those with most of SSOR residues identified by prior NOAA surveys and RPI (2007) survey sites most likely to have SSOR based on 1990–92 surveys, using the same methodology as the 2007 survey.	382	334	28	20	SSOR found at 10 sites. Six of the 8 pits with HOR were at 2 sites.

[a] Short et al., 2004
[b] Short et al., 2002
[c] Rice, 2002
[d] Michel et al., 2006
[e] Fredericks, 2006
[f] Short et al., 2006
[g] Short et al., 2007a
[h] Michel et al., 2010
[i] Zevenbergen, 2010
[j] Taylor and Reimer, 2008
[k] Neff et al., 2006
[l] Boehm et al., 2008; Page et al., 2008

Hydrocarbon chemistry data for sediment samples from 2002 to 2008 showed that all remaining deposits of SSOR had undergone natural oil loss and most were highly weathered by 2007 (Chapter 8; Boehm et al., 2008; Page et al., 2008). Grab-sample sites in 2001 and 2005 also showed extensive weathering (Short et al., 2007a). Where heavy deposits of SSOR are in a sequestered form, the degree of weathering can be low (Short et al., 2007a), but these situations are rare and usually in the upper shore, away from the biologically productive lower intertidal zone. For lightly weathered SSOR to be present after 20 years, it must be environmentally inaccessible and, therefore, not bioavailable (Chapters 7 and 16).

6.3.3.4 SSOR deposits persist at well-known locations predicted in 1991

The 2001–08 surveys clearly show that the 1991 MAYSAP survey was an excellent predictor of where SSOR was likely to persist beyond 2000: locations with heavy SSOR after 2000 correspond well to locations with SSOR deposits from 1991 site reports

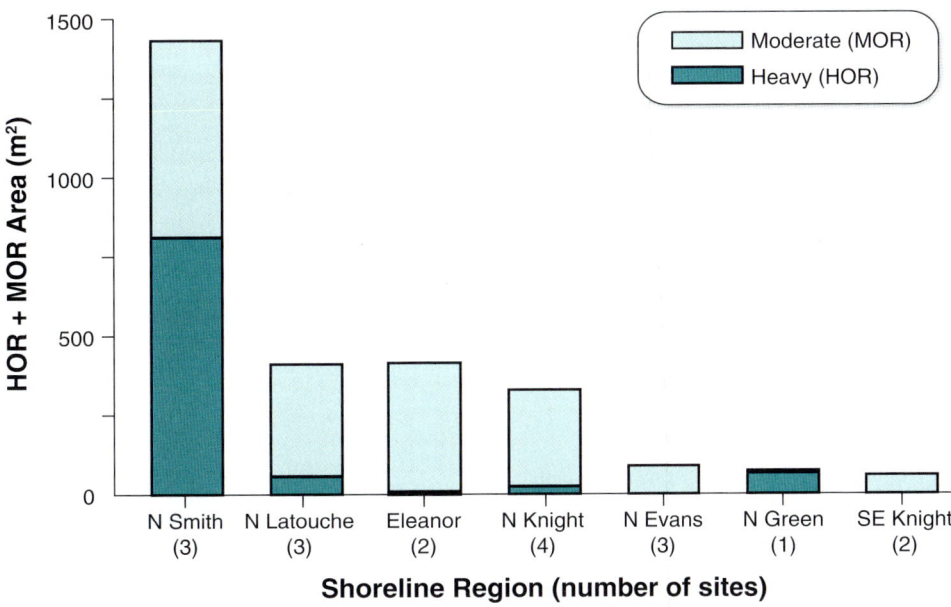

Figure 6.6 Moderate and heavy SSOR found by NOAA in 2001 was restricted to a small number of well-known areas on north- and east-facing shorelines on islands that were heavily oiled in 1989. The number of NOAA 2001 sites in each area is given in parentheses. Of the 91 sites surveyed by NOAA in 2001, only the 18 represented here had moderate and heavy levels of SSOR. (Adapted from Michel *et al.*, 2006)

(Exxon Company, USA, 1991). Moreover, no post-2000 survey found any SSOR not identified by earlier surveys: sites where heavy SSOR was found in 2001 were the same north- and east-facing locations that were heavily oiled in 1989 (Chapter 4). The extent of moderate and heavy SSOR found in 2001 at seven of these areas is shown in Figure 6.6. Together, they account for ~70% of the total moderate and heavy SSOR mapped by MAYSAP in 1991 and over 90% of moderate and heavy SSOR mapped by NOAA in 2001. The potential for moderate and heavy SSOR to persist long-term in PWS is limited to a small number of worst-case sites already well known in 1991.

6.3.3.5 SSOR persists at certain shoreline types

As discussed above, shoreline type plays a key role in determining SSOR persistence (Hayes, 1980; Hayes and Michel, 1999), as confirmed by surveys from the *Exxon Valdez* spill zone (Owens *et al.*, 2008). The vast majority of moderate and heavy SSOR found in the NOAA 2001 survey was located at sites in nine boulder/cobble/gravel shoreline locations (Fig. 6.7).

Two exposed, armored sites on southeastern Knight Island and northern Smith Island had similar amounts of moderate and heavy SSOR in 1991. By 2001, however, there was a ~99% decrease in the areal extent of SSOR at the Knight Island site, compared to a ~60% decrease at the Smith Island site. The Knight Island site has few intertidal bedrock outcrops, whereas the Smith Island site has a large transverse bedrock ridge in the middle of the beach. Most of the SSOR in 2001 was in the upper beach behind this ridge (Fig. 6.7b): the ridge impedes tidal-water flow, promoting SSOR persistence (Page *et al.*, 2008; Chapters 7 and 8).

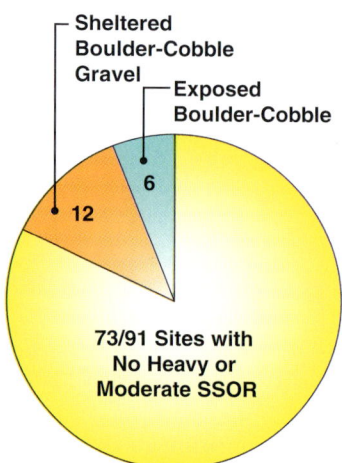
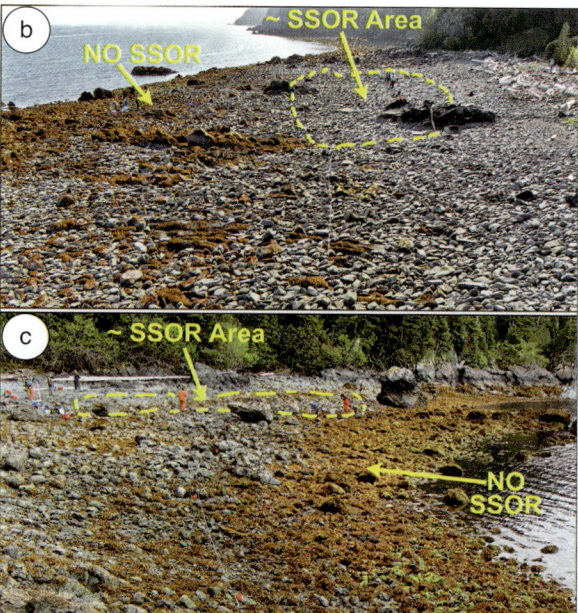

Figure 6.7 (a) The 2001 NOAA survey found that most of the persistent moderate and heavy SSOR deposits were in the upper shore at 18 well-known, exposed and sheltered boulder/cobble/gravel locations (Fig. 6.6). (b) The exposed boulder/cobble/gravel beach at a northern Smith Island site (SM006B) had more moderate and heavy SSOR than any other site surveyed in 2001. (c) Sheltered boulder/cobble/gravel beach in eastern Herring Bay, northern Knight Island, indicating the approximate zone of SSOR. (Photos: David S. Page)

6.3.3.6 SSOR persists predominantly in the upper shore

Comparing results of the 1991 surveys with those of the 2001–08 surveys confirms that SSOR deposits were not being transported down the shore, but remained in place over time while undergoing natural oil loss.

The 2003 NOAA study (Fredericks, 2006; Short et al., 2006) surveyed the intertidal zone between −0.2 m and +4.8 m. From mostly sheltered areas on northern Knight Island that were heavily oiled in 1989, they randomly chose 23 sites and resurveyed nine sites from the 2001 survey. NOAA found few pits (~5%) with SSOR, most in the middle- to upper-intertidal zone (Fig. 6.8). In the biologically productive lower tide zone, only two pits had moderate oil residues and none had heavy residues.

The Exxon 2002–08 surveys covered the intertidal zone at tide heights ranging from ≤ 0.0 m to +3 m and confirmed NOAA's observations (Neff et al., 2006; Boehm et al., 2007, 2008; Taylor and Reimer, 2008): SSOR in the biologically productive lower-tide zone was rare, and never occurred in the lower intertidal zone at locations where animals dig for prey (Neff et al., 2011; Chapters 15 and 16).

The most studied of these unusual, lower tidal zone SSOR locations is an armored beach at the head of the east arm of Northwest Bay on Eleanor Island (see Map 3, p. vii). Discontinuous SSOR patches were found in the lower tidal zone in 2002–08. Their persistence is most likely related to this shoreline's 1989 oiling and cleanup history. Facing northwest, it was heavily oiled in March 1989 and received extensive cleanup, including storm-berm relocation. In April 1989, during the height of cleanup in the area, the east arm of Northwest Bay was boomed off as an oil collection point

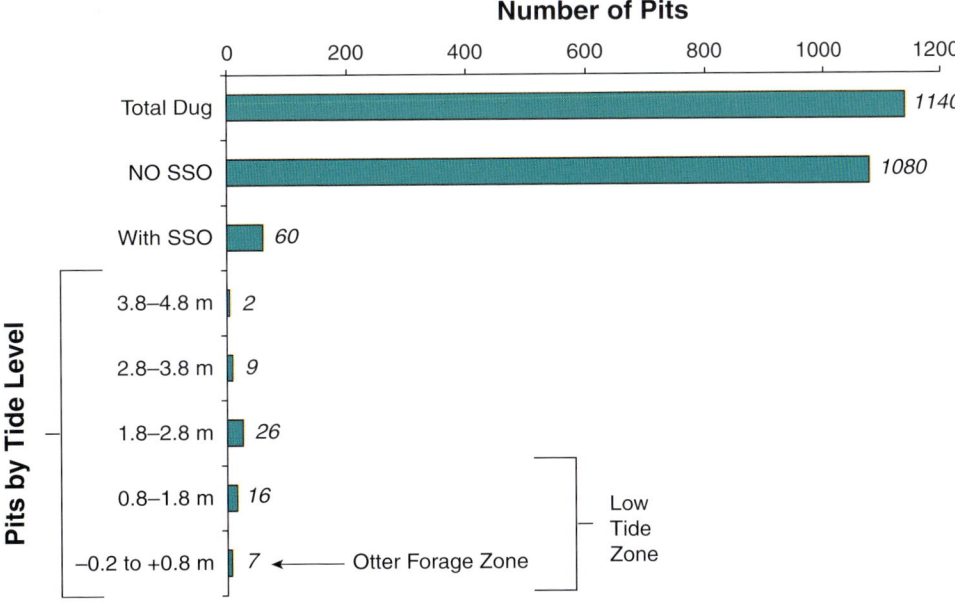

Figure 6.8 NOAA 2003 shoreline survey results, showing total pits dug and tide-zone distribution of pits containing any level of SSOR. The results confirm other surveys (Boehm et al., 2007, 2008) that found little SSOR in the biologically productive low-tide zone (data from Fredericks, 2006). The numbers of pits are given in italics.

(Erich R. Gundlach, 1989, personal observation and photographs), making this an unusual worst-case location. This example underscores the importance of knowing the oiling and cleanup history of sites when evaluating long-term survey results.

6.3.3.7 The occurrence of heavy SSOR in the spill zone is rare

The survey data can also help to estimate the frequency of occurrence of moderate and heavy SSOR deposits after 2001 throughout the spill zone. Four shoreline surveys (the NOAA 2001 survey, 2003 survey, 2005 survey, and the RPI 2007 survey) recorded the visual level of oil at randomly selected pits that had been dug. The sites were primarily from known oiled areas and at different tidal elevations (Table 6.2). Figure 6.9 shows the low frequency of moderate and heavy SSOR (1.5%) after 2000, even when survey sites were expected to have persistent SSOR based on pre-1992 surveys.

6.3.3.8 Synthesis of 2001–08 shoreline survey results

Surveys from 2001 to 2008 found limited SSOR and documented a continued decline (Fig. 6.5). Just as important, the post-2000 surveys did not find sites with SSOR that were not known from earlier surveys and predicted by the 1991 MAYSAP survey. About 75% of the ~51 000 m² of moderate and heavy SSOR mapped in 1991 was again surveyed in 2001–08 (Fig. 6.10).

The major conclusion is that, after 20 years, spill remnants do persist as inaccessible, sequestered SSOR deposits at a few shoreline locations in PWS. These locations correspond to shoreline types where experience from other spills has shown that SSOR deposits are likely to persist. They were more specifically predicted from where SCAT surveys in 1990–92 found heavy SSOR deposits, and are now well known. This underscores the predictive value of comprehensive surveys immediately following a spill.

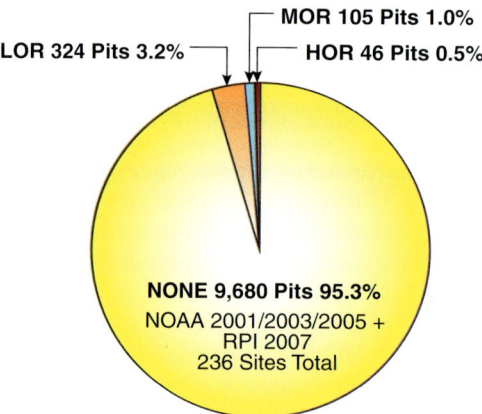

Figure 6.9 Distribution of number of pits at each visual oiling level based on data from the NOAA 2001, 2003, and 2005 surveys and the RPI 2007 survey (Table 6.2). Oiling-level abbreviations: Heavy = HOR; Moderate = MOR; Light = LOR.

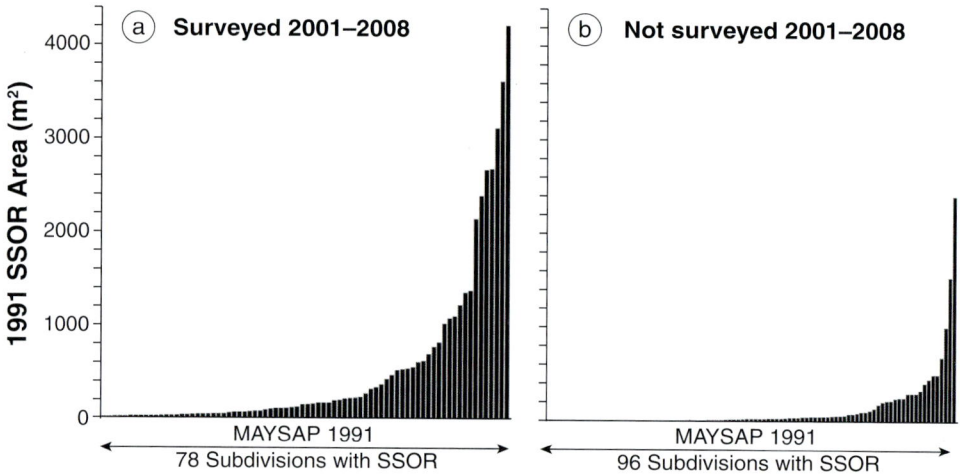

Figure 6.10 Each bar in the histograms corresponds to the 1991 SSOR area in a given shoreline subdivision surveyed by MAYSAP in 1991. (a) Areas surveyed in 2001–08 contained most (75%) of the heavy and moderate SSOR mapped by MAYSAP in 1991. (b) Shoreline areas surveyed by MAYSAP in 1991 but not surveyed in 2001–08 contained much less SSOR in 1991. This means that most remaining deposits of SSOR have been identified. (1991 SSOR area data source: http://www.valdezsciences.com/maysap.cfm).

6.4 Lessons learned

- A thorough understanding of how physical and chemical factors influence the persistence of shoreline oil residues is possible only when comprehensive, carefully designed, and well-documented SCAT surveys begin at the onset of the spill and continue in a systematic manner over multiple years.
- The 1990–92 SCAT program was an excellent framework and predictor for later studies. SSOR have not been found where they were not located by the 1990–92 SCAT surveys.

- The study of the fate of oil on the shoreline by multidisciplinary survey teams (chemists, geologists, geomorphologists, microbiologists, biologists, hydrologists, petroleum reservoir engineers, etc.) has been most fruitful. The whole has proved to be very much more than the sum of its parts.

- The form and location of SSOR are easily understood in the context of physical and chemical processes. SSOR will persist longer in settings where it is physically sequestered from natural weathering processes.

- The persistence of *Exxon Valdez* SSOR for more than 20 years indicates that they are not being released into the environment to become bioavailable, as confirmed by direct measurement of PAH concentrations in mussel tissues. Therefore, the environmental significance of persistent SSOR is negligible.

- Fully understanding the form, location, and bioavailability of shoreline oil at a given time requires concurrent sampling and analysis of water, sediments, and biota. These samples must address intertidal distribution and relative degree of oiling.

- Rigorous and objective science requires studying a range of sites in a spill area, both oiled and not; random sampling; unoiled reference sites that are similar to oiled sites; and accounting for hydrocarbon sources other than the spill.

- Differing perceptions about the long-term status of PWS shorelines underscore the importance of viewing an oil-spill zone as a whole and not as a few worst-case locations.

- The fate of shoreline oil from the *Exxon Valdez* spill is consistent with that of other spills and the established scientific principles about the fate of oil.

REFERENCES

Boehm, P.D., P.J. Mankiewicz, R. Hartung, J.M. Neff, D.S. Page, E.S. Gilfillan, J.E. O'Reilly, and K.R. Parker (1996). Characterization of mussel beds with residual oil and the risk to foraging wildlife four years after the *Exxon Valdez* oil spill. *Environmental Toxicology and Chemistry* **15**(8): 1289–1303.

Boehm, P.D., D.S. Page, J.S. Brown, J.M. Neff, J.R. Bragg, and R.M. Atlas (2008). Distribution and weathering of crude oil residues on shorelines 18 years after the *Exxon Valdez* spill. *Environmental Science & Technology* **42**(24): 9210–9216.

Boehm, P.D., D.S. Page, J.S. Brown, J.M. Neff, and W.A. Burns (2004). Polycyclic aromatic hydrocarbon levels in mussels from Prince William Sound, Alaska, USA, document the return to baseline conditions. *Environmental Toxicology and Chemistry* **23**(12): 2916–2929.

Boehm, P.D., D.S. Page, E.S. Gilfillan, A.E. Bence, W.A. Burns, and P.J. Mankiewicz (1998). Study of the fates and effects of the *Exxon Valdez* oil spill on benthic sediments in two bays in Prince William Sound, Alaska. 1: Study design, chemistry, and source fingerprinting. *Environmental Science & Technology* **32**(5): 567–576.

Boehm, P.D., D.S. Page, E.S. Gilfillan, W.A. Stubblefield, and E.J. Harner (1995). Shoreline Ecology Program for Prince William Sound, Alaska, following the *Exxon Valdez* oil spill: Part 2 – Chemistry. In Exxon Valdez *Oil Spill: Fate and Effects in Alaskan Waters*. P.G. Wells, J.N. Butler, and J.S. Hughes, eds. Philadelphia, PA, USA: American

Society for Testing and Materials; ASTM Special Technical Publication 1219; ISBN-10: 0803118961; pp. 347–397.

Boehm, P.D., D.S. Page, J.M. Neff, and C.B. Johnson (2007). Potential for sea otter exposure to remnants of buried oil from the *Exxon Valdez* oil spill. *Environmental Science & Technology* **41**(19): 6860–6867.

Bragg, J.R. and S.H. Yang (1995). Clay-oil flocculation and its role in natural cleansing in Prince William Sound following the *Exxon Valdez* oil spill. In Exxon Valdez *Oil Spill: Fate and Effects in Alaskan Waters*. P.G. Wells, J.N. Butler, and J.S. Hughes, eds. Philadelphia, PA, USA: American Society for Testing and Materials; ASTM Special Technical Publication 1219; ISBN-10: 0803118961; pp. 178–214.

Carls, M.G., P.M. Harris, and S.D. Rice (2004). Restoration of oiled mussel beds in Prince William Sound, Alaska. *Marine Environmental Research* **57**: 359–376.

DeMicco, E., P.A. Schuler, T. Omer, and B. Baca (2011). Net environmental benefit analysis (NEBA) of dispersed oil on nearshore tropical ecosystems: Tropics – the 25th year research visit. In *Proceedings of the 2011 International Oil Spill Conference (Promoting the Science of Spill Response), May 24–26, 2011, Portland, Oregon, USA*. Washington DC, USA: American Petroleum Institute.

Exxon Company, USA (1991). *May Shoreline Assessment Program (MAYSAP) Survey*. Anchorage, Alaska, USA: Exxon Company, USA. Available from Alaska Resources Library and Information Service (ARLIS), Anchorage, AK, USA. [www.valdezsciences.com]

Fredericks, B.S. (2006). *Materials Pertaining to NOAA 2003 Shoreline Program Provided Pursuant to Freedom of Information Act Request 2004–0131*. Juneau, AK, USA: National Oceanic and Atmospheric Administration, National Marine Fisheries Service, Auke Bay Laboratory.

Galt, J.A., W.J. Lehr, and D.L. Payton (1991). Fate and transport of the *Exxon Valdez* oil spill. *Environmental Science & Technology* **25**(2): 202–209.

Gundlach, E.R., P.D. Boehm, M. Marchand, R.M. Atlas, D.M. Ward, and D.A. Wolfe (1983). The fate of *Amoco Cadiz* oil. *Science* **221**(4606): 122–129.

Gundlach, E.R. and M.O. Hayes (1978a). Classification of coastal environments in terms of potential vulnerability to oil spill damage. *Marine Technology Society Journal* **12**(4): 18–27.

Gundlach, E.R. and M.O. Hayes (1978b). Investigation of beach processes. In *The Amoco Cadiz Oil Spill, A Preliminary Scientific Report*. W.N. Hess, ed. Boulder, CO, USA: National Oceanic and Atmospheric Administration and US Environmental Protection Agency, Environmental Research Laboratories; Section 4; pp. 85–196. [http://www.gpo.gov/fdsys/pkg/CZIC-gc1321-a46–1978/html/CZIC-gc1321-a46–1978.htm]

Gundlach, E.R., C.H. Ruby, M.O. Hayes, and A.E. Blount (1978). The *Urquiola* oil spill, La Coruña, Spain: impact and reaction on beaches and rocky coasts. *Environmental Geology* **2**(3): 131–143.

Gundlach, E.R., E.A. Pavia, C. Robinson, and J. Gibeaut (1991). Shoreline surveys at the *Exxon Valdez* oil spill: the State of Alaska response. In *Proceedings of the 1991 International Oil Spill Conference (Prevention, Behavior, Control, Cleanup), March 4–7, 1991,*

San Diego, California. Washington DC, USA: American Petroleum Institute; Publication 4529; pp. 519–529.

Hayes, M.O. (1980). Oil spill vulnerability, coastal morphology, and sedimentation of Outer Kenai Peninsula and Montague Island. In *Outer Continental Shelf Environmental Assessment Program, Final Reports of Principal Investigators, Vol. 51*. Anchorage, AK, USA: National Oceanic and Atmospheric Administration; NTIS No. PB87198867; December 1986; pp. 419–583.

Hayes, M.O. and J. Michel (1998). *Evaluation of the Condition of Prince William Sound Shorelines following the Exxon Valdez Oil Spill and Subsequent Shoreline Treatment: 1997 Geomorphological Monitoring Survey*. Seattle, WA, USA: National Oceanic and Atmospheric Administration, National Ocean Service; NOAA Technical Memorandum NOS ORCA 126.

Hayes, M.O. and J. Michel (1999). Factors determining the long-term persistence of *Exxon Valdez* oil in gravel beaches. *Marine Pollution Bulletin* **38**(2): 92–101.

Hayes, M.O., J. Michel, and D.V. Betenbaugh (2010). The intermittently exposed, coarse-grained gravel beaches of Prince William Sound, Alaska: comparison with open-ocean gravel beaches. *Journal of Coastal Research* **26**(1): 4–30.

Hoff, R.Z. and G. Shigenaka (1999). Lessons from 10 years of post-*Exxon Valdez* monitoring on intertidal shorelines. In *Proceedings of the 1999 International Oil Spill Conference (Beyond 2000: Balancing Perspective), March 8–11, 1999, Seattle, Washington*. Washington DC, USA: American Petroleum Institute; Publication 4686B; pp. 111–117.

Irvine, G.V., D.H. Mann, and J.W. Short (2006). Persistence of 10-year old *Exxon Valdez* oil on Gulf of Alaska beaches: the importance of boulder-armouring. *Marine Pollution Bulletin* **52**(9): 1011–1022.

Kennedy, D.M. (1991). *Review of the Status of Prince William Sound Shorelines following Two Years of Treatment by Exxon*. Seattle, WA, USA: National Oceanic and Atmospheric Administration, Hazardous Materials Response Branch, Spill Response Program; unpublished report for Rear Admiral D.E. Ciancaglini. Anchorage, AK, USA: United States Coast Guard, Federal On Scene Coordinator *Exxon Valdez* Archive; Document No. F119; David M. Kennedy (NOAA) to David E. Ciancaglini (USCG), March 15, 1991.

Kvenvolden, K.A., F.D. Hostettler, P.R. Carlson, J.B. Rapp, C.N. Threlkeld, and A. Warden (1995). Ubiquitous tarballs with a California-source signature on the shorelines of Prince William Sound, Alaska. *Environmental Science & Technology* **29**(10): 2684–2694.

Lauenstein, G.G. and K.D. Daskalakis (1998). US long-term coastal contaminant temporal trends determined from mollusk monitoring programs, 1965–1993. *Marine Pollution Bulletin* **37**(1–2): 6–13.

Long, E.R., L.J. Field, and D.D. MacDonald (1998). Predicting toxicity in marine sediments with numerical sediment quality guidelines. *Environmental Toxicology and Chemistry* **17**(4): 714–727.

Michel, J. and M.O. Hayes (1993). Persistence and weathering of *Exxon Valdez* oil in the intertidal zone: 3.5 years later. In *Proceedings of the 1993 International Oil Spill*

Conference (Prevention, Preparedness, Response), March 29–April 1, 1993, Tampa, Florida. Washington DC, USA: American Petroleum Institute; Publication 4580; pp. 279–286.

Michel, J., M.O. Hayes, W.J. Sexton, J.C. Gibeaut, and C. Henry (1991). Trends in natural removal of the *Exxon Valdez* oil spill in Prince William Sound from September 1989 to May 1990. In *Proceedings of the 1991 International Oil Spill Conference (Prevention, Behavior, Control, Cleanup), March 4–7, 1991, San Diego, California*. Washington DC, USA: American Petroleum Institute; Publication 4529; pp. 181–187.

Michel, J., Z. Nixon, and L. Cotsapas (2006). *Evaluation of Oil Remediation Technologies for Lingering Oil from the* Exxon Valdez *Oil Spill in Prince William Sound, Alaska*. Juneau, AK, USA: National Oceanic and Atmospheric Administration, National Marine Fisheries Service; *Exxon Valdez* Oil Spill Restoration Project 050778 Final Report. [http://www.evostc.state.ak.us/Files.cfm?doc=/Store/FinalReports/2005-050778-Final.pdf&]

Michel, J., Z. Nixon, M.O. Hayes, J. Short, G. Irvine, D. Betenbaugh, C. Boring, and D. Mann (2010). *Distribution of subsurface oil from the* Exxon Valdez *oil spill*. Juneau, AK, USA: National Oceanic and Atmospheric Administration; *Exxon Valdez* Oil Spill Restoration Project 070801 Final Report. [http://www.evostc.state.ak.us/Files.cfm?doc=/Store/FinalReports/2007-070801-Final.pdf&]

National Oceanic and Atmospheric Administration (2007). *Prince William Sound, Alaska: July 2000, Environmental Sensitivity Index Maps, Digital Data Re-Release, April 2007*. Seattle, WA, USA: National Oceanic and Atmospheric Administration, Office of Response and Restoration, Emergency Response Division. [http://response.restoration.noaa.gov/maps-and-spatial-data/esi-coverage-alaska.html]

National Research Council (1985). *Oil in the Sea: Inputs, Fates, and Effects*. Washington DC, USA: National Research Council, National Academy Press; ISBN-10: 0309078350, ISBN-13: 9780309078351.

Neff, J.M., A.E. Bence, K.R. Parker, D.S. Page, J.S. Brown, and P.D. Boehm (2006). Bioavailability of PAH from buried shoreline oil residues thirteen years after the *Exxon Valdez* oil spill: a multispecies assessment. *Environmental Toxicology and Chemistry* **25**(4): 947–961.

Neff, J.M., E.H. Owens, S.W. Stoker, and D.M. McCormick (1995). Shoreline oiling conditions in Prince William Sound following the. *Exxon Valdez* oil spill. In Exxon Valdez *Oil Spill: Fate and Effects in Alaskan Waters*. P.G. Wells, J.N. Butler, and J.S. Hughes, eds. Philadelphia, PA, USA: American Society for Testing and Materials; ASTM Special Technical Publication 1219; ISBN-10: 0803118961; pp. 312–346.

Neff, J.M., D.S. Page, and P.D. Boehm (2011). Exposure of sea otters and harlequin ducks in Prince William Sound, Alaska, USA, to shoreline oil residues 20 years after the *Exxon Valdez* oil spill. *Environmental Toxicology and Chemistry* **30**(3): 659–672.

O'Clair, C.E., J.W. Short, and S.D. Rice (1996). Contamination of intertidal and subtidal sediments by oil from the *Exxon Valdez* in Prince William Sound. In *Proceedings of the* Exxon Valdez *Oil Spill Symposium*. S.D. Rice, R.B. Spies, D.A. Wolfe, and B.A. Wright, eds. Bethesda, MD, USA: American Fisheries Society; Symposium 18; ISBN-10: 0913235954; ISSN: 08922284; pp. 61–93.

Oudot, J. and F. Chaillan (2010). Pyrolysis of asphaltenes and biomarkers for the fingerprinting of the *Amoco Cadiz* oil spill after 23 years. *Comptes Rendus Chimie* **13**(5): 548–552.

Owens, E.H. (1991). Shoreline conditions following the *Exxon Valdez* spill as of fall 1990. In *Proceedings Fourteenth Arctic and Marine Oilspill Program (AMOP) Technical Seminar, June 12–14, 1991, Vancouver, British Columbia, Canada*. Ottawa, ON, Canada: Environment Canada; pp. 579–606.

Owens, E.H. and G.A. Sergy (2005). Time series observations of marsh recovery and pavement persistence at three *Metula* spill sites after 30½ years. In *Proceedings Twenty-Eighth Arctic and Marine Oilspill Program (AMOP) Technical Seminar, June 7–9, 2005, Calgary, Alberta, Canada*. Ottawa, ON, Canada: Environment Canada; pp. 463–472.

Owens, E.H., E. Taylor, and B. Humphrey (2008). The persistence and character of stranded oil on coarse-sediment beaches. *Marine Pollution Bulletin* **56**(1): 14–26.

Page, D.S., P.D. Boehm, G.S. Douglas, and A.E. Bence (1995). Identification of hydrocarbon sources in the benthic sediments of Prince William Sound and the Gulf of Alaska following the *Exxon Valdez* oil spill. In Exxon Valdez *Oil Spill: Fate and Effects in Alaskan Waters*. P.G. Wells, J.N. Butler, and J.S. Hughes, eds. Philadelphia, PA, USA: American Society for Testing and Materials; ASTM Special Technical Publication 1219; ISBN-10: 0803118961; pp. 41–83.

Page, D.S., P.D. Boehm, G.S. Douglas, A.E. Bence, W.A. Burns, and P.J. Mankiewicz (1996). The natural petroleum hydrocarbon background in subtidal sediments of Prince William Sound, Alaska. *Environmental Toxicology and Chemistry* **15**(8): 1266–1281.

Page, D.S., P.D. Boehm, G.S. Douglas, A.E. Bence, W.A. Burns, and P.J. Mankiewicz (1997). An estimate of the annual input of natural petroleum hydrocarbons to seafloor sediments in Prince William Sound, Alaska. *Marine Pollution Bulletin* **34**(9): 744–749.

Page, D.S., P.D. Boehm, G.S. Douglas, A.E. Bence, W.A. Burns, and P.J. Mankiewicz (1999). Pyrogenic polycyclic aromatic hydrocarbons in sediments record past human activity: A case study in Prince William Sound Alaska. *Marine Pollution Bulletin* **38**(4): 247–260.

Page, D.S., P.D. Boehm, J.S. Brown, J.M. Neff, W.A. Burns, and A.E. Bence (2005). Mussels document loss of bioavailable polycyclic aromatic hydrocarbons and the return of baseline conditions for oiled shorelines in Prince William Sound, Alaska. *Marine Environmental Research* **60**(4): 422–436.

Page, D.S., P.D. Boehm, and J.M. Neff (2008). Shoreline type and subsurface oil persistence in the *Exxon Valdez* spill zone of Prince William Sound, Alaska. In *Proceedings of the Thirty-First Arctic and Marine Oilspill Program (AMOP) Technical Seminar, Environmental Contamination and Response, June 3–5, 2008, Calgary, AB, Canada*. Ottawa, ON, Canada: Environment Canada; pp. 545–564.

Page, D.S., P.D. Boehm, W.A. Stubblefield, K.R. Parker, E.S. Gilfillan, J.M. Neff, and A.W. Maki (2002). Hydrocarbon composition and toxicity of sediments following the *Exxon Valdez* oil spill in Prince William Sound, Alaska. *Environmental Toxicology and Chemistry* **21**(7): 1438–1450.

Page, D.S., J.S. Brown, P.D. Boehm, A.E. Bence, and J.M. Neff (2006). A hierarchical approach measures the aerial extent and concentration levels of PAH-contaminated shoreline sediments at historic industrial sites in Prince William Sound, Alaska. *Marine Pollution Bulletin* **52**(4): 367–379.

Page, D.S., J.C. Foster, P.M. Fickett, and E.S. Gilfillan (1988). Identification of petroleum sources in an area impacted by the *Amoco Cadiz* oil spill. *Marine Pollution Bulletin* **19**(3): 107–115.

Page, D.S., D.W. Mayo, J.F. Cooley, E. Sorenson, E.S. Gilfillan, and S.A. Hanson (1979). Hydrocarbon distribution and weathering characteristics at a tropical oil spill site. In *Proceedings of the 1979 International Oil Spill Conference (Prevention, Behavior, Control, Cleanup), March 99–22, 1979, Los Angeles, California*. Washington DC, USA: American Petroleum Institute; pp. 709–712.

Payne J.R., J.R. Clayton, G.D. McNabb, and B.E. Kirstein (1991). *Exxon Valdez* oil weathering fate and behavior: Model prediction and field observation. In *Proceedings of the 1991 International Oil Spill Conference (Prevention, Behavior, Control, Cleanup), March 4–7, 1991, San Diego, California*. Washington DC, USA: American Petroleum Institute; Technical Publication 4529; pp. 641–654.

Payne, J.R., W.B. Driskell, J.W. Short, and M.L. Larsen (2008). Long term monitoring for oil in the *Exxon Valdez* spill region. *Marine Pollution Bulletin* **56**(12): 2067–2081.

Peacock, E.E., R.K. Nelson, A.R. Solow, J.D. Warren, J.L. Baker, and C.M. Reddy (2005). The West Falmouth oil spill: 100 kg of oil found to persist decades later. *Environmental Forensics* **6**(3): 273–281.

Prince, R.C., E.H. Owens, and G.A. Sergy (2002). Weathering of an Arctic oil spill over 20 years: The BIOS experiment revisited. *Marine Pollution Bulletin* **44**(11): 1236–1242.

Reed, M., E. Gundlach, and T. Kana (1989). A coastal oil spill model: Development and sensitivity studies. *Oil and Chemical Pollution* **5**(6): 411–449.

Rice, S.D. (2002). *Materials Pertaining to NOAA 2001 Shoreline Program Provided Pursuant to Freedom of Information Act Request 02–133*. Juneau, Alaska, USA: National Oceanic and Atmospheric Administration, National Marine Fisheries Service, Auke Bay Laboratory.

Schrope, M. (2010). The lost legacy of the last great oil spill. *Nature* **466**(7304): 304–305.

Semple, K.T., K.J. Doick, K.C. Jones, P. Burauel, A. Craven, and H. Harms (2004). Defining bioavailability and bioaccessibility of contaminated soil and sediment is complicated. *Environmental Science & Technology* **38**(12): 228A–231A.

Short, J.W., G.V. Irvine, D.H. Mann, J.M. Maselko, J.J. Pella, M.R. Lindeberg, J.M. Payne, W.B. Driskell, and S.D. Rice (2007a). Slightly weathered *Exxon Valdez* oil persists in Gulf of Alaska beach sediments after 16 years. *Environmental Science & Technology* **41**(4): 1245–1250.

Short, J.W., J.J. Kolak, J.R. Payne, and G.K. Van Kooten (2007b). An evaluation of petrogenic hydrocarbons in northern Gulf of Alaska continental shelf sediments: The role of coastal oil seep inputs. *Organic Geochemistry* **38**(4): 643–670.

Short J.W., M.R. Lindeberg, P.M. Harris, J. Maselko, J.J. Pella, and S.D. Rice (2004). Estimate of oil persisting on the beaches of Prince William Sound 12 years after the *Exxon Valdez* oil spill. *Environmental Science & Technology* **38**(1): 19–25.

Short, J.W., M.R. Lindeberg, P.M. Harris, J. Maselko, and S.D. Rice (2002). Vertical oil distribution within the intertidal zone 12 years after the *Exxon Valdez* oil spill in Prince William Sound, Alaska. In *Proceedings of the Twenty-Fifth Arctic and Marine Oilspill*

Program (AMOP) Technical Seminar, Environmental Contamination and Response, June 11–13, 2002, Calgary, AB, Canada. Ottawa, ON, Canada: Environment Canada; pp. 57–72.

Short, J.W., J. Maselko, M.R. Lindeberg, P.M. Harris, and S.D. Rice (2006). Vertical distribution and probability of encountering intertidal *Exxon Valdez* oil on shorelines of three embayments within Prince William Sound. *Environmental Science & Technology* **40**(12): 3723–3729.

Taylor, E. and D. Reimer (2008). Oil persistence on beaches in Prince William Sound: a review of SCAT surveys conducted from 1989 to 2002. *Marine Pollution Bulletin* **56**(3): 458–474.

Wolfe, D.A., M.J. Hameedi, J.A. Galt, G. Watabayashi, J. Short, C. O'Claire, S. Rice, J. Michel, J.R. Payne, J. Braddock, S. Hanna, and D. Sale (1994). The fate of the oil spilled from the *Exxon Valdez. Environmental Science & Technology* **28**(13): 561A–568A.

Wooley, C. (2002). The myth of the "pristine environment": Past human impacts in Prince William Sound and the Gulf of Alaska. *Spill Science and Technology Bulletin* **7**(1–2): 89–104.

Zevenbergen, M. (2010). *Materials Related to the Michel 2007–2010* Exxon Valdez *Oil Spill Trustee Council Restoration Project 070801: Assessment of Areal Distribution and Amount of Lingering Oil in Prince William Sound and the Gulf of Alaska. Pursuant to Freedom of Information Act Requests NOAA-2008–0046 (October 29, 2007) and NOAA-2009–00040 (October 17, 2008).* Seattle, WA, USA: US Department of Justice; National Oceanic and Atmospheric Administration, Damage Assessment.

CHAPTER SEVEN

Understanding subsurface contamination using conceptual and mathematical models

Gary A. Pope, Kimberly D. Gordon, and James R. Bragg

7.1 Introduction

Petroleum spills and other sources of hydrocarbon contamination represent risks for society. Regardless of whether oil is stranded on a shoreline, spilled from a pipeline, or leaked from underground storage tanks, the same basic physical and chemical principles characterize exposure levels of contaminants. The purpose of this chapter is to explain and illustrate these principles. In particular, we use these principles to explain the apparent paradox of how oil residues persist at some shorelines of Prince William Sound (PWS) as isolated subsurface patches, but yet pose little if any exposure risk to the local ecology. We resolve this apparent paradox using well-established scientific and engineering tools.

One of the biggest challenges of any study of a contaminated site is identifying the most important questions and the most important observations and data needed to answer these questions. This challenge is discussed in this chapter in both a general way and for the PWS study in particular. One of the key lessons learned from this study was the need for experts in multiphase flow in contaminated sediments to be a central part of the team addressing these questions. Our goal is to convey a coherent understanding and perspective that brings all of the observations and measurements by various environmental experts of different scientific disciplines into a consistent explanation.

The tools of conceptual and mathematical modeling have previously been applied to organize and integrate individual site properties, physical and chemical parameters, principles of multiphase flow and transport, and other factors affecting contaminated terrestrial sites for the purpose of characterization and remedial assessment. However, these tools have only recently been applied to shoreline contamination from marine oil

Oil in the Environment: Legacies and Lessons of the Exxon Valdez *Oil Spill*, ed. J. A. Wiens. Published by Cambridge University Press. © Cambridge University Press 2013.

spills (Li and Boufadel, 2010; Pope *et al.*, 2011a). We will show how the models developed from these tools integrate the various types of data generated by experts of several disciplines to foster an understanding of the factors affecting persistence and environmental risk.

7.2 Models

A model represents a simplified conceptualization of some phenomenon, system, or process. Scientific models can be conceptual or mathematical. Conceptual models can be simple qualitative descriptions or drawings or complex, semiquantitative process flows. Mathematical models can be simple analytical tools that hold many parameters as constants or numerical simulations with multidimensional parameter distributions. The development and degree of simplification of appropriate models for any given problem depend on the objectives of the environmental managers, available resources, available field data, and the legal and regulatory framework of the problem. Consistent conceptual, analytical, and numerical models can be used as tools to understand the physical and chemical processes that govern the behavior of water, air, and oil in the subsurface and free-flowing nearshore water. Numerical simulation can be used to provide quantitative estimates of exposure levels.

The development of these conceptual and mathematical models typically requires a multidisciplinary approach (e.g., chemists, hydrogeologists, engineers, biologists) to identify important measurements and to answer critical questions associated with risk. We use studies performed for selected PWS shorelines to illustrate this approach and show how the general principles of multiphase flow and transport have been, and can be, adapted to other sites.

7.2.1 Characterization for model development

The first step in model development involves general site reconnaissance and gathering information on the site layout, history, and records of management (US Environmental Protection Agency, 1988). As described by Bear *et al.* (1992), the physical and chemical components of a conceptual model provide the basis for a mathematical model and may include:

- Geometry and boundaries; nature of the solid media accounting for heterogeneity, anisotropy, fractures, etc.;
- Flow characteristics of the media (gradients, directions, multiphase parameters including saturation and relative permeability);
- Sources and sinks of water and/or relevant components with reference to their estimation as point or distributed sources or sinks;
- Initial conditions (e.g., starting water table, source distributions, recharge conditions, pumping);
- Stresses or perturbation of boundaries that show the interaction of the system with the environment; and
- Fluid properties (e.g., saturation, density, viscosity, component water solubilities).

The relative importance of these components varies for individual sites depending on spill conditions and remediation goals. The spatial and temporal distribution of these properties is also important. In particular, spatial variations in porosity and permeability (heterogeneity) strongly influence subsurface contaminant transport. Poeter and Gaylord (1990) show that a change in permeability of a single order of magnitude causes a significant change in the contaminant spatial distribution, as compared to a homogeneous medium. These parameters can be obtained through a desktop review of available data, but they usually require direct field measurements and laboratory analyses to adequately characterize the processes described by the conceptual model. Following this phase of conceptual model development, the results can be evaluated for overall site characterization and risk assessment, and data needs can be reassessed for the development of remedial strategies (US Environmental Protection Agency, 1988). The need for mathematical models can also be determined at this point.

As context for the development of multiphase models, a brief review of some characterization parameters is shown in Table 7.1. Knowledge of these parameters is critical to the understanding of how and why oil and other fluids persist in the subsurface. Each parameter is readily determined through existing methods that are as simple as direct observations or as complex as detailed laboratory studies. The level of detail required to define each parameter varies with the level of site risk and the required accuracy and precision for individual sites.

7.3 The conceptual model

Conceptual models of fluid flow are qualitative descriptions of the hydrogeologic system that incorporate information about contaminant sources and their fate and transport. They provide a systematic assessment of processes at contaminated sites and can include hydrogeological, biological, and ecological components. They can also be used for understanding and illustrating processes or making predictions. *Guidance for Conducting Remedial Investigations and Feasibility Studies Under CERCLA* (US Environmental Protection Agency, 1988) gives practical guidance for the development of site conceptual models. Conceptual models provide the assumptions for, and directly impact, the quality of mathematical or numerical models developed for the purpose of quantifying site conditions. They are commonly used throughout all phases of a site investigation and remediation planning (Bear *et al.*, 1992).

The development of models is iterative: as more information is gained and study and remediation goals change, models are updated. Figure 7.1 shows how a conceptual model evolves. Beginning as a basic site drawing (or map), two- and three-dimensional details are added, as are key processes and features that govern the flow of fluids at the site. Residual oil[1] is trapped in the subsurface of the shoreline in low-permeability sediments. Seawater infiltration occurs at the surface with changes in tidal elevation. Fresh groundwater and surface water from streams comes from inland sources. The transient interaction of these fluids governs how quickly the subsurface oil degrades and how components of the oil are released into the surrounding environment.

Schwille (1975) discusses the need to understand the behavior of the oil phase in the subsurface to evaluate and apply remedial alternatives for removing or isolating contaminants. Physical and chemical properties of each site are unique, but the basic

---

[1] Because we are primarily concerned with the effects of oil stranded in shoreline sediments in this chapter, we will use the terms oil and nonaqueous phase liquids (NAPL) interchangeably.

principles that govern the fate and transport of hydrocarbons in the subsurface are the same. Feenstra and Cherry (1988) identify factors that influence subsurface oil migration and persistence: volume of oil released, infiltration area, time and duration of the release, properties of the oil, properties of the media, and subsurface flow conditions. Saenton *et al.* (2002) stress the importance of understanding multiphase-flow problems where components of the oil will be distributed between free-phase oil, oil adsorbed on solids, components dissolved in the aqueous phase, or components volatilized into the gaseous phase. There may be ongoing sources of soluble or volatile components for as long as the free-phase oil and/or adsorbed oil persist in the subsurface. The magnitude and distribution of these components in the aqueous phase or the gas phase depends on the properties of the individual components of the oil and fluid interactions with the oil.

The basic conceptual model that applies to many sites contaminated with oil has been described by Abriola (1989), Mercer and Cohen (1990), and Newell *et al.* (1995). As shown in Figure 7.2, when light NAPL (LNAPL) (such as oil that is less dense than water) is released either at the surface or in the subsurface, it will migrate downward under the force of gravity through the unsaturated zone, where a residual fraction of the oil will be trapped in the pores by capillary forces (surface tension effects). This capillary trapping gradually depletes the oil flow until there is no longer sufficient mass for it to continue to move under natural gradients. Within the unsaturated zone, there are three separate phases in the sediments: air, water, and oil. Capillary forces also retain some water in the soil or sediment even under unsaturated conditions. Fluids that are trapped by these forces are said to be at residual saturation.

If the spill is large enough, the oil will continue to migrate downward under gravitational forces until it encounters finer-grained sediments that impede flow because they have smaller pores (and thus a high capillary pressure), requiring a higher pressure for the oil to enter them. Lateral spreading will then occur until the oil reaches coarser-grained sediments that allow it to again flow downward (or until it is removed by tidal flow). Overall, mobile oil becomes increasingly trapped by capillary forces in the pores of the sediment. Eventually, all of the oil may become immobile because of the capillary trapping. Volatile components of the oil may evaporate and form a gas plume that can migrate in the subsurface.

If the oil reaches the water table and is less dense than the water (as was the case with oil from the *Exxon Valdez*), it will spread laterally along the upper boundary of the water table in the direction of lower water-table elevation (down-hydraulic gradient) until a balance is reached between pressure, gravity, and capillary forces. Capillary and gravity forces can also draw some oil below the water table until the oil reaches vertical equilibrium: the overall amount of oil in the pore space below the water table generally decreases with depth. Meanwhile, soluble components of the oil dissolve into the groundwater and will follow groundwater-flow gradients and form an aqueous plume of dissolved components.

When the oil components are relatively insoluble in water and only slightly volatile in the air phase, the oil can persist for a long time. Soluble or volatile components of oil partition into the aqueous or air phase, respectively, where they are exposed to natural microbial processes and can be transported in the saturated zone (in the groundwater) or unsaturated zone (above the water table). Thus, understanding the mass transfer of hydrocarbons to the water and the air is essential to making scientifically based

Table 7.1 Key parameters for multiphase flow models.

Parameter	Narrative	References
Boundary conditions	Site geometry and boundaries, topography, morphology, and stratigraphy. Key features such as bedrock outcrops, streams and other recharge features, significant vegetation, and surface infrastructure such as buildings, roads, parking lots. Sources: site histories; field investigations; ground-surface elevation surveys; borehole logs; core descriptions; hand-dug pits and trenches.	Anderson and Woessner, 1992
Porosity and hydraulic conductivity	Porosity is the volume of the pores of a rock or sediment divided by the total volume. Hydraulic conductivity describes the ease with which fluids can flow through the pore spaces of the porous medium and is described mathematically by Darcy's law. Sources: literature values; core floods; soil column studies; grain-size data; well testing; permeameters; infiltrometers; tracer testing.	van Bavel and Kirkham, 1948 Theis et al., 1963 Bouwer and Rice, 1976 Bear, 1979 Freeze and Cherry, 1979 US Environmental Protection Agency, 1988 Shepherd, 1989 Poeter and Gaylord, 1990 Domenico and Schwartz, 1998
Flow characteristics	Hydraulic head distributions; recharge or discharge from stream flow, tidal effects, precipitation, evaporation; tidal boundaries. Sources: fluid levels in wells; piezometers.	Anderson and Woessner, 1992
Fluid saturation	The saturation of a fluid is the volume of the fluid in the pore space divided by the total fluid volume. Oil saturation is the volume of oil divided by the total pore volume. Residual saturations are the saturations at which a fluid becomes immobile as a result of capillary forces (interfacial tension effects). Sources: geophysical methods; tensiometers; total organic mass in a sample, taking adsorption into account; geostatistical methods.	Mariner et al., 1997 Mercer and Cohen, 1990
Relative permeability	Relative permeability is the permeability of a fluid divided by a reference permeability such as the permeability when only water is present. The water-relative permeability decreases sharply in a very nonlinear way as oil saturation increases. Sources: standard core floods; soil column experiments.	Muskat, 1949 Brooks and Corey, 1964 van Genuchten, 1980 Lake, 1989 Delshad and Pope, 1989 Domenico and Schwartz, 1998 Fetter, 1993
Density	Density is mass per unit volume and varies with temperature, composition, and pressure. Fluids with densities less than that of water migrate down until they are depleted or reach the water table and then migrate along the top of the water table. Fluids with densities greater than water migrate down until they are depleted or reach the bottom of the water table where bedrock, clay, or some impermeable barrier stops flow. Sources: literature values; laboratory determinations.	Mackay et al., 1985
Viscosity	Viscosity is a fluid's internal friction that causes resistance to flow. Viscosity can be thought of in terms of low-viscosity "thin" fluids that flow rapidly and high-viscosity "thick" fluids that flow slowly. Viscosity changes over time with weathering as lighter components are dissolved or volatilized. Sources: literature values; laboratory determinations.	Mercer and Cohen, 1990
Solubility	Aqueous solubility of a chemical component is the maximum concentration that will dissolve into water at a given temperature and pressure. Influenced by the mix of components in the oil and the dissolved components already in	Domenico and Schwartz, 1998 Pankow and Cherry, 1996

Table 7.1 (*cont.*)

Parameter	Narrative	References
	the aqueous solution. Spatial distributions of dissolved components in groundwater can provide some indication of the location of a nonaqueous phase liquid (NAPL) source zone. Sources: literature values; laboratory determinations.	Fetter, 1993
Wettability	Wettability is the tendency of a fluid to preferentially adhere to a solid surface. It affects the movement and trapping of fluids in the subsurface. The interfacial tension between fluids and how the fluids wet the sediment determine the distribution of the fluids in the sediment, the capillary pressure, the residual saturations, and the relative permeabilities. Sources: literature values; laboratory determinations.	Dwarakanath *et al.*, 2002
Other parameters	Groundwater quality measurements such as pH, total dissolved solids, and contaminant concentration are needed to determine exposure via groundwater and to define a contaminant plume. Sources: field sampling; field or laboratory measurements.	Farr *et al.*, 1990

predictions of their persistence. The amount of a particular component transferred to other phases is directly proportional to the solubility or volatility of the components and the component concentration in the oil. Highly soluble or volatile components will quickly transfer to other phases from the oil, while less soluble or volatile components remain behind in the oil phase.

Groundwater-flow and contaminant-transport modeling have been used with varying degrees of success in conjunction with the study of contaminated sites. In particular, for sites that contain persistent sources of petroleum hydrocarbons, modeling can assess the risks from hydrocarbons and their dissolved or volatilized components. Modeling can also help to identify treatment options. Cygan *et al.* (2007) provide a summary of locations where various US agencies have used models to evaluate transport at a wide variety of spill sites. As risk-assessment tools, models give environmental managers and scientists the ability to predict the rate and direction of contaminant migration and transport.

7.3.1 The *Exxon Valdez* oil spill case study

The concepts of capillarity and multiphase fluid flow in permeable media (as described above) explain how oil got into the shoreline sediments in PWS following the *Exxon Valdez* oil spill, why only a very small amount remains, and why only very low background concentrations of dissolved components of the oil are found in the groundwater and surface water at locations known to have subsurface oil residue.

We conducted a series of investigations to construct conceptual models of selected shoreline sites within PWS based on common characterization practices used to assess subsurface contamination at other sites. We then selected two sites (Fig. 7.3) to illustrate the important factors that cause oil to persist at those locations and the effects of weathered oil on the groundwater at those sites. The National Oceanic and Atmospheric Administration (NOAA) identified shoreline segments EL056C on Eleanor Island and SM006B on Smith Island (see Map 3, p. vii) as two of the sites containing significant amounts of subsurface oil residue (Michel *et al.*, 2006). Both sites were heavily oiled following the spill, and both sites were among the most well-characterized

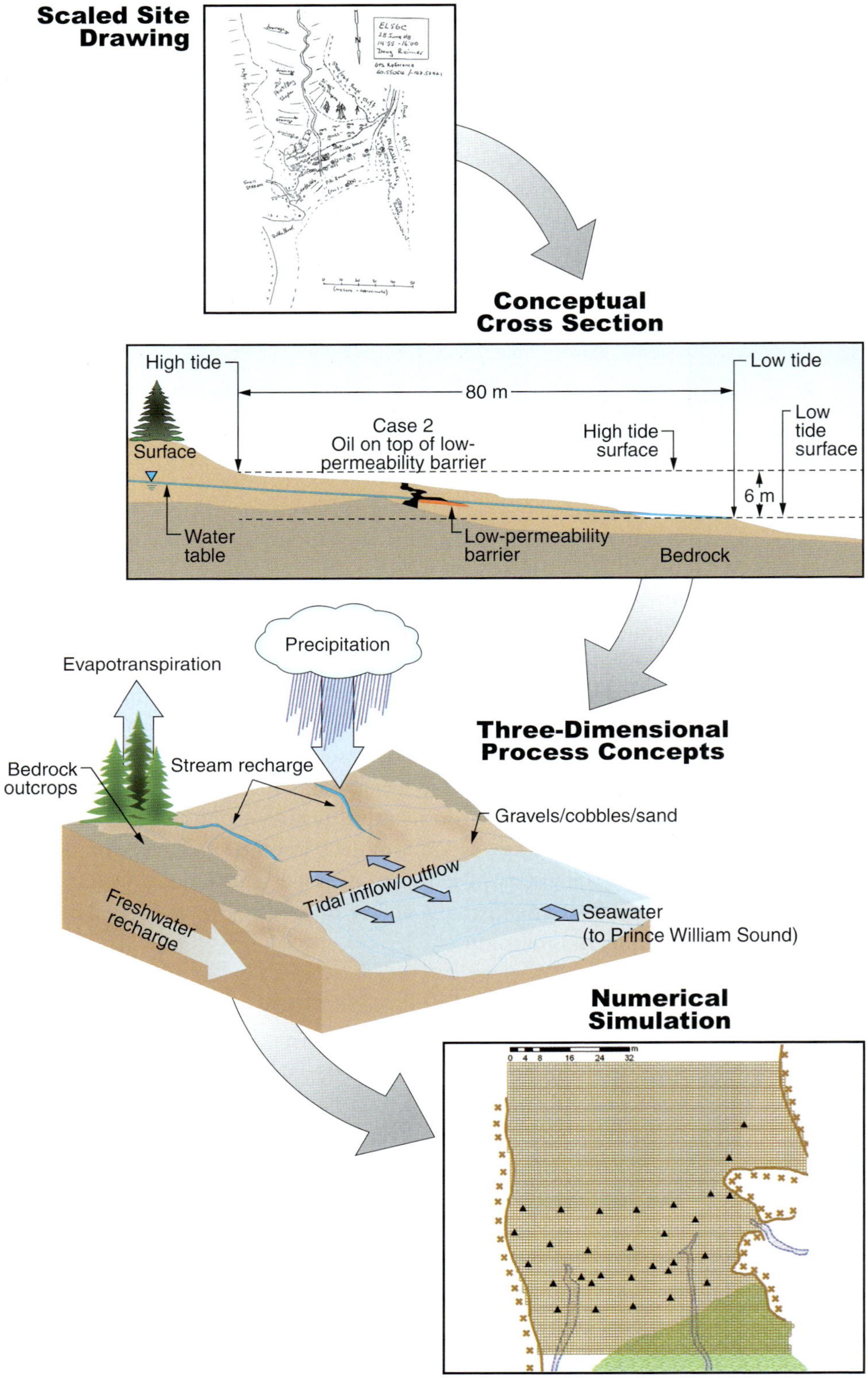

Figure 7.1 The evolution (shown by arrows) of models from qualitative conceptualizations to quantitative numerical models. From upper left: detailed site sketch with measurement locations noted (for illustrative purposes only; see Fig. 7.4 for site details); schematic cross section depicting how oil may have infiltrated into the subsurface; three-dimensional conceptualization that includes boundary conditions and geomorphology; and numerical model plan view.

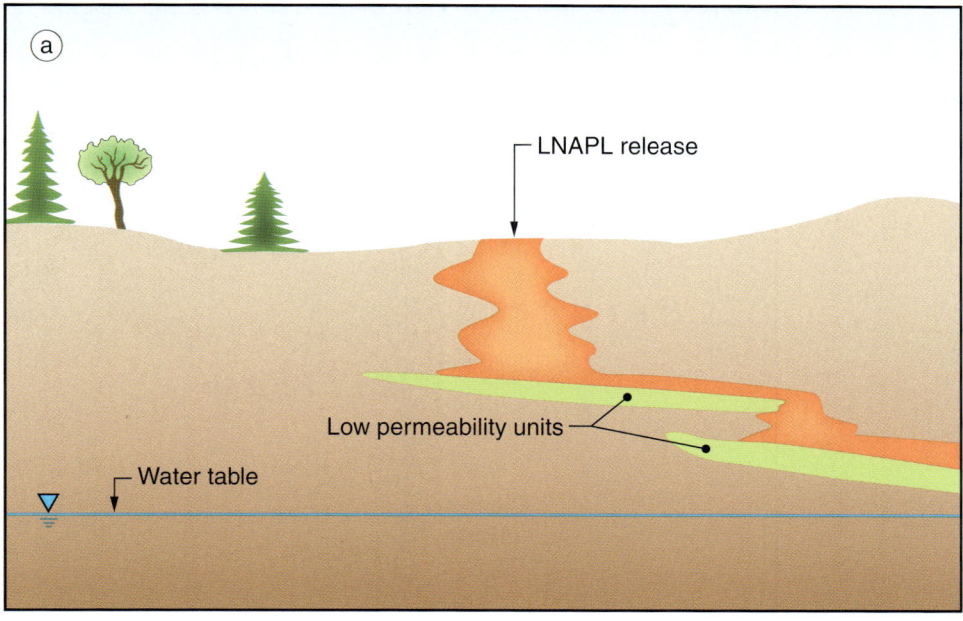

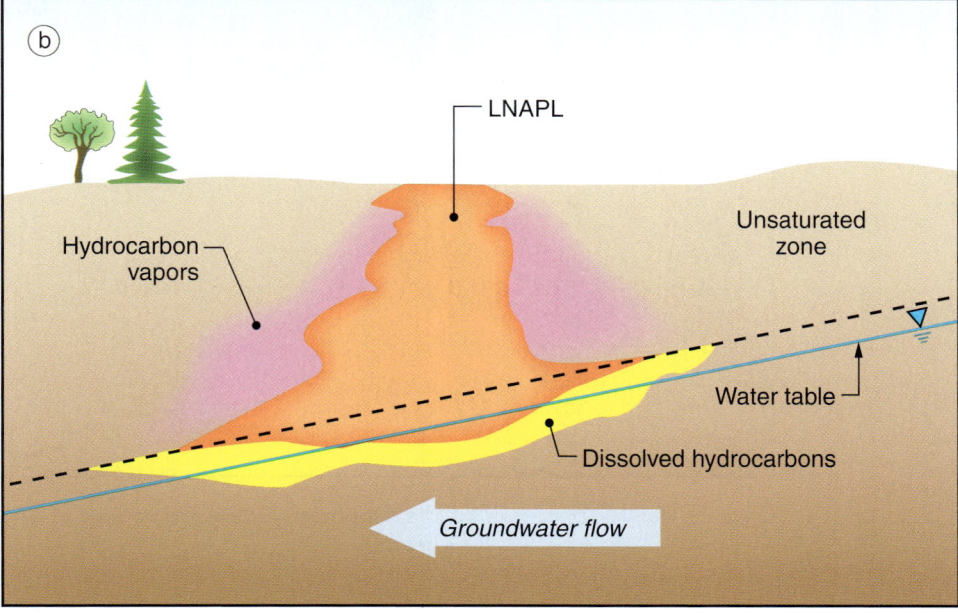

Figure 7.2 Simplified conceptual models for light nonaqueous phase liquids (LNAPL) release into the subsurface and migration. (a) LNAPL migrates downward under the force of gravity until it reaches a low permeability zone that it cannot penetrate or until it reaches the water table, where it spreads laterally down gradient. (b) Weathering occurs through volatilization of constituents in the unsaturated zone and dissolution in the saturated zone.

sites within PWS. Pope *et al.* (2011a, b) provide a description of this work and detailed discussions about these sites.

The two principal objectives of our study were to use both conceptual and mathematical models to (1) improve the understanding and visualization of the hydrogeologic processes that caused some of the oil to be sequestered at some sites but not others, and

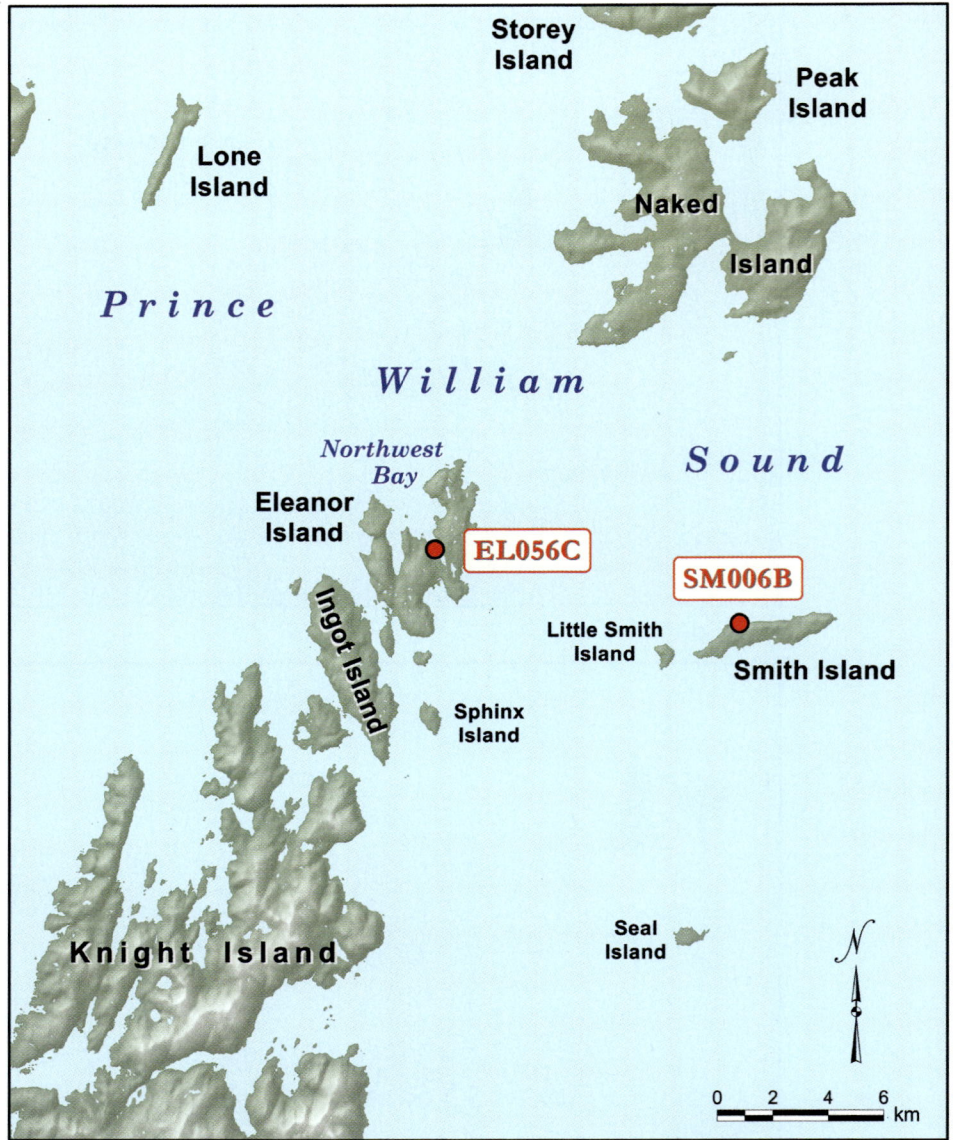

Figure 7.3 Map showing the study locations in this chapter (EL0056C and SM006B).

(2) quantify the polycyclic aromatic hydrocarbon (PAH) concentrations dissolved in the water in the intertidal zone. Models of two shorelines were constructed using detailed data from field programs conducted from shortly after the spill in 1989 to 2008 and 2009, when samples were collected specifically to define the most recent model parameters for this study. Our discussion provides the general context for generic model development and illustrates how these principles were applied to develop conceptual and numerical models for PWS shorelines. As shown in Figure 7.1, we start with individual components of site characterization for conceptual model development, then follow with the integration of the qualitative components of the conceptual model into the numerical simulation model.

7.3.2 Conceptual model of oil migration in the subsurface

Short *et al.* (2004, p. 24) noted that very little oil remained in the subsurface – "the oil remaining is only about 0.14–0.28% of the volume originally beached" – and estimated that it would decrease at an annual rate of ~4% after 2001 (Chapter 6). Boehm *et al.* (2008) further demonstrated the extent to which the oil that persists in the subsurface exists as isolated patches. Based on field and laboratory investigations, these findings can be supported by conceptual models that describe the migration of oil in the subsurface of PWS shorelines. Sediment heterogeneity and extreme contrast in permeability between sediment zones controlled how the oil was able to penetrate the subsurface initially and why it persisted in the sediments at some locations but not others in the years following the spill. Field sampling and measurements focused on determining sediment permeability from grain-size and permeameter measurements and on geomorphological surveys of impermeable features such as outcrops and subsurface bedrock that control the flow at these sites. For the PWS case studies reported here, a 1000-fold variation in the hydraulic conductivity was found between the top sediment layers and the bottom sediment layers (Pope *et al.*, 2011b). With such a large contrast between layers, the groundwater velocity in the lower sediment layers is extremely low.

Shortly after stranding on the shoreline, the oil infiltrated shallow, high-permeability sediments during low tides, when these coarse sediments were not water-saturated. As tides rose, downward migration of the oil was slowed as it encountered the sediments saturated with denser water. Some of this oil was removed through tidal action and cleanup. Some of it drained slowly downward into less permeable sediments, where a small fraction of it still remains in discontinuous lenses trapped within the sediments.

During the wave action of storm activity and the daily tides, more of the oil washed out of the high-permeability zones, while some of the oil remained trapped by capillary forces in small pores. Eventually, the oil trapped in the low-permeability zones slowly weathered, changing composition and properties and forming oil residues. In these low-permeability sediments where this residual oil is sequestered, it is most weathered in more permeable sediment zones and least weathered in less permeable zones, where the groundwater flow is very low (Atlas and Bragg, 2009). Weathering and removal of the residual oil at PWS shorelines occurs through natural processes that include biodegradation (Chapter 8), dissolution of soluble components into groundwater, and reworking of sediments by natural events that include storms and stream action. The term used in this book for this weathered oil is oil residue. "Oil residue" does not have the same meaning as "residual oil," which is the well-established scientific term used by reservoir engineers and environmental scientists for the oil remaining in a pore owing to capillary forces, whether the oil is weathered or not.

Although the conceptual model for oil migration within the subsurface of PWS shorelines can be generalized, subsurface conditions for a particular site determine whether or not subsurface oil persists. Both PWS locations selected for our case study exhibited extremely large permeability contrasts that explained the persistence of the oil in the low-permeability sediments. Most PWS sites lack the highly impermeable sediment layers that lead to such extreme permeability contrasts and, as a result, little if any oil persists at those sites (Chapter 6). It has been estimated that approximately 1% of the 780 km of shorelines originally impacted by the spill contained subsurface oil residue in 2001 (Short *et al.*, 2004). The most impacted shorelines have been studied extensively. Field studies delineated the distribution of residual oil and weathering extent for these sites and found that oil residues persisted in discontinuous lenses below the surface (Boehm *et al.*, 2008;

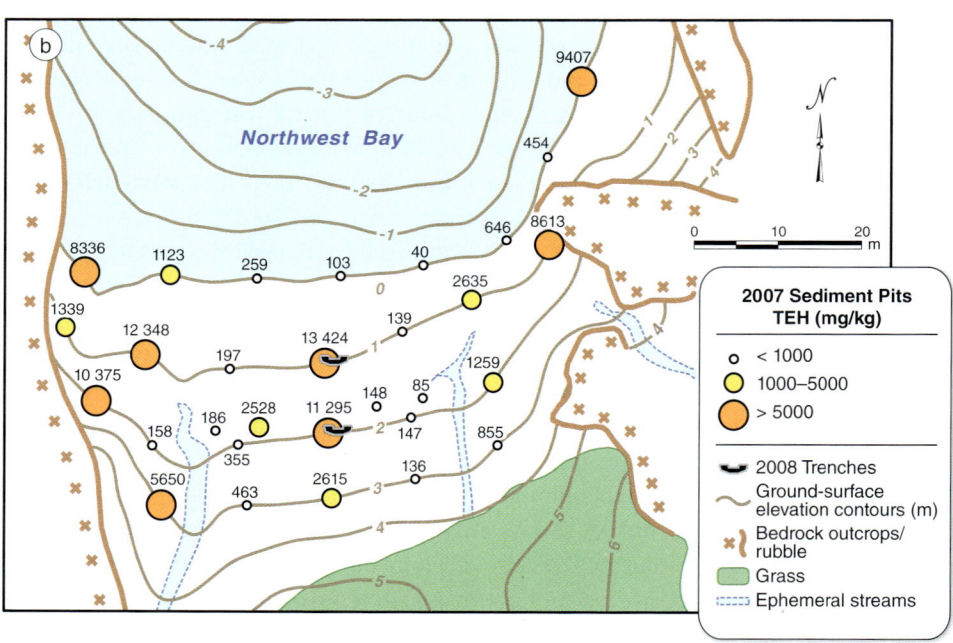

Figure 7.4 (a) EL056C site photo. (b) EL056C site drawing showing surface features, measurement locations, and ground-surface elevation contours (m). (Photo: A.E. Bence)

Page *et al.*, 2008; Taylor and Reimer, 2008). Carls *et al.* (2008) have suggested that the significant risk from the oil is not from the oil residue, but rather from slightly soluble PAH components that dissolve in groundwater in contact with the oil residue.

EL056C is a sheltered (low wave energy) cobble and gravel shoreline within Northwest Bay of Eleanor Island (Fig. 7.4a). The shoreline is approximately 80 × 80 m at low tide and bounded by bedrock outcrops that define and separate the site from adjacent shorelines (Fig. 7.4b). Poorly consolidated coarse sands, gravels, cobbles, and boulders exist below the

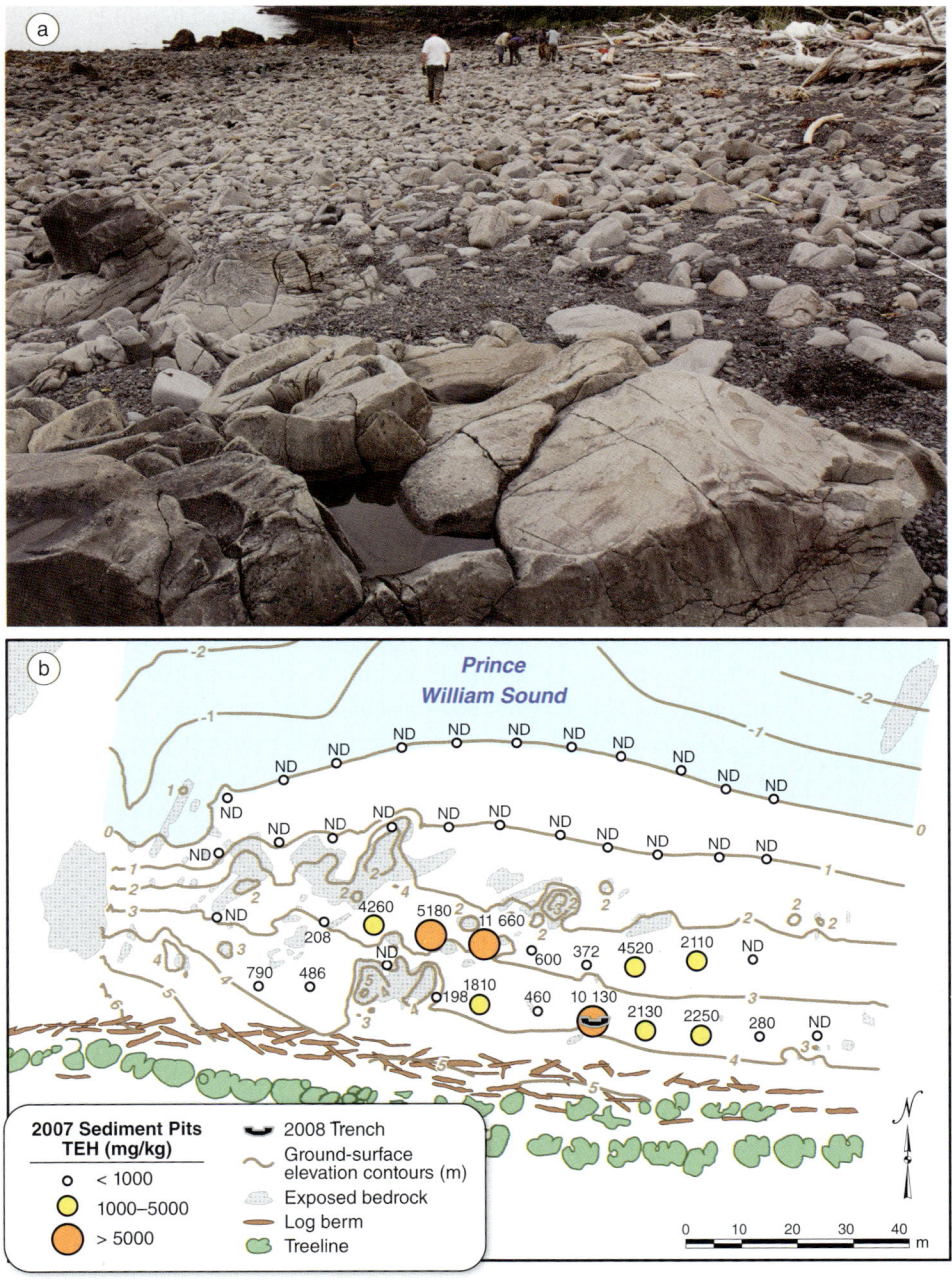

Figure 7.5 (a) SM006B site photo. (b) SM006B site drawing showing surface features, measurement locations, and ground-surface elevation contours (m). (Photo: David S. Page)

surface sediments. Underlying these sediments are more compacted heterogeneous fine sand, gravel, and cobbles. Layered beach structures are common to many mid- and high-latitude coastal environments (Owens *et al.*, 2008). However, this distinctive change in sediment layers at this particular site resulted primarily from more permeable sediments, including cobble and gravel, being deposited on top of subtidal sands and silts that were uplifted during the 1964 earthquake (see Chapter 1, Fig. 1.2; Stanley, 1968).

SM006B is an exposed (high wave energy) boulder/cobble shoreline on northwest Smith Island (Fig. 7.5a). The boulders and cobbles on this shoreline exist as a veneer that overlies more boulders and cobbles intermixed with gravel, sands, and silts above shallow bedrock (Fig. 7.5b). Bedrock that is exposed mainly in the upper intertidal zone creates significant hydrogeologic barriers to groundwater and surface-water flow.

Although the general principles and methods are standardized (US EPA, 2008), characterization of contaminated sites requires the development of site-specific data and measurement techniques. In fact, the unique characteristics and hydrogeologic features of these two sites are the primary reasons for the subsurface-oil persistence at these sites and why only limited amounts of PAH leach from oiled sediments. We performed extensive field and laboratory measurements and collected new types of experimental data for these sites (Pope et al., 2011a). These data include measurements of the solubility of PAH in seawater from column experiments using oiled sediment collected from the sites, as well as estimates of hydraulic conductivity based on measurements collected at multiple sites. These PAH-solubility (partitioning) tests are, to our knowledge, the first to measure PAH concentrations in seawater both at equilibrium and while flowing through sediments oiled by a coastal spill.

In 2007, the lateral extent of subsurface oil at both EL056C and SM006B was mapped using gridded pit surveys (Boehm et al., 2008; Page et al., 2008): evenly spaced pits were dug along constant-elevation transects, and sediment samples were collected and analyzed for total extractable hydrocarbons (TEH) and individual oil components, including PAH. Figures 7.4b and 7.5b show the general locations of the pits and the TEH concentration measured for each pit where samples were collected in 2007.

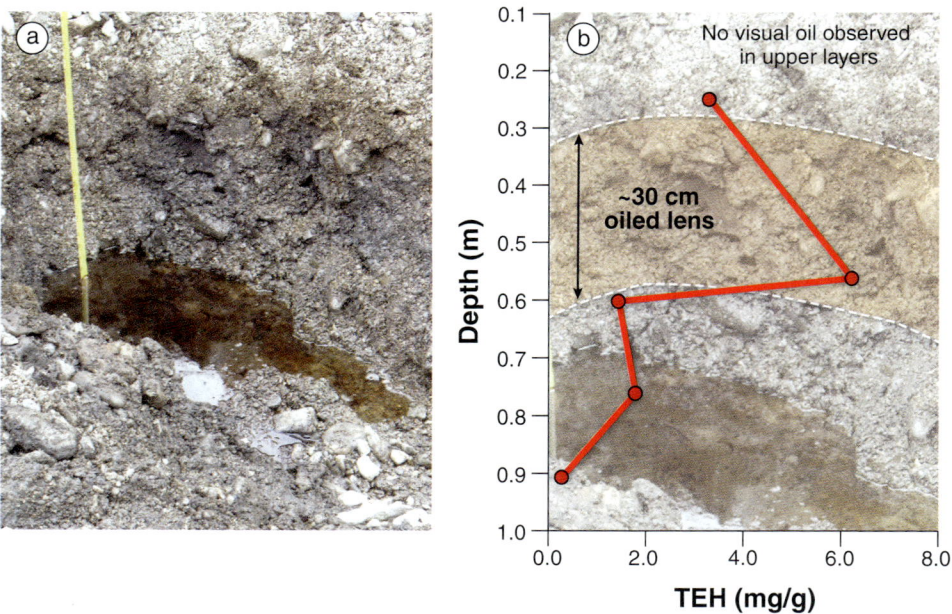

Figure 7.6 (a) Photo of trench dug at EL056C. (b) Trench sidewall with an overlay of measured TEH sediment concentrations showing occurrence of typical oiling in the subsurface. The oiled sediments shown in this figure are typical of oil residues found in the subsurface on shorelines with persistent oil. (Photo: A.E. Bence)

For both locations, no oil was observed in the very permeable surface sediments. Where residual oil was observed within the underlying sediments, it was associated with finer-grained sediments rather than coarse overlying sediments. During subsequent studies in 2008, trenches were excavated at both sites to observe the vertical profiles of oil content in the subsurface. Observations in these trenches indicate that vertical migration of the oil in the subsurface was limited by the permeability of the underlying materials, either finer-grained sediments or bedrock. Figure 7.6 shows an overlay of measured TEH concentration (in mg/kg of sediment) with a photograph of a trench sidewall where an approximately 30-cm thick lens of sediments containing weathered residual oil was found. The oiled sediments shown in the photograph illustrate the weathered oil that exists as an immobile fraction typical of those trenches and pits where oil residue is found (Michel et al., 2006; Taylor and Reimer, 2008).

Based on these data, we developed three-dimensional heterogeneous models for the sites. These models included three-dimensional realizations of the parameters characterized during the field studies. Sediment descriptions and permeability measurements were correlated using geostatistics, which uses statistics developed for spatial and/or temporal datasets. Three-dimensional hydraulic-conductivity distributions were developed for both shoreline segments that represented the unique variability in permeability for each site. Similarly, spatial distributions of total PAH (TPAH) in the oil residue at both sites (Fig. 7.7) were developed using both the TEH from the sediment samples and the TPAH found in the extracted oil fraction (Pope et al., 2011a). We then used these models to expand on previous studies (Li and Boufadel, 2010; Guo et al., 2010; Xia et al., 2010) to better understand the persistence of the oil and its impact on the groundwater.

The ways in which oil is sequestered and persists in the subsurface are locally controlled because of the extreme variability of individual sites. How oil is sequestered explains both why oil persists in some locations and why slow dissolution of PAH from this weathered oil into groundwater results in negligible concentrations in the near-shore water column. Results of the column studies show that PAH components of the *Exxon Valdez* oil remaining today have very low solubilities in water (< 10 µg TPAH/L; Pope et al., 2011a) and, therefore, very low bioavailability.

Figure 7.8 compares the TPAH content in weathered oil on sediments collected at site EL056C and used in column tests with the TPAH content in water reaching equilibrium with the oil residue. Concentrations in water are very low because of the very low solubility of the PAH remaining in the oil residue. The ratio of TPAH in the oil to the TPAH in water gives the partition coefficient that is input into the numerical model to predict the TPAH in groundwater flowing through the oiled sediment. Numerous surveys conducted within PWS show that TPAH concentrations dissolved in the near-surface groundwater and nearshore surface water were at background levels ranging from nondetect (< 0.001 µg/L) to 0.94 µg/L as early as 1990 (Boehm et al., 2007). Indirect TPAH intertidal water-concentration estimates derived from mussels (*Mytilus trossulus*) (average 0.001 µg/L; Boehm et al., 2007) are consistent with these findings.

The difference between the TPAH concentrations in pore water in contact with oil residue and the background-level TPAH concentrations observed in the near-surface groundwater and nearshore surface-water field measurements is readily explained by the conceptual model. Estimates of hydraulic conductivity show that fast groundwater-flow paths exist in the near-surface sediments compared to the underlying sediments.

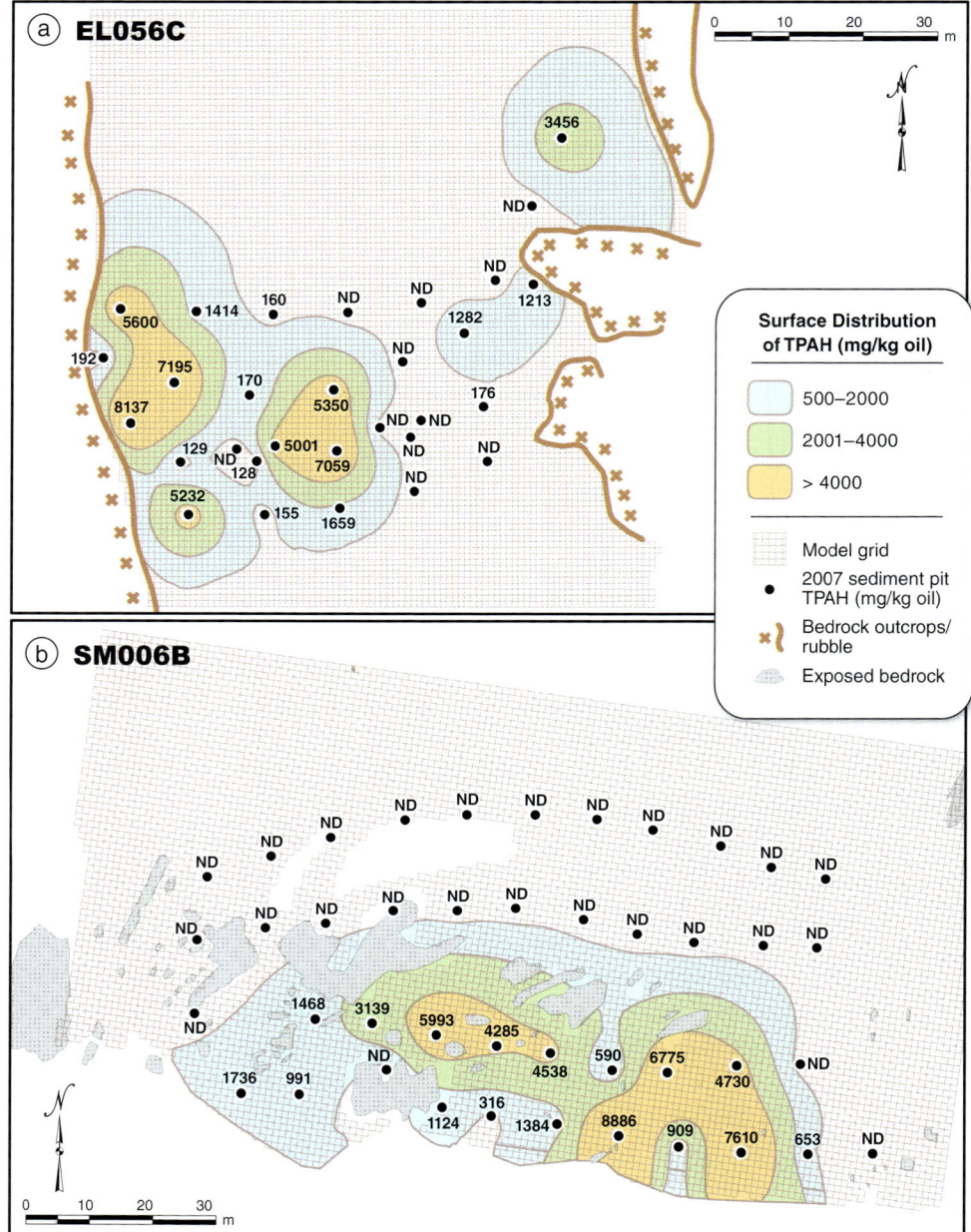

Figure 7.7 Subsurface distribution of TPAH concentrations (mg TPAH/kg oil) used to assign aqueous TPAH concentrations to (a) EL056C and (b) SM006B numerical model constant concentration cells. Aqueous TPAH concentrations were assigned to each grid block based on measured TPAH concentrations in the oil (shown here) and partition coefficients estimated from the column studies. ND = TPAH not detected.

The bulk of the water flowing in and out of the shorelines during each tidal cycle is within these fast-flow paths. As PAH components slowly leach from the adjacent, low-permeability lenses containing residual oil, the dissolved concentrations are rapidly diluted by tidal waters flowing in the shallow, high-permeability pathways. If the contact time between the water and oil is limited, the amount of each PAH component

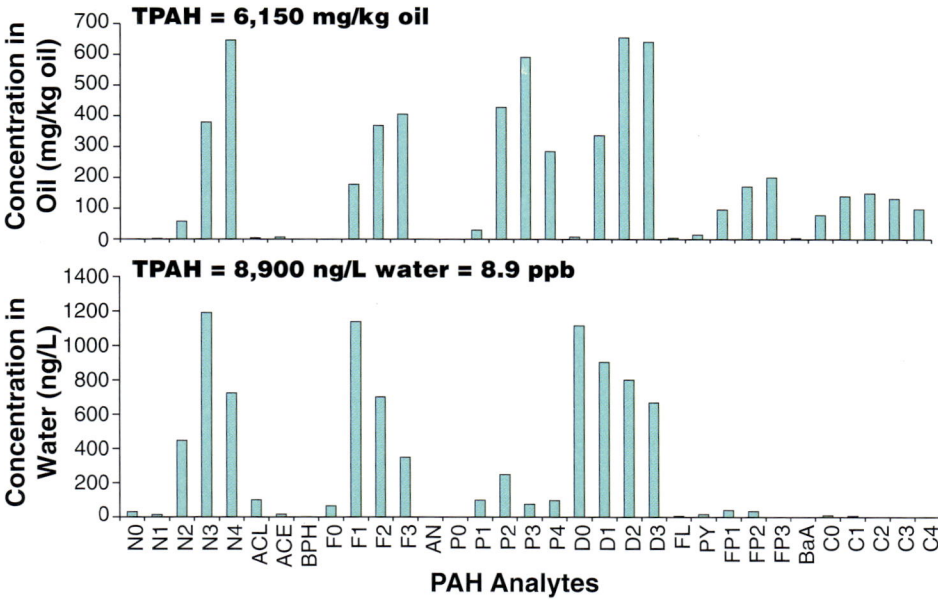

Figure 7.8 Concentration of PAH analytes in oil residue on EL056C sediments in column test and in equilibrated water. Note concentration scale change. Analytes heavier than dibenzothiophenes (D0-D3) are so insoluble that concentrations in water are negligible. For PAH analyte abbreviations see Table 7.2 in Atlas and Bragg (2009).

that dissolves into the water will be less than its equilibrium concentration because there will not be sufficient time for the PAH components to diffuse through and out of the oil into the flowing pore water. This is called *rate-limited mass transfer*, and it slows down the dissolution of the PAH components in the oil. In our simulations, however, we used the equilibrium partition coefficients measured from the column experiments. Hence, the simulated TPAH concentration may be higher than the actual TPAH concentration in the pore water.

7.4 The mathematical models

Mathematical models are quantitative models based on one or more equations that describe the conceptual model that includes the biology, chemistry, geology, and physics of an idealized system, subject to specified assumptions. In our case, some of these equations are differential equations with associated initial and boundary conditions. If these differential equations are simple enough that they can be integrated analytically, then the model is said to be an *analytical model*. Charbeneau (2000) developed and applied analytical models to groundwater hydraulics and pollutant transport. A more complete description will generally lead to a set of coupled, nonlinear partial differential equations that must be integrated numerically: this is a *numerical model*.

Analytical models are useful tools for the study of interrelationships among the important variables identified by the conceptual model. Many analytical models can be as simple as a single algebraic equation or as complex as a spreadsheet analysis. Simple calculations can sometimes be used to estimate how far oil has penetrated into the subsurface following a spill, how far the oil might move laterally, and how long it might take the oil to degrade. As an example, we used an analytical model as an aid in understanding the persistence of patches of oil residues at some PWS shorelines and the partitioning of PAH components from these residues to the pore water.

Numerical modeling of oil reservoirs has a long history, starting in the 1950s (Douglas *et al.*, 1959) and reaching maturity by the 1980s (Coats, 1982). These numerical models were based on the theory of multiphase flow in permeable media developed by Muskat (1949) and others to predict gas and oil production from petroleum reservoirs. Three-dimensional, multiphase, contaminant-transport models have been developed based on these same principles, but include special features needed to predict flow and transport of contaminants in groundwater (Abriola and Pinder, 1985; Pruess and Narasimhan, 1985; Abriola, 1989; Delshad *et al.*, 1996; Pruess, 2004; Panday and Huyakorn, 2008).

7.4.1 The analytical model

Although we used numerical simulations (described below) as the primary modeling tool, we also used analytical models to understand subsurface processes pertinent to the PWS shorelines. We illustrate this approach with a simple model used to help us understand the partitioning of PAH components from oil residue to seawater.

It was not computationally practical to simulate the partitioning of all of the PAH components over the thousands of tidal cycles since the spill. However, it was easy to use the analytical solution to provide quantitative insight for the entire time period. The analytical solution can also be used for long-term predictions. It also has the advantage that it is independent of the details of the hydrogeology and all of the uncertainties in the measurements and models, as well as the extreme heterogeneity within and among sites.

The analytical model can be used to calculate approximate aqueous concentrations for each of the PAH components, based on widely different partition coefficients measured from laboratory column experiments assuming equilibrium partitioning between the oil and water. The equilibrium partition coefficient K_i for component i is defined as

$$K_i = C_{io}/C_{iw}$$

where C_{io} is the concentration of a component in the oil phase and C_{iw} is the concentration of a component in the aqueous phase.

With a few reasonable approximations, such as local equilibrium, a simple mass balance yields the following equation using the K_i and the oil saturation, S_o, to determine the fraction of PAH component i (F_{in}) remaining in the oil after n extractions or n tidal cycles (two times the number of days):

$$F_{in} = \left(1 - \frac{1}{K_i S_o}\right)^n$$

For large K_i values, a good approximation to this equation is:

$$F_{in} = \exp\left(\frac{-n}{K_i S_o}\right)$$

The fraction depleted is $1 - F_{in}$.

Table 7.2 Comparison of measured values of remaining oil fractions using sediment from SM006B with values calculated from an analytical model.

Component	PAH conc. in sediment oil, C_{io} (mg/kg oil)	PAH conc. in water, C_{iw} (ng/L)	Partition coefficient $K_i = C_{io}/C_{iw}$	PAH conc. in Exxon Valdez crude oil (mg/kg oil)	Measured remaining fraction in the oil, F_{in}	Calculated remaining fraction in the oil, F_{in}
Naphthalene	10	55	178 000	820	0.008	0.065
C1-Naphthalenes	34	111	303 000	1670	0.015	0.200
C2-Naphthalenes	183	818	223 000	2200	0.061	0.113
C3-Naphthalenes	696	1271	548 000	1690	0.303	0.411
C4-Naphthalenes	784	482	1 620 000	986	0.560	0.741
Fluorene	23	229	101 000	100	0.164	0.008
C1-Fluorenes	145	532	273 000	258	0.396	0.168
C2-Fluorenes	333	426	780 000	372	0.630	0.536
C3-Fluorenes	473	282	1 675 000	431	0.773	0.748
Phenanthrene	87	278	313 000	301	0.204	0.212
C1-Phenanthrenes/Anthracenes	426	463	922 000	662	0.454	0.590
C2-Phenanthrenes/Anthracenes	776	241	3 220 000	774	0.706	0.860
C3-Phenanthrenes/Anthracenes	651	84	7 767 000	548	0.836	0.939
C4-Phenanthrenes/Anthracenes	377	77	4 872 000	309	0.859	0.905
Dibenzothiophene	90	422	213 000	261	0.243	0.102
C1-Dibenzothiophenes	395	405	975 000	527	0.527	0.607
C2-Dibenzothiophenes	748	292	2 561 000	682	0.773	0.827
C3-Dibenzothiophenes	741	188	3 952 000	599	0.871	0.884
Pyrene	19	12	1 612 000	15	0.903	0.739
C1-Fluoranthenes/Pyrenes	111	38	2 923 000	86	0.901	0.847
TPAH	8285	6778	1 222 000	14 342	0.407	0.423

The results shown in Table 7.2 illustrate the model calculations for the PAH components measured in the column experiment using sediment from SM006B collected in 2009. Because this sediment sample was taken 20 years after the spill, $n = 14\,600$ (20 years × 365 days/year × 2 tides/day). A value of 0.03 (typical of that measured for PWS sediment samples from this site) was used for the oil saturation in this example calculation. For comparison with the values computed from the model, the experimental values for the remaining fraction of each PAH component in the oil sample compared to the original *Exxon Valdez* crude oil are also shown in Table 7.2. The experimental values were normalized by the measured concentration of C_{29}R-stigmastane, which is a conserved marker (Atlas and Bragg, 2009; Chapter 8 and Box 8.1) in both *Exxon Valdez* crude and the sediment sample.

Actual remaining fractions of PAH vary from the predictions for several reasons: heterogeneity of the original oil stranded on the shorelines could cause differences in the overall composition in the oil from location to location even on the same shoreline; biodegradation in the water might in some instances enhance the rate of dissolution; or rate-limited mass transfer can occur either by nonequilibrium mass transfer from the oil to the pore water or by incomplete transport from the pore water to the surface water. The experimental values for the aqueous concentrations are also uncertain because of the very low solubilities of the individual PAH components. These factors can be used to explain why some observed values are smaller than the calculated values and why some are larger.

The fraction of the TPAH remaining at any given time is the sum of $C_{io} F_{in}$ divided by the sum of C_{io}. For the example shown in Table 7.2, the remaining fraction of TPAH calculated using the analytical model is 0.42. Considering all of the approximations in the model and all of the uncertainties in the data, this value is remarkably close to the observed value of 0.41.

The partition coefficient for the TPAH has increased with time, as the oil residue has become partially or totally depleted of more soluble PAH components. Thus, the TPAH solubility in water is now much lower than the original value. As the more soluble components of the oil residue continue to be depleted, the PAH concentrations in water will continue to decrease. Numerous measurements on samples from different sites taken at different times and measured by both batch and flow experiments clearly established this trend. The model can also be used to quantitatively predict this trend.

The analytical model provides support for the partition coefficient values selected for use in the numerical model, and thus for the very important conclusions based on the numerical simulation results. If the partition coefficient for the TPAH were, for example, ten times smaller, then the solubility in seawater would be ten times larger. If the dissolved PAHs based on the smaller K values either biodegraded in the water or were transported to the surface water, then a calculation using the analytical model would show that the same sediment sample (Table 7.2) would have been depleted of all but 1.8% of the original TPAH, which is far lower than the observed value even taking into account the large uncertainty in the measurements. A factor of ten is much larger than the uncertainty in the measured partition coefficients, and a smaller factor does not make enough difference to change the conclusions based on the numerical simulations. As shown below, the numerical simulations using these K values explain why the TPAH concentration in the surface water is extremely low and thus the oil residue is effectively sequestered.

7.4.2 The numerical model

The development of numerical models involves the integration of qualitative information from conceptual models with quantitative data measurements. A general numerical model of flow and transport in permeable media requires a set of partial differential equations, as well as a set of auxiliary equations representing the chemical and physical property relationships needed to describe the particular phenomena of interest. Initial and boundary conditions are needed for a complete solution of these equations. These equations are highly nonlinear for multiphase flow, and thus numerical methods must be used to solve them. Variations in geologic properties (e.g., permeability and porosity) and fluid properties (e.g., density and viscosity) add to the complexity of the equations. They must be accounted for, however, as these variations have a significant influence on flow and transport. Distributed-parameter models allow realistic distributions of these properties. Numerical methods are used to determine the solutions to the differential equations using finite-element or finite-difference methods (Anderson and Woessner, 1992).

Hydrogeologic systems are complex, three-dimensional, and heterogeneous. Variability in the system strongly influences groundwater flow and transport, so the hydrogeologic parameters like those listed in Table 7.1 should be carefully and adequately described. Because of this heterogeneity, however, uncertainty about the properties and boundary conditions always exists. Numerical uncertainty is introduced through effects such as discretization of the modeled domain and the solution technique for the equations. Unrecognized errors and uncertainty in the modeling process can affect the quality of the solution. The best modelers appreciate the limitations of their models and account for this when applying them to field problems by understanding the qualifying assumptions, the uncertainties of the data, and the errors associated with numerical simulation.

The first step in the modeling process involves the determination of whether or not numerical simulation is actually needed. Can simple analytical models represent the problem, or is a more complex mathematical representation needed? Hill (2006) encourages practitioners to start simple and add complexity when developing numerical models. Questions to ask before the simulation process begins might include:

- What is the level of complexity required?
- What are the important drivers of the flow and transport processes?
- What are the important drivers in terms of risks?
- What type of simulator is needed?
- What is the uncertainty associated with the model parameters?

For special cases such as immobile oil in a saturated zone, the problem is greatly simplified because the numerical model only needs to include single-phase flow of water. Several authors have discussed the formulation of numerical methods (e.g., Hill, 2006). Mercer and Faust (1980) provide discussions of finite-element and finite-difference methods and various solution techniques for the system of partial differential equations needed to model flow and transport in a permeable medium.

Understanding the uncertainties and assumptions associated with numerical modeling allowed us to develop the most appropriate realizations for selected PWS shorelines. Because of the morphology of the shorelines and the heterogeneity of the sediments and the oil residue itself, a three-dimensional representation was needed. Because rising and falling tides create levels of variable saturation in the subsurface, we needed a multiphase model. However, because all observations indicate that the oil remaining in the subsurface at the two locations selected for numerical modeling is immobile or at residual saturation, we can model the oil as a fixed source rather than a separate phase.

Simulators have been developed to model flow and transport in permeable media. We used MODFLOW–SURFACT as the simulator for the PWS studies. It is a finite-difference numerical simulator for variably saturated flow and transport in three dimensions with a full set of essential multiphase flow features. MODFLOW–SURFACT is based on the US Geological Survey's groundwater flow and transport code MODFLOW and includes additional modules that allow for more robust and efficient numerical solution of multiphase flow and transport equations for three-dimensional, highly heterogeneous porous media. The simulator is capable of handling the time-variant boundary conditions associated with rising and falling tides, and variable grid specifications allow for realistic geometries. MODFLOW–SURFACT includes robust rewetting and drying schemes, adaptive time-stepping solutions, and Newton–Raphson iterative solutions of highly nonlinear equations (Panday and Huyakorn, 2008). Pope *et al.* (2011a) provide a more complete discussion of the model development and results for the PWS case studies described.

7.4.3 The *Exxon Valdez* oil spill case study
7.4.3.1 Mathematical models of PAH dissolution in the subsurface

One of the objectives of these case studies was to predict quantitatively the PAH concentrations dissolved in the water in the intertidal zone. Based on experimental field measurements, dissolved PAH in intertidal waters were determined to be at background concentrations (Boehm *et al.*, 2004). However, oil residue was known to be present at some locations, including EL056C and SM006B, where these background-level measurements were collected. The conceptual model describes the processes whereby the PAH components are dissolved from the oil residue and rapidly diluted by tidal infiltration. Quantifying the amount and distribution of dissolved PAH depends on the spatial variability in the sediments and the oil residue itself. Because of spatial variability, an adequate representation of the site conditions at EL056C and SM006B requires a three-dimensional simulation with spatially variable properties that include ground-surface and stratigraphic elevations, hydraulic conductivity, distributions of projected dissolved PAH concentrations, and irregular boundary features.

Based on the estimates of hydraulic conductivity, the conceptual model of groundwater flow included faster, more permeable groundwater-flow paths in the near-surface sediments as compared to the underlying sediments. The simulated groundwater velocities showed a large contrast between upper and lower sediments for the Eleanor (EL056C) and Smith Island (SM006B) sites, consistent with observations and measurements showing that oil in the more permeable sediments washed away while patches of oil residue persisted in less-permeable sediments. In contrast, sites that do not have such large variation in permeability have been observed to have very little if any oil residue remaining, even though they were heavily oiled in 1989 (e.g., Point Helen, southeastern Knight Island; see Map 3, p. vii; Chapter 6).

Figure 7.9 shows the relative magnitudes of the average groundwater velocity vectors in a model cross-section for EL056C. Minimum groundwater velocities calculated in the least permeable sediments were more than 100 000 times less than maximum velocities calculated in the most permeable sediments, which indicates significant water stagnation zones in the subsurface. The simulations for SM006B showed similar contrasts in the modeled flow velocities between the cells representing oiled sediments and the surface sediments.

In addition to the permeability contrasts between the overlying and underlying sediments, impermeable features such as bedrock and large boulders control the groundwater flow. Although this control is observed at many locations throughout PWS (Hayes and Michel, 1999; Page *et al.*, 2008), it is well illustrated by the flow-modeling results for SM006B. Figure 7.10 shows simulated water saturations during two different tidal elevations at SM006B. Bedrock shadow zones create areas of low

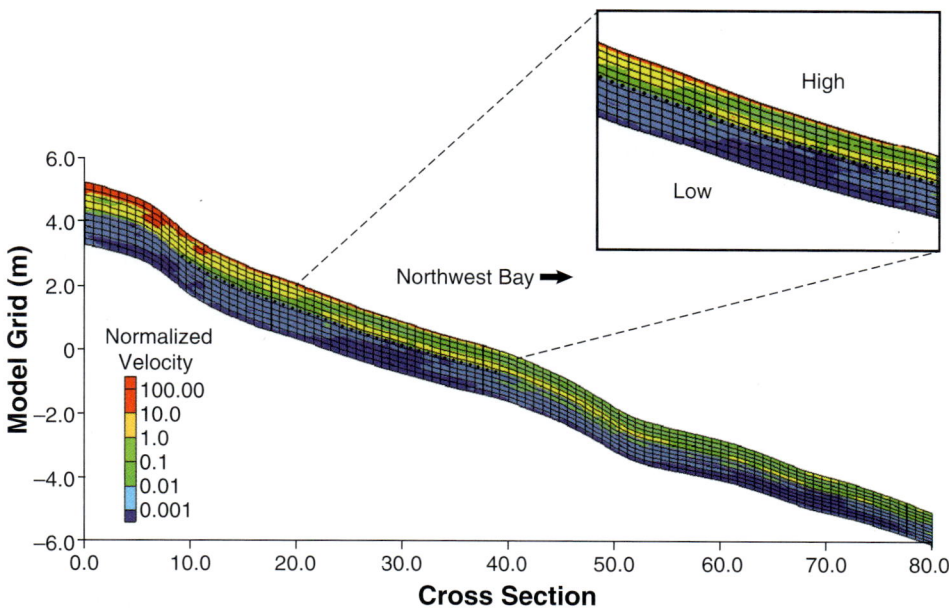

Figure 7.9 EL056C model cross section along the Y-axis of the model grid showing extreme variations in groundwater velocity normalized to the average velocity in the subsurface cells (water cells are hidden). Higher velocities are simulated for the upper layers where more permeable sediments occur. Lower velocities are simulated in the underlying low-permeability sediments where oil residue persists. Note that the groundwater velocities are not zero in these layers, suggesting the potential for weathering to continue. (m = meters above mean sea level.)

energy that drain slowly and are sheltered from tidal action that reworks sediments and washes away residual oil trapped in those sediments.

These simulations showed the effects of extreme heterogeneity within sediments observed at both locations. The three-dimensional complexity at these sites demands this type of model to properly demonstrate how fluid movement is governed.

Simulated transport of PAH dissolved in groundwater also shows the effect of extreme permeability contrasts. Seawater infiltration during the daily tidal cycles rapidly dilutes any dissolved PAH that leach from the residual oil lenses into the groundwater. The lateral extent of the oil residue lenses remaining at EL056C and SM006B has been well characterized by numerous surveys (Boehm *et al.*, 2008; Page *et al.*, 2008; Taylor and Reimer, 2008). However, the depth of these lenses is not as well characterized, so a constant depth from the ground surface to the top of the oiled zone was assumed for both models. As shown in a cross section for EL056C (Fig. 7.11a), with some concentrations greater than 4000 mg TPAH/kg of extracted oil (see Fig. 7.7a), simulated dissolved TPAH concentrations around 7 µg/L adjacent to subsurface oiled sediments (near the fourth layer shown on the section) approached the maximum solubility derived from the column experiments (around 10 µg/L) for oiled sediments collected at EL056C. However, in areas near the surface and in the lower intertidal zone where organisms could be affected by dissolved PAH, the simulated daily time-weighted mean TPAH groundwater concentration was 0.08 µg/L, which is within the background range for TPAH concentrations found in the PWS nearshore water column. The model clearly shows very high dilution, consistent with both direct and indirect measurements of dissolved TPAH observed at the ground surface and in nearshore waters.

Saturation (fraction) at 1 m-tide elevation (falling)

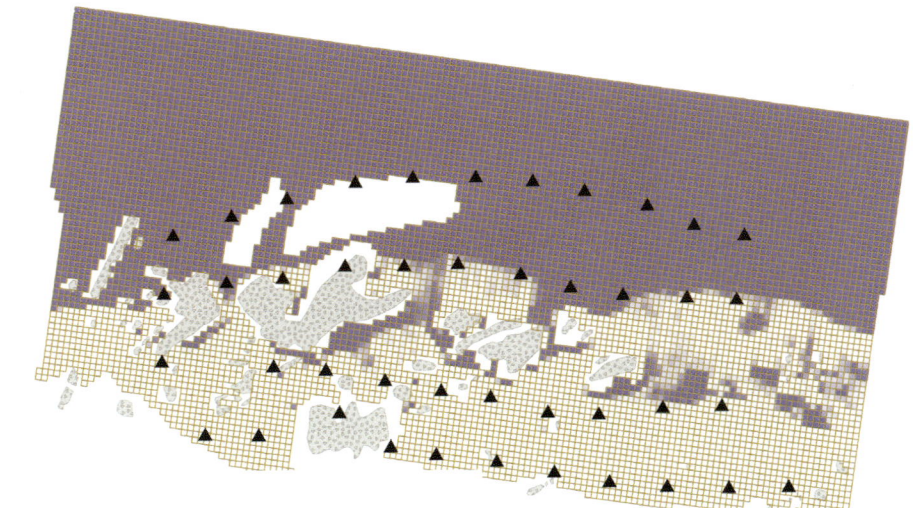

Saturation (fraction) at 0 m-tide elevation (falling)

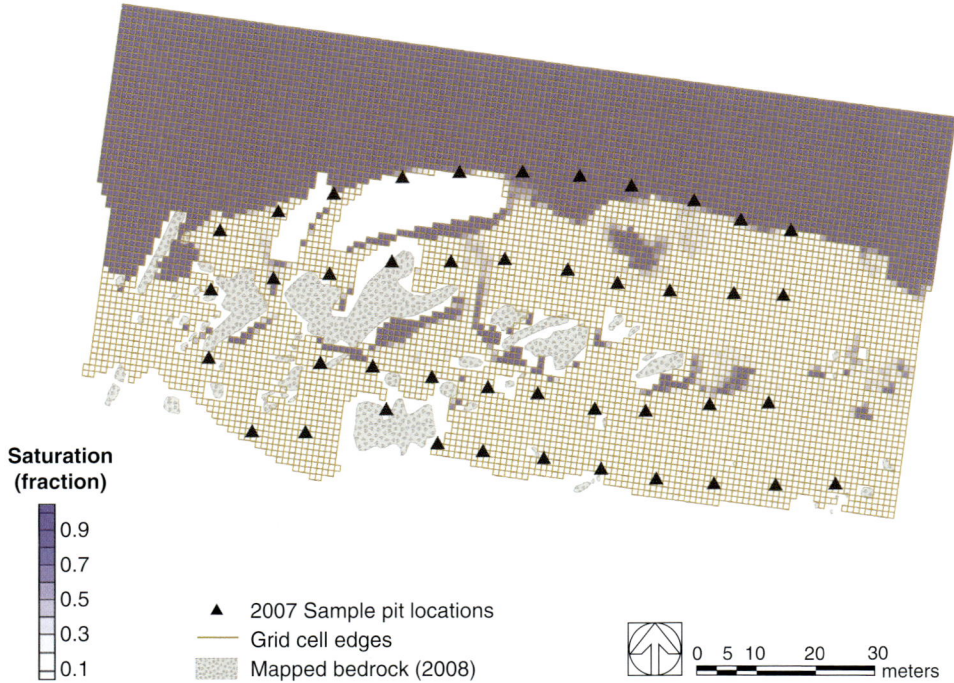

Figure 7.10 Contour maps of SM006B water saturation for two different tidal elevations at +1 m and at 0 m. During low tides, groundwater drains more slowly from behind bedrock outcrops. The outcrops form tidal-break "shadow zones" that establish a lower energy groundwater-flow environment.

Simulation results also indicated cyclical variability in the TPAH concentrations, where dissolved concentrations increased as tidal elevations decrease (Fig. 7.11b).

A further reduction in the concentrations of dissolved PAH would be expected from natural biodegradation, which was not included in these simulations to allow a more

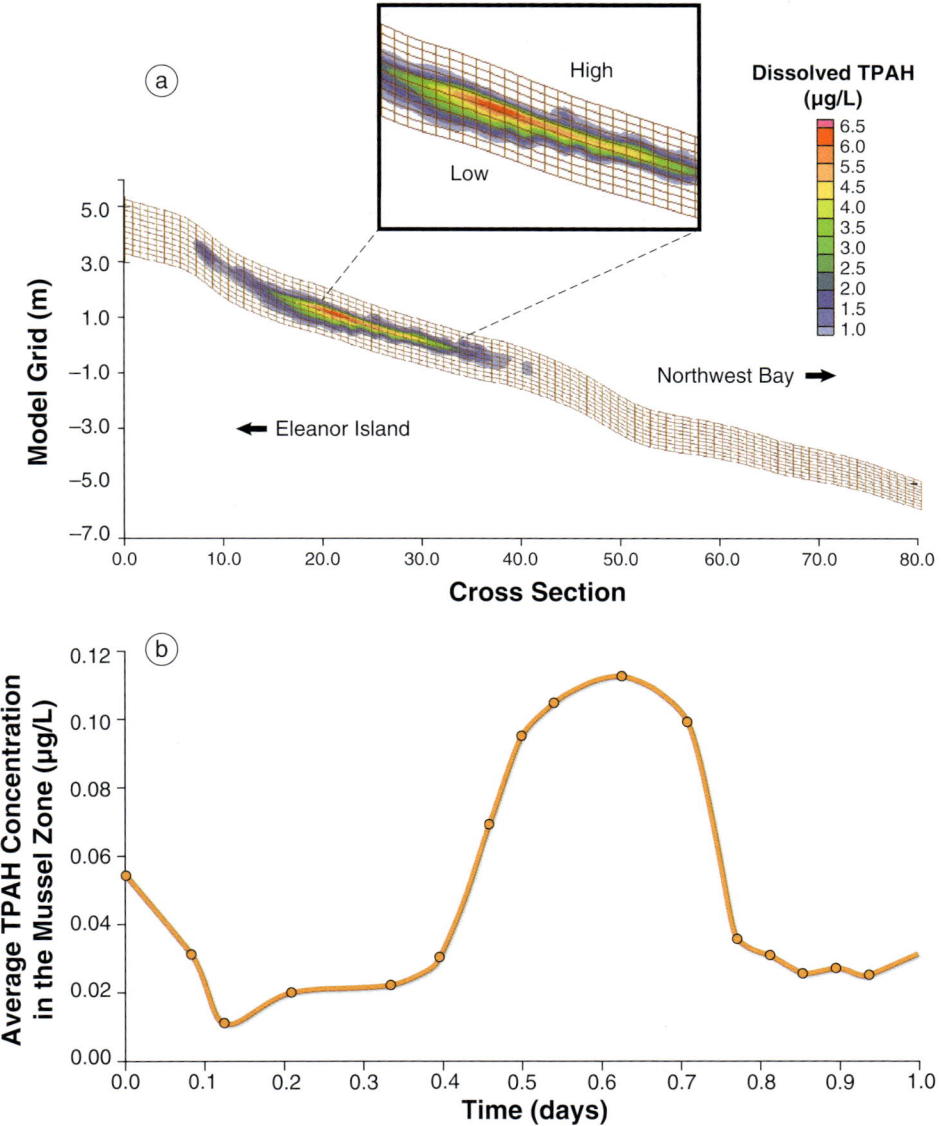

Figure 7.11 (a) Model cross section along the Y-axis of the model grid of EL056C simulated aqueous TPAH (mg/L) averaged over a single tidal cycle (1 day). The model plan view is not shown because dissolved concentrations at ground surface are < 0.5 mg/L. (m = meters above mean sea level.) (b) Average simulated aqueous TPAH (mg/L) for grid blocks between 1.0 m and 1.5 m elevation showing transient variability over the period of a single day's tidal cycle. Simulated dissolved TPAH concentrations in the cross section are highest adjacent to subsurface oiled sediments and are within background range in areas near the surface and in the lower intertidal zone.

conservative approach. Many flow-and-transport simulators do include biodegradation (Delshad *et al.*, 1996; Chapter 8). Water flowing through the high-permeability layers already contains sufficient natural nutrients and dissolved oxygen to support active biodegradation of PAH if the oil residue were to be in contact with the groundwater. Studies conducted by Venosa *et al.* (2010) confirm that the oil would be rapidly degraded without needing application of exogenous nutrients (Chapter 8).

Biodegradation of the PAH in weathered oil is slowed by the limited solubility of PAH in water. However, once dissolved, the PAH reaching the water flowing through the high-permeability zone containing the necessary nutrients and dissolved oxygen would be rapidly consumed by hydrocarbon-degrading bacteria.

Dilution of dissolved TPAH concentrations that leach from oiled source zones was also observed in the simulations for SM006B. These simulations illustrated the hydraulic control that bedrock outcrops exhibit over the groundwater movement (Fig. 7.12a). The highest simulated time-weighted mean TPAH concentrations occurred in the shadow zones behind bedrock outcrops (the break in the model grid blocks on the cross section; Fig. 7.12b). The results show that dissolved TPAH concentrations are diluted by almost two orders of magnitude, from greater than 5.0 µg/L aqueous TPAH near buried oiled-subsurface sediments in the upper intertidal zone to concentrations less than background concentrations (< 0.06 µg/L aqueous TPAH) in the lower intertidal zone.

7.4.3.2 Sensitivity analyses

Sensitivity of the model results to various parameters was determined by varying the model input. These parameters included absolute differences in permeability of sediments, heterogeneity of sediments, amounts of recharge from streams and inland sources, density differences between fresh and saline groundwater, and concentrations of PAH dissolving from the oil residue. Results of these simulations showed that values for the parameters used in the model were within reasonable ranges and that the controlling factors were the permeability contrasts, volume of water from the tidal influx, and the amount of PAH dissolving from the subsurface oil. Although there are uncertainties associated with some data, the likely ranges of uncertainty do not change the final answer significantly and are in agreement with field measurements.

7.4.3.3 Conclusions from numerical simulations

The physical principles applied to develop the conceptual and mathematical models that explain why oil residue remains at some shorelines in PWS are well-understood concepts that have been applied under many scenarios for many reservoir engineering and environmental problems. Each of the models (conceptual, analytical, and numerical) supports the conclusions that residual oil is trapped in low-permeability sediments and that dissolution of PAH from that oil residue is minimal and rapidly diluted as they leach into more permeable sediments. The numerical simulation results provide a quantitative explanation for why oil can persist in some locations and the TPAH concentrations are nevertheless at or near background levels in the lower intertidal zone (Boehm *et al.*, 2007), where organisms could be affected by dissolved PAH. Oil residue remains trapped at residual oil saturations in low-permeability sediments. Like recovery processes in petroleum reservoirs that mobilize oil and also strip lighter components from residual oil, altering its original composition, natural processes that remove and weather oil are affected by the rate and amount of groundwater that contacts the oil residue on a daily basis. Thus, over time these natural attenuation processes remove oil from more permeable sediments at a much faster rate than from less permeable sediments.

Groundwater velocities from the numerical simulations support this conceptualization. Subsurface oil residue remains in areas where limited groundwater circulation occurs. However, modeling results show that groundwater velocities in these zones are not zero. This means that exchange between oil residue and the groundwater is

Conceptual and mathematical models of subsurface contamination 169

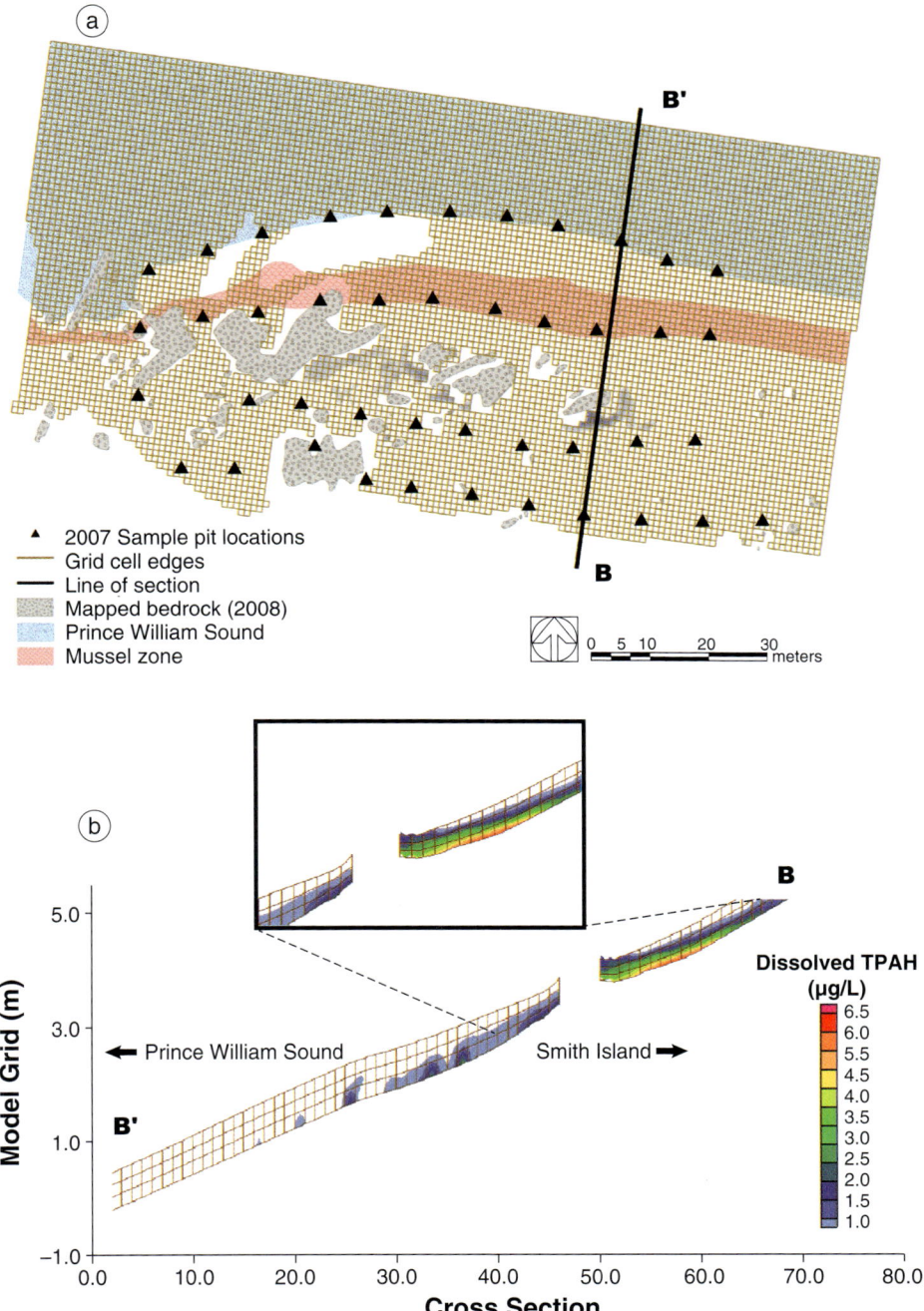

Figure 7.12 Model plan view of SM006B groundwater in (a) ground surface model layer and (b) model cross section from B to B′ along the Y-axis of the model grid of simulated aqueous TPAH concentrations (mg/L) averaged over a single tidal cycle (1 day). Higher dissolved TPAH concentrations are simulated within low-energy "shadow zones" behind bedrock outcrops where subsurface oil residue persists. (m = meters above mean sea level.)

still occurring and natural attenuation is ongoing. A simple analytical model was used to calculate the remaining fraction of each PAH component from a sample collected from SM006B in 2009 and compared with the values measured from a column experiment. The agreement between the measured and calculated results is consistent with the conclusion that a slow leaching of the PAH from the oil residue is occurring and will continue to occur for several more decades, although at very low and ever-diminishing rates. Simulation results and field observations indicate that the overall effect of this exchange is small because of high oil–water partition coefficients and the dilution of aqueous TPAH that occurs as large volumes of water are exchanged between the surface water and groundwater on a daily basis.

The numerical model briefly described in this chapter proved to be very useful, as it provided quantitative answers to long-pending questions, such as the relative importance of subsurface processes including dilution and heterogeneity controls. Initial sensitivity studies using the numerical model clearly indicated a need for additional and more accurate PAH aqueous solubility data. Sediment samples were collected specifically for this purpose and column experiments were then conducted to yield the required data under the required subsurface conditions. The development of a more complete hydrogeologic understanding led to important model results: in areas near the surface and in the lower intertidal zone where organisms could be affected by dissolved PAH, the simulated daily time-weighted mean TPAH groundwater concentrations were within the background range found in PWS nearshore waters.

7.5 Conclusions

As with most major coastal oil spills occurring over the past 50 years, in the aftermath of the *Exxon Valdez* spill, shoreline geomorphologists, chemists, and other scientists dug pits, analyzed sediment samples, and performed numerous tests that allowed them to identify the oil concentrations, extent of weathering, and locations of oil residues, some of which have persisted for many years. This work provided the necessary foundation for the conceptual and mathematical models we have applied in our own studies of contaminated PWS shorelines.

Although the conceptualization that the remaining oil was sequestered, and therefore not environmentally available, had been posed and shown through scientific measurements, the application of the fundamental principles of fluid flow in porous media and appropriate models provided a quantitative scientific basis for this conceptualization. Scientists knowledgeable in multiphase flow in the subsurface were needed to determine the kind of measurements affecting fluid flow that are necessary to characterize the sites with sequestered oil residue and to construct appropriate mathematical models based on more complete characterization and the updated conceptual model. These scientists were then able to use these models to make quantitative predictions. Such quantitative predictions followed from the natural progression of scientific studies of PWS shorelines.

7.6 Lessons learned

- For the PWS cases studied, the most important parameters affecting persistence of oil residues and low concentrations of PAH in nearshore waters were the permeability variations of the sediment, the geomorphology of bedrock boundaries, the composition of the oil residues, and the dissolution of the PAH into seawater.

- Technically sound principles based on hydrogeology and reservoir engineering applied to fluid flow can explain field observations and serve as a guide to field investigations. Previous experimental measurements of the concentrations of PAH dissolved in nearshore seawater indicated background levels at locations where subsurface oil residues persist. However, additional hydrogeologic data and flow-and-transport models were needed to explain the low PAH concentrations quantitatively.

- An integrated, interdisciplinary team implementing a multidisciplinary approach to the development of conceptual models and numerical simulation is a requisite. Modeling and remediation of hydrocarbon sites often include complex processes that incorporate geologic heterogeneities, multiphase flow, and biological and chemical processes. Communication of modeling results to the full body of scientists studying a spill (as well as other interested parties) can often provide answers to questions seemingly unrelated to the goals of the modeling exercise itself.

- The same basic methodology applies to characterizing different hydrocarbon spill sites, although each site will have unique characteristics and risks. Critical questions and important measurements to answer these questions can be identified according to site-specific needs. Conceptualization methods, however, are adaptable and can be applied to many problems regardless of scale, and they can be updated as new data are gathered.

REFERENCES

Abriola, L.M. (1989). Modeling multiphase migration of organic chemicals in groundwater systems: a review and assessment. *Environmental Health Perspectives* **83**: 117–143.

Abriola, L.M. and G.F. Pinder (1985). A multiphase approach to the modeling of porous media contamination by organic compounds. 2: Numerical simulation. *Water Resources Research* **21**(1): 19–26.

Anderson, M.P. and W.W. Woessner (1992). *Applied Groundwater Modeling*. San Diego, CA, USA: Academic Press. ISBN-10: 0120594854; ISBN-13: 9780120594856.

Atlas, R. and J.R. Bragg (2009). Evaluation of PAH depletion of subsurface *Exxon Valdez* oil residues remaining in Prince William Sound in 2007–2008 and their likely bioremediation potential. In *Proceedings of the 32nd Arctic and Marine Oilspill Program (AMOP) Technical Seminar, June 9–11, 2009, Vancouver, BC, Canada*. Ottawa, ON, Canada: Environment Canada; pp. 723–747.

Bear, J. (1979). *Hydraulics of Groundwater*. New York, NY, USA: McGraw-Hill. ISBN-10: 0486453552; ISBN-13 9780486453552. Also reprinted (2007); Mineola, NY, USA: Dover Publications.

Bear, J., M.S. Beljin, and R.R. Ross (1992). *Fundamentals of Ground-Water Modeling*. Washington DC, USA: US Environmental Protection Agency, Office of Research and Development, Office of Solid Waste and Emergency Response; EPA/540/S-92/005. [http://babel.hathitrust.org/cgi/pt?id=umn.3195d00762214d#page/n0/mode/1up]

Boehm, P.D., J.M. Neff, and D.S. Page (2007). Assessment of polycyclic aromatic hydrocarbon exposure in the waters of Prince William Sound after the *Exxon Valdez* oil spill: 1989–2005. *Marine Pollution Bulletin* **54**(3): 339–356.

Boehm, P.D., D.S. Page, J.S. Brown, J.M. Neff, J.R. Bragg, and R.M. Atlas (2008). Distribution and weathering of crude oil residues on shorelines 18 years after the *Exxon Valdez* spill. *Environmental Science & Technology* **42**(24): 9210–9216.

Boehm, P.D., D.S. Page, J.S. Brown, J.M. Neff, and W.A. Burns (2004). Polycyclic aromatic hydrocarbon levels in mussels from Prince William Sound, Alaska, document the return to baseline conditions. *Environmental Toxicology and Chemistry* **23**(12): 2916–2929.

Bouwer, H. and R.C. Rice (1976). A slug test for determining hydraulic conductivity of unconfined aquifers with completely or partially penetrating wells. *Water Resources Research* **12**(3): 423–428.

Brooks, R.H. and A.T. Corey (1964). *Hydraulic Properties of Porous Media*. Fort Collins, CO, USA: Colorado State University, Civil Engineering Department; Hydrology Paper 3.

Carls, M.G., L. Holland, M. Larsen, T.K. Collier, N.L. Scholz, and J.P. Incardona (2008). Fish embryos are damaged by dissolved PAHs, not oil particles. *Aquatic Toxicology* **88**(2): 121–127.

Charbeneau, R.J. (2000). *Ground Water Hydraulics and Pollutant Transport*. Upper Saddle River, NJ, USA: Prentice-Hall. ISBN-10: 0139756167; ISBN-13:9780139756160.

Coats, K. (1982). Reservoir simulation: state of the art. *Journal of Petroleum Technology* **34**(8): 1633–1642; SPE Paper 10020.

Cygan, R.T., C.T. Stevens, R.W. Puls, S.B. Yabusaki, R.D. Wauchope, C.J. McGrath, G.P. Curtis, M.D. Siegel, L.A. Veblen, and D.R. Turner (2007). Research activities at US government agencies in subsurface reactive transport modeling. *Vadose Zone Journal* **6**(4): 805–822.

Delshad, M. and G.A. Pope (1989). Comparison of the three-phase oil relative permeability models. *Transport in Porous Media* **4**(1): 59–83.

Delshad, M., G.A. Pope, and K. Sepehrnoori (1996). A compositional simulator for modeling surfactant enhanced aquifer remediation. 1. Formulation. *Journal of Contaminant Hydrology* **23**(4): 303–327.

Domenico, P.A. and W. Schwartz (1998). *Physical and Chemical Hydrogeology*, 2nd edn. Indianapolis, IN, USA: Wiley. ISBN-10: 0471597629; ISBN-13: 9780471597629.

Douglas, J., D.W. Peaceman, and H.H. Rachford (1959). A method for calculating multidimensional immiscible displacement. *Transactions of the American Institute of Mechanical Engineers* **216**: 297–308.

Dwarakanath, V., R.E. Jackson, and G.A. Pope (2002). Influence of wettability on the recovery of NAPLs from alluvium. *Environmental Science & Technology* **36**(2): 227–231.

Farr, A.M., R.J. Houghtalen, and D.B. McWhorter (1990). Volume estimation of light nonaqueous phase liquids in porous media. *Ground Water* **28**(1): 48–56.

Feenstra, S. and J.A. Cherry (1988). Subsurface contamination by dense non-aqueous phase liquid (DNAPL) chemicals. In *Proceedings of the International Groundwater Symposium, International Association of Hydrogeologists, May 1–5, 1988, Halifax, Nova*

Scotia. C.L. Lin, ed. Goring, Reading, UK: International Association of Hydrogeologists; pp. 62–69.

Fetter, C.W. (1993). *Contaminant Hydrogeology*. New York, NY, USA: Macmillan Publishing Company. ISBN-10: 0023371358; ISBN-13: 9780023371356.

Freeze, R.A. and J.A. Cherry (1979). *Groundwater*. Englewood Cliffs, NJ, USA: Prentice Hall. ISBN-10 0133653129; ISBN-13: 9780133653120.

Guo. Q., H. Li., M.C. Boufadel, and Y. Sharifi (2010). Hydrodynamics in a gravel beach and its impact on the *Exxon Valdez* oil spill. *Journal of Geophysical Research, Oceans* **115**: C12077; DOI:10.1029/2010JC006169.

Hayes, M.O. and J. Michel (1999). Factors determining the long-term persistence of *Exxon Valdez* oil in gravel beaches. *Marine Pollution Bulletin* **38**(2): 92–101.

Hill, M.C. (2006). The practical use of simplicity in developing ground water models. *Ground Water* **44**(6): 775–781.

Lake, L.W. (1989). *Enhanced Oil Recovery*. Old Tappan, NJ, USA: Prentice-Hall. ISBN-10: 0132816016; ISBN-13: 978155563305–9. Also reprinted (2010): Allen, TX, USA: Society of Petroleum Engineers (original edition).

Li, H. and M.C. Boufadel (2010). Long-term persistence of oil from the *Exxon Valdez* spill in two-layer beaches. *Nature Geoscience* **3**(2): 96–99.

Mackay, D.M., P.V. Roberts, and J.A. Cherry (1985). Transport of organic contaminants in groundwater. *Environmental Science & Technology* **19**(5): 384–392.

Mariner, P.E., M. Jin, and R.E. Jackson (1997). An algorithm for the estimation of NAPL saturation and composition from typical soil chemical analyses. *Ground Water Monitoring & Remediation* **17**(2): 122–129.

Mercer, J.W. and R.W. Cohen (1990). A review of immiscible fluids in the subsurface: Properties, models, characterization and remediation. *Journal of Contaminant Hydrology* **6**(2): 107–163.

Mercer, J.W. and C.R. Faust (1980). Ground-water modeling: an overview. *Ground Water* **18**(3): 212–227.

Michel, J., Z. Nixon, and L. Cotsapas (2006). *Evaluation of Oil Remediation Technologies for Lingering Oil from the* Exxon Valdez *Oil Spill in Prince William Sound*. Juneau, AK, USA: National Oceanic and Atmospheric Administration National Marine Fisheries Service; *Exxon Valdez* Oil Spill Restoration Project 050778 Final Report. [http://www.evostc.state.ak.us/Files.cfm?doc=/Store/FinalReports/2005-050778-Final.pdf&]

Muskat, M. (1949). *Physical Principles of Oil Production*. New York, NY, USA: McGraw-Hill.

Newell, C.J., S.D. Acree, R.R. Ross, and S.G. Huling (1995). *Light Non-Aqueous Phase Liquids*. Ada, OK, USA: US Environmental Protection Agency, Office of Research and Development, Robert S. Kerr Environmental Research Laboratory; EPA/540/5-95/500. [http://www.epa.gov/superfund/remedytech/tsp/download/lnapl.pdf]

Owens, E.H., E. Taylor, and B. Humphrey (2008). The persistence and character of stranded oil on coarse-sediment beaches. *Marine Pollution Bulletin* **56**(1): 14–26.

Page, D.S., P.D. Boehm, and J.M. Neff (2008). Shoreline type and subsurface oil persistence in the *Exxon Valdez* spill zone of Prince William Sound, Alaska. In

Proceedings of the 31st Arctic and Marine Oilspill Program (AMOP) Technical Seminar, June 3–5, 2008, Calgary, Alberta, Canada. Ottawa, ON, Canada: Environment Canada; pp. 545–563.

Panday, S. and P.S. Huyakorn (2008). MODFLOW SURFACT: a state-of-the-art use of vadose zone flow and transport equations and numerical techniques for environmental evaluations. *Vadose Zone Journal* **7**(2): 610–631.

Pankow, J.F. and J.A. Cherry (1996). *Dense Chlorinated Solvents and other DNAPLs in Groundwater*. Portland, OR, USA: Waterloo Press. ISBN-10: 0964801418; ISBN-13: 9780964801417.

Poeter, E. and D.R. Gaylord (1990). Influence of aquifer heterogeneity on contaminant transport at the Hanford site. *Ground Water* **28**(6): 900–909.

Pope, G.A., K.D. Gordon, and J.R. Bragg (2011a). Fundamental reservoir engineering principles explain lenses of shoreline oil residue twenty years after the *Exxon Valdez* oil spill. In *Proceedings of the Society of Petroleum Engineers' Americas E&P Health, Safety, Security, and Environmental Conference, March 21–23, 2011, Houston, Texas*. Houston, TX, USA: Society for Petroleum Engineers; SPE Paper 141809.

Pope, G.A., K.D. Gordon, and J.R. Bragg (2011b). Using fundamental practices to explain field observations twenty-one years after the *Exxon Valdez* oil spill. In *Proceedings of the 2011 International Oil Spill Conference (Promoting the Science of Spill Response), May 24–26, 2011, Portland, Oregon, USA*. Washington DC, USA: American Petroleum Institute.

Pruess, K. (2004). The TOUGH codes: a family of simulation tools for multiphase flow and transport processes in permeable media. *Vadose Zone Journal* **3**(3): 738–746.

Pruess, K. and T.N. Narasimhan (1985). A practical method for modeling fluid and heat flow in fractured porous media. *Society of Petroleum Engineers Journal* **25**: 14–26.

Saenton, S., T.H. Illangasekare, K. Soga, and T. Saba (2002). Effects of source zone heterogeneity on surfactant enhanced NAPL dissolution and resulting remediation end-points. *Journal of Contaminant Hydrology* **59**(1–2): 27–44.

Schwille, F. (1975). Groundwater pollution by mineral oil products. In *Groundwater Pollution Symposium, Proceedings of the Moscow Symposium, August 1971*. Washington DC, USA: International Association of Hydrological Sciences; IAHS-AISH Publication No. **103**; pp. 226–240.

Shepherd, R.G. (1989). Correlations of permeability and grain-size. *Ground Water* **27**(5): 633–638.

Short, J.W., M.R. Lindeberg, P.M. Harris, J.M. Maselko, J.J. Pella, and S.D. Rice (2004). Estimate of oil persisting on beaches of Prince William Sound, 12 years after the *Exxon Valdez* oil spill. *Environmental Science & Technology* **38**(1): 19–25.

Stanley, K.W. (1968). *Effects of the Alaska Earthquake of March 27, 1964, on Shore Processes and Beach Morphology*. Denver, CO, USA: US Geological Survey, Information Services; USGS Professional Paper 543-J. [http://pubs.usgs.gov/pp/0543j/report.pdf]

Taylor, E. and P.D. Reimer (2008). Oil persistence on beaches in Prince William Sound: a review of SCAT surveys conducted from 1989 to 2002. *Marine Pollution Bulletin* **56**(3): 458–474.

Theis, C.V., R.H. Brown, and R.R. Meyer (1963). Estimating the transmissivity of aquifers from the specific capacity of wells. In *Methods of Determining Permeability, Transmissibility and Drawdown: Ground-Water Hydraulics*. R. Bentall, compiler. Alexandria, VA, USA: US Geological Survey, Distribution Branch; Geological Survey Water-Supply Paper 1536-I; pp. 331–341. [http://pubs.usgs.gov/wsp/1536i/report.pdf]

US Environmental Protection Agency (1988). *Guidance for Conducting Remedial Investigations and Feasibility Studies under CERCLA*. Washington DC, USA: US Environmental Protection Agency, Office of Emergency and Remedial Response; Final Interim; EPA/540/G-89/004, OSWER Directive 9355.3–01; October 1988. [http://www.epa.gov/superfund/policy/remedy/pdfs/540g-89004-s.pdf]

van Bavel, C.H.M. and D. Kirkham (1948). Field measurement of soil permeability using auger holes. *Soil Science Society of America Journal* **13**(C): 90–96.

van Genuchten, M.Th. (1980). A closed-form equation for predicting the hydraulic conductivity of unsaturated soils. *Soil Science Society of America Journal* **44**(5): 892–898.

Venosa, A.D., P. Campo, and M.T. Suidan (2010). Biodegradability of lingering crude oil 19 years after the *Exxon Valdez* oil spill. *Environmental Science & Technology* **44**(19): 7613–7621.

Xia, Y., H. Li., M.C. Boufadel, and Y. Sharifi (2010). Hydrodynamic actors affecting the persistence of the *Exxon Valdez* oil in a shallow bedrock beach. *Water Resources Research* **46**: W10528; DOI:10.1029/2010WR009179.

CHAPTER EIGHT

Removal of oil from shorelines: biodegradation and bioremediation

Ronald M. Atlas and James R. Bragg

8.1 Introduction

Many microorganisms have evolved the ability to feed on naturally occurring petroleum hydrocarbons,[1] which they use as sources of carbon and energy to make new microbial cells. Most of the tens of thousands of chemical compounds that make up crude oil can be attacked by bacterial populations indigenous to marine ecosystems. A consortium of different bacterial species rather than any single species acts together to break hydrocarbons down into carbon dioxide, water, and inactive residues. Even toxic oil residues, including highly toxic polycyclic aromatic hydrocarbons (PAH), can be detoxified. Microorganisms do not accumulate hydrocarbons as they consume and degrade them, so they are not a conduit for transferring hydrocarbons into the food web. In fact, microorganisms grown on hydrocarbons can be a potential source of protein for animal and human food (Shennan, 1984).

For many years before the *Exxon Valdez* oil spill, the US Environmental Protection Agency (EPA), the National Oceanic and Atmospheric Administration (NOAA), and other governmental agencies had supported research on microbial degradation of oil in marine environments – *biodegradation* – and on ways to enhance and accelerate it – *bioremediation*.[2] These studies showed that, while in many cases biodegradation can mitigate toxic impacts of spilled oil without causing ecological harm, environmental conditions for it to happen rapidly are not always ideal (Atlas, 1995). If water carrying sufficient amounts of oxygen and nutrients cannot reach the oil, rates of biodegradation will be severely limited: oil incorporated into, or on, sediment above the tidal zone, oil buried in low-permeability sediments

---

[1] See Chapter 1, Box 1.1 for a discussion of the chemical composition of petroleum and the hydrocarbons found in North Slope crude oil.
[2] For general overviews of petroleum biodegradation and bioremediation, see reports of the National Research Council (2003) and the American Academy of Microbiology (2011).

Oil in the Environment: Legacies and Lessons of the Exxon Valdez *Oil Spill*, ed. J. A. Wiens. Published by Cambridge University Press. © Cambridge University Press 2013.

(Chapter 7), and thick oil layers and tarballs that are not intimately in contact with flowing water are especially resistant to biodegradation.

Because oil-degrading microorganisms are widespread throughout marine environments, including Prince William Sound (PWS) (Atlas, 1975, 1995; Button *et al.*, 1981; Chianelli *et al.*, 1991), bioremediation employing fertilizers was considered for use following the *Exxon Valdez* oil spill. Bioremediation complemented the physical cleanup of oil and was applied to surface and subsurface porous sediments (e.g., boulder/cobble/gravel shorelines).[3] Seawater is a poor source of the required nutrients nitrogen and phosphorus, so adding them to the shoreline surface had the potential to accelerate biodegradation and thereby reduce the overall ecological effects of the spill.

The *Exxon Valdez* spill was the first time a full-scale, microbial-treatment process was developed using bioremediation. In all, 48 400 kg of nitrogen and 5200 kg of phosphorus were applied from 1989 to 1991, involving 2237 separate shoreline applications of fertilizer (each application was one time at one site). This represents the largest use of bioremediation ever undertaken.

In this chapter, we discuss how and why bioremediation is safe and effective for cleaning up petroleum hydrocarbons at the time of a spill. We also evaluate its potential use on oil residues that still remain many years after the end of cleanup (e.g., more than 20 years in the case of the *Exxon Valdez* spill).

8.2 Initial background studies

8.2.1 A cooperative approach to bioremediation

A month after the *Exxon Valdez* oil spill, EPA convened a panel of scientists to consider bioremediation for oil cleanup on PWS shorelines. The panel recommended that EPA conduct a field test to examine whether fertilizer addition could stimulate the rates of oil biodegradation in PWS (Pritchard and Costa, 1991). Exxon scientists had already begun laboratory studies to determine whether the biodegradation of North Slope crude oil by indigenous PWS microorganisms could be accelerated (Chianelli *et al.*, 1991). On June 2, 1989, Exxon and EPA agreed to test the effectiveness of bioremediation in PWS cooperatively. Their overall objectives were to:

1. Determine the rate and extent of natural biodegradation on oiled shorelines;
2. Determine whether adding fertilizers could significantly increase rates of biodegradation;
3. Develop methods for large-scale application of fertilizers to effects on shorelines;
4. Evaluate the environmental risk of adding fertilizers; and
5. Establish methods to monitor the efficacy and potential ecological effects of bioremediation (Pritchard *et al.*, 1991).

The cooperation between Exxon and EPA leveraged the strengths of both organizations, improving the evaluation and approval process. Recommendations for action were reached jointly.

---

[3] About 26% of shorelines in the PWS spill path are sedimentary (mixed pebble/gravel and boulder/cobble/gravel) (Chapter 6). Oil penetration was not an issue at the dominant shoreline types, which are bedrock/rubble, so bioremediation was not considered for them.

8.2.2 Exxon laboratory testing of bioremediation

Starting in April 1989, Exxon scientists initiated research that demonstrated in the lab that bacteria from sediment samples in PWS could degrade both aliphatic and aromatic hydrocarbons in North Slope crude oil. This research also showed that adding nutrients could accelerate biodegradation (Chianelli et al., 1991). Although these studies measured numbers of naturally occurring, oil-degrading bacteria, they paid little attention to identifying specific bacteria, which was deemed less important than determining the aggregate's ability to degrade hydrocarbons based on changes in oil composition.

With extra nutrients and dissolved oxygen added to flasks, microbes degraded up to 90% of alkanes and about 36% of the initial total oil mass in 20–60 days. This represents a three-fold enhancement of the biodegradation rate compared to unfertilized controls (Chianelli et al., 1991; Bragg et al., 1992).

8.2.3 Proposals for adding oil-degrading microorganisms

Exxon received several proposals in the spring and summer of 1989 claiming that specific commercial bioremediation agents, including cultures of microorganisms, would be effective for cleanup. None of the products, however, had an established scientific basis for application to oiled shorelines, and certainly not to those in PWS. Additionally, the State of Alaska opposed the introduction of nonindigenous microorganisms into PWS.

Even so, and to make sure that no useful commercial technologies were overlooked, EPA established the National Environmental Technology Applications Corporation to evaluate and facilitate the application and commercialization of bioremediation technologies. Thirty-nine candidate technologies, most of which involved adding specific microorganisms, were evaluated; these included technologies developed by the government and by biotechnology companies.

Laboratory tests were conducted on 10 technologies, and field tests were performed on two (Zhu et al., 2004). The tests failed to demonstrate that any of the products were effective. Given the failure of microbial seed agents to increase rates of oil biodegradation under real-world conditions, EPA judged the use of such agents for treating oil spills as dubious (Venosa et al., 1992).

8.2.4 Toxicity testing

Before bioremediation employing fertilizer application could be used in PWS, extensive toxicological studies were required to determine whether any fertilizer components might be harmful to the organisms living on or near the shoreline (Table 8.1) (Pritchard et al., 1991). Test results showed ammonia from the fertilizer to be the most toxic component to marine life, but because of how diluted it would be in the spill zone, EPA did not expect it to cause harm. EPA concluded that 2-butoxyethonol, a component of the liquid fertilizer Inipol EAP 22®, posed a negligible risk to wildlife if inhaled – but that there was a brief period after application when wildlife could be harmed by eating it. To prevent the risk of ingestion during Inipol® field trials, wildlife deterrents (e.g., helium-filled balloons to deter birds or mammals) were used during and for 2 days after fertilizer applications. EPA and Exxon monitored wildlife for toxic effects throughout and following field trials (Pritchard et al., 1991).

Table 8.1 Toxicity tests on Inipol® and Customblen®.

Organisms tested	Fertilizer tested	Results
Herring	Inipol	Because of high mortality in controls, the herring tests were invalid.
Mussel larvae	Inipol	The LC50[a] for mussel larvae was 55 μL/L in the presence of 8 μL/L weathered oil.
Salmon smolts	Inipol	Mussel larvae were more sensitive than salmon smolts.
Stickleback		LC50 = 131 ppm
Pandalid shrimp	Inipol	LC50 = 358 ppm
Oyster larvae	Inipol	LC50 = 41 ppm; field tests directly above fertilizer application showed less than 70% mortality – within 24 h no toxicity due to dilution
Mysids	Inipol	LC50 = 14.8 ppm
Black-necked stilts (a representative shorebird)	Customblen	The birds sometimes picked up the fertilizer pellets but did not ingest them
Bobwhite quail	Customblen	Based upon force-feeding of fertilizer pellets the calculated LD50[b] was 1 g/quail (40–60 pellets) or 5 g/kg quail. When offered fertilizer pellets, bobwhite only consumed one or at most a very few pellets.

[a]LC50 is the concentration that resulted in 50% mortality.
[b]LD50 is the total dosage that resulted in 50% mortality.

8.3 EPA field trials (1989)

8.3.1 Snug Harbor

In early June 1989, EPA began field trials along shorelines in Snug Harbor on Knight Island (see Map 3, p. vii) (Pritchard and Costa, 1991). The Snug Harbor shoreline was typical of moderately oiled shorelines: large cobblestones atop a mixed sand and gravel base. The trials involved separately testing three different types of fertilizers:

1. A water-soluble fertilizer, typical of that used in gardens;
2. A solid, slow-release fertilizer that would gradually release nutrients (similar to that used on lawns): Customblen® 28–8–0, manufactured by Sierra Chemicals of California; and
3. An oleophilic liquid fertilizer, designed to adhere to oil: Inipol®, manufactured by Elf Aquitaine of France.

These three fertilizers were chosen for testing based on application strategies and attendant logistical issues for large-scale application (e.g., the ability to get necessary quantities of fertilizer to isolated shorelines and to minimize the number of personnel needed to apply them), commercial availability, and the product's ability to deliver nitrogen and phosphorus to surface and subsurface microbial communities for sustained periods.

In the field trials, EPA scientists applied fertilizer to the shoreline surface and then allowed tidal waters to disperse the nutrients to surface and subsurface oil. About 2 weeks after fertilizer application, there was a visible reduction in the amount of oil on rock surfaces where the oleophilic fertilizer had been applied, but this visual loss of oil was not observed with the other fertilizer treatments (Pritchard and Costa, 1991). The oleophilic-fertilizer-treated areas even looked clean from the air, which was important for gaining public and political support, but it was not enough to meet scientific standards. More precise analytical methods were needed to prove that bioremediation could work. Using detailed chemical analyses (Box 8.1), EPA was able to quantitatively confirm the reduction in surface and subsurface oil (Pritchard and Costa, 1991).

Box 8.1 Measuring oil biodegradation

To know whether bioremediation has been effective, scientists must be able to measure rates of increased biodegradation quantitatively. This necessitates accurate and precise chemical analyses. Such analyses are complicated, require adequate quality assurance and control, and are costly; only a few laboratories are able to perform them in a reliable and repeatable manner. Given the magnitude of the *Exxon Valdez* spill, hundreds of analyses were performed on numerous shoreline samples, allowing statistical treatment of the data.

To determine the amount and composition of the residual oil, a measured quantity of sediment is extracted with a solvent and the extract is analyzed for hydrocarbons by gas chromatography with flame ionization detection (GC-FID) and gas chromatography-mass spectrometry (GC-MS). There is also a residual fraction that includes asphaltenes and high-molecular-weight resins that cannot be analyzed by these methods. However, this fraction is weighed to determine its percentage of the entire oil sample. Since this fraction contains components highly resistant to biodegradation, it increases as a percentage of the whole oil sample as the oil becomes more biodegraded.

In laboratory tests, rates of oil loss can easily be determined by measuring the change in weight (gravimetric analysis). By comparing samples treated with biocide (so there is no microbial activity) to samples with active microbes, scientists can determine the amount of hydrocarbon loss due to biodegradation. Also, the products of biodegradation (such as carbon dioxide and partially oxidized molecules) can be captured and quantified to obtain a mass balance, i.e., to account for what has happened to all of the components in the oil. The mass balance reveals what percentage of starting hydrocarbons was biodegraded, what percentage was partially degraded to byproducts, and what remains undegraded.

But in a dynamic, shoreline field experiment, such determinations are complicated by other oil-removal processes, such as physical cleanup and waves. Changes in the composition of oil residues must be used instead. By tracking the change in the ratio of a hydrocarbon or a group of hydrocarbons (such as alkanes, PAH, etc.) to a stable (or "conserved") component in the oil not subject to degradation, the rates of biodegradative change of specific hydrocarbons or groupings of hydrocarbons can be determined. Such stable components are called "conserved biomarkers."

Identifying suitable conserved biomarkers takes time, since candidates must be watched to see if they degrade for the length of time necessary for a given analysis (e.g., over 20 years in the case of the *Exxon Valdez*). Some components in oil are known to biodegrade more rapidly than others; e.g., straight-chain alkanes usually degrade more rapidly than branched alkanes. In 1989, the ratios of straight-chain alkanes (heptadecane or octadecane) to branched chain alkanes (pristane or phytane) were investigated. They proved unreliable, however, because indigenous bacteria were able to degrade pristane and phytane at significant rates.

Other conserved biomarkers, such as hopane, are especially resistant to biodegradation and thus are suitable for measuring biodegradative changes in oil. EPA, Environment Canada, and others have adopted their use as a best practice for measuring weathering loss (Wang et al., 1998; Zhu et al., 2001). C_{30}-hopane was used extensively for PWS bioremediation analyses in the years immediately after the spill (Pritchard et al., 1992; Prince et al., 1993, 1994; Bragg et al., 1994; Venosa et al., 1997). Later assessments found $C_{29}R$-stigmastane to be somewhat more stable and the best for tracking long-term degradation of *Exxon Valdez* oil residues in PWS (Atlas and Bragg, 2007, 2009a, b).

Atlas, R. and J. Bragg (2007). Assessing the long-term weathering of petroleum on shorelines: Uses of conserved components for calibrating loss and bioremediation potential. In *Proceedings of the Thirtieth Arctic and Marine Oilspill Program (AMOP) Technical Seminar, June 5–7, Edmonton, Alberta, Canada*. Ottawa, ON, Canada: Environment Canada; pp. 263–290.

Atlas, R.M. and J.R. Bragg (2009a). Bioremediation of marine oil spills: When and when not – the *Exxon Valdez* experience. *Microbial Biotechnology* 2(2): 213–221.

Atlas, R.M. and J.R. Bragg (2009b). Evaluation of PAH depletion of subsurface *Exxon Valdez* oil residues remaining in Prince William Sound in 2007–2008 and their likely bioremediation potential. In *Proceedings of the 29th Arctic and Marine Oilspill Program (AMOP) Technical Seminar, June 5–7, Edmonton, Alberta, Canada*. Ottawa, ON, Canada: Environment Canada; pp. 723–748.

Bragg, J.R., R.C. Prince, and R.M. Atlas (1994). Effectiveness of bioremediation for oiled intertidal shorelines. *Nature* 368(6470): 413–418.

Prince, R.C., J.R. Clark, J.E. Lindstrom, E.L. Butler, E.J. Brown, G. Winter, W.G. Steinhauer, G.S. Douglas, J.R. Bragg, J.E. Harner, and R.M. Atlas (1993). Bioremediation of the *Exxon Valdez* oil spill: Monitoring safety and efficacy. In *Bioremediation of Chlorinated and Polycyclic Aromatic Hydrocarbon Compounds*. R.E. Hinchee, B.C. Alleman, R.E. Hoeppel, and R.N. Miller, eds. Boca Raton, FL, USA: Lewis Publishers. ISBN-10: 0873719832; ISBN-13: 9780873719834; pp. 107–124.

Prince, R.C., D.L. Elmendorf, J.R. Lute, C.S. Hsu, C.E. Haith, J.D. Senlus, G.J. Gary, G.J. Dechert, G.S. Douglas, and E.L. Butler (1994). 17α(H),21β(H)-Hopane as a conserved internal marker for estimating the biodegradation of crude oil. *Environmental Science & Technology* 28(1): 142–145.

Pritchard, P.H., J.G. Mueller, J.C. Rogers, F.V. Kremer, and J.A. Glaser (1992). Oil spill bioremediation: Experiences, lessons, and results from the *Exxon Valdez* oil spill in Alaska. *Biodegradation* 3(2–3): 315–335.

Venosa, A.D., M.T. Suidan, D.W. King, and B.A. Wrenn (1997). Use of hopane as a conservative biomarker for monitoring the bioremediation effectiveness of crude oil contaminating a sandy beach. *Journal of Industrial Microbiology & Biotechnology* 18(2–3): 131–139.

Wang, Z., M. Fingas, S. Blenkinsopp, G. Sergy, M. Landriault, L. Sigouin, J. Foght, K. Semple, and W. Westlake (1998). Comparison of oil composition changes due to biodegradation and physical weathering in different oils. *Journal of Chromatography A* 809(1–2): 89–107.

Zhu, X., A.D. Venosa, T. Suidan, and K. Lee (2001). *Guidelines for the Bioremediation of Marine Shorelines and Freshwater Wetlands*. Cincinnati, OH, USA: US Environmental Protection Agency, Office of Research and Development, National Risk Management Research Laboratory, Land Remediation and Pollution Control Division. [http://www.epa.gov/oem/docs/oil/edu/bioremed.pdf]

8.3.2 Passage Cove

EPA continued field trials at Passage Cove on northern Knight Island (see Map 3, p. vii). They tested applying the oleophilic fertilizer and the slow-release fertilizer together, as well as applying water-soluble fertilizer via a sprinkler system. EPA wanted to test a sprinkler system since it potentially could deliver water-soluble fertilizer in regulated doses. This was in contrast to Exxon's focus on using oleophilic and slow-release fertilizers because of logistical issues involved in moving fertilizers, application equipment, and crews to remote shorelines for large-scale applications.

Within a month of beginning the tests at Passage Cove, concentrations of subsurface oil declined in all treated plots, regardless of the fertilizer type. Compositional changes in the oil confirmed that oleophilic and slow-release fertilizer treatments produced comparable enhancements to biodegradation (Pritchard and Costa, 1991).

8.3.3 EPA recommendations for using bioremediation

Based primarily on results of the field trials, EPA supported the use of bioremediation (US Environmental Protection Agency, 1990) and recommended:

1. Application of both oleophilic fertilizer and slow-release soluble fertilizer on cobble and mixed sand and gravel shorelines;
2. Physical washing before fertilization on moderately and heavily oiled shorelines; and
3. Rates of application such that the oleophilic fertilizer covered oiled areas completely in a thin coat and the slow-release fertilizer released nitrogen (as ammonia or nitrate) and phosphate at rates of 1–10 and 0.1–0.5 mg/L/day per 100 g of granules, respectively, for up to 40 days (US Environmental Protection Agency, 1989a).

Additional guidelines were given for monitoring potential toxicity due to fertilizer application, with the mandate that fertilizer application would be terminated if any adverse environmental effects were detected.

The proposal was approved by the Regional Response Team, chaired by the US Coast Guard (USCG).

8.4 Exxon's use of bioremediation (1989)

Exxon began applying fertilizer on a large scale on August 1, 1989, even while EPA continued its scientific studies. By the time EPA field trials ended, Exxon had treated 119 km of shoreline under its provisional authorization – the largest application of bioremediation in history. Exxon used Inipol® and Customblen® (Bragg et al., 1992). Application rates were based on EPA's calculations of the highest concentrations of ammonia that could be released, with worst-case minimal dilution, while still maintaining a safety margin below concentrations considered toxic by EPA water-quality standards (US Environmental Protection Agency, 1989b).

Although field trials showed that increasing fertilizer application frequency and dosage could produce greater increases in the rates of biodegradation, the decision was made to limit concentrations of fertilizer to levels estimated to be safe for organisms living on or near the shoreline so as to ensure that there were no adverse ecological consequences. The goal of bioremediation was to achieve an increase in biodegradation rate of more than two-fold.

The normal rates of oil biodegradation, based on field measurements from April 1989–May 1990 at three sites on Knight Island, ranged from ~0.6 to 2.8 g oil/kg sediment per year for surface oil and ~0.2 to 3.0 g oil/kg sediment per year for subsurface oil (Bragg et al., 1992). This equated to a mean loss in the mass of residual oil of about 28% per year for surface oil and 12% per year for subsurface oil. The rate of oil removal from biodegradation contributed to the total rate of oil removed from shorelines by all factors (physical cleanup, waves, storms, etc.). The total rate of oil removal from 1989 through the winter of 1990 was 75–90% (Koons and Jahns, 1993; Wolfe et al., 1994).

8.5 Winter laboratory testing and monitoring (1989–90)

Although large-scale applications of fertilizers in 1989 showed that they were safe and effective on surface oil, questions still remained about their effectiveness on subsurface oil. Because cleanup activities for the *Exxon Valdez* spill were suspended during winter for safety and weather reasons, there was time for laboratory work that could inform the following season's bioremediation program. In less extreme environments, such time would likely not have been available. The additional research, conducted in EPA and Exxon laboratories, aimed to:

1. Address the potential effectiveness of subsurface bioremediation;
2. Gain a better understanding of the mechanisms by which Inipol® worked;
3. Determine the products formed by oil biodegradation; and
4. Evaluate potential chronic and acute toxicity to additional organisms (Pritchard et al., 1991; Bragg et al., 1992).

8.5.1 Exxon laboratory testing

Exxon scientists sought to:

1. Determine the specific hydrocarbon compounds microbes were degrading (and the products formed by oil biodegradation) under conditions where nutrients and oxygen were not limited;
2. Evaluate various fertilizers on a small scale, where nutrients, oxygen, and sediments had good contact; and
3. Simulate, on a large scale, nutrient and oxygen transport to subsurface sediments by tidal flow, when the fertilizer was applied only on the surface.

These tests included large-scale experiments that were designed to model conditions prevailing on most shorelines in 1990, where water freely moved through sediment; they therefore did not address possible sequestered oil residues. The goal was to determine whether nutrients applied to shoreline surfaces at low tide could penetrate permeable subsurface sediments to depths of about 1 m.

Plexiglas® columns were packed with about 50 L of sediment. For over 4 months under favorable laboratory conditions, these microcosms experienced water flow simulating highly permeable PWS shorelines (Bragg et al., 1992). Water entering the base of a column (rising tide) or the top of a column (falling tide) was saturated with dissolved oxygen, as most shoreline water would be. Columns were either unfertilized (natural attenuation controls), had biocide added to the seawater to eliminate all microbial activity (negative controls), or had Inipol®, Customblen®, or water-soluble inorganic fertilizers (potassium phosphate and ammonium nitrate) applied to the surface in the same proportion as had been used in 1989 field applications.

Oxygen consumption, which indicates that biodegradation is occurring, increased soon after one application of Inipol®, but diminished over the first month (Fig. 8.1). A second application, this time with Customblen®, caused oxygen consumption to increase again. This suggests that fertilizer nutrients from the earlier application of Inipol® had been depleted, meaning that to achieve efficient bioremediation, sediment

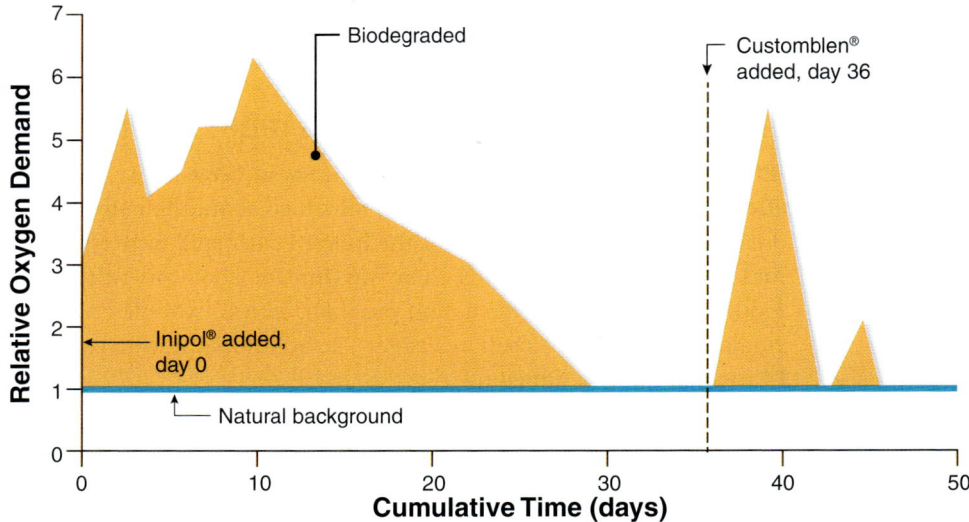

Figure 8.1 Ratio of fertilized oxygen demand to oxygen demand in control (natural attenuation) for a microcosm initially treated with Inipol®. Oxygen demand (rate of oxygen consumption by bacteria degrading hydrocarbons) is proportional to biodegradation rate and is useful for tracking ongoing biodegradation without disturbing the experiment. Oxygen depletion relative to control increased 5-fold over the first two weeks and an average of 3.7-fold over the first month after application. Oxygen levels reached background levels after 28 days. When the same amount of nitrogen was reapplied as Customblen® at day 36, consumption again increased, but less so, indicating that prior loss of easily degraded hydrocarbons had slowed biodegradation rate. The increase in oxygen consumption is proportional to the amount of hydrocarbon consumed and reflects the effectiveness of adding nutrients to enhance biodegradation.

fertilizer concentrations should be monitored to determine appropriate schedules for repeat treatments.

After 4 months, samples of sediments were collected at various depths in the columns and the residual oil mass and composition were quantified. Results showed that, for columns tested with both Inipol® and Customblen®, about 24% of the total oil mass had been consumed based on mass loss. Analyses based on changes in ratios of hydrocarbons (e.g., PAH) to the conserved biomarker C_{30}-hopane (Box 8.1) indicated that 26% of total oil mass had been lost. That these numbers are in close agreement indicates that, in the field where measuring changes in oil mass would be impossible, scientists can estimate changes in oil composition by using the ratio of hydrocarbons to C_{30}-hopane (Box 8.1). The results also show that both Inipol® and Customblen® produce comparable increases in biodegradation rates – about five-fold for alkanes and about three-fold for total oil mass over the first month – relative to natural rates.

Although these experiments were valuable in understanding bioremediation of subsurface oil residues (SSOR) when those SSOR are in direct contact with nutrient- and oxygen-containing water, it must be emphasized that they do not apply to conditions where SSOR are sequestered from contact with water by impermeable or nearly impermeable sediment (Chapter 7).

Field surveys of over 20 shorelines in PWS and the Gulf of Alaska (GOA) showed that oil concentrations continued to decline over this first winter (Owens, 1991). Scientific evidence supported continuing bioremediation in the spring; additional efforts by EPA, Exxon, and the Alaska Department of Environmental Conservation (ADEC) to communicate this information prior to the summer fertilizer application period helped to ensure public support and governmental approval for continuing bioremediation in 1990 (Box 8.2).

Box 8.2 Consultations with governmental agencies and Alaska citizens

In October and November 1989 and February 1990, EPA, NOAA, and Exxon met to review results and discuss the potential for further use of bioremediation. At these meetings, Exxon presented results from laboratory tests indicating that the oil remaining after the 1989 cleanup was still amenable to bioremediation and that fertilizer application could enhance rates of subsurface biodegradation to a depth of at least 0.9 m. The data helped to convince governmental officials of the usefulness of bioremediation.

To gain further acceptance of bioremediation, Exxon and EPA conducted a series of public meetings in Alaska during the spring and summer of 1990 with municipal officials, Alaska Native groups, and environmental organizations to discuss the results of the 1989 program and the winter monitoring and laboratory experiments. ADEC helped organize these meetings, which served an educational function and helped overcome popular misperceptions about bioremediation. These meetings were critical for gaining acceptance of continuing the use of bioremediation in 1990.

8.6 Joint EPA, ADEC, and Exxon field trials (summer 1990)

Additional field trials were conducted jointly by EPA, ADEC, and Exxon during the summer of 1990 at three sites on Knight Island: a shoreline in the Bay of Isles with surface and subsurface oil that was relatively sheltered from waves (low energy), a sheltered shoreline in Herring Bay with only surface oil, and a shoreline with primarily subsurface oil that was exposed to wave action (high energy) (Prince *et al.*, 1993; Bragg *et al.*, 1994). The trials were designed to produce data within 6 weeks to assist in cleanup decisions. They assessed:

1. The ability of nutrients to penetrate sediments and remain within the water pores;
2. The ability of microorganisms to degrade residual hydrocarbons;
3. Changes in microbial populations in response to fertilizer;
4. Changes in the amount and composition of the oil residues in response to fertilizer; and
5. Potential adverse toxicological or ecological effects of fertilizer.

The field trials involved oil residues a little more than a year old. At this time, most of the surface and subsurface oil residues were in contact with nutrient- and oxygen-containing water and not sequestered in water-impermeable sediment layers.

The trials confirmed that the rate of oil degradation under these conditions was critically dependent on the ratio of nitrogen to biodegradable oil (Bragg *et al.*, 1994). Biodegradation rates for PAH could increase by a factor of two, and for aliphatic hydrocarbons by a factor of five, with fertilizer.[4] Measurements of dissolved oxygen in subsurface pore water showed it was lower as compared to the surface, but was not

---

[4] The slow-release, solid fertilizers had to be applied at low tide. In one test application, much of the solid fertilizer floated away at first incoming tide and did not deliver sufficient nutrients. Other applications at low tide were successful.

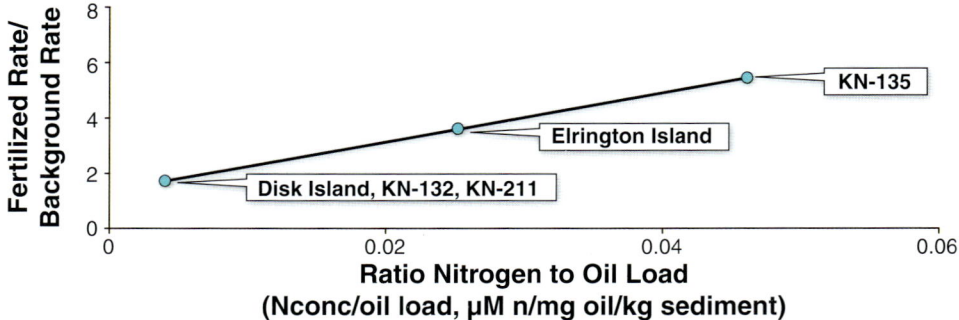

Figure 8.2 Increases in biodegradation rates of total gas chromatographically detectable hydrocarbons with added nutrients over natural background for bioremediation tests in PWS, 1989–90. The correlation shown was based on measured average oil loadings and average nitrogen concentrations measured in the pore water at each test site regardless of the nutrient application methods and site locations.

rate limiting. Oil biodegradation slowed once the more readily degradable components, such as the normal alkanes and lower molecular weight aromatics, were depleted, even when fertilizer was reapplied.

Weekly monitoring showed no adverse ecological consequences of fertilizer application (US Environmental Protection Agency, 1990). Nearshore waters collected following fertilizer application showed no toxicity to mysid shrimp (*Mysidopsis bahia*). There was no evidence of algal blooms from the fertilizer (no increase in chlorophyll) nor that fertilizer caused mobilization of hydrocarbons[5] into nearshore waters (no increase in hydrocarbons detected in caged mussels, *Mytilus trossulus*) (US Environmental Protection Agency, 1990).

After reviewing the results of these trials, the Federal On Scene Coordinator (FOSC)[6] concluded that bioremediation should continue throughout PWS; from his perspective, bioremediation had been shown to be effective and safe for cleanup (Leschine *et al.*, 1993).

8.6.1 EPA field tests (1990)

EPA continued independent testing on Elrington and Disk Islands (see Map 3, p. vii) (Pritchard and Costa, 1991). These tests focused on the effects of using different doses of Customblen®. They showed no difference between control and treated plots. Monitoring later showed that this was due to inadequate concentrations of nutrients delivered to pore waters in contact with oil residues.

As a consequence, others later developed simple ways to measure nutrient levels on site during bioremediation to ensure that adequate nutrient levels were applied (Prince *et al.*, 2003). The amount of nutrients delivered to pore water relative to the amount of biodegradable oil proved to be the critical factor affecting rates of hydrocarbon biodegradation (Fig. 8.2).

---

[5] Oleophilic fertilizers are capable of acting like detergents and physically moving into nearshore waters.
[6] The FOSC is the point of contact for the coordination of federal efforts with those of the local response community. The FOSC monitors, provides technical assistance, and/or directs federal and potentially responsible party resources.

Figure 8.3 (a) Bottles of slow release Customblen® fertilizer and Inipol EAP22® oleophilic fertilizer. (b) Work crew spraying Inipol® on shoreline. (Photos: (a) James R. Bragg; (b) Exxon.)

8.7 Full-scale bioremediation (1990–92)

Large-scale applications of Inipol® and Customblen® fertilizer (Fig. 8.3) during summer 1990 targeted sites recommended by Shoreline Cleanup Assessment Technique (SCAT) surveys (Owens, 1991; Chapter 4). Over 1400 sites in PWS and 378 in the GOA were selected (Bragg *et al.*, 1992).

Logistical challenges, especially in remote areas where weather can prevent access to shorelines, were an important consideration in designing the large-scale bioremediation program. Water-soluble nutrients were rejected because of the logistical issues involved in applying and monitoring continuous application across many kilometers of shoreline. Because Inipol® could be applied from a backpack system (Fig. 8.3) and solid pellets of Customblen® could be broadcast by an individual worker, with either fertilizer supplied by barge, only small teams were needed. Inipol® was applied to surface oil and the slow-release Customblen® pellets were used for subsurface oil; where both existed, both were applied. Generally, shoreline segments were given 1–3 treatments over several months.

Ecological monitoring was conducted concurrently with fertilizer application (Pritchard *et al.*, 1991; Prince *et al.*, 1993). The potentially toxic ammonia concentration never exceeded 1.9 ppm in intertidal water (measured at a depth of 0.5 m above the sediment), well below EPA's 9.8-ppm limit for short-term exposure at 15°C and salinity of 2% (US Environmental Protection Agency, 1989b). Caged mussels, which were used as sentinels to monitor whether undegraded oil or biodegradation products were accumulating in the environment, showed no evidence of this in their tissues (Pritchard *et al.*, 1991). EPA concluded, therefore, that the oleophilic fertilizer, which can have a dispersant action, was not washing oil away from shorelines.

As in the joint trials, no algal blooms were detected and chlorophyll measurements confirmed that the added nutrients were not causing eutrophication. Toxicity tests on mussels, Pacific herring (*Clupea pallasii*), coho salmon smolts (*Onchorhynchus kisutch*), Pacific oyster larvae (*Crassotrea gigas*), pandalid shrimp (*Pandalus danae*), mysids, and threespine stickleback fish (*Gasterosteus aculeatus*), showed no adverse effects from intertidal water after the application of oleophilic fertilizer (Pritchard *et al.*, 1991). Oleophilic fertilizer could be applied at levels that would be effective without causing ecological harm.

The next year, in 1991, fertilizer was applied to only about 220 sites. This was because residual oil at the surface and subsurface of shorelines had declined rapidly. In 1992, bioremediation was used on only a few small and scattered sites that still showed significant concentrations of SSOR.

A survey in May–June 1992 found that the vast majority of the oil had been removed from shorelines (Neff *et al.*, 1995). On June 10, 1992, the USCG officially declared the cleanup over (Leschine *et al.*, 1993). Even though there was no objective standard for "how clean is clean," the Federal and State On Scene Coordinators concluded that further cleanup, either physical or through bioremediation, would not provide a net environmental benefit; indeed, it could cause harm. A net environmental benefit analysis compares and ranks the net environmental benefit associated with multiple management alternatives, including allowing the oil to remain without further treatment. In the case of the *Exxon Valdez* oil spill, by 1992 any residual oil was thought to pose a minimal risk to biota and the risk would continue to diminish as the oil continued to biodegrade naturally.

A 1992 NOAA status report, quoted in the FOSC's report (Leschine *et al.*, 1993, p. 196, note 69), stated with respect to subsurface oil that:

> As the "pockets" of subsurface oil become smaller and more discontinuous with each passing year, the overriding concern must shift to minimizing the disturbance to the ecosystem as a whole. Reworking of shoreline sediments, either manually
> or mechanically, may increase the availability, and therefore the toxicity, of the oil to the surrounding biological community thereby reducing the benefit of treatment.

A final response update released by the Seventeenth USCG District commander, also quoted in the FOSC's report, stated that:

> There is still some oil remaining on the shorelines impacted by the *Exxon Valdez* oil spill. The oil left is generally oil mousse, which is very high in water content, weathered and has lost most of its toxicity. This oil is primarily located in areas protected from the elements, behind rocks and boulders or is below the surface. Algae, mussels, periwinkles and other marine life are recolonizing in strength on these shorelines. Removal of the remaining weathered, generally benign oil, would require invasive cleanup measures which would disrupt the environmental recovery process that is well underway. The consensus and judgment of state and federal agencies involved in the spill response is that additional cleanup would cause unacceptable environmental harm. Accordingly, the FOSC has determined cleanup 'complete' (Leschine *et al.*, 1993, p. 198).

8.8 Oil on shorelines more than a decade after the end of bioremediation and cleanup

In 2001, 2003, and 2005, NOAA carried out surveys (Chapter 6) to determine where and how much residual oil remained on PWS shorelines. The surveys used statistically based, random site selection to choose the sites to sample; the sites chosen were mostly sedimentary shoreline sites that had been oiled in 1989 and were still heavily or moderately oiled some time during 1990–92 according to SCAT surveys (Short et al., 2004, 2006). Of the 42 shoreline segments surveyed in 2001, about 42% of the segments contained some heavy or moderate oil residues (Michel et al., 2006). In 2002, 2005, and 2007, scientists contracted by Exxon also surveyed sites that NOAA had reported to be the most heavily oiled, focusing on worst-case scenarios (Chapter 6; Boehm et al., 2008; Page et al., 2008; Taylor and Reimer, 2008).

NOAA and Exxon scientists analyzed the PAH components of SSOR. But, unlike the Exxon studies, NOAA did not measure C_{30}-hopane and other conserved biomarker components that could be used as internal standards for calculating the extent of weathering (Box 8.1). Instead, NOAA initially relied on a "weathering index" (Short et al., 2007). Later, they used the sum of C_2–C_4 chrysenes as an internal standard, which caused them to underestimate the extent of weathering compared to using C_{29}R-stigmastane, which proved to be the most conserved internal standard (Atlas and Bragg, 2007).

Because of the differences in oil-analysis methods used following the field surveys, NOAA and Exxon came to very different conclusions about the extent of weathering (particularly, how much PAH had been depleted) and whether SSOR remained an ecological risk. Both showed similar masses of oil residue remaining. Some researchers working for NOAA (Short et al., 2007) reported that most of the oil sampled in 2001–05 was only slightly weathered, even though both NOAA and Exxon found only a few sites where SSOR had undergone minimal weathering (< 50% total PAH loss). When compared on the same basis (as shown in Fig. 8.4, which includes all NOAA data), NOAA and Exxon samples collected over 2001–03 had lost almost the same amounts of total PAH (TPAH). Average TPAH depletion for NOAA samples was 74% versus 72.1% for Exxon samples (computed using sum of chrysenes as conserved marker, which actually understates true TPAH depletion by about 9%). Later surveys by Exxon in 2007–08 showed that most of the residual oil was highly weathered (> 70% TPAH mass loss) (Fig. 8.5) (Atlas and Bragg, 2009a, b).

Based on NOAA's report that substantial amounts of unweathered SSOR remained, some scientists suggested that bioremediation might again be employed to remove remaining oil residues (Michel et al., 2006; Boufadel and Bobo, 2011; Boufadel and Michel, 2011). But does the remaining SSOR pose a risk to biota – would there be a benefit to removing it? And if there were a risk, could bioremediation still be effective in cleaning up SSOR over 20 years after the spill (Atlas and Bragg, 2009a)?

To answer the first question, scientists measured whether toxic components from the SSOR were *bioavailable* – being taken up by the surrounding intertidal biota (Chapter 7). All evidence suggests that the toxic components are too sequestered to be bioavailable; hence, they are not of ecological concern (Neff et al., 2006; Pope et al., 2011a, b; Chapters 7 and 16) and there would be no net environmental benefit to their removal. SSOR in PWS will continue to degrade slowly without posing risk.

One way to answer the second question is to determine whether the oil still contains enough degradable hydrocarbons for bioremediation to work (an oil-residue composition question). A second approach examines whether geology and hydrogeology that restrict access of long-sequestered SSOR to water containing nutrients makes bioremediation as previously applied in PWS ineffective (a flow-and-transport question, covered in Chapter 7).

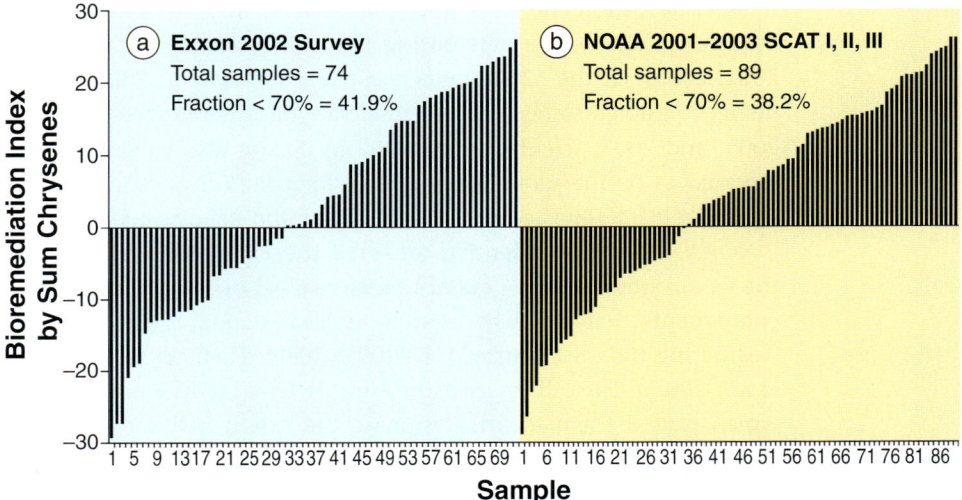

Figure 8.4 Comparison of percentage depletion of TPAH computed using the sum of chrysenes as "conserved" marker for (a) 2002 Exxon samples and (b) NOAA 2001–03 samples. TPAH depletion is relative to spilled *Exxon Valdez* oil. Results are comparable in extent of TPAH depletion. NOAA did not measure more conserved biomarkers such as C_{30}-hopane or $C_{29}R$-stigmastane. Note: this plot underestimates the true TPAH depletion by at least 9% since chrysenes are also degraded (Atlas and Bragg, 2007). Only samples with total PAH > 500 ng/g sediment are included.

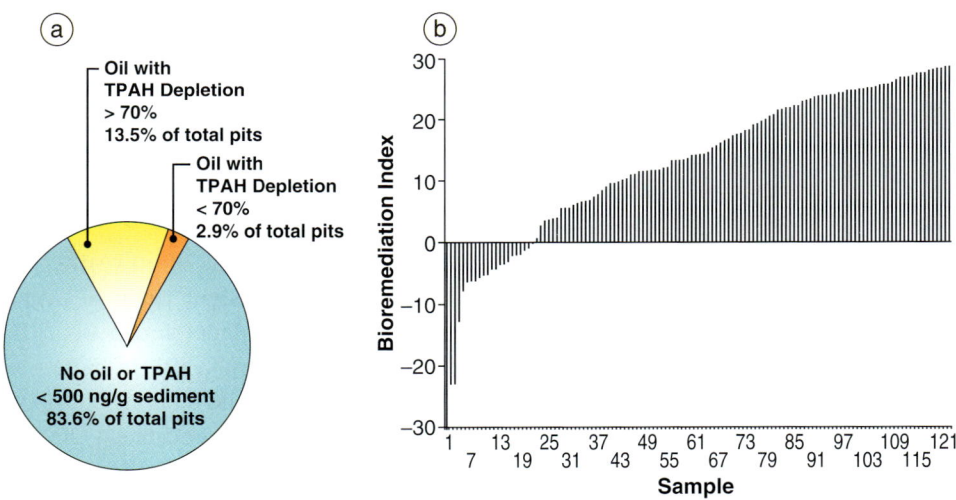

Figure 8.5 (a) Extent of PAH weathering measured in pits sampled in 2007–08. Less than 3% of the 761 total pits had SSOR that were less than 70% depleted of TPAH. (b) Bioremediation indices for 2007–08 samples based on $C_{29}R$-stigmastane. Only 125 samples (16.4% of 761 total pits) had enough oil to quantify TPAH depletion. Of those samples, 82.4% had positive bioremediation indices.

To more easily assess the potential for bioremediation of the remaining SSOR, Atlas and Bragg (2009b) proposed using a "bioremediation index," which is the extent of PAH depletion (as a percentage) minus 70%. The 70% is a scientifically based (but subjective) estimate of the extent of PAH degradation above which addition of nutrients would be unlikely to significantly accelerate biodegradation over natural

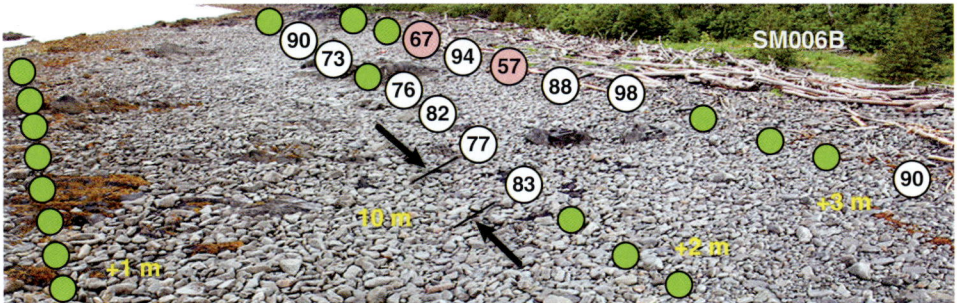

Figure 8.6 Photo of shoreline on Smith Island, which had the highest concentrations of SSOR identified in the 2001 NOAA survey, showing the extent of depletion of TPAH of oil in pits dug in 2007 at 10-m intervals along transects at indicated elevations above mean low tide. Green dots indicate PAH below a concentration too low for accurate quantification (<500 ng/g sediment). Remaining oil residue is spotty in location and most of the oil is highly degraded. From Atlas and Bragg, 2009a.

background biodegradation (Atlas and Bragg, 2009a, b).[7] The bioremediation index would be computed and plotted for each sediment sample, providing a ranking of the bioremediation potential for the entire site.

Bioremediation index analyses suggest that, even if the remaining sequestered oil residues could easily be contacted by added nutrients and dissolved oxygen in seawater, most are already too weathered for biodegradation rates to be significantly enhanced by adding fertilizer (Atlas and Bragg, 2009a, b). Consequently, the composition and patchy distribution of the remaining oil residues (for example, see Fig. 8.6) are not the major factors arguing against the use of bioremediation. The main reason is that the SSOR are sequestered in low-permeability layers (often 1000 times less permeable than surrounding sediments) (Boufadel *et al.*, 2010; Li and Boufadel, 2010; Pope *et al.*, 2011a, b; Chapter 7). Because pore water in those layers is nearly stagnant, it is depleted of nutrients and oxygen. Shoreline surveys in 2006–08 showed that the pore waters adjacent to sequestered SSOR have enough natural nitrogen and dissolved oxygen to support biodegradation, if only the water were able to flow through the SSOR (Atlas and Bragg, 2009b). So, if water had continued to flow through the sequestered SSOR, those residues would have been biodegraded years ago.

The low-permeability problem was confirmed in laboratory experiments with PWS sediments containing 20-year-old SSOR (Venosa *et al.*, 2010). Removing the sediment samples from the shoreline effectively unsequestered the oil. The samples were kept in intimate contact with water containing nutrients and dissolved oxygen. Biodegradation with no added fertilizers was almost as rapid as biodegradation when very high nitrogen concentrations were added.

All of the field and laboratory tests are consistent – the remaining oil would naturally biodegrade fairly rapidly if the oil were not sequestered in low-permeability sediment layers. The microcosm experiments conducted by Venosa and colleagues did not demonstrate a net environmental benefit nor that nutrients could reach the sequestered oil residues without physical disruption of the sediments in which they are sequestered.

---

[7] Almost all alkanes had been lost in residues by 2001–03, so the bioremediation index only considered PAH, which are the components of highest toxicological concern.

Results from hydrodynamic simulations (Boufadel et al., 2010; Pope et al., 2011a, b; Chapter 7) show that adding nutrients to the surface of shorelines at low tide (as practiced in PWS in 1989–92) would not significantly increase nitrogen concentrations in low-permeability sediment layers containing the SSOR. Any water, whether natural or provided by some piping system, will take the path of least resistance through sediments: it will flow around impermeable layers, supplying no nutrients or oxygen to sequestered SSOR. For nutrients to reach the sequestered oil residues, physical disruption (e.g., tilling) of the low-permeability sediments would be needed.

8.9 Lessons learned

- Bioremediation can be an effective technology for oil spill cleanup. In the case of the *Exxon Valdez* spill, it was possible to speed up the rates of natural biodegradation by adding fertilizers to the surfaces of oiled shorelines. Accelerated rates of three to five times were achieved without any toxicity to biota or any other adverse environmental effects.

- Efficacy and safety of bioremediation must be scientifically demonstrated in the laboratory and in the field before large-scale application to shorelines. Rigorous chemical analyses were needed to establish rates of biodegradation. Laboratory tests provided critical scientific information, but were considered inadequate for ensuring that bioremediation was applicable to the actual shorelines impacted by oil from the *Exxon Valdez* spill. Field testing was critical for establishing efficacy and safety.

- A cooperative approach is critical for establishing credibility and gaining public acceptance for the use of bioremediation. Having EPA, Exxon, and ADEC working together – and conducting joint outreach to the public by the governments and Exxon – was important for gaining approval to apply fertilizers to stimulate microbial oil degradation.

- Bioremediation and natural oil biodegradation have limitations and are not effective in all environments. Bioremediation was shown to be effective in highly porous shorelines where nutrients and oxygenated seawater could reach the surface and subsurface oil residue. However, it will be no more effective than natural biodegradation if oil is sequestered from the significant water flow needed to transport nutrients and oxygen.

- Bioremediation will not result in the complete removal of all of the oil. Some oil components, such as asphaltenes and high-molecular-weight resins, are resistant to biodegradation and will remain as inert residues, and some residues will remain in sequestered SSOR.

- Naturally occurring hydrocarbon-degrading bacteria are widespread and introducing new bacteria is not necessary. Nonnative bacteria that work well in the laboratory might not be useful for real-world application to an oil spill; their effectiveness would have to be scientifically demonstrated in the field; and government and public concerns about the introduction of nonindigenous microorganisms would need to be addressed.

- Scaling-up is a critical factor that must be considered in a real-world application of bioremediation. Full-scale application of bioremediation required major logistical considerations and monitoring to ensure effectiveness. Practical logistical constraints generally dictated that fertilizers applied be slow-release or oleophilic.

- The decision to use bioremediation should be based on a net environmental benefit analysis. If residual oil poses no ecological risk, it should be left to undergo natural biodegradation.
- Bioremediation lessons learned from the *Exxon Valdez* spill are applicable to other marine shorelines. Site-specific differences, however, will require additional considerations.

REFERENCES

American Academy of Microbiology (2011). *Microbes & Oil Spills*. Washington DC, USA: American Society for Microbiology. [http://academy.asm.org/images/stories/documents/Microbes_and_Oil_Spills.pdf]

Atlas, R.M. (1975). Microbial degradation of petroleum in marine environments. In *Proceedings of the First Intersectional Congress of the International Association of Microbiological Societies, September 1–7, 1974, Tokyo, Japan*. T. Hasegawa, ed. Tokyo, Japan: Science Council of Japan; Volume 2; pp. 527–531.

Atlas, R.M. (1995). Petroleum biodegradation and oil spill bioremediation. *Marine Pollution Bulletin* **31**(4–12): 178–182.

Atlas, R.M. and J.R. Bragg (2007). Assessing the long-term weathering of petroleum on shorelines: Uses of conserved components for calibrating loss and bioremediation potential. In *Proceedings of the Twenty-Ninth Arctic and Marine Oilspill Program (AMOP) Technical Seminar, June 5–7, Edmonton, Alberta, Canada*. Ottawa, ON, Canada: Environment Canada; pp. 263–290.

Atlas, R.M. and J.R. Bragg (2009a). Bioremediation of marine oil spills: when and when not – the *Exxon Valdez* experience. *Microbial Biotechnology* **2**(2): 213–221.

Atlas, R.M. and J.R. Bragg (2009b). Evaluation of PAH depletion of subsurface *Exxon Valdez* oil residues remaining in Prince William Sound in 2007–2008 and their likely bioremediation potential. In *Proceedings of the 29th Arctic and Marine Oilspill Program (AMOP) Technical Seminar, June 5–7, Edmonton, Alberta, Canada*. Ottawa, ON, Canada: Environment Canada; pp. 723–748.

Boehm, P.D., D.S. Page, J.S. Brown, J.M. Neff, J.R. Bragg, and R.M. Atlas (2008). Distribution and weathering of crude oil residues on shorelines 18 years after the *Exxon Valdez* spill. *Environmental Science & Technology* **42**(24): 9210–9216.

Boufadel, M.C., Y. Harifi, B. Van Aken, B. Wrenn, and K. Lee (2010). Nutrient and oxygen concentrations within the sediments of an Alaskan beach polluted with the *Exxon Valdez* oil spill. *Environmental Science & Technology* **44**(19): 7418–7424.

Boufadel, M.C. and A.M. Bobo (2011). Feasibility of high pressure injection of chemicals into the subsurface for the bioremediation of the *Exxon Valdez* oil. *Ground Water Monitoring and Remediation* **31**(1): 59–67.

Boufadel, M. and J. Michel (2011). *Pilot Studies of Bioremediation of the Exxon Valdez Oil in Prince William Sound Beaches*. Anchorage, Alaska, USA: *Exxon Valdez* Oil Spill Trustee Council Restoration Project 11100836. [http://www.evostc.state.ak.us/Projects/ProjectInfo.cfm?project_id=2189]

Bragg, J.R., R.C. Prince, and R.M. Atlas (1994). Effectiveness of bioremediation for oiled intertidal shorelines. *Nature* **368**(6470): 413–418.

Bragg, J.R., R.C. Prince, J.B. Wilkinson, and R.M. Atlas (1992). *Bioremediation for Shoreline Cleanup following the 1989 Alaskan Oil Spill*. Houston, Texas, USA: Exxon Company, USA. [http://www.valdezsciences.com]

Button, D.K., B.R. Robertson, and K.S. Craig (1981). Dissolved hydrocarbons and related microflora in a fjordal seaport: Sources, sinks, concentrations, and kinetics. *Applied and Environmental Microbiology* **42**(4): 708–719.

Chianelli, R.R., T. Aczel, R.E. Bare, G.N. George, M.W. Genowitz, M.J. Grossman, C.E. Haith, F.J. Kaiser, R.R. Lessard, R. Liotta, R.L. Mastracchio, V. Minak-Bernero, R.C. Prince, W.K. Robbins, E.I. Stiefel, J.B. Wilkinson, S.M. Hington, J.R. Bragg, S.J. McMillen, and R.M. Atlas (1991). Bioremediation technology development and application to the Alaskan spill. In *Proceedings of the 1991 International Oil Spill Conference (Prevention, Behavior, Control, Cleanup), March 4–7, 1991, San Diego, California*. Washington DC, USA: American Petroleum Institute Technical Publication 4529; pp. 549–558.

Koons, C.B. and H.O. Jahns (1993). The fate of oil from the *Exxon Valdez*: A perspective. *Marine Technology Society Journal* **26**(3): 61–69.

Leschine, T.M., J. McGee, R. Gaunt, A. van Emmerik, D.M. McGuire, R. Travis, and R. McCready (1993). T/V Exxon Valdez *Oil Spill: Federal On Scene Coordinator's Report*. Washington DC, USA: United States Department of Transportation, United States Coast Guard; Report DOT-SRP-94-1; National Technical Information Service Order Number PB94-121845; Volume 1; pp. 198–200.

Li, H. and M.C. Boufadel (2010). Long-term persistence of oil from the *Exxon Valdez* spill in two-layer beaches. *Nature Geoscience* **3**(2): 96–99.

Michel, J., Z. Nixon, and L. Cotsapas (2006). *Evaluation of Oil Remediation Technologies for Lingering Oil from the* Exxon Valdez *Oil Spill in Prince William Sound*. Juneau, AK, USA: National Oceanic and Atmospheric Administration, National Marine Fisheries Service: *Exxon Valdez* Oil Spill Restoration Project 050778 Final Report. [http://www.evostc.state.ak.us/Files.cfm?doc=/Store/FinalReports/2005–050778-Final.pdf&]

National Research Council (2003). *Oil in the Sea III: Inputs, Fates, and Effects*. Washington DC, USA: National Academy Press; ISBN-10: 0309084385. [http://www.nap.edu/openbook.php?record_id=10388&page=R1]

Neff, J.M., A.E. Bence, K.R. Parker, D.S. Page, J.S. Brown, and P.D. Boehm (2006) Bioavailability of polycyclic aromatic hydrocarbons from buried shoreline oil residues thirteen years after the *Exxon Valdez* oil spill: a multispecies assessment. *Environmental Toxicology & Chemistry* **25**(4): 947–961.

Neff, J.M., E.H. Owens, S.W. Stoker, and D. McCormick (1995). Shoreline oiling conditions in Prince William Sound following the *Exxon Valdez* oil spill. In Exxon Valdez *Oil Spill: Fate and Effects in Alaskan Waters*. P.G. Wells, J.N. Butler, and J.S. Hughes, eds; Philadelphia, PA, USA: American Society for Testing and Materials; ASTM Special Technical Publication 1219; ISBN-10: 0803118961; pp. 312–346.

Owens, E.H. (1991). Shoreline conditions following the *Exxon Valdez* oil spill as of fall 1990. In *Proceedings of the 14th Arctic and Marine Oilspill Program (AMOP) Technical*

Seminar, June 12–14, 1991, Vancouver, British Columbia, Canada. Ottawa, ON, Canada: Environment Canada; pp. 579–606.

Page, D.S., P.D. Boehm, and J.M. Neff (2008). Shoreline type and subsurface oil persistence in the *Exxon Valdez* spill zone of Prince William Sound, Alaska. In *Proceedings of the 31st Arctic and Marine Oilspill Program (AMOP) Technical Seminar, June 3–5, 2008, Calgary, Alberta, Canada*. Ottawa, ON, Canada: Environment Canada; pp. 545–563.

Pope, G.A., K.D. Gordon, and J.R. Bragg (2011a). Fundamental reservoir engineering principles explain lenses of shoreline oil residue twenty years after the *Exxon Valdez* oil spill. In *Proceedings of the Society of Petroleum Engineers' Americas E&P Health, Safety, Security, and Environmental Conference, March 21–23, 2011, Houston, Texas*. Houston, TX, USA: Society for Petroleum Engineers; SPE Paper 141809.

Pope, G.A., K.D. Gordon, and J.R. Bragg (2011b). Using fundamental practices to explain field observations twenty-one years after the *Exxon Valdez* oil spill. In *Proceedings of the 2011 International Oil Spill Conference (Promoting the Science of Spill Response), May 24–26, 2011, Portland, Oregon, USA*. Washington DC, USA: American Petroleum Institute.

Prince, R.C., R.E. Bare, R.M. Garrett, M.J. Grossman, C.E. Haith, L.G. Keim, K. Lee, G.J. Holtom, P. Lambert, G.A. Sergy, E.H. Owens, and C.C. Guénette (2003). Bioremediation of stranded oil on an Arctic shoreline. *Spill Science & Technology Bulletin* 8(3): 303–312.

Prince, R.C., J.R. Clark, J.E. Lindstrom, E.L. Butler, E.J. Brown, G. Winter, W.G. Steinhauer, G.S. Douglas, J.R. Bragg, J.E. Harner, and R.M. Atlas (1993). Bioremediation of the *Exxon Valdez* oil spill: Monitoring safety and efficacy. In *Bioremediation of Chlorinated and Polycyclic Aromatic Hydrocarbon Compounds*. R.E. Hinchee, B.C. Alleman, R.E. Hoeppel, and R.N. Miller, eds. Boca Raton, FL, USA: Lewis Publishers. ISBN-10: 0873719832; ISBN-13: 9780873719834; pp. 107–124.

Pritchard, P.H. and C.F. Costa (1991). EPA's Alaska oil spill bioremediation project. Part 5. *Environmental Science & Technology* 25(3): 372–379.

Pritchard, P.H., C.F. Costa, and L. Suit (1991). *Alaska Oil Spill Bioremediation Project, Science Advisory Board Draft Report*. Gulf Breeze, FL, USA: US Environmental Protection Agency, Office of Research and Development, Environmental Research Laboratory; EPA Report EPA/600/9–91/046a. [http://nepis.epa.gov/Exe/ZyPURL.cgi?Dockey=2000C9BB.txt *and* http://nepis.epa.gov/Exe/ZyPURL.cgi?Dockey=2000C9JQ.txt]

Shennan, J.L. (1984). Hydrocarbons as substrates in industrial fermentations. In *Petroleum Microbiology*. R. Atlas, ed. New York, NY, USA: Macmillan Publishing Company; ISBN-10: 0029490006; pp. 643–683.

Short, J.W., G.V. Irvine, D.H. Mann, J.M. Maselko, J.J. Pella, M.R. Lindeberg, J.M. Payne, W.B. Driskell, and S.D. Rice (2007). Slightly weathered *Exxon Valdez* oil persists in Gulf of Alaska beach sediments after 16 years. *Environmental Science & Technology* 41(4): 1245–1250.

Short, J.W., M.R. Lindeberg, P.M. Harris, J.M. Maselko, J.J. Pella, and S.D. Rice (2004). Estimate of oil persisting on beaches of Prince William Sound, 12 years after the *Exxon Valdez* oil spill. *Environmental Science & Technology* 38(1): 19–25.

Short, J.W., J.M. Maselko, M.R. Lindeberg, P.M. Harris, and S.D. Rice (2006). Vertical distribution and probability of encountering intertidal *Exxon Valdez* oil on shorelines of three embayments within Prince William Sound. *Environmental Science & Technology* **40**(12): 3723–3729.

Taylor, E. and D. Reimer (2008). Oil persistence on beaches in Prince William Sound: a review of SCAT surveys conducted from 1989 to 2002. *Marine Pollution Bulletin* **56**(3): 458–474.

US Environmental Protection Agency (1989a). *Bioremediation of Exxon Valdez Oil Spill*. Washington DC, USA: US Environmental Protection Agency, Office of Research and Development; Press Release, July 31, 1989; Letter to K.T. Koonce, Exxon Corporation, July 26, 1989. [http://www.epa.gov/history/topics/valdez/01.html]

US Environmental Protection Agency (1989b). *Ambient Water Quality Criteria for Ammonia (Salt Water): 1989*. Narragansett, RI, USA: US Environmental Protection Agency, Office of Research and Development, Environmental Research Laboratory; EPA 440/5–88–004. [http://water.epa.gov/scitech/swguidance/standards/upload/2001_10_12_criteria_ambientwqc_ammoniasalt1989.pdf]

US Environmental Protection Agency (1990). *Alaskan Oil Spill Bioremediation Project*. Washington DC, USA: US Environmental Protection Agency, Office of Research and Development. National Service Center for Environmental Publications; EPA/600/8–89/073. [http://nepis.epa.gov/Exe/ZYPURL.cgi?Dockey=30001ISV.txt]

Venosa, A.D., J.R. Haines, and D.M. Allen (1992). Efficacy of commercial inocula in enhancing biodegradation of weathered crude oil contaminating a Prince William Sound beach. *Journal of Industrial Microbiology & Biotechnology* **10**(1): 1–11.

Venosa, A.D., P. Campo, and B.A. Wrenn (2010). Biodegradability of lingering crude oil 19 years after the *Exxon Valdez* oil spill. *Environmental Science & Technology* **44**(19): 7613–7621.

Wolfe, D.A., M.J. Hameedi, J.A. Galt, D. Watabayashi, J. Short, C. O'Clair, S. Rice, J. Michel, J.R. Payne, J. Braddock, S. Hanna, and D. Sale (1994). Fate of the oil spilled from the T/V *Exxon Valdez* in Prince William Sound, Alaska. *Environmental Science & Technology* **28**(13): 561A–568A.

Zhu, X., A.D. Venosa, and T. Suidan (2004). *Literature Review of the Use of Commercial Bioremediation Agents for Cleanup of Oil-Contaminated Estuarine Environments*. Cincinnati, OH, USA: US Environmental Protection Agency, Office of Research and Development, National Risk Management Research Laboratory. National Service Center for Environmental Publications; EPA/600/R-04/075. [http://nepis.epa.gov/Adobe/PDF/2000E76J.pdf]

PART III

BIOLOGICAL EFFECTS

INTRODUCTION

As oil spreads through the marine environment and undergoes the changes described in Part II, concerns are raised by the general public, scientists, and regulators about possible effects on biological resources. These concerns may be most obvious when they relate to important subsistence and commercially valuable resources such as salmon or herring, but they extend as well to a variety of wildlife. In the days and weeks following an oil spill, speculations and hyperbole abound. Documenting whether there are actual effects, how long they last, and whether the effects are due to the oil spill or something else requires rigorous science. The chapters in this section describe how science was brought to bear on assessing potential injury to natural resources and what was learned in the process, both about the effects of the *Exxon Valdez* spill and about the challenges of conducting scientific investigations in a harsh and variable environment with much at stake.

We begin with two chapters that provide essential background on several analytical and design issues that reappear in subsequent chapters. In Chapter 9, James Oris and Aaron Roberts provide a cautionary review of the use of biomarkers – biochemical and molecular responses of organisms to chemicals and other stressors in the environment – as indicators of exposure. They focus on cytochrome P450 1A (CYP1A), which was used to assay exposure of several species to aromatic hydrocarbons from fossil fuels following the *Exxon Valdez* spill. Because CYP1A can be mobilized in response to a wide array of hydrocarbons, elevated concentrations do not provide unambiguous evidence of exposure to a specific hydrocarbon source, nor do they link exposure to potential injuries, so they must be interpreted with care.

Careful interpretation is also a theme in Chapter 10, in which Keith Parker, John Wiens, Robert Day, and Stephen Murphy bring statistics and scientific methodology to bear in discussing the confounding influences of temporal and spatial variation on study design and statistical assessments of environmental accidents such as oil spills. Because different locations in a spill path may suffer different levels of oiling that are superimposed on pre-existing environmental differences, and because both oiling and environmental conditions change over time in different ways in different locations, attributing any observed effects to an oil spill (or any environmental disruption) without incorporating differences in space and time into analyses and considering alternative causes is difficult.

The remaining chapters in this section consider how scientific studies were conducted to evaluate the potential effects of the *Exxon Valdez* spill on key elements of the biota of Prince William Sound: shoreline organisms (Chapter 11), pink salmon (Chapter 12), Pacific herring (Chapter 13), marine birds (Chapter 14), and sea otters (Chapter 15). All of the studies described in these chapters were confronted with the issues of study design and analysis mentioned above, as well as the challenges of dealing with conflicting results from different studies. They dealt with these issues and challenges in different ways, reflecting the vast differences in life histories among the target taxa, their habitats, and their exposure and potential vulnerability to acute and chronic oiling effects. Despite these differences, some common themes emerged: the importance of concurrent measures of the biota, chemistry, and habitat; the value of employing multiple approaches to gauge spill effects and recovery; the imperative of defining recovery in operational terms with realistic targets; the need to disentangle the effects of sampling from those of oiling; and the importance of ensuring that postulated oiling effects are based on well-established and biologically plausible exposure pathways rather than inference, to name a few.

CHAPTER NINE

Cytochrome P450 1A (CYP1A) as a biomarker in oil spill assessments

James T. Oris and Aaron P. Roberts

9.1 Introduction

More than 30 years ago, scientists began measuring biochemical and molecular responses in organisms as a way to understand pathways of exposure to chemicals in the environment. These responses, termed *biomarkers*, help screen for the presence or absence of classes of chemicals (e.g., aromatic hydrocarbons, metals) and other stressors (e.g., temperature, oxidative stress). They can also indicate possible mechanisms or pathways of potential toxic outcomes and provide direction for additional, more detailed analysis of the effects of exposures.

Biomarkers have been used extensively in studies of oil spills (Anderson and Lee, 2006). Investigations following the *Exxon Valdez* spill considered biomarkers for many species, including sea otters (*Enhydra lutris*), river otters (*Lontra canadensis*), harlequin ducks (*Histrionicus histrionicus*), Barrow's goldeneye (*Bucephala islandica*), black oystercatchers (*Haematopus bachmani*), pigeon guillemots (*Cepphus columba*), intertidal fish, rockfish (*Sebastes* spp.), bottom fish, pink salmon embryos (*Oncorhynchus gorbuscha*), and mussels (*Mytilus* spp.).

No biomarker has received more attention than the Cytochrome P450 1A (CYP1A) enzyme system. Tens of thousands of papers have been published on using the CYP1A system as evidence of exposure to aromatic hydrocarbons found in fossil fuels and industrial chemicals. However, there are conflicting opinions in the literature on using the CYP1A system as a measure of low-level oil exposure when multiple sources of aromatic hydrocarbons are present. There are also conflicting opinions on whether the CYP1A system can be used as an indicator of both exposure and of effect or injury.

In this chapter, we focus on the use and utility of specific components of the CYP1A system as biomarkers of exposure to oil and oil components. We also discuss whether these components can be useful biomarkers of harmful effects in oil-spill assessments.

Oil in the Environment: Legacies and Lessons of the Exxon Valdez *Oil Spill*, ed. J. A. Wiens. Published by Cambridge University Press. © Cambridge University Press 2013.

9.2 Introduction to biomarkers

A *biomarker* is an attribute of a living system that can be used to indicate a condition or change in response to an internal or external stimulus. In other words, it is a **bio**logical measure that **marks** a condition or change. Biomarkers are used to characterize chemical and stressor exposure and effects in a wide variety of organisms (Schlenk *et al.*, 2008).

Scientists can use biomarkers at any scale of biological organization:

- *Molecules and biochemical pathways:* abnormal genetic information (DNA) in a developing fetus can be a biomarker of developmental abnormalities (e.g., an extra chromosome 21 is a biomarker for Down syndrome).
- *Cells, tissues, and organs:* the presence of specific chemicals in bodily fluids can be a biomarker of disease (e.g., a high level of prostate-specific antigen in blood is a biomarker for prostate cancer).
- *Whole organism:* elevated body temperature (i.e., fever) is a biomarker of bacterial or viral infection.
- *Populations:* Rates or incidence of disease in groups of organisms (e.g., prevalence of heart disease in obese compared to normative-weight individuals) is a biomarker.
- *Communities, ecosystems, and landscapes:* Walking along a path and making note of the different kinds of plants and animals you see can be a biomarker (e.g., hardwood forest versus a desert).

In order to use a biomarker, studies must be conducted to establish ranges of levels that would be considered physiologically normal or expected. Thus, what makes a biomarker a biomarker is its ability to discriminate when deviations from normal or expected conditions occur and can be detected. "Normal" or "expected" are determined using scientific evidence and best professional judgment. Levels of detectable change are determined by the statistical characteristics of the attribute measured as the biomarker, including variability, sources of measurement error, and sample sizes.

In the context of this chapter, biologists use the term "biomarker" differently than petroleum chemists and geologists. Petroleum or geochemical biomarkers are suites of organic compounds (derived from organisms that died) accumulated in deposits and formed into fossil fuels (Wang and Stout, 2007). These compounds have been used for many years to investigate the properties, sources, and origins of petroleum and fossil fuels (see Chapter 1, Box 1.1).

9.3 Categories of biomarkers

There are three categories of biomarkers that correspond to information necessary to conduct ecological risk assessments (Schlenk, 1999): exposure, effects, and susceptibility.

9.3.1 Exposure

Biomarkers of exposure respond as a result of the uptake of a chemical or chemicals from the environment. They are typically molecules or biochemicals that are activated after chemical or stressor exposure. They respond rapidly, metabolizing and excreting the foreign chemical or repairing its damage. When the exposure stops and the metabolism or repair process is complete, the biomarker turns off and returns to normal, background levels.

Biomarkers of exposure have gained fairly widespread acceptance in the scientific community in the context of ecological risk assessment (Schlenk et al., 2008). Most biomarkers of exposure respond to specific classes of chemical or stressor, and thus can help identify what was available in the environment at levels sufficient enough to trigger a biological response. In addition, biomarkers of exposure correlate relatively well with the magnitude of the exposure, providing some indication of the amount of biologically available chemical or stressor present in the environment.

Components of the CYP1A system are commonly used as biomarkers of exposure. It is a metabolic enzyme system that is turned on upon exposure to aromatic hydrocarbons. Other common biomarkers of exposure include a sequestering protein induced upon exposure to heavy metals (metallothionein) and an egg-yolk protein induced upon exposure to estrogen-like chemicals (vitellogenin). All three of these biomarkers respond to broad classes of chemicals; thus, without complementary studies, these biomarkers cannot decipher to which specific aromatic hydrocarbon, metal, or estrogen-like chemical an organism was exposed. Additional studies to maximize the effectiveness of biomarkers of exposure include:

- Environmental analysis, which can provide information about the specific chemicals present that triggered the biomarker response.

- Testing and validation under laboratory conditions prior to use in field situations (McClain et al., 2003). Local populations can acclimate to long-term exposure, which alters how readily biomarkers of exposure are turned on (Meyer et al., 2003). It is therefore important to conduct exposure assessments using organisms that are naïve to the site under consideration, such as organisms caged and raised in the lab or collected from areas known not to have the target chemicals present (McClain et al., 2003; Roberts et al., 2005, 2006).

- Parallel studies that compare sites known to be contaminated with those known not to be contaminated. This helps account for environmental variability in biomarker responses in field situations (Roberts et al., 2005, 2006).

9.3.2 Effect

Biomarkers of effect indicate damage or dysregulation of a biological process that results in adverse effects. They can be measured at any scale, from molecular to ecosystem, and may vary considerably in their specificity (Schlenk, 1999).

Measures made at the suborganismal level focus on physiological processes that, if disrupted, may lead to tissue or organ failure. For example, inhibition of an enzyme responsible for nerve function (acetylcholinesterase) can be used as a biomarker of insecticide poisoning. Inhibition of an enzyme that plays a role in red blood cell formation (delta-aminolevulinate dehydratase) can be used as a biomarker of lead poisoning.

At the tissue or organ level, adverse effects can be shown through microscopic examinations of pathological conditions. Conditions such as necrosis (uncontrolled cell death due to membrane breakdown) and apoptosis (controlled cell death due to DNA damage), or the presence of precancerous or cancerous lesions, are commonly used as biomarkers to define the type and extent of adverse, tissue level effects.

Biomarkers of effect at the population level include cancer incidence, reduced abundance, reduced reproduction, or increased mortality.

At the community and ecosystem scales, changes in species diversity or changes in nutrient flux can be used as biomarkers of effect.

9.3.3 Susceptibility

Biomarkers of susceptibility indicate individual differences in sensitivity to a stressor within a population. Individual variability within populations is a hallmark of biology; it is the basis for adaptation and evolution. More apparent examples include sensitivity to sunburn in fair-skinned individuals or increased risk of osteoporosis in postmenopausal women of Caucasian or Asian descent. Different sensitivities can be identified by measuring elements of metabolic pathways that lead to increased or decreased rates of cellular damage or repair. For example, scientists have identified specific genetic variants of enzymes that are strongly correlated to an individual's susceptibility to alcohol dependence (e.g., Mulligan *et al.*, 2003) or to their susceptibility to a wide range of chemically induced diseases and cancers (e.g., Walraven *et al.*, 2008).

9.4 Uses of biomarkers in ecological toxicology assessments

In ecological studies, biomarkers can detect exposure to and potential effects of chemicals and stressors in plants and animals (Fig. 9.1). While in protecting human health the concern is for individual organisms, in protecting ecosystems the focus is primarily on effects at the population, community, or ecosystem level.

Figure 9.1 Scientists working in PWS on a project to examine biomarkers of contaminant exposure in juvenile coho salmon (*Oncorhynchus kisutch*) (Roberts *et al.*, 2006). (a) Juvenile fish obtained from a fish hatchery (and thus with no exposure history to chemicals in PWS) were placed in cages at field sites for 48 hours. Image illustrates collection of a cage at the end of an exposure period. (b) Juvenile fish recovered from exposure cage. (c) Scientists dissect liver and gill tissue from fish, place tissues in vials, and immerse vials in liquid nitrogen to preserve tissues for subsequent biomarker analysis. (Photos: James T. Oris)

A significant amount of research in ecological toxicology has been conducted on developing highly sensitive, rapid-response biomarkers at the individual level that can predict effects at the population level and beyond; however, results have been mixed (Schlenk *et al.*, 2008). Linking short-term, individual responses to broader effects has met with limited success and considerable criticism (e.g., Forbes *et al.*, 2006; Emlen and Springman, 2007).

Environmental scientists have taken advantage of recent advances in clinical science and are now exploring the use of large numbers of biomarkers simultaneously to link suborganismal responses to population-scale responses (cf., Hoffmann and Oris, 2006; Hoffmann *et al.*, 2006; Ankley *et al.*, 2007). Despite these advances, however, there are still critics of using biomarkers for anything other than showing mechanisms or assessing exposure at the individual level (Forbes *et al.*, 2008).

9.5 The Cytochrome P450 (CYP) family of enzymes

Like other enzymes, CYPs are proteins that catalyze biochemical reactions. Their name – Cytochrome P450 – gives a clue about their chemical characteristics.

- A *cytochrome* is a membrane-bound protein that contains iron in the form of a heme group and functions in electron-transport reactions. The heme group in CYPs can bind and release oxygen – similar to hemoglobin in the blood. CYPs can also transfer electrons from one chemical to another; in their functional form, CYPs are bound to cellular membranes.
- P450 means that, in a test tube, if the heme group is saturated with electrons and oxygen is removed using carbon monoxide, CYPs absorb light with a characteristic absorption peak at 450 nanometers.

Thus, CYPs are **CY**tochromes with an absorption **P**eak at **450** nm. The protein (i.e., enzyme) form of CYP is written in plain font (CYP1A), whereas the gene on the DNA is represented in italics (*CYP1A*).

CYPs are found in all living things, including bacteria, fungi, plants, and animals (Nebert, 2005). All CYPs catalyze the same basic reaction of adding an oxygen atom to a chemical substrate (Box 9.1). The original CYPs evolved in bacterial ancestors about 3 billion years ago and likely played a role in the synthesis and maintenance of the cell membrane (Nelson, 1998). Since then, the general reaction that CYPs catalyze has found thousands of uses.

While CYP is the superfamily of enzymes, an Arabic numeral adds further specificity by designating the family. CYPs with an amino acid sequence similarity of 40% or

Box 9.1 The basic CYP reaction

The basic reaction that CYP enzymes catalyze is given by:

$$R-H + O_2 + 2H^+ + 2e^- \rightarrow R-OH + H_2O$$

R represents an organic molecule of varied structure (e.g., steroid, fatty acid, drug, environmental chemical). A hydrogen atom (H) on one part of the molecule is replaced with an atom of oxygen bound to a hydrogen (OH). The second atom of oxygen is converted into water (H_2O). The reaction requires two additional hydrogen ions (H^+) and two electrons (e^-) that are supplied by an electron donor.

greater are put into the same family. There are at least 37 families of CYPs (CYP1–CYP37). Within a family, CYPs with an amino acid sequence similarity of 55% or greater are put into the same subfamily, designated with a capital letter (e.g., CYP1A). Finally, based on the specific reaction they catalyze, individual enzymes are numbered with an additional Arabic numeral (e.g., CYP1A1). All CYP families from CYP5 and higher catalyze a few specific biosynthetic reactions. CYP1–CYP4, however, are more diverse and catalyze both normal biosynthetic reactions as well as reactions that metabolize drugs and environmental chemicals (Nebert, 2005).

All organisms are challenged by the accumulation of foreign chemicals absorbed directly from the environment or taken in through their diets. If allowed to accumulate unchecked, any chemical will eventually become toxic. An organism's primary mechanism for eliminating foreign chemicals is excretion, or egestion, in a watery medium. CYPs ability to add oxygen to chemicals serves two purposes. First, it increases the chemical's water solubility, promoting excretion. Second, the oxygen itself serves as a reactive "handle," allowing other enzyme systems to react with the oxygenated chemical. This makes it extremely water soluble, nearly always abolishing toxicity. Adding oxygen is termed a *Phase I* reaction, since it is typically the first step in eliminating a foreign chemical. *Phase II* reactions further detoxify and allow easy excretion.

9.5.1 Cytochrome P450 1A (CYP1A)

The CYP1A system has been characterized in organisms from bacteria to humans. Overall, in studies of the biochemical effects of chemicals in the natural environment, various specific components of the CYP1A system are probably the most commonly used biomarkers. The CYP1A system has both a normal physiological (i.e., endogenous) function and a function in the metabolism of planar aromatic hydrocarbons (i.e., exogenous).

It is normal to have CYP1A activity in the absence of an exogenous exposure. Although CYP1A's endogenous pathways and reactions are not fully known, scientists do know that CYP1A is induced or suppressed at specific times during embryonic development or during the reproductive cycle of many vertebrates. The endogenous function of CYP1A may involve the normal biochemical processing of a large group of chemicals important in development, reproduction, and normal cellular maintenance (Nebert and Karp, 2008). In organisms with dysfunctional *CYP1A* genes, a variety of developmental syndromes occur even without exposure to foreign chemicals. In mice, *CYP1A*-deficient individuals display reduced viability and cardiac abnormalities (Nebert and Karp, 2008). Similar syndromes have been observed in fish (e.g., Goldstone et al., 2009).

The CYP1A protein is a Phase I enzyme in the metabolism and detoxification of small planar aromatic hydrocarbons. Oxygen added to planar aromatic hydrocarbons increases polarity and enhances reactivity with Phase II enzymes. Phase II enzymes connect large, water-soluble groups to the added oxygen, which completes the detoxification process and allows the planar aromatic hydrocarbons to be excreted.

9.5.2 Cellular regulation of CYP1A

The CYP1A enzymes are regulated in a multistep, negative feedback loop through a receptor called the *aromatic hydrocarbon receptor* (AHR). The AHR is a protein in cells that binds to small, planar (i.e., flat) molecules such as polycyclic aromatic hydrocarbons (PAH), chlorinated dioxins (e.g., TCDD), and certain polychlorinated biphenyls (PCBs) (Nebert and Karp, 2008).

AHRs are cellular monitors for PAH. When a PAH enters the cell, it binds to the AHR. This complex moves from the cell's cytoplasm into the cell's nucleus. Once in the nucleus, the complex forms a *transcription factor* that can bind to DNA at specific locations near the *CYP1A* gene. These binding locations are called *aromatic hydrocarbon response elements* (AHREs).

Upon binding to the AHRE, *CYP1A* is turned on and transcribed into messenger RNA (mRNA). The mRNA moves out to the cytoplasm, where it is processed into the sequence of amino acids that form the CYP1A enzyme. After additional processing and activation, the CYP1A enzyme can add oxygen to PAH (Box 9.1). The PAH can then undergo additional reactions with the CYP1A enzyme or it can be acted upon by other enzymes in Phase II reactions, resulting in the metabolism and excretion of the PAH.

As PAH molecules are metabolized, there are fewer AHR–PAH complexes to interact with the DNA and make new CYP1A enzyme. Therefore, when the cell is no longer exposed to PAH, the CYP1A system is turned down to background levels over a relatively short period of time (hours to days). These background levels are controlled by an endogenous pathway that is always present to maintain a low level of CYP1A activity. To emphasize, a cell may show measurable levels of CYP1A enzyme activity, even absent any exposure to a contaminant.

When the CYP1A system is turned on, so too are a series of Phase I, Phase II, and other enzyme systems. The system of coordinated control of these enzymes through the AHR activation pathway has been referred to as the *CYP Gene Battery* (Nebert, 2005). The CYP Gene Battery results in pathways where rapid metabolism, detoxification, and excretion of PAH and other aromatic hydrocarbons predominate.

Any compound that binds to the AHR will turn on the CYP1A system, including endogenous and foreign compounds. Thus, comparisons of CYP1A induction between potentially exposed and unexposed organisms can only inform the investigator that exposure to a compound that binds to the AHR (e.g., PAH, PCB, or dioxins in general) has occurred. Additional information, such as analytical measurements of specific chemicals in the environment or tissues of an organism, is required to provide evidence of specific chemical exposure. Concluding that an organism has been exposed to oil just because CYP1A activity is present or elevated, in the absence of environmental or tissue chemical analyses, is unjustified. Nonetheless, some investigators continue to make conclusions concerning sources of exposure in the absence of chemical-specific analyses and exposure pathways (e.g., Esler *et al.*, 2010, 2011; see Chapter 16).

9.5.3 Is CYP1A a detoxification pathway or an activation pathway?

CYP1A enzyme reactions typically detoxify and eliminate chemicals (Nebert *et al.*, 2004; Puga *et al.*, 2005). This is not to be confused with toxic syndromes that can develop through the reaction of AHR with, primarily, other *CYP* genes (such as *CYP2A* or *CYP1B*) and other non-*CYP* genes (e.g., growth regulating and immune system genes). Studies in cell cultures have shown that certain PAH compounds can be metabolized into forms that bind to DNA, causing mutations that lead to tumors and cancers.

Studies with PCB or dioxins, in which the CYP1A system is strongly activated but does not metabolize or excrete the exogenous chemical readily, caused initial concern about the role of CYP1A. Even short-term exposure to PCB or dioxins results in a long-term elevation of the AHR-mediated pathways that can lead to toxic outcomes. But it is now generally accepted that other genes are responsible.

PAH exposure may still cause toxicity: prolonged exposure to PAH or exposure during sensitive life stages has been shown to cause toxic outcomes in organisms. However, the effects are mediated through both AHR-dependent and independent pathways (Incardona *et al.*, 2005, 2009; Hicken *et al.*, 2011).

9.5.4 Measuring CYP1A: as a protein, as catalytic activity, and as mRNA

Various, specific components of the CYP1A system can be measured in cells or tissues, and it is these measures that are used as biomarkers.

CYP1A can be measured as the protein in terms of "is it physically there?" using several techniques, including enzyme-linked immuosorbent assays (ELISAs), staining electrophoresis gels (Western blots), and immunohistochemical staining (IHC). In IHC, antibodies bind to a specific part of the CYP1A protein, producing a color change or stain in tissue samples mounted on slides. Based on the extent and intensity of staining, researchers either subjectively score how much CYP1A has been triggered (e.g., $1 =$ low through $5 =$ high) or use spectroscopy to determine optical density of the staining. As the results are not quantitative, IHC can neither be used to determine the concentration of PAH exposure nor to compare results among experiments (Oris and Roberts, 2007).

CYP1A can also be measured in terms of catalytic activity. **Ethoxyresorufin-O-deethylase (EROD)** measures chemical reactions being carried out by CYP1A protein as a specific rate or activity. EROD is the most common method for measuring CYP1A and, arguably, the best studied (Whyte *et al.*, 2000). It was used extensively in assessments of the *Exxon Valdez* oil spill (e.g., Trust *et al.*, 2000; Huggett, *et al.*, 2003, 2006; Carls *et al.*, 2005; Roberts *et al.*, 2006; Springman *et al.*, 2008a, b). The EROD assay is semiquantitative in that the total amount of CYP1A protein can be estimated in a cell or tissue. While EROD activity correlates relatively well to the level of PAH, PCB, and dioxin exposures in a laboratory, field studies must establish control (reference) or nonexposed field sites that closely match the suspected exposed sites (Roberts *et al.*, 2005, 2006; Oris and Roberts, 2007).

Other techniques can measure how much CYP1A mRNA is being made, as an indirect measure of the CYP1A itself. In general, the more CYP1A mRNA being made, the more CYP1A protein the cell will make. **Reverse transcriptase–polymerase chain reaction (RT-PCR)** – sometimes called realtime PCR or quantitative PCR – is the most common and modern way (Oris and Roberts, 2007). Like EROD, RT-PCR is semiquantitative: the reaction will produce DNA copies of the desired mRNA in proportion to the amount of mRNA in the original sample. To our knowledge, however, no investigation has been able to quantitatively relate CYP1A RT-PCR results to PAH concentration in the environment, especially in field studies and oil-spill assessments.

Failure to follow appropriate quality-control procedures in RT-PCR can lead to erroneous results and conclusions about exposure to oil. For example, Bodkin *et al.* (2002) and Ballachey *et al.* (2002) reported RT-PCR results for sea otter tissues collected from different sites in Prince William Sound (PWS) oiled by the *Exxon Valdez* spill. They concluded that there was evidence of oil exposure due to elevated CYP1A mRNA. However, RNA purity, RNA integrity, and primer specificity were not reported

(Hook et al., 2008). Further examination of their procedures indicated that the step to remove DNA from the initial extraction was likely omitted and that the amplified product from the PCR step was most closely associated with a noncoding region of DNA unrelated to CYP1A mRNA (Hook et al., 2008). Quality assurance concerns rendered the conclusions invalid; however, some investigators still cite these studies as evidence of oil exposure in PWS sea otters (e.g., Esler et al., 2010, 2011; see discussion in Chapter 16).

9.5.5 Use of CYP1A as a biomarker: what is "significant"?

Because the CYP1A system is normally active in the background, using it as a biomarker for PAH exposure requires measuring the difference in CYP1A levels between exposed and nonexposed organisms. What is a "significant" triggering of CYP1A is defined by each investigator based on the type of CYP1A measurement used and on statistical significance or best professional judgment. With EROD activity, most investigators rely on the statistical standard of a Type I error of 5% (i.e., $\alpha = 0.05$). For mRNA measurements, it is most common to compare exposed and nonexposed organisms in a relative manner and use professional judgment to decide whether they are different. An informal survey of scientists and the literature (James T. Oris, unpublished) indicates that two- to five-fold increases in mRNA levels are considered biologically significant. Other scientists use the threshold value for the number of copy cycles in the amplification step of PCR required to detect the presence of the mRNA (i.e., the C_t value) and analyze with standard statistics, assuming – incorrectly – that statistical significance equates to biological significance. Statistically significant changes in CYP1A, however, have never been shown to directly correspond with biological significance. Statistical significance is related to the design of an experiment (i.e., the number of replicates and the variability of the responses). Professional judgment in the absence of studies to show biological significance will always be subjective.

Statistical versus biological significance can be illustrated by the study of Golet et al. (2002) on CYP1A levels in pigeon guillemot adults and chicks. A decade after the *Exxon Valdez* oil spill, EROD activity in adults from previously oiled sites was significantly greater (statistically) than that in adults from unoiled (reference) sites. EROD activity was 1.6-fold higher in livers of adults from oiled sites than from unoiled sites tens of kilometers away (3.1 ± 0.4 pmol/min/mg, compared to 1.9 ± 0.2 [mean $\pm$ SE]). EROD activity in livers of chicks, however, was higher at unoiled sites than at oiled sites (4.7 ± 0.5, compared to 4.1 ± 0.4). The informal survey described above and the analysis by Oris and Roberts (2007; see below) indicate that a 1.6-fold difference is too small to be biologically relevant. However, because the results were statistically significant, Golet et al. (2002) suggested that "adults, but not chicks, were exposed to residual petroleum hydrocarbons at the oiled site a decade after the spill," albeit at low levels. They offered no explanation as to why chicks showed an opposite trend in EROD activity. Given the range and very low (in our judgment, background) level of EROD activity across all sites and ages studied, it seems reasonable that the small difference in EROD activity between oiled and unoiled adults and chicks should be considered biologically insignificant – especially in the absence of tissue PAH analyses. The large sample size (11–14 individuals per site within each age class) made the analysis very sensitive to small changes in EROD activity, and a post hoc analysis of the statistical power of the test indicates that even a modest reduction in sample size (e.g., to eight individuals) would negate the statistical difference.

9.6 Biomarker studies often fail to address variation

The significance of biomarker studies of PAH exposure and effects in oil-spill assessments is confounded by multiple sources of exposure, biogeochemical characteristics of home ranges, age of organisms, species, sex, temperature, and season (Anderson and Lee, 2006).

Oris and Roberts (2007) conducted an analysis of over 100 published CYP1A biomarker experiments in fish following the *Exxon Valdez* spill. They found a high degree of variation in CYP1A responses among fish species (differences in species sensitivity), measurement method (differences in technical sensitivity), and among or within laboratories (human error). Given current measurement methods, CYP1A cannot serve as a stand-alone, quantitative measure of exposure to a specific concentration of PAH. Rather, as a biomarker of exposure, CYP1A must be evaluated in the context of appropriate controls (unexposed fish, fish caged at reference sites, etc.), viewed as a semi-quantitative description of *relative* PAH exposure, and corroborated with multiple lines of investigation (e.g., measuring oil residues on shorelines, measuring exposures in mussels).

Additional variation happens at the site and individual levels. Differences in field conditions, life history, and physiological status (e.g., Anderson and Lee, 2006) can influence the sensitivity of the CYP1A response in individuals or populations. Natural chemicals in the diet of many of the organisms can affect biomarker response. For example, ingestion of dietary cyanobacteria triggers EROD activity in birds (Pašková et al., 2008). Numerous studies have demonstrated a suppressed CYP1A response in fish chronically exposed to PAH-contaminated environments. Collection and measurement of background CYP1A response of these populations would reveal nothing remarkable; however, chemical analysis of their habitat would demonstrate significant PAH concentrations.

A good example of a rigorous approach – although not related to an oil-spill assessment – was demonstrated in a study of seaducks (Steller's eiders, *Polysticta stelleri*, and harlequin ducks) in the Aleutian Islands (Miles et al., 2007). The study established clearly defined field sites and correlated PAH and PCB levels in mussel and seaduck tissues with levels of EROD activity in those ducks. By using multiple lines of evidence, this study demonstrated the relative bioavailability and importance of different sources of contaminants in the food chain that resulted in exposure and uptake in ducks.

This combination of biomarker and analytical chemistry measurements is rare, however. The assumption that, in the absence of information from analytical chemistry, induction of a biomarker such as CYP1A can be used as an accurate surrogate to identify the source and measure the environmental concentration of oil exposure is "doomed to failure" (Handy et al., 2003). Most field studies on biomarkers of exposure have used fish (Whyte et al., 2000; Oris and Roberts, 2007), but often the results are not easily interpreted or are not scientifically valid:

- Using fish caged on site cannot account for the multiple environmental factors that may influence CYP1A response. Roberts et al. (2006) report CYP1A mRNA expression in hatchery-reared fish caged at oiled and at unoiled sites in PWS. The study used juvenile, hatchery-reared salmon obtained from a single source to decrease variability associated with physiology and populations. The study did report increased CYP1A expression at a site oiled by the *Exxon Valdez* spill but also found increased CYP1A expression at an unoiled site that included an abandoned cannery where hydrocarbons were present. The CYP1A biomarker response itself was unable to distinguish between different PAH sources in the system.

- Exposing fish in a lab to concentrated doses of toxin does not mimic the long time period over which organisms would be exposed in the wild. Short *et al.* (2008) and Springman *et al.* (2008a, b) put semipermeable membrane devices (SPMDs) in pits at several sites in PWS where the SPMDs were exposed to subsurface oil residues, and let the SPMDs accumulate PAH. Extracts were taken from the SPMDs, injected into laboratory-reared fish, and CYP1A response was measured. The authors observed an increased CYP1A response, which they attributed to oil exposure. However, the SPMDs had accumulated PAH over several weeks. Although wild fish might be exposed over the same time period, they are unlikely ever to receive such a concentrated exposure all at once. In the wild, Phase I and II metabolic pathways in wild fish would clear the PAH as it accumulated in real time, as evidenced by the generally short-lived CYP1A response observed in wild fish following oil spills (Anderson and Lee, 2006). It is difficult to link CYP1A response following SPMD extract injection to any ecologically meaningful exposure or effect pathway.

As with the fish studies, it is common among nearly all bird and marine mammal studies that either biomarker levels or contaminant residues (but not both) were measured and that proper laboratory or field controls, including well-characterized contaminated and reference sites, were not included.

Marine bird biomarker studies from the *Exxon Valdez* oil spill have focused on EROD activity and CYP1A mRNA induction, primarily in harlequin ducks (e.g., Trust *et al.*, 2000; Miles *et al.*, 2007; Esler and Iverson, 2010; Esler *et al.*, 2010; Ricca *et al.*, 2010), Barrow's goldeneye (Trust *et al.*, 2000; Esler *et al.*, 2011), and pigeon guillemot (Golet *et al.*, 2002).

Less biomarker work has been done with marine mammals. The logistics and ethical considerations associated with collecting (legally protected) organisms and tissue samples preclude conducting detailed laboratory studies. This prevents scientists from establishing the required dose–response relationships and limits studies primarily to field-collected samples with limited scope and small sample sizes. A few laboratory studies have used mink (*Neovison vison*) as a model organism (e.g., Schwartz *et al.*, 2004; Bowen *et al.*, 2007) or conducted field studies with river otters (Ben-David *et al.*, 2001) and sea otters (e.g., Ballachey *et al.*, 2002; Bodkin *et al.*, 2002; Kannan and Perrotta, 2008; Bowen *et al.*, 2012). Studies of other marine mammals are even more limited in scope. Most involve field-collected samples and have been primarily descriptive. These include studies on several species of seals (Wolkers *et al.*, 1998; Nyman *et al.*, 2000; Miller *et al.*, 2005; Assunção *et al.*, 2007), dolphins (Montie *et al.*, 2008), and whales (Wilson *et al.*, 2005; Godard-Codding *et al.*, 2011).

In each case, conclusions about why biomarkers (CYP1A in particular) changed in one geographic area compared to another are speculative and not backed up by appropriate controls or analytical chemistry. This is especially true of *Exxon Valdez* studies conducted 5–10 years or more after the spill, when aromatic hydrocarbons from the spill were at or near background levels compared to other sources of chemicals or influences that may affect biomarker pathways.

Using biomarkers in wildlife is complicated and other factors can obscure the ability to describe specific, quantitative, and predictable dose–response relationships between hydrocarbon exposure and CYP1A induction (Ben-David et al., 2001). Diet and pyrogenic sources (e.g., from shipping or commercial or recreational fishing), historic human sources (e.g., abandoned fuel storage tanks; Page et al., 1999; Wooley, 2002; Roberts et al., 2006; Chapter 6), and natural sources (e.g., seeps; Page et al., 1996; Chapter 6) should have been considered in more studies following the *Exxon Valdez* oil spill.

9.7 Evidence of harm: CYP1A as a biomarker of exposure but not of injury

Only a few investigators have suggested that CYP1A is a biomarker of injury. In fact, Nebert et al. (2004) suggest that using CYP1A as a biomarker of potential effect in drug studies is not correct. In the area of oil-spill-impact assessment, however, it has been suggested that CYP1A can serve as a biomarker of both exposure and effect. This has been an area of vigorous debate in recent years, but new information on PAH-toxicity pathways indicates that CYP1A is a useful biomarker of exposure but not of effect (injury), even at the individual level. Incardona et al. (2005, p. 1761) concluded that

> [f]or many years, measuring CYP1A in field-collected samples has been the basis for assessing ecological damage and recovery after oil spills or remediation efforts in urbanized watersheds. However, CYP1A appears to play a protective rather than a causal role in petrogenic PAH toxicity. This greatly reduces the significance of CYP1A as a biomarker of PAH effect.

Hicken et al. (2011) also refer to CYP1A activity as being protective of low-level exposure to oil.

Despite the preponderance of evidence to the contrary, the idea that CYP1A can be used as an indicator of harm persists. For example, Carls et al. (2005) contended that "CYP1A is not only a biomarker, but is a bioindicator of population level PAH toxicity." Balk et al. (2011) refer to CYP1A as being an "effect biomarker," but in the sense of a physiological response to oil exposure rather than a biomarker of harm, and they studied a suite of other biomarkers in addition to CYP1A in their experiments. Curtis et al. (2011) cite liver EROD activity as an indicator of PAH exposure and make a correlation to DNA adducts in the blood, but they do not conclude that EROD is a biomarker of harm.

9.8 The future of biomarkers in oil-spill-impact assessment

Over the past several years, there has been an explosion of technical progress related to the field of biomarkers. In the 1990s, it was common to examine more than one biomarker (e.g., gene) at a time, but most often this was done with just a few. This was due to technical limitations. The deciphering of the human genome, and then the genomes of laboratory species; the development of tools to sequence genes rapidly and determine their function; the development of ultrasensitive and high-resolution detection methods; the development of miniaturized measurement techniques; and the development of the computing and database methods required to handle huge amounts of information have allowed scientists to measure more and more biomarkers simultaneously using genomic techniques. Three fields of genomic biomarker studies have emerged: transcriptomics (the study of multiple DNA biomarkers), proteomics (the study of multiple protein/enzyme biomarkers), and metabolomics (the study of patterns of metabolites from physiological or detoxification pathways). In transcriptomics, gene chips or DNA arrays containing hundreds to tens of thousands of DNA biomarkers have been developed. In proteomics and metabolomics, scientists use sophisticated, high-resolution analytical chemistry techniques to measure tens to hundreds of biochemicals that may be present in tissues. All three techniques examine specific *patterns* of change in biomarkers to decipher biological pathways or potential effects of chemical exposure. They rarely rely on the change of a single biomarker to come to conclusions about these pathways or effects.

The use of these "omic" techniques in regulatory ecotoxicology and risk assessment is still in the development and exploratory stages. Ankley *et al.* (2007, pp. 2–5) describe genomic biomarkers as offering the

> potential to effectively address a number of data needs and uncertainties currently confronting risk assessors and regulators. However, exactly how toxicogenomics might be incorporated into regulatory programs is uncertain. […] Ongoing research will, in the long term, serve to obviate limitations related to the global identification of gene products, proteins, and metabolites in test species relevant to ecological risk assessments.

Shorter-term research to develop molecular "profiles" or "fingerprinting" techniques could be used to support regulatory decision-making in ecotoxicology (Ankley *et al.*, 2007), but these studies need to be done on a case-by-case basis with targeted test species. A recent study by Hook *et al.* (2010) demonstrated the difficulty of using single-gene analysis as a diagnostic biomarker approach in assessment of oil spills. Most studies since the mid-2000s have relied on a suite of biomarkers to examine exposure to and effects of oil (e.g., Ramos-Gomez *et al.*, 2008; Bilbao *et al.*, 2010; Binelli *et al.*, 2010; Carvalho and Lettieri, 2011). Much research is still required in order to use these techniques to determine exposure and effect relationships for wild species.

9.9 Lessons learned

- The most commonly used biomarker to assess exposure to aromatic hydrocarbons in both terrestrial and aquatic organisms has been the CYP1A system. The system is clearly responsive to a wide variety of aromatic hydrocarbons and can indicate even relatively low levels of exposure to the general class of compounds.
- It is now generally accepted that background levels of CYP1A activity are normal, that the induction of the CYP1A system in response to aromatic hydrocarbons is adaptive and protective, and that pathways leading to toxicity of aromatic hydrocarbons include key components that cannot be identified simply by measuring CYP1A.
- In the absence of other biochemical or physiological measures, the conclusion that induction of CYP1A is evidence of harm is unfounded.
- If exposures to single chemicals or well-characterized mixtures are measured in the lab or in the field along with the CYP1A biomarker, the biomarker can be a good indicator of bioavailability and exposure. However, in the absence of analytical chemistry, the induction of the CYP1A system provides only a first-tier screening-level tool to ascertain

that an organism has been exposed to one or more of hundreds of compounds from many potential sources (Roberts *et al.*, 2005).

- The use of proper laboratory techniques (e.g., Hook *et al.*, 2008); selection of appropriate reference sites; inclusion of positive controls or known sites of contamination; examination of statistical versus biological relevance of measurements; and detailed consideration of physiological condition, sex, age, season, and location of origin of organisms must be part of a study designed to assess contaminant exposure in the field. These considerations are especially critical following oil spills, where environmental damage assessments inform decisions about safety, cleanup, and responsibility.

- As oil concentrations return to background levels in the environment impacted by a spill, biomarker studies must be especially rigorous about separating the sources of aromatic hydrocarbon exposures and making conclusions about oiled versus unoiled sites.

REFERENCES

Anderson, J.W. and R.F. Lee (2006). Use of biomarkers in oil spill risk assessment in the marine environment. *Human and Ecological Risk Assessment* **12**(6): 1192–1222.

Ankley, G.T., A.L. Miracle, E.J. Perkins, and G.P. Daston, eds (2007). *Genomics in Regulatory Ecotoxicology: Applications and Challenges*. Boca Raton, FL, USA: CRC Press; ISBN-10: 142006682X; ISBN-13: 9781420066821.

Assunção, M.G.L., K.A. Miller, N.J. Dangerfield, S.M. Bandiera, and P.S. Ross (2007). Cytochrome P450 1A expression and organochlorine contaminants in harbour seals (*Phoca vitulina*): evaluating a biopsy approach. *Comparative Biochemistry and Physiology Part C: Toxicology & Pharmacology* **145**(2): 256–264.

Balk, L., K. Hylland, T. Hansson, M.H.G. Berntssen, J. Beyer, G. Jonsson, A. Melbye, M. Grung, B.E. Torstensen, J.F. Børseth, H. Skarphedinsdottir, and J. Klungsøyr (2011). Biomarkers in natural fish populations indicate adverse biological effects of offshore oil production. *PLoS ONE* **6**(5): e19735; DOI:10.1371/journal.pone.0019735.

Ballachey, B.E., J.J. Stegeman, P.W. Snyder, G.M. Blundell, J.L. Bodkin, T.A. Dean, L. Duffy, D. Esler, G. Golet, S. Jewett, L. Holland-Bartels, A.H. Rebar, P.E. Seiser, and K.A. Trust (2002). Oil exposure and health of nearshore vertebrate predators in Prince William Sound following the 1989 *Exxon Valdez* oil spill. In *Mechanisms of Impact and Potential Recovery of Nearshore Vertebrate Predators following the 1989* Exxon Valdez *Oil Spill*. L.E. Holland-Bartels, ed. Anchorage, AK, USA: US Geological Survey, Alaska Biological Science Center; *Exxon Valdez* Oil Spill Restoration Project 99025 Final Report; Volume 1, 2.1–2.35. [http://www.evostc.state.ak.us/Files.cfm?doc=/Store/FinalReports/1999-99025-Final.pdf&]

Ben-David, M., T. Kondratyuk, B.R. Woodin, P.W. Snyder, and J.J. Stegeman (2001). Induction of cytochrome P450 1A1 expression in captive river otters fed Prudhoe Bay crude oil: Evaluation by immunohistochemistry and quantitative RT-PCR. *Biomarkers* **6**(3): 218–235.

Bilbao, E., D. Raingeard, O. Diaz de Cerio, M. Ortiz-Zarragoitia, P. Ruiz, U. Izagirre, A. Orbea, I. Marigómez, M.P. Cajaraville, and I. Cancio (2010). Effects of exposure to

Prestige-like heavy fuel oil and to perfluorooctanesulfonate on conventional biomarkers and target gene transcription in the thicklip grey mullet *Chelon labrosus*. *Aquatic Toxicology* **98**(3): 282–296.

Binelli, A., D. Cogni, M. Parolini, and A. Provini (2010). Multi-biomarker approach to investigate the state of contamination of the R. Lambro/R. Po confluence (Italy) by zebra mussel (*Dreissena polymorpha*). *Chemosphere* **79**(5): 518–528.

Bodkin, J.L., B.E. Ballachey, T.A. Dean, A.K. Fukuyama, S.C. Jewett, L. McDonald, D.H. Monson, C.E. O'Clair, and G.R. VanBlaricom (2002). Sea otter population status and the process of recovery from the 1989 "Exxon Valdez" oil spill. *Marine Ecology Progress Series* **241**: 237–253.

Bowen, L., A.K. Miles, M. Murray, M. Haulena, J. Tuttle, W. Van Bonn, L. Adams, J.L. Bodkin, B. Ballachey, J. Estes, M.T. Tinker, R. Keister, and J.L. Stott (2012). Gene transcription in sea otters (*Enhydra lutris*); development of a diagnostic tool for sea otter and ecosystem health. *Molecular Ecology Resources* **12**(1): 67–74.

Bowen, L., F. Riva, C. Mohr, B. Aldridge, J. Schwartz, A.K. Miles, and J.L. Stott (2007). Differential gene expression induced by exposure of captive mink to fuel oil: a model for the sea otter. *EcoHealth* **4**(3):298–309.

Carls, M.G., R.A. Heintz, C.D. Marty, and S.D. Rice (2005). Cytochrome P450 1A induction in oil-exposed pink salmon *Oncorhynchus gorbuscha* embryos predicts reduced survival potential. *Marine Ecology Progress Series* **301**: 253–265.

Carvalho, R.N. and T. Lettieri (2011). Proteomic analysis of the marine diatom *Thalassiosira pseudonana* upon exposure to benzo(a)pyrene. *BMC Genomics* **12**: 159. [http://www.biomedcentral.com/1471-2164/12/159]

Curtis, L., C.B. Garzon, M. Arkoosh, T. Collier, M.S. Myers, J. Buzitis, and M.E. Hahn (2011). Reduced cytochrome P450 1A activity and recovery from oxidative stress during subchronic benzo[a]pyrene and benzo[e]pyrene treatment of rainbow trout. *Toxicology and Applied Pharmacology* **254**(1): 1–7.

Emlen, J.M. and K.R. Springman (2007). Developing methods to assess and predict the population level effects of environmental contaminants. *Integrated Environmental Assessment and Management* **3**(2): 157–165.

Esler, D., B.E. Ballachey, K.A. Trust, S.A. Iverson, J.A. Reed, A.K. Miles, J.D. Henderson, B.R. Woodin, J.J. Stegeman, M. McAdie, D.M. Mulcahy, and B.W. Wilson (2011). Cytochrome P450 1A biomarker indication of the timeline of chronic exposure of Barrow's goldeneyes to residual *Exxon Valdez* oil. *Marine Pollution Bulletin* **62**(3): 609–614.

Esler, D., K.A. Trust, B.E. Ballachey, S.A. Iverson, T.L. Lewis, D.J. Rizzolo, D.M. Mulcahy, A.K. Miles, B.R. Woodin, J.J. Stegeman, J.D. Henderson, and B.W. Wilson (2010). Cytochrome P450 1A biomarker indication of oil exposure in harlequin ducks up to 20 years after the *Exxon Valdez* oil spill. *Environmental Toxicology and Chemistry* **29**(5): 1138–1145.

Esler, D. and S.A. Iverson (2010). Female harlequin duck winter survival 11 to 14 years after the *Exxon Valdez* oil spill. *Journal of Wildlife Management* **74**(3): 471–478.

Forbes, V.E., P. Calow, and R.M. Sibly (2008). The extrapolation problem and how population modeling can help. *Environmental Toxicology and Chemistry* **27**(10): 1987–1994.

Forbes, V.E., A. Palmqvist, and L. Bach (2006). The use and misuse of biomarkers in ecotoxicology. *Environmental Toxicology and Chemistry* **25**(1): 272–280.

Godard-Codding, C.A.J., R. Clark, M.C. Fossi, L. Marsili, S. Maltese, A.G. West, L. Valenzuela, V. Rowntree, I. Polyak, J.C. Cannon, K. Pinkerton, N. Rubio-Cisneros, S.L. Mesnick, S.B. Cox, I. Kerr, R. Payne, and J.J. Stegeman (2011). Pacific Ocean-wide profile of CYP1A1 expression, stable carbon and nitrogen isotope ratios, and organic contaminant burden in sperm whale skin biopsies. *Environmental Health Perspectives* **119**(3): 337–343.

Goldstone, J.V., M.E. Jönsson, L. Behrendt, B.R. Woodin, M.J. Jenny, D.R. Nelson, and J.J. Stegeman (2009). Cytochrome P450 1D1: A novel CYP1A-related gene that is not transcriptionally activated by PCB126 or TCDD. *Archives of Biochemistry and Biophysics* **482**(1–2): 7–16.

Golet, G.H., P.E. Seiser, A.D. McGuire, D.D. Roby, J.B. Fischer, K.J. Kuletz, D.B. Irons, T.A. Dean, S.C. Jewett, and S.H. Newman (2002). Long-term direct and indirect effects of the 'Exxon Valdez' oil spill on pigeon guillemots in Prince William Sound, Alaska. *Marine Ecology Progress Series* **241**: 287–304.

Handy, R.D., T.S. Galloway, and M.H. Depledge (2003). A proposal for the use of biomarkers for the assessment of chronic pollution and in regulatory toxicology. *Ecotoxicology* **12**(1–4): 331–343.

Hicken, C.E., T.L. Linbo, D.H. Baldwin, M.L. Willis, M.S. Myers, L. Holland, M. Larsen, M.S. Stekoll, S.D. Rice, T.K. Collier, N.L. Scholz, and J.P. Incardona (2011). Sublethal exposure to crude oil during embryonic development alters cardiac morphology and reduces aerobic capacity in adult fish. *Proceedings of the National Academy of Sciences* **108**(17): 7086–7090.

Hoffmann, J.L. and J.T. Oris (2006). Altered gene expression: A mechanism for reproductive toxicity in zebrafish exposed to benzo[a]pyrene. *Aquatic Toxicology* **78**(4): 332–340.

Hoffmann, J.L., S.P. Torontali, R.G. Thomason, D.M. Lee, J.L. Brill, B.B. Price, G.J. Carr, and D.J. Versteeg (2006). Hepatic gene expression profiling using genechips in zebrafish exposed to 17 alpha-ethynylestradiol. *Aquatic Toxicology* **79**(3): 233–246.

Hook, S.E., M. Cobb, J. Oris, and J. Anderson (2008). Gene sequences for Cytochrome P450 1A1 and 1A2: The need for biomarker development in sea otters (*Enhydra lutris*). *Comparative Biochemistry and Physiology – Part B: Biochemistry and Molecular Biology* **151**(3): 336–348.

Hook, S.E., M.A. Lampi, E.J. Febbo, J.A. Ward, and T.F. Parkerton (2010). Temporal patterns in the transcriptomic response of rainbow trout, *Oncorhynchus mykiss*, to crude oil. *Aquatic Toxicology* **99**(3): 320–329.

Huggett, R.J., J.M. Neff, J.J. Stegeman, B. Woodin, K.R. Parker, and J.S. Brown (2006). Biomarkers of PAH exposure in an intertidal fish species from Prince William Sound, Alaska: 2004–2005. *Environmental Science & Technology* **40**(20): 6513–6517.

Huggett, R.J., J.J. Stegeman, D.S. Page, K.R. Parker, B. Woodin, and J.S. Brown (2003). Biomarkers in fish from Prince William Sound and the Gulf of Alaska: 1999–2000. *Environmental Science & Technology* **37**(18): 4043–4051.

Incardona, J.P., M.G. Carls, H.L. Day, C.A. Sloan, J.J. Bolton, T.K. Collier, and N.L. Scholz (2009). Cardiac arrhythmia is the primary response of embryonic Pacific herring (*Clupea pallasi*) exposed to crude oil during weathering. *Environmental Science & Technology* 43(1): 201–207.

Incardona, J.P., M.G. Carls, H. Teraoka, C.A. Sloan, T.K. Collier, and N.L. Scholz (2005). Aryl hydrocarbon receptor-independent toxicity of weathered crude oil during fish development. *Environmental Health Perspectives* 113(12): 1755–1762.

Kannan, K. and E. Perrotta (2008). Polycyclic aromatic hydrocarbons (PAHs) in livers of California sea otters. *Chemosphere* 71(4): 649–655.

McClain, J.S., J.T. Oris, G.A. Burton, and D. Lattier (2003). Laboratory and field validation of multiple molecular biomarkers of contaminant exposure in rainbow trout (*Oncorhynchus mykiss*). *Environmental Toxicology and Chemistry* 22(2): 361–370.

Meyer, J.N., D.M. Wassenberg, S.I. Karchner, M.E. Hahn, and R.T. Di Giulio (2003). Expression and inducibility of aryl hydrocarbon receptor pathway genes in wild-caught killifish (*Fundulus heteroclitus*) with different contaminant-exposure histories. *Environmental Toxicology and Chemistry* 22(10): 2337–2343.

Miles, A.K., P.L. Flint, K.A. Trust, M.A. Ricca, S.E. Spring, D.E. Arrieta, T. Hollmen, and B.W. Wilson (2007). Polycyclic aromatic hydrocarbon exposure in Steller's eiders (*Polysticta stelleri*) and harlequin ducks (*Histrionicus histrionicus*) in the eastern Aleutian Islands, Alaska, USA. *Environmental Toxicology and Chemistry* 26(12): 2694–2703.

Miller, K.A., M.G.L. Assunção, N.J. Dangerfield, S.M. Bandiera, and P.S. Ross (2005). Assessment of cytochrome P450 1A in harbour seals (*Phoca vitulina*) using a minimally-invasive biopsy approach. *Marine Environmental Research* 60(2): 153–169.

Montie, E.W., P.A. Fair, G.D. Bossart, G.B. Mitchum, M. Houde, D.C.G. Muir, R.J. Letcher, W.E. McFee, V.R. Starczak, J.J. Stegeman, and M.E. Hahn (2008). Cytochrome P450 1A1 expression, polychlorinated biphenyls and hydroxylated metabolites, and adipocyte size of bottlenose dolphins from the Southeast United States. *Aquatic Toxicology* 86(3): 397–412.

Mulligan, C.J., L.G. Goldfarb, R.W. Robin, N. Sambuughin, M.V. Osier, R.A. Kittles, D. Goldman, D. Hesselbrock, and J.C. Long (2003). Allelic variation at alcohol metabolism genes (ADH1b, ADH1c, ALDH2) and alcohol dependence in an American Indian population. *Human Genetics* 113(4): 325–336.

Nebert, D.W (2005). Role of host susceptibility to toxicity and cancer caused by pesticides: cytochromes P450. *Journal of Biochemical and Molecular Toxicology* 19(3): 184–186.

Nebert, D.W. and C.L. Karp (2008). Endogenous functions of the aryl hydrocarbon receptor (AHR): Intersection of cytochrome P450 1 (CYP1)-metabolized eicosanoids and AHR biology. *Journal of Biological Chemistry* 283(52): 36061–36065.

Nebert, D.W., T.P. Dalton, A.B. Okey, and F.J. Gonzalez (2004). Role of aryl hydrocarbon receptor-mediated induction of the CYP1 enzymes in environmental toxicity and cancer. *Journal of Biological Chemistry* 279(23): 23847–23850.

Nelson, D.R (1998). Metazoan cytochrome P450 evolution. *Comparative Biochemistry and Physiology – Part C: Toxicology and Pharmacology* 121(1–3): 15–22.

Nyman, M., H. Raunio, and O. Pelkonen (2000). Expression and inducibility of members in the cytochrome P4501 (CYP1) family in ringed and grey seals from polluted and less polluted waters. *Environmental Toxicology and Pharmacology* **8**(4): 217–225.

Oris, J.T. and A.P. Roberts (2007). Statistical analysis of cytochrome P450 1A biomarker measurements in fish. *Environmental Toxicology and Chemistry* **26**(8): 1742–1750.

Page, D.S., P.D. Boehm, G.S. Douglas, A.E. Bence, W.A. Burns, and P.J. Mankiewicz (1996). The natural petroleum hydrocarbon background in subtidal sediments of Prince William Sound, Alaska. *Environmental Toxicology and Chemistry* **15**(8): 1266–1281.

Page, D.S., P.D. Boehm, G.S. Douglas, A.E. Bence, W.A. Burns, and P.J. Mankiewicz (1999). Pyrogenic polycyclic aromatic hydrocarbons in sediments record past human activity: A case study in Prince William Sound Alaska. *Marine Pollution Bulletin* **38**(4): 247–260.

Pašková, V., O. Adamovský, J. Pikula, B. Skočovská, H. Band'ouchová, J. Horáková, P. Babica, B. Maršálek, and K. Hilscherová (2008). Detoxification and oxidative stress responses along with microcystins accumulation in Japanese quail exposed to cyanobacterial biomass. *Science of the Total Environment* **398**(1–3): 34–47.

Puga, A., C.R. Tomlinson, and Y. Xia (2005). Ah receptor signals cross-talk with multiple developmental pathways. *Biochemical Pharmacology* **69**(2): 199–207.

Ramos-Gómez, J., M.L. Martin-Diaz, A. Rodríguez, I. Riba, and T.Á. DelValls (2008). In situ evaluation of sediment toxicity in Guadalete Estuary (SW Spain) after exposure of caged *Arenicola marina*. *Environmental Toxicology* **23**(5): 643–651.

Ricca, M.A., A.K. Miles, B.E. Ballachey, J.L. Bodkin, D. Esler, and K.A. Trust (2010). PCB exposure in sea otters and harlequin ducks in relation to history of contamination by the *Exxon Valdez* oil spill. *Marine Pollution Bulletin* **60**(6): 861–872.

Roberts, A.P., J.T. Oris, and W.A. Stubblefield (2006). Gene expression in caged juvenile coho salmon (*Oncorhynchys kisutch*) exposed to the waters of Prince William Sound, Alaska. *Marine Pollution Bulletin* **52**(11): 1527–1532.

Roberts, A.P., J.T. Oris, G.A. Burton, and W.H. Clements (2005). Gene expression in caged fish as a first-tier indicator of contaminant exposure in streams. *Environmental Toxicology and Chemistry* **24**(12): 3092–3098.

Schlenk, D. (1999). Necessity of defining biomarkers for use in ecological risk assessments. *Marine Pollution Bulletin* **39**(1–12): 48–53.

Schlenk, D., W.H. Benson, S. Steinert, R. Handy, and M. Depledge (2008). Biomarkers. In *The Toxicology of Fishes*. R.T. DiGiulio and D.E. Hinton, eds. Boca Raton, FL, USA: CRC; ISBN-10: 041524868X; ISBN-13: 9780415248686; pp. 683–723.

Schwartz, J.A., B.M. Aldridge, B.L. Lasley, P.W. Snyder, J.L. Stott, and F.C. Mohr (2004). Chronic fuel oil toxicity in American mink (*Mustela vison*): Systemic and hematological effects of ingestion of a low concentration of bunker C fuel oil. *Toxicology and Applied Pharmacology* **200**(2): 146–158.

Short, J.W., K.R. Springman, M.R. Lindeberg, L.G. Holland, M.L. Larsen, C.A. Sloan, C. Khan, P.V. Hodson, and S.D. Rice (2008). Semipermeable membrane devices link site-specific contaminants to effects: Part II – A comparison of lingering *Exxon Valdez* oil with other potential sources of CYP1A inducers in Prince William Sound, Alaska. *Marine Environmental Research* **66**(5): 487–498.

Springman, K.R., J.W. Short, M.R. Lindeberg, J.M. Maselko, C. Khan, P.V. Hodson, and S.D. Rice (2008a). Semipermeable membrane devices link site-specific contaminants to effects: Part I – Induction of CYP1A in rainbow trout from contaminants in Prince William Sound, Alaska. *Marine Environmental Research* **66**(5): 477–486.

Springman, K.R., J.W. Short, M.R. Lindeberg, and S.D. Rice (2008b). Evaluation of bioavailable hydrocarbon sources and their induction potential in Prince William Sound, Alaska. *Marine Environmental Research* **66**(1): 218–220.

Trust, K.A., D. Esler, B.R. Woodin, and J.J. Stegeman (2000). Cytochrome P450 1A induction in sea ducks inhabiting nearshore areas of Prince William Sound, Alaska. *Marine Pollution Bulletin* **40**(5): 397–403.

Walraven, J.M., J.O. Trent, and D.W. Hein (2008). Structure-function analyses of single nucleotide polymorphisms in human N-acetyltransferase 1. *Drug Metabolism Reviews* **40**(1): 169–184.

Wang, Z. and S.A. Stout, eds (2007). *Oil Spill Environmental Forensics: Fingerprinting and Source Identification*. Burlington, MA, USA: Academic Press; ISBN-13: 9780123695239; ISBN-10: 0123695236.

Whyte, J.J., R.E. Jung, C.J. Schmitt, and D.E. Tillitt (2000). Ethoxyresorufin-O-deethylase (EROD) activity in fish as a biomarker of chemical exposure. *Critical Reviews in Toxicology* **30**(4): 347–570.

Wilson, J.Y., S.R. Cooke, M.J. Moore, D. Martineau, I. Mikaelian, D.A. Metner, W.L. Lockhart, and J.J. Stegeman (2005). Systemic effects of arctic pollutants in beluga whales indicated by CYP1A1 expression. *Environmental Health Perspectives* **113**(11): 1594–1599.

Wolkers, J., R.F. Witkamp, S.M. Nijmeijer, I.C. Burkow, E.M. de Groene, C. Lydersen, S. Dahle, and M. Monshouwer (1998). Phase I and phase II enzyme activities in ringed seals (*Phoca hispida*): Characterization of hepatic cytochrome P450 by activity patterns, inhibition studies, mRNA analyses, and western blotting. *Aquatic Toxicology* **44**(1–2): 103–115.

Wooley, C. (2002). The myth of the "pristine environment": Past human impacts in Prince William Sound and the Gulf of Alaska. *Spill Science and Technology Bulletin*, **7**(1–2): 89–104.

CHAPTER TEN

Assessing effects and recovery from environmental accidents

Keith R. Parker, John A. Wiens, Robert H. Day, and Stephen M. Murphy

10.1 Introduction

When an environmental accident such as an oil spill occurs, several things happen. There can be immediate efforts to contain the damage to natural ecosystems or human structures or livelihoods. Steps can be taken to provide relief to the people or environments most immediately affected. If the accident is sufficiently large, media accounts can fuel responses by the broader public. For damages resulting from human-caused accidents, claims and counter-claims of the magnitude and extent of the accident and its consequences will be made and then amplified, often followed (in the United States, at least) by litigation. All of these consequences require knowledge of what really happened – where did the oil go, what natural resources or services were affected, and how persistent were these effects? As the chapters in this book amply demonstrate, the need for careful, rigorous, and objective science is paramount.

Just because there is a need for careful, rigorous, and objective science, however, does not mean that it is easily attainable. Because of demands for immediate action and the heightened emotions following an environmental accident, attempts to document the ecological effects and subsequent recovery run the risk of being hastily developed and inadequately designed. This can foster never-ending arguments about conclusions. Such accidents occur against a complex and dynamically varying environmental background, so they cannot be treated as traditional experiments to be analyzed with straightforward statistical procedures that are planned in advance. There is only a single replicate of the "treatment" (i.e., the accident), there are no pre-established controls, and a welter of other factors with varying degrees of intercorrelation confounds attempts to attribute observed changes to the environmental accident. After all, the real world is messy! Designing field studies and analyses that are quantitative, objective, and scientifically rigorous under such circumstances is difficult – yet it is essential.

Oil in the Environment: Legacies and Lessons of the Exxon Valdez *Oil Spill*, ed. J. A. Wiens. Published by Cambridge University Press. © Cambridge University Press 2013.

In this chapter, we discuss the elements of study design, analysis, and interpretation that are necessary to meet this objective. We distill and synthesize the lessons we have learned from our experience with the *Exxon Valdez* oil spill to provide guidance to resource managers, field biologists, litigants, corporate managers, and policy makers – those involved in responding to environmental accidents and developing the regulations that govern such responses. We begin by describing the challenge of conducting observational studies to assess impact (i.e., effects)[1] and recovery in a highly variable environment. We then address several aspects of how one goes about determining cause–effect relationships between an environmental accident and its consequences and establishing when recovery occurs. Finally, we consider the analytical aspects of assessing the effects of an environmental accident: statistical power, study duration, reducing variation, and interpreting the results of the scientific studies. We draw on examples from the *Exxon Valdez* oil spill, but our comments and the lessons we derive apply more broadly to any large-scale environmental perturbation, regardless of whether it is accidental or natural. The points we make in this chapter provide a framework for considering what was (or was not) done following the *Exxon Valdez* spill, as detailed in the following chapters in this section.

10.2 An environmental accident is not an experiment

In studying the effects of an environmental accident, it is tempting to think of the event as akin to an experiment to which standard approaches of experimental design and statistical analysis can be applied to ensure scientific rigor. However, because it is by definition unplanned and because experimental and control areas are not identical ecologically, studies of an environmental accident lack those features of experimental design that control for the confounding effects of other causal factors. Natural variation in habitats and populations further obscures the relationship between the accident and possible effects on natural resources. Logistical constraints, inclement weather, and a host of unforeseen complications can erode a well-planned study design. Consequently, one must consider these limitations and aim to assess cause-and-effect relationships objectively. Put simply, one should not attempt to interpret statistical results as if they were derived from experiments in a situation where only observations can be made. To see why this is so, it is useful to contrast observational studies with what they are not: planned statistical experiments (see Eberhardt and Thomas, 1991).

A *planned statistical experiment* is used to infer cause–effect relationships by eliminating causal factors other than the treatment under study. Hypothetically, had the *Exxon Valdez* spill been a planned experiment, the application of a known level of oil (the treatment) would have been randomized among, say, half the sampling sites. Those sites that had not received oil would serve as control sites. Even though sampling sites would differ in factors (e.g., wave exposure, shoreline substrate) that affect levels of the natural resource being assessed, oiled and control sites would, *on average*, be affected similarly by those factors other than oil. Because the potential biases of confounding natural factors with the effects of oiling would have been controlled by randomization, differences in levels of the resource (e.g., abundance) between oiled and control sites could be inferred to be due solely to oiling.

---

[1] Our use of "impact," "effect," "injury," and "recovery" in this book is detailed in Chapter 1, Box 1.3.

In contrast, in an *observational study*, causal natural factors other than the perturbation are not controlled. As a consequence, inferences about effects of the perturbation cannot be made without considering the confounding effects of these other natural factors. Oil from the *Exxon Valdez* was not distributed randomly among potential sampling sites, the treatment itself (the oil spill) was not replicated, and the quantity of oil varied among parts of the treatment area. Where, when, and how much oil was deposited on shorelines was a function of wind, currents, proximity to the spill, and geographic features such as protective islands and peninsulas (Chapters 4 and 6). Generally northeasterly winds and waves carried oil southwesterly and on to north- and northeasterly facing shorelines, which had distinctive environmental features. Nearby, nonoiled shorelines typically did not face to the north and northeast and often had different environmental features such as wave exposure. Consequently, oiled and nonoiled shorelines differed in the other natural factors that affected the biota. Hence, natural factors and oiling were confounded, challenging attempts to establish unbiased cause–effect relationships between oiling and biota (Wiens and Parker, 1995).

10.3 Things do not stay the same

Environments vary, populations ebb and flow, and communities change over time in manifold ways. The notion that this temporal flux and spatial heterogeneity simply represent constrained variation around some unchanging long-term mean is a myth (Hutchings *et al.*, 2000; Milly *et al.*, 2008). Consequently, analytical approaches for determining the consequences of an environmental perturbation necessarily make assumptions about background environmental variation and its effects on biological resources. For example, Pacific herring (*Clupea pallasii*) biomass in Prince William Sound (PWS) changes seasonally owing to migratory patterns and differs spatially owing to a host of poorly understood factors that are referred to collectively as "habitat quality." When an environmental accident occurs, its effects are superimposed on this natural variation, making it difficult to separate the effects of the perturbation from those of the varying natural factors. To interpret the results of observational studies, in which natural variation may obscure or confound the effects of the perturbation, one must make assumptions about how environmental variation has affected levels of the resource. Depending on study design, one of three ecological assumptions is made about variation in time, space, or time and space together (Fig. 10.1) (Wiens and Parker, 1995; Parker and Wiens, 2005).

The assumption of *steady-state equilibrium* means that, in the absence of the perturbation, the system would have varied around a stable long-term mean; there would be no systematic changes over time (i.e., stationarity; Milly *et al.*, 2008). In the case of oil spills, this assumption underlies the definitions of "effect" or "injury" as a departure from prespill conditions and "recovery" as a return to those conditions (*Exxon Valdez* Oil Spill Trustee Council [hereafter, "Trustees"], 2002; see Chapter 1, Box 1.3). The assumption of *spatial equilibrium* implies that the natural factors affecting levels of a resource do not differ systematically among locations at a given time – impacted and reference areas are assumed to be equivalent in all respects except for the occurrence of the disturbance. The assumption of *dynamic equilibrium* relaxes the previous assumptions by recognizing that conditions may vary in time and space – absent the perturbation, locations may differ from one another and conditions vary over time. However,

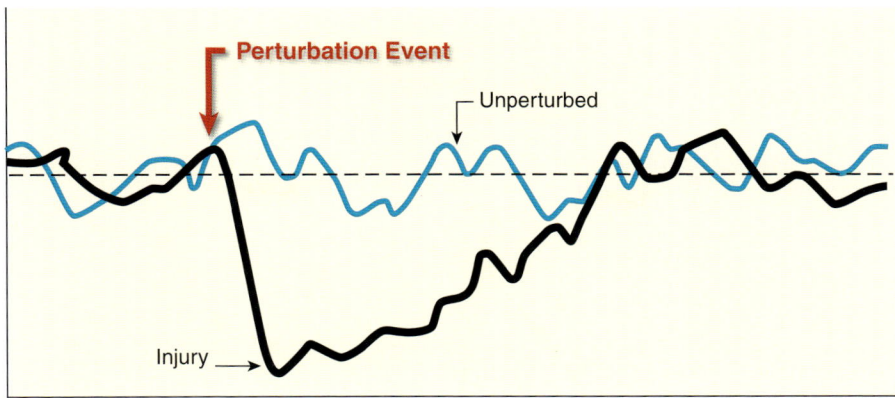
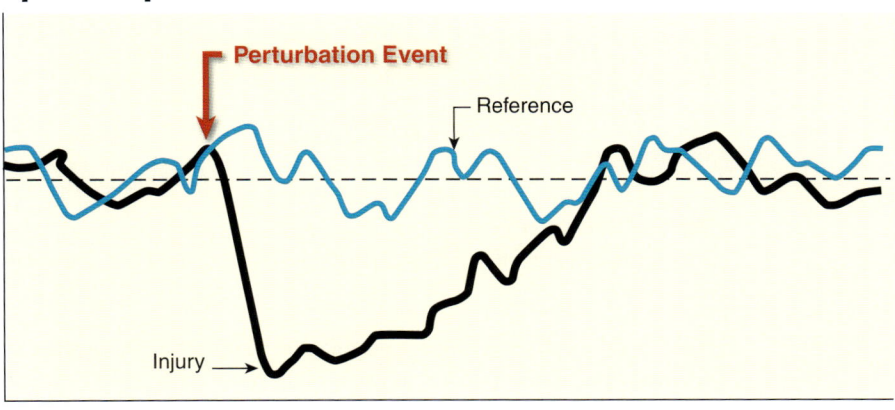
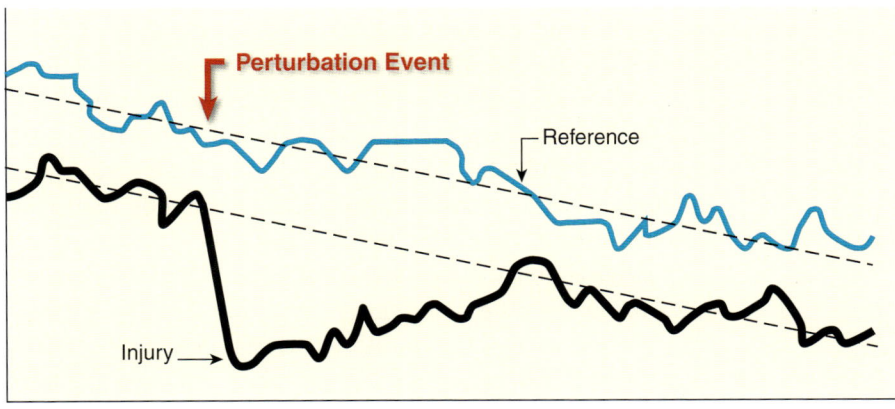

Figure 10.1 Hypothetical illustration of three equilibrium assumptions. In steady-state equilibrium, the unperturbed system (blue line) would vary about a stable long-term mean; an injury following a perturbation event would be recognized by a departure from the normal range of variation (black line), and recovery would occur when the system returned to vary about the long-term mean. In spatial equilibrium, an injury to the system would be recognized by a departure from the conditions in an appropriate reference area and recovery by a return to those conditions. In dynamic equilibrium, the impacted and reference areas differ under unperturbed conditions (and are trending toward lower values in the example shown). An affected system departs from the conditions in the reference area; recovery is indicated when the dynamics in the affected and reference areas are parallel, although the levels will differ.

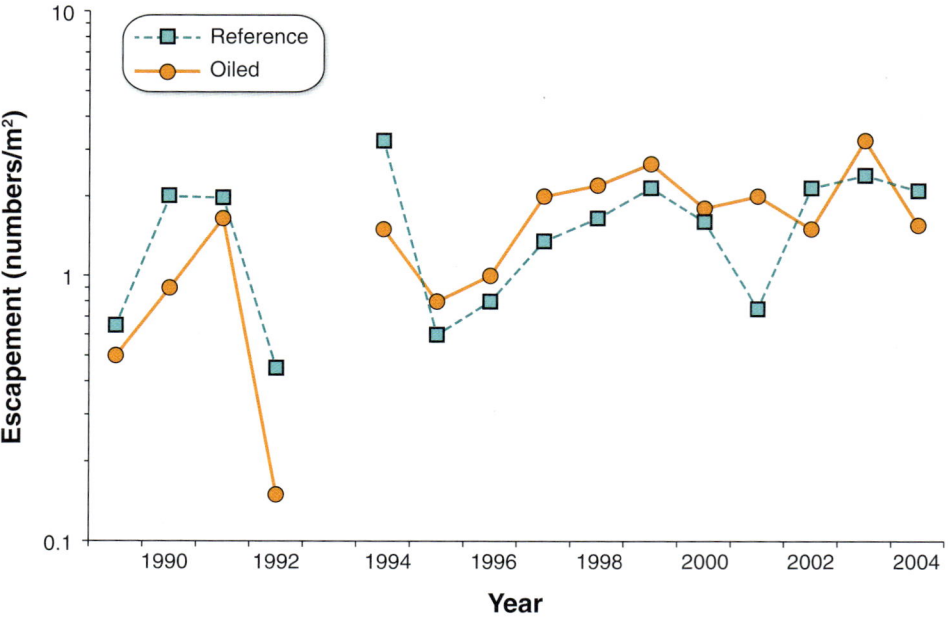

Figure 10.2 Profiles of geometric mean whole-stream escapement (numbers/m^2) of spawning pink salmon for oiled and reference streams from 1989 to 2004 (from Brannon et al., 2006, with permission from Elsevier).

the factors of interest (e.g., population size) vary in similar ways, so the *differences* between the areas remain similar over time. The forces driving these dynamics are therefore assumed to be similar in the different areas. This assumption is most likely (but not inevitable) if the areas are located near one another.

These ecological assumptions can be illustrated using the example of mean escapement (returning spawners that "escaped" the fishery) of pink salmon (*Oncorhynchus gorbuscha*) in five oiled and five reference streams from 1989 through 2004 (Brannon et al., 2006). Over the 16-year period, mean escapement densities (numbers/m^2) for both oiled and reference streams varied by a factor of 10 (Fig. 10.2). The assumption of steady-state equilibrium clearly does not hold. Neither does the assumption of spatial equilibrium hold: reference streams had consistently higher spawning densities than did oiled streams from 1989 to 1994, then the opposite held true from 1995 to 1999, and patterns of relative densities for oiled and reference streams were mixed from 2000 onward. On the other hand, mean spawning densities at oiled and reference streams generally (albeit imperfectly) showed similar patterns of variation over time; densities approximated a dynamic equilibrium. Thus, although the factors affecting spawning density changed over time, they affected oiled and reference streams in similar ways.

These ecological assumptions about the nature of equilibrium are important in different ways, depending on the form of the environmental perturbation. When a perturbation is large, the effects of spatial and temporal variation in natural factors may be overwhelmed and the assumptions of steady-state or spatial equilibrium can be temporarily accepted, even though they may be false. As the effects of the perturbation diminish over time, however, temporal and spatial environmental variation may increasingly overwhelm the diminishing effects of the perturbation; violations of the assumptions therefore become increasingly important.

As the scale in time or space increases, the magnitude of variations in biological resources is likely to increase and mean values will become less stable. Under such conditions, the assumptions of steady-state and spatial equilibrium are eroded and one is left to deal with the assumption of dynamic equilibrium. In such a situation, short-term studies can provide only a snapshot of a system in continual flux, and sampling from only a few locations will not capture the true extent of spatial heterogeneity. Ideally, studies of the effects of environmental perturbations on biological resources should be conducted in a standardized way over multiple years and multiple sites so that the dynamics of interest (effects and recovery) can be distinguished from all else that is going on. Although they are logistically more demanding, study designs that assume dynamic equilibrium and incorporate multiple sampling times to compare areas subjected to different levels of disturbance (e.g., impact-level-by-time designs; Wiens and Parker, 1995) may be less sensitive to the effects of natural temporal and spatial variation than simpler study designs. No matter how a study of biological resources is designed, however, the underlying assumptions about the temporal and spatial dynamics of the system must be explicitly acknowledged at the onset of an assessment. Their potential effects on the results and interpretations should be evaluated and studies designed with an appropriate spatial extent and temporal duration needed to address the assumptions. Problems with interpretation become particularly difficult for species for which information on natural history and prespill data are lacking. For such species one may need to evaluate results with regard to similar (but well-studied) species and accept the weakness of conclusions about cause and effect.

These ecological assumptions often are ignored or are not explicitly recognized and evaluated. Longstanding beliefs in equilibrium, stationarity, and the "balance of nature" (Wiens, 1984; Rohde, 2005; Milly *et al.*, 2008) may lead one to accept the assumption of steady-state equilibrium uncritically, and inattention to the spatial heterogeneity of habitats at multiple scales may lead one to ignore differences among study locations and accept the assumption of spatial equilibrium. These ecological assumptions about equilibrium are not just abstract theoretical considerations, however. They affect study designs and statistical analyses and cannot be wished away when one interprets results. Even if one documents statistically significant differences between, say, mean population sizes in impacted and reference areas or significant differences disappear over time, the results and their interpretation may be misleading or invalid if the study design or statistical analysis is vulnerable to violations of these equilibrium assumptions.

10.4 Things do not happen in isolation

Environments are multivariate mosaics, full of factors that affect both each other and the resources of interest in various ways and to varying degrees. Some of these factors may vary in tandem with the disturbance, producing effects that may parallel or obscure those of the disturbance itself and confounding the results of statistical analyses. One cannot pretend that these other environmental factors do not matter or are somehow averaged away by the study design, as they can be in experiments. One instead must play the hand of a nonrandom and nonreplicated event that one has been dealt. If such potentially confounding factors are not included in an analysis, leaving the perturbation as the only causal factor considered, one may conclude that an effect is due to the perturbation when it actually is a consequence of the other factors.

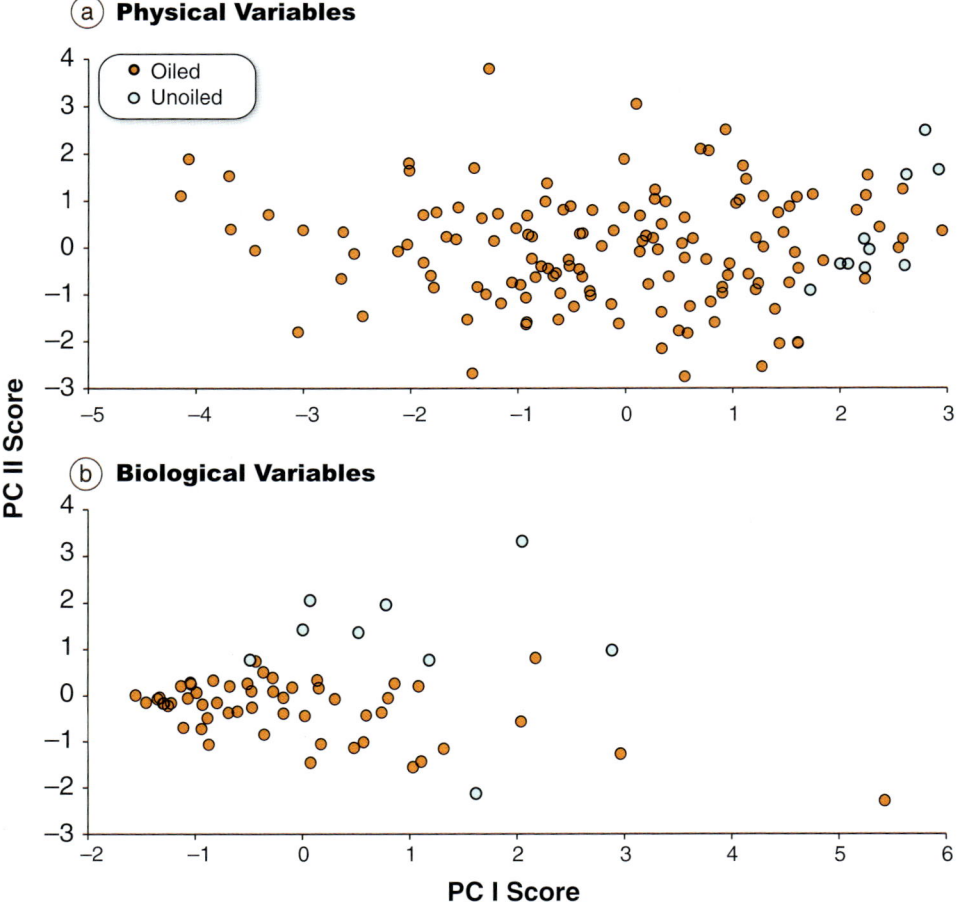

Figure 10.3 Principal component analysis results for differences in physical habitat variables (a) and biological habitat variables (b) between shoreline sections on Montague Island (unoiled reference area) and shoreline sections on the Knight Island group and the northwestern part of the spill-affected area (oiled area). The positions of oiled and unoiled symbols indicate that shoreline and nearshore habitats differ significantly between oiled and unoiled areas. Had habitat features been similar, the oiled and unoiled symbols would be thoroughly mixed. For the physical variables, important features of principal component I (PC I) are percentage bedrock shoreline, islands/km shoreline, and percentage boulder/cobble shoreline; important features for principal component II (PC II) are percentage bedrock/rubble shoreline, shallowness index, and islets/km shoreline. For the biological variables, important features of PC I are salmon runs/km of shoreline and percentage of shoreline with seagrass; important features for PC II are percentage of shoreline with mussels and percentage of shoreline with *Fucus* cover. (From Wiens et al., 2010, with permission from Springer Science and Business Media)

The potential for confounding factors to cloud analyses is especially great in comparisons of "treatment" (e.g., oiled) areas and "reference" (e.g., unoiled) areas. Several studies of the effects of the *Exxon Valdez* spill on marine birds (e.g., Esler et al., 2000, 2002) used data from Montague Island as unoiled reference samples in comparisons with nearby Green and Knight Islands (see Map 3, p. vii), parts of which were oiled (Fig. 1.3). Such a comparison rests on the assumption that the areas are equivalent in all respects except their oiling history (or that any differences do not matter). Montague Island differs from the oiled areas in several bathymetric, oceanographic, and shoreline habitat features (Fig. 10.3), so attributing differences in marine bird abundance, distribution, or other features to oiling effects alone is suspect (Wiens et al., 2001, 2010).

Similar systematic differences between oiled and reference shorelines affect broader comparisons among regions of PWS. Pacific herring spawn on kelp along rocky shorelines throughout northern PWS, where spawning success is subject to a wide range of conditions that are difficult to observe and measure, including fluctuating water temperature and predation by seabirds and fish. To assess the effects of oiling on Pacific herring eggs, Pearson *et al.* (1995; Chapter 13) flew recently spawned eggs-on-kelp back to the laboratory, where assessments of mortality and morbidity were observed under similar and measureable environmental conditions.

The potential confounding effects of differences among compared areas can be reduced if measures of important environmental factors are included as covariates in the analyses. For example, the effects of *Exxon Valdez* oiling on mortality of salmon embryos disappeared when Brannon *et al.* used sample timing as a covariate (Chapter 12). Gilfillan *et al.* (1995; see Chapter 11) reduced the confounding influences of wave exposure and sediment grain size on shoreline biota by using them as covariates. Conducting analyses with and without covariates can provide insights into the importance of multiple environmental factors. Esler *et al.* (2010) tested with and without covariates of sex, age, and body mass of harlequin ducks (*Histrionicus histrionicus*) to evaluate the potential effects of these factors on mean levels of a liver enzyme (CYP1A; see Chapter 9) between oiled and reference areas of PWS. The downside of using covariates in such analyses is that they reduce degrees of freedom (effective sample size for testing and estimation) and, therefore, statistical power. Accordingly, the advantages of using covariates to reduce confounding effects should be balanced against the reduction in power.

10.5 Hypotheses must be carefully framed

Except in obvious cases, one seldom knows whether or by how much a resource has been affected. Effects and recovery must be inferred from samples, which represent only a subset of the injured resource, not a complete census of it. In such situations, hypothesis testing and statistical analyses provide formal means for dealing with uncertainty and quantifying the strength of inferences.

Scientific hypotheses are falsifiable assertions that lend rigor to the process of inferring cause–effect relationships (Popper, 1959; Pickett *et al.*, 1994). A statistical null hypothesis proposes a default or null position (typically the absence of a particular effect), much like the legal precedent of innocent until proven guilty. Hence, for assessing injuries, the appropriate null hypothesis is that there were no effects.

If the null hypothesis is one of no effects, the alternative hypothesis logically is one of an effect (i.e., injury). Intuitively, one might expect an oil spill to have only adverse effects, in which case one can use a one-tailed statistical test to determine whether the null hypothesis has been falsified and the conclusion of a negative effect is (statistically) warranted. In some cases, however, the effect may be either negative or positive (e.g., black-legged kittiwakes, *Rissa tridactyla*; see Table 14.1). In this case, a two-tailed test must be used, which is less powerful than a one-tailed test if the direction of effect is known. Such positive effects would be missed if only negative effects were considered as an alternative hypothesis. Of course, a failure to reject the null hypothesis may tempt one to conclude that there were no effects. This inference is incorrect (statistically, at least): one simply has failed to demonstrate significance (i.e., an effect) at a specified α-level (a Type I error, also known as a false positive, which occurs when a statistical test rejects a true null hypothesis). The appropriate α-level and the form of the statistical test must be determined *before* hypothesis tests are conducted; a post facto adjustment of the α-level or the decision of whether to use a one-tailed or two-tailed test to make the results fit one's expectations is scientifically indefensible.

Testing hypotheses about recovery is different from testing hypotheses about effects (Parker and Wiens, 2005). A resource can recover only if it has been previously affected (i.e., only if the null hypothesis of no injury has been previously rejected). By this measure, recovery is the disappearance of a previously documented effect; it is recognized when the null hypothesis of no effect can no longer be rejected (Chapter 1, Box 1.3). The recovery process can occur over varying lengths of time, depending on the magnitude of the initial effects, the presence and magnitude of any continuing stressors, and the biological attributes (e.g., recruitment rates, immigration rates) of the injured resource. As time passes following a perturbation, the effects of the event on the resource may become progressively clouded by other, unrelated events or by natural variability in the environment, which complicates efforts to detect when recovery has occurred. If the recovery potential of a resource is intrinsically low and/or the environment is highly variable, it may become impossible to distinguish a continuing signal of effects of the perturbation from the accumulating noise of environmental variability, precluding definitive statements about when or if recovery occurs.

Proper framing of the null hypothesis is important. It may be hard to avoid the expectation that a major environmental disruption such as an oil spill must surely have had major effects. In this case, the null hypothesis of no effect is sometimes replaced by the presumption that the accident had major negative effects (e.g., Irons *et al.*, 2000; Lance *et al.*, 2001). In this case, the null hypothesis is not really null. For example, McKnight *et al.* (2008; see Chapter 14) tested for seabird recovery by comparing trends in density over time (1989–2006) between oiled and unoiled areas in PWS, reasoning that if a species was recovering, the trend slope for the oiled area would be significantly more positive than that for the unoiled area. They tested the null hypothesis of equality of slopes in the two areas against the alternative of a greater positive slope in the oiled area. A rejection of the null hypothesis was taken as evidence that the species was recovering. If the null hypothesis was not rejected, however, the species was considered still to be affected. Although McKnight *et al.* (2008) considered only species that were previously thought to be negatively affected by the spill, the analysis could not detect recovery for species that were never affected or that were positively affected. Consequently, such species would be considered (incorrectly) as still being affected.

Some (e.g., Burnham and Anderson, 2002) have argued that hypothesis testing is overrated in ecology. Although that view may be correct in some situations, it is essential to have clear thinking in assessing effects and recovery from an environmental accident such as an oil spill, and the process of framing a hypothesis helps one do that. Without an explicit null hypothesis, preconceptions about possible effects can exert a powerful pull, potentially leading to poor design, biased tests, and erroneous conclusions. Moreover, the statistical tests that are applied to gauge the significance of the results of a hypothesis are constrained by the way the hypothesis is framed. Selecting the appropriate hypotheses to test for the effects of an environmental perturbation requires careful thought because it implicitly sets the direction for design, implementation, and (most importantly for observational studies) interpretation of results.

10.6 Improve statistical power to detect effects through sampling and analysis

Statistical analyses are at the heart of hypothesis testing. Ideally, one strives for statistical tests with the greatest power to detect an effect, thereby bolstering one's confidence in the results. Statistical power increases as the effect size (magnitude of effect) or sample size increases and/or as the level of statistical significance (α-level) becomes more permissive (e.g., $\alpha = 0.20$ versus $\alpha = 0.05$). Statistical power can be low when any of these components is inadequate, resulting in a failure to reject the null hypothesis of no effect even when an effect or injury actually occurred. If the effect size is small, for example, the difference in means (e.g., for oiled and reference samples) may be insufficient to pass the a priori significance threshold. This was unlikely in the case of the *Exxon Valdez* spill (at least initially), so statistical power should not have been constrained by this factor.

Power may also be low, however, because the sample size is too small to achieve adequate precision for rejecting the null hypothesis. Attaining adequate sample size is likely to be challenging in any observational study, especially if the affected area is large and samples are not easily obtained. Bue *et al.* (1998) optimized power for a two-factor model (oiling levels and tidal heights) by blocking (i.e., grouping) streams within oiling level, thus reducing the effects of the considerable variation among streams on tests of hypotheses. In another example, McDonald *et al.* (1995) improved the power of a stratified random design by using a paired *t*-test on oiled and reference sites that were matched on features of shoreline habitat.

The third component of statistical power is the α-level selected to judge significance. Statistical tests of hypotheses will always produce a *P*-value, but whether or not the results are deemed statistically significant (i.e., the null hypothesis is rejected) depends on the α-level. Traditionally, the α-level in statistical tests of hypothesis tests is set at 0.05; less commonly used are $\alpha = 0.10$, which increases the probability of rejecting the null hypothesis (i.e., of finding an effect, although at an increased risk of finding an effect when there is none), and $\alpha = 0.01$, which makes the hypothesis test more stringent. Aside from tradition, however, there are no compelling a priori reasons to use these α-levels in observational studies of environmental perturbations. The level of α used in tests should be selected (*before* the analysis is conducted) based on what is appropriate to a given situation. If one expects low power – typically because of low sample size – an α-level greater than 0.05 should be considered. For example, because they wished to err on the side of documenting effects of the *Exxon Valdez* spill on seabirds, even if such effects did not actually occur, Day *et al.* (1995, 1997a, b) used $\alpha = 0.05$ (strong evidence of an effect), $\alpha = 0.10$ (moderate evidence), and $\alpha = 0.20$ (weak evidence) (Chapter 14). Because sample sizes were small and variances typically were high, the use of higher α-levels increased statistical power to detect an effect. Because $\alpha = 0.05$ is so widely accepted as the standard for significance, however, the use of a different α-level should be justified.

10.7 Use pre-impact data to reduce confounding factors

Pre-impact data taken at both impacted and reference areas can be used to make assessments that reduce the confounding effects of temporally and/or spatially varying natural factors among sites. Murphy *et al.* (1997) categorized 10 bays in PWS as either oiled (moderate or heavy oiling; i.e., "impacted") or unoiled (light oiling or unoiled; i.e., "control") and then compared nearshore bird densities with those recorded before the spill in 1984 by Irons *et al.* (1988) by using a Before–After/Control–Impact (BACI[2]) design (Skalski and McKenzie, 1982; Stewart-Oaten *et al.*, 1986; Wiens and Parker, 1995). Erikson (1995) also used a BACI design to assess spill effects on the abundance of murres (*Uria* spp.). Klosiewski and Laing (1994) used a similar analytical approach to assess effects on birds following the *Exxon Valdez* spill but used different before and after data sets from PWS for their analyses. In a BACI analysis, one assumes that whatever environmental changes occurred between the before and after surveys would affect the impacted and reference sites similarly (i.e., sampling sites were in dynamic equilibrium).

A BACI analysis compares *relative* changes in abundance between impacted and control (reference) sites; any significant differences are ascribed to the perturbation. BACI analyses are useful because they deal with environmental variation in an uncomplicated manner and because they contribute to a weight-of-evidence approach to impact assessment. As stand-alone studies, however, BACI analyses implemented after the *Exxon Valdez* spill were not as sensitive as other approaches to assessing effects. In the particular studies of Murphy *et al.* (1997) and Klosiewski and Laing (1994), the reduced sensitivity may have resulted from several factors, including unreplicated sampling on each survey, treating oiling as a categorical variable, low statistical power, flawed prespill baseline data, and/or poor selection of reference sites. These are all factors that should be considered in designing a BACI study, although some (e.g., the quality of the preperturbation data) are beyond the control of the investigator.

10.8 Use replication to increase precision and statistical power

A consistent theme in many of the chapters in this book is the pervasive influence of numerous, complex sources of environmental variation on everything from the fate of oil (Chapter 6) to our ability to detect oiling effects on (and recovery of) the biota (Chapters 11–15). Understanding environmental variation and accounting for it in study and analytical designs are critical for developing quality scientific assessments. In the immediate aftermath of the *Exxon Valdez* spill, Day *et al.* (1995, 1997a, b) used replication to help account for variation in bird distributions associated with tides, weather, and other attributes of the highly variable marine environment and to improve statistical power. A measure of the effectiveness of replicate sampling can be inferred by comparing the results of the initial assessment of spill effects in 1989 reported by Day *et al.* (1997b; 3–5 replicates/survey yielded 19 species classified as negatively affected) with those reported by Klosiewski and Laing (1994) (one replicate/survey yielded nine taxa classified as negatively affected). Although there were other differences between the two studies besides replication (e.g., how oiling levels were classified), Day *et al.*'s sampling design detected more than twice the number of initial effects as that of Klosiewski and Laing. As a corollary, Day *et al.* (1995) also were

---

[2] In other formulations, replicates were taken over time within before and after periods. In the following examples, replicates were taken at different sampling sites at approximately the same time. "Reference" is more appropriate than "control" because oiling was not randomized among sites.

able to document recovery of affected species (Chapter 14, Fig. 14.4) more effectively than were Klosiewski and Laing (1994), who used single-replicate methods.

In addition to improving statistical power to detect effects, replication provides an opportunity to document and evaluate unevenness in the distribution or abundance of species among replicates – clustering – and how it can affect our understanding of natural history and, as a consequence, the interpretation of statistical results (Ryan, 2011). For example, Day et al. (1995) observed 5200 gulls (mostly glaucous-winged, *Larus glaucescens*) feeding on seasonal salmon spawn during one replicate survey in Galena Bay. On a winter cruise, Day et al. also observed 132–133 common mergansers (*Mergus merganser*) on back-to-back replicate surveys of Drier Bay (see Map 3, p. vii), evidence that mergansers cluster and reside within protective bays during the winter season. These counts of gulls and mergansers were at least two orders of magnitude greater than those typically recorded. These experiences illustrate the degree of widespread variability in bird counts and the usefulness of replicate surveys to capture variability and observe the effects of factors such as season, location, tide, or weather. High variability from intense clustering can increase the likelihood of obtaining nonsignificant results when there is, in fact, a real effect (i.e., a Type II error).

Despite the obvious benefits of increased replication, however, there are tradeoffs. Assuming that one has limited resources, the downside of replicate sampling of individual sites is that fewer sites can be sampled, which may result in less extensive coverage of the impacted and reference areas than would be achieved by more widespread, single-sample surveys. Whether one opts for extensive sampling at the expense of replication, or for increased replication and information about clustering in more intensive sampling of fewer sites, depends on the study objectives. In general, extensive sampling may be more appropriate in early phases of an oil spill (Chapter 2), when determining the extent of potential effects is important, whereas more intensive, replicated sampling may be more suitable for assessing effects in localized areas. The Day et al. (1995) study, for example, was focused on 10 study locations (bays) that were selected to represent a quantitatively defined gradient of levels of oiling from the *Exxon Valdez* spill.

10.9 Use multiple-year studies to assess recovery

Multiple-year studies can be essential for detecting recovery of biota from an environmental perturbation. The longer the time to recovery, the more essential multiple-year studies are. Although the rate of recovery and status of injured bird populations continues to be a source of debate (Chapter 14), virtually all observers anticipated that in most instances recovery from the *Exxon Valdez* oil spill would take years. Conducting multiple-year surveys allowed Day et al. (1995, 1997b) to document decreasing effects and the trajectory of recovery in habitat use by marine birds through time (Chapter 14, Fig. 14.4). Although Day et al. (1995, 1997b) concluded that most affected species had recovered in 2–3 years, it took 7 years to document recovery in habitat use for all bird species studied (Wiens et al., 2004). In an example requiring a multiple-year time series of data to detect effects (rather than recovery), Matkin et al. (2008) analyzed 21 years of survey data (1984–2004) on killer whales (*Orcinus orca*; a long-lived species with a low reproductive rate) to show that the unexpectedly slow recovery of the AB pod in PWS likely resulted from abnormal mortalities of juveniles and reproductive females in the year following the spill. As a counter-example, Gilfillan *et al.* (1995) conducted a single-year study of shoreline ecology following the *Exxon Valdez* spill. Surveys in 1990 indicated that shoreline biota had recovered within the range of 73% to 91%; because of this rapid recovery, broad-scale studies were discontinued after 1990. Consequently, there is no way to estimate if and when full shoreline recovery occurred or whether the apparent increases in 1990 continued in subsequent years. Multiple-year studies would have helped to resolve such questions, although the benefits of the answers would need to be balanced against the costs of additional years of (expensive) shoreline sampling (Chapter 11).

10.10 Statistics is not enough

If a test of a hypothesis is significant, the immediate question is, "Why?" Statistics alone does not provide an answer. For a well-designed experiment in which the confounding effects of sampling, natural factors, and variation have been controlled, the "why" is logically inferred: the rejection of the null hypothesis provides evidence of a perturbation effect. Determining "why" in an observational study, however, is not so straightforward. A statistically significant difference between impacted and reference areas could be the result of the perturbation, but it could instead have been caused by differences in other factors (if the study design is incorrect), or it might simply reflect spatial or temporal variation in an inadequately sampled system. On the other hand, a nonsignificant test outcome could indicate that there were indeed no effects, but it could also result if differences in natural factors mask a real effect of the perturbation (i.e., statistical power is too low to achieve significance at the designated α-level).

Because environmental accidents often occur in complex, dynamic environments where much else is going on, it is easy to presume that complex explanations of the observed patterns are necessary. Bear in mind, however, that Occam's Razor ("the simplest explanation is most likely the correct one") may cut just as finely for assessing the consequences of an environmental accident as it does for other areas of science. Consider an example. Salmon egg mortality in oiled streams was found to be significantly higher from 1989 to 1993 and, after a 3-year hiatus, again in 1997 (Chapter 12). To explain this pattern in mortality, Rice *et al.* (2001) and Craig *et al.* (2002) proposed that a shift in stream channels uncovered deposits of weathered oil, over which salmon spawned, exposing eggs to high-molecular-weight polycyclic aromatic hydrocarbons (PAH) in the newly exposed sediments. There were no data to indicate that PAH concentrations had in fact increased. Field experiments, however, showed that the higher egg mortality at oiled streams was a consequence of the timing of sampling rather than of cascading effects of oiling history (Chapter 12; Brannon *et al.* 2001, 2012) – a simpler explanation and one relying on direct measurements rather than inferred relationships.

10.11 Interpret data with an eye toward corroborating evidence

We have advocated a statistically rigorous approach to assessing the effects of environmental perturbations. Statistics alone, however, can only tell us what is and what is not significantly different, with a specified probability, as revealed by the data analyzed.

The data, however, are only the numbers that we have obtained through (it is hoped) well-designed sampling and appropriate measurements. Interpreting the statistical results requires one to step beyond the data and analyses to consider whether statistically significant results are also biologically significant and whether the cause-effect pathways supported by the statistical analyses are in fact plausible. Biological significance means that a statistically significant relationship must translate into something that affects key features of the biology of the resource of concern. For example, the demonstration that levels of an enzyme (CYP1A; Chapter 9) that is induced by exposure to hydrocarbons were significantly higher in individuals in areas that were oiled years before by the *Exxon Valdez* spill than in individuals from an unoiled area (e.g., Esler *et al.*, 2010) may not be important if the higher levels do not translate into decreased individual or population performance (e.g., survival, reproduction, habitat selection). Plausibility requires that there be a clear biological mechanism and pathway to link factors that are statistically correlated or interpretations that build on multiple lines of evidence. Thus, suggestions that harlequin ducks are exposed to the hydrocarbons that induce CYP1A by foraging close to sites where sea otters (*Enhydra lutris*) have been digging subtidal pits, potentially releasing residual *Exxon Valdez* hydrocarbons from buried deposits (Short *et al.*, 2004), do not seem plausible; harlequin ducks are rarely seen near sea otters (Harwell *et al.*, 2011), much less foraging near foraging otters (Chapter 16). In addition, sea otters generally do not forage at locations with residual oil (Boehm *et al.*, 2011; Neff *et al.*, 2011).

Just as the inclusion of covariates in an analysis can help to tease apart the effects of potential confounding factors, the use of multiple analyses in a weight-of-evidence approach can strengthen (or weaken) one's confidence in a conclusion. Complementary results among several analyses lend support to conclusions about an effect or its absence, whereas dissimilar results lead to weaker conclusions. For example, a "sediment triad" approach (Long and Chapman, 1985; Chapman, 1990) has been used to understand the linkages between three important components of an exposure pathway: (1) elevated levels of specific chemicals, which result in (2) elevated toxicity, which in turn (3) leads to injuries to biota. To assess possible effects of the *Exxon Valdez* spill on shoreline ecology in 1990 and 1991, Page *et al.* (1995) sampled sediments for total PAH (TPAH), conducted bioassays of toxicity to benthic amphipods (*Rhepoxynius abronius*), and collected core samples of interstitial biota. Using these data, Boehm *et al.* (1995) found a positive relationship between TPAH concentration and toxicity. Results from analyses of core samples (Gilfillan *et al.*, 1995) were consistent with those of Boehm *et al.* (1995): effects on biological resources were positively correlated with hydrocarbon contamination and toxicity to amphipods. Using these data, Page *et al.* (2002) found that biodiversity decreased with increasing TPAH concentration above the toxicity threshold. Collectively, these lines of evidence provided stronger support for inferences about effects and recovery than would have any single one alone.

A weight-of-evidence approach need not be restricted to the results of statistical tests. An examination of the data or results of preliminary analyses may help one interpret and understand test results. For example, Day *et al.* (1995) and Wiens *et al.* (2004) compared graphical scatter-plots of the frequency of occurrence of marine-bird species over several sampling periods to see whether the overall patterns supported the conclusions about recovery derived from statistical analyses. Such descriptive tools can supplement tests of hypothesis.

Assessing effects and recovery with observational studies never is simple and clear-cut, and data analysis involves more than simply running statistical software and unquestioningly accepting the results of tests. Interpreting results relies on a close examination of data and analyses, questioning whether there are hidden biases in a study design or in the data, evaluating assumptions, and gauging the strength of results. It's not just science. Interpretation is really an art, and it is best done by a team. Many eyes are better at seeing than a few, and insights emerge from discussions among people with different skill sets and experiences. Teamwork is critical to resolving questions about effects and recovery from a severe perturbation that extends over a large geographic area, such as the *Exxon Valdez* spill.

10.12 Adapt to changing circumstances

There is an additional factor that complicates observational studies: "The best-laid schemes o' mice an' men gang aft agley" (the best-laid plans often go awry) syndrome. Unlike designed experiments, in which the researcher has control over levels of treatments and numbers of subjects and replicates, it often is difficult in observational studies to follow one's initial study design, no matter how much care and effort are expended in planning. The locations sampled and sample sizes are likely to be less than planned for, initially balanced designs can become unbalanced, potentially confounding habitat variables go unmeasured or are recognized only in retrospect, and so on. In studies of salmon following the *Exxon Valdez* spill, for example, it was necessary to sample streams at low tide at both oiled and reference locations that were some distance apart, limiting the number of locations that could be sampled in any tide window. Priorities for surveying sea otters versus shoreline birds had to be adjusted depending on the availability of suitable vessels and qualified personnel. Vagaries of weather required constant modifications of sampling strategies. In an observational study, one must be willing to adjust the study design and analyses to suit the realities of fieldwork and the available data. At the same time, one has to be vigilant about potential biases introduced to data sets because of uncontrolled sources of variation in field sampling.

Designing an efficient and focused, yet flexible, field study to assess the effects of a perturbation on a natural resource requires an understanding of its life history and how variations in natural factors affect that resource. With this foundation, one can make a good first cut at which measures of the resource and concomitant variables should be taken and develop a concept of how to analyze the data. Even with a team of experts, however, there will be a learning curve to climb. Logistic and budgetary constraints, inclement weather, and other unforeseen events will tarnish the beauty of a well-designed study plan. In addition, issues may emerge only years after initial studies have been completed (e.g., biomarkers, Chapter 9; the effect of sample timing on salmon egg mortality, Chapter 12), necessitating reanalysis or reinterpretation of data. The results of other research may also bring new factors into the interpretive mix. For example, the effects of humpback whale (*Megaptera novaeangliae*) predation or juvenile pink salmon competition on population dynamics of Pacific herring (Chapter 13) or of killer whale predation on sea otters (Chapter 15) were realized only more than a decade after the *Exxon Valdez* spill. The imperative is to remain flexible.

10.13 Lessons learned

With some notable exceptions – e.g., the eruption of Mt. St. Helens in 1980 (Dale *et al.*, 2005), the Yellowstone National Park fires of 1988 (Turner *et al.*, 2003) – few investigations of disturbances (especially human-caused environmental accidents) are carried on long enough to obtain a sufficient understanding of the temporal dynamics of the resources or their habitats. Investigations of the consequences of the *Exxon Valdez* oil spill continued for so long in part because the continuing litigation and the high financial stakes drove a need to find out what the short- and long-term effects really were. Because the scale of the oil spill was large, the logistics difficult and expensive, and the overtones of ongoing litigation and emotional responses to the spill ever-present, the need to be attentive to the details of study design and analysis was especially great. Drawing on the studies covered in Chapters 11–16 and our own experience with the *Exxon Valdez* spill, we can offer some thoughts about how to assess the effects of oil spills or other environmental perturbations, be they anthropogenic or natural.

- The single most important lesson from the *Exxon Valdez* oil spill is that there are no shortcuts. Although determining the immediate, acute effects of an environmental accident may seem straightforward and logistically manageable, assessing the subsequent dynamics of the system, evaluating possible chronic effects, determining when recovery occurs, or whether "recovery" is even a useful or operational concept, is complex and demanding. It takes time and a great deal of thought.

- Environmental accidents cannot be evaluated as if they were experiments. The treatment itself (the accident) is not replicated and, because the effects of the accident are not randomized, there are no true controls. Environmental accidents instead entail observational studies, in which the replicates are samples in statistical analyses, so one must rely on reference samples rather than true controls.

- Study designs necessarily make assumptions about the nature of environmental equilibrium in the system under study. These assumptions must be evaluated carefully and explicitly to determine which are reasonable and which are not under the circumstances.

- Evaluating the consequences of an environmental perturbation requires that terms such as "impact," "effect," "injury," and "recovery" be operationally defined in ways that are appropriate to the system attributes being considered. It is generally easier to detect effects or injuries than recovery, but doing so becomes more problematic as more time passes after the perturbation. Recovery should be assessed by using criteria that are amenable to statistical tests; defining recovery as a return to some previous (but usually unknown) state ignores the realities of system variability in time and space.

- Even though it may seem counterintuitive, the appropriate null hypothesis should generally be one of no effect. Framing the hypothesis in this way reduces potential biases that may arise from preconceptions about perturbation effects.

- Evaluation of plausible alternative causal hypotheses should be part of the study design. When the direction of perturbation is known (either positive or negative effect), one-tailed tests are more powerful that two-tailed tests at the same α-level.

- A failure to reject the null hypothesis of no effect does not mean that there actually were no effects, only that no effects are statistically detectable given the power of the statistical test. Because sample sizes are often small, the power of a test to detect effects can be enhanced by increasing the α-level to 0.10 or 0.20. Slavish adherence to an α-level of 0.05 should be avoided when assessing the consequences of environmental accidents.

- Because environments vary in space and time, it may be difficult to determine whether an observed effect is due to the perturbation or to other factors. The signal of the perturbation may be obscured by the noise of other things, and the likelihood of such confounding effects increases as the spatial or temporal scale of analysis is increased. The effects of confounding factors can be reduced by incorporating them as covariates into the study design and statistical analyses.

- Replicate sampling within a sampling period represents an effective way to reduce variance, which in turn will result in more powerful statistical tests and a greater ability to evaluate effects and recovery. Although there is a tradeoff between surveying fewer sites repeatedly and conducting single surveys of more sites, our experience indicates that the advantages of replicate sampling generally outweigh the disadvantages.

- Multiple postspill years are often needed to assess recovery. Severity of effects, lingering risks, climatic conditions, natural history, and population dynamics often extend effects beyond a single year, potentially delaying recovery for several years.

- Be wary about constructing a complex explanation for results when a simpler one will do, especially if the complex explanation involves steps that have not been (or cannot be) empirically documented.

- Single statistical tests may yield ambiguous results, especially if they are sensitive to the influences of confounding environmental factors. A weight-of-evidence approach that incorporates multiple tests, including inspection of patterns in the underlying data and qualitative results, may lend strength to conclusions.

- Do not put the interpretive cart before the analytical horse. Frame the hypotheses, collect the data, conduct the analyses, and only then interpret the results. This is how good science is done. Jumping to conclusions about the possible effects of an environmental accident before the data are available and analyzed can produce misinformation that feeds advocacy, not science.

- Because accidents occur in a temporally dynamic and spatially variable environment, unforeseen factors can erode the tightness and balance of an initial study design. One should be willing to adjust study designs or analyses as conditions and experiences in the field warrant.

- Crafting a powerful study design and developing a thoughtful interpretation of results can benefit from the multiple insights of a team of scientists and nonscientists. Dealing with a large environmental perturbation in a complex and dynamic environment requires teamwork.

REFERENCES

Boehm, P.D., D.S. Page, E.S. Gilfillan, W.A. Stubblefield, and E.J. Harner (1995). Shoreline Ecology Program for Prince William Sound, Alaska, following the *Exxon Valdez* oil spill: Part 2 – Chemistry and toxicology. In Exxon Valdez *Oil Spill: Fate and Effects in Alaskan Waters*. P.G. Wells, J.N. Butler, and J.S. Hughes, eds. Philadelphia,

PA, USA: American Society for Testing and Materials; ASTM Special Technical Publication 1219; ISBN-10: 0803118961; pp. 347–397.

Boehm, P.D., D.S. Page, J.M. Neff, and J.S. Brown (2011). Are sea otters being exposed to subsurface intertidal oil residues from the *Exxon Valdez* oil spill? *Marine Pollution Bulletin* **62**(3): 581–589.

Brannon, E.L., K. Collins, L.L. Moulton, M.A. Cronin, A.W. Maki, and K.R. Parker (2012). Review of the *Exxon Valdez* oil spill effects on pink salmon in Prince William Sound, Alaska. *Reviews in Fisheries Science* **20**(1): 20–60.

Brannon, E.L., K.C.M. Collins, M.A. Cronin, L.L. Moulton, and K.R. Parker (2001). Resolving allegations of oil damage to incubating pink salmon eggs in Prince William Sound. *Canadian Journal of Fisheries and Aquatic Sciences* **58**(6): 1070–1076.

Brannon, E.L., A.W. Maki, L.L. Moulton, and K.R. Parker (2006). Results from a sixteen year study on the effects of oiling from *Exxon Valdez* on adult pink salmon returns. *Marine Pollution Bulletin* **52**(8): 892–899.

Bue, B.G., S. Sharr, and J.E. Seeb (1998). Evidence of damage to pink salmon populations inhabiting Price William Sound, Alaska, two generations after the *Exxon Valdez* oil spill. *Transactions of the American Fisheries Society* **127**(1): 35–43.

Burnham, K.P. and D.R. Anderson (2002). *Model Selection and Multimodel Inference: A Practical Information–Theoretic Approach, 2nd edn*. New York, NY, USA: Springer-Verlag; ISBN-10: 0387953647; ISBN-13: 9780387953649.

Chapman, P.M. (1990). The sediment quality triad approach to determining pollution-induced degradation. *Science of the Total Environment* **97–98**: 815–825.

Craig, A.K., T.M. Willette, D.G. Evans, and B.G. Bue (2002). *Injury to Pink Salmon Embryos in Prince William Sound: Field Monitoring*. Cordova, Soldotna, Anchorage, AK, USA: Alaska Department of Fish and Game, Division of Commercial Fisheries; *Exxon Valdez* Oil Spill Restoration Project Restoration Project 98191A-1 Final Report. [http://www.evostc.state.ak.us/Files.cfm?doc=/Store/FinalReports/1998–98191A1-Final.pdf&]

Dale, V.H., F.J. Swanson, and C.M. Crisafulli, eds (2005). *Ecological Responses to the 1980 Eruption of Mount St. Helens*. New York, NY, USA: Springer-Verlag; ISBN-13: 9780387238685 (hardcover); ISBN-13: 9780387238500 (softcover).

Day, R.H., S.M. Murphy, J.A. Wiens, G.D. Hayward, E.J. Harner, and L.N. Smith (1995). Use of oil-affected habitats by birds after the *Exxon Valdez* oil spill. In *Exxon Valdez Oil Spill: Fate and Effects in Alaskan Waters*. P.G. Wells, J.N. Butler, and J.S. Hughes, eds. Philadelphia, PA, USA: American Society for Testing and Materials; ASTM Special Technical Publication 1219; ISBN-10: 0803118961; pp. 726–761.

Day, R.H., S.M. Murphy, J.A. Wiens, G.D. Hayward, E.J. Harner, and B.E. Lawhead (1997a). Effects of the *Exxon Valdez* oil spill on habitat use by birds along the Kenai Peninsula, Alaska. *The Condor* **99**(3): 728–742.

Day, R.H., S.M. Murphy, J.A. Wiens, G.D. Hayward, E.J. Harner, and L.N. Smith. (1997b). Effects of the *Exxon Valdez* oil spill on habitat use by birds in Prince William Sound, Alaska. *Ecological Applications* **7**(2): 593–613.

Eberhardt, L.L. and J.M. Thomas (1991). Designing environmental field studies. *Ecological Monographs* **61**(1): 53–73.

Erikson, D.E. (1995). Surveys of murre colony attendance in the northern Gulf of Alaska following the *Exxon Valdez* oil spill. In Exxon Valdez *Oil Spill: Fate and Effects in Alaskan Waters*. P.G. Wells, J.N. Butler, and J.S. Hughes, eds. Philadelphia, PA, USA: American Society for Testing and Materials; ASTM Special Technical Publication 1219; ISBN-10: 0803118961; pp. 780–819.

Esler, D., T.D. Bowman, K.A. Trust, B.E. Ballachey, T.A. Dean, S.C. Jewett, and C.E. O'Clair (2002). Harlequin duck population recovery following the "Exxon Valdez" oil spill: progress, process, and constraints. *Marine Ecology Progress Series* **241**: 271–286.

Esler, D., J.A. Schmutz, R.L. Jarvis, and D.M. Mulcahy (2000). Winter survival of adult female harlequin ducks in relation to history of contamination by the "Exxon Valdez" oil spill. *Journal of Wildlife Management* **64**(3): 839–847.

Esler, D., K.A. Trust, B.E. Ballachey, S.A. Iverson, T.L. Lewis, D.J. Rizzolo, D.M. Mulcahy, A.K. Miles, B.R. Woodin, J.J. Stegeman, J.D. Henderson, and B.W. Wilson (2010). Cytochrome P450 1A biomarker indication of oil exposure in harlequin ducks up to 20 years after the *Exxon Valdez* oil spill. *Environmental Toxicology and Chemistry* **29**(5): 1138–1145.

Exxon Valdez Oil Spill Trustee Council (2002). *Exxon Valdez Oil Spill Restoration Plan Update on Injured Resources and Services, August, 2002*. Anchorage, AK, USA: *Exxon Valdez* Oil Spill Trustee Council. [http://www.evostc.state.ak.us/Universal/Documents/Publications/2002IRSUpdate.pdf]

Gilfillan, E.S., D.S. Page, E.J. Harner, and P.D Boehm (1995). Shoreline Ecology Program for Prince William Sound, Alaska, following the *Exxon Valdez* oil spill. Part 3: Biology. In Exxon Valdez *Oil Spill: Fate and Effects in Alaskan Waters*. P.G. Wells, J.N. Butler, and J.S. Hughes, eds. Philadelphia, PA, USA: American Society for Testing and Materials; ASTM Special Technical Publication 1219; ISBN-10: 0803118961; pp. 398–443.

Harwell, M.A., J.H. Gentile, K.R. Parker, S.M. Murphy, R.H. Day, A.E. Bence, J.M. Neff, and J.A. Wiens (2011). Quantitative assessment of current risks to harlequin ducks in Prince William Sound, Alaska, from the *Exxon Valdez* oil spill. *Human and Ecological Risk Assessment* **18**(2): 261–328.

Hutchings, M.J., E.A. John, and A.J.A. Stewart, eds (2000). *The Ecological Consequences of Environmental Heterogeneity: 40th Symposium of the British Ecological Society*. Cambridge, UK: Cambridge University Press; ISBN-10: 0521549353; ISBN-13: 9780521549356.

Irons, D.B., S.J. Kendall, W.P. Erickson, L.L. McDonald, and B.K. Lance (2000). Nine years after the "Exxon Valdez" oil spill: effects on marine bird populations in Prince William Sound, Alaska. *The Condor* **102**(4): 723–737.

Irons, D.B., D.R. Nysewander, and J.L. Trapp (1988). *Prince William Sound Waterbird Distribution in Relation to Habitat Type*. Anchorage, AK, USA: US Fish and Wildlife Service.

Klosiewski, S.P. and K.K. Laing (1994). *Marine Bird Populations of Prince William Sound, Alaska, Before and After the* Exxon Valdez *Oil Spill*. Anchorage, AK, USA: US Fish and Wildlife Service; *Exxon Valdez* Oil Spill State/Federal Natural Resources Damage Assessment Bird Study 2 Final Report. [http://www.evostc.state.ak.us/Files.cfm?doc=/Store/FinalReports/1989-B02-Final.pdf&]

Lance, B.K., D.B. Irons, S.J. Kendall, and L.L. McDonald (2001). An evaluation of marine bird population trends following the *Exxon Valdez* oil spill, Prince William Sound, Alaska. *Marine Pollution Bulletin* **42**(4): 298–309.

Long, E.R. and P.M. Chapman (1985). A sediment quality triad: measures of sediment contamination, toxicity, and infaunal community composition in Puget Sound. *Marine Pollution Bulletin* **16**(10): 405–415.

Matkin, C.O., E.L. Saulitis, G.M. Ellis, P. Olesiuk, and S.D. Rice (2008). Ongoing population-level impacts on killer whales *Orcinus orca* following the "Exxon Valdez" oil spill in Prince William Sound, Alaska. *Marine Ecology Progress Series* **356**: 269–281.

McDonald, L.L, W.P. Erickson, and M.D. Strickland (1995). Survey design, statistical analysis, and basis for inferences in coastal habitat injury assessment: *Exxon Valdez* oil spill. In Exxon Valdez *Oil Spill: Fate and Effects in Alaskan Waters*. P.G. Wells, J.N. Butler, and J.S. Hughes, eds. Philadelphia, PA, USA: American Society for Testing and Materials; ASTM Special Technical Publication 1219; ISBN-10: 0803118961; pp. 296–311.

McKnight, A., K.M. Sullivan, D.B. Irons, S.W. Stephensen, and S. Howlin (2008). *Prince William Sound Marine Bird Surveys, Synthesis and Restoration*. Anchorage, AK, USA: US Fish and Wildlife Service; *Exxon Valdez* Oil Spill Restoration Project 080751 Final Report. [http://www.evostc.state.ak.us/Files.cfm?doc=/Store/FinalReports/2008-080751-Final.pdf&]

Milly, P.C.D., J. Betancourt, M. Falkenmark, R.M. Hirsch, Z.W. Kundzewicz, D.P. Lettenmaier, and R.J. Stouffer (2008). Stationarity is dead: whither water management? *Science* **319**(5863): 573–574.

Murphy, S.M., R.H. Day, J.A. Wiens, and K.R. Parker (1997). Effects of the *Exxon Valdez* oil spill on birds: comparisons of pre- and post-spill surveys in Prince William Sound, Alaska. *The Condor* **99**(2): 299–313.

Neff, J.M., D.S. Page, and P.D. Boehm (2011). Exposure of sea otters and harlequin ducks in Prince William Sound, Alaska, USA, to shoreline oil residues 20 years after the *Exxon Valdez* oil spill. *Environmental Toxicology and Chemistry* **30**(3): 659–672.

Page, D.S., P.D. Boehm, W.A. Stubblefield, K.R. Parker, E.S. Gilfillan, J.A. Neff, and A.W. Maki (2002). Hydrocarbon composition and toxicity of sediments following the *Exxon Valdez* oil spill in Prince William Sound, Alaska, USA. *Environmental Toxicology and Chemistry* **21**(7): 1438–1450.

Page, D.S., E.S. Gilfillan, P.D Boehm, and E.J. Harner (1995). Shoreline Ecology Program for Prince William Sound, Alaska, following the *Exxon Valdez* oil spill. Part 1: Study design and methods. In Exxon Valdez *Oil Spill: Fate and Effects in Alaskan Waters*. P.G. Wells, J.N. Butler, and J.S. Hughes, eds. Philadelphia, PA, USA: American Society for Testing and Materials; ASTM Special Technical Publication 1219; ISBN-10: 0803118961; pp. 263–295.

Parker, K.R. and J.A. Wiens (2005). Assessing recovery following environmental accidents: environmental variation, ecological assumptions, and strategies. *Ecological Applications* **15**(6): 2037–2051.

Pearson, W.H., E. Moksness, and J.R. Skalski (1995). A field and laboratory assessment of oil spill effects on survival and reproduction of Pacific herring following the *Exxon*

Valdez oil spill. In Exxon Valdez *Oil Spill: Fate and Effects in Alaskan Waters*. P.G. Wells, J.N. Butler, and J.S. Hughes, eds. Philadelphia, PA, USA: American Society for Testing and Materials; ASTM Special Technical Publication 1219; ISBN-10: 0803118961; pp. 626–661.

Pickett, S.T.A., J. Kolasa, and C.G. Jones (1994). *Ecological Understanding*. San Diego, CA, USA: Academic Press, Inc.; ISBN 0-12-554720-X.

Popper, K.R. (1959). *The Logic of Scientific Discovery*. London, UK: Routledge.

Rice, S.D., R.E. Thomas, M.G. Carls, R.A. Heintz, A.C. Wertheimer, M.L. Murphy, J.W. Short, and A. Moles (2001). Impacts to pink salmon following the *Exxon Valdez* oil spill: persistence, toxicity, sensitivity, and controversy. *Reviews in Fisheries Science* **9**(3): 165–211.

Rohde, K. (2005). *Nonequilibrium Ecology*. Cambridge, UK: Cambridge University Press; ISBN-13: 9780521674553.

Ryan, M.J. (2011). Replication in field biology: the case of the frog-eating bat. *Science* **334**(6060): 1229–1230.

Short, J.W., M.R. Lindeberg, P.M. Harris, J.M. Maselko, J.J. Pella, and S.D. Rice (2004). Estimate of oil persisting on beaches of Prince William Sound 12 years after the *Exxon Valdez* oil spill. *Environmental Science & Technology* **38**(1): 19–25.

Skalski, J.R. and D.H. McKenzie (1982). A design for aquatic monitoring programs. *Journal of Environmental Management* **14**(3): 237–251.

Stewart-Oaten, A., W.W. Murdoch, and K.R. Parker (1986). Environmental impact assessment: "pseudoreplication" in time? *Ecology* **67**(4): 929–940.

Turner, M.G., W.H. Romme, and D.B. Tinker (2003). Surprises and lessons from the 1988 Yellowstone fires. *Frontiers in Ecology and the Environment* **1**(7): 351–358.

Wiens, J.A. (1984). On understanding a nonequilibrium world: Myth and reality in community patterns and processes. In *Ecological Communities: Conceptual Issues and the Evidence*. D.R. Strong, Jr,. D. Simberloff, L.G. Abele, and A.B. Thistle, eds. Princeton, NJ, USA: Princeton University Press; ISBN-10: 0691083401; ISBN-13: 9780691083407; pp. 439–457.

Wiens, J.A., R.H. Day, S.M. Murphy, and M.A. Fraker (2010). Assessing cause–effect relationships in environmental accidents: harlequin ducks and the *Exxon Valdez* oil spill. *Current Ornithology* **17**: 131–189.

Wiens, J.A., R.H. Day, S.M. Murphy, and K.R. Parker (2001). On drawing conclusions nine years after the *Exxon Valdez* oil spill. *The Condor* **103**(4): 886–892.

Wiens, J.A., R.H. Day, S.M. Murphy, and K.R. Parker (2004). Changing habitat and habitat use by birds after the *Exxon Valdez* oil spill, 1989–2001. *Ecological Applications* **14**(6): 1806–1825.

Wiens, J.A. and K.R. Parker (1995). Analyzing the effects of accidental environmental impacts: approaches and assumptions. *Ecological Applications* **5**(4): 1069–1083.

CHAPTER ELEVEN

Shoreline biota

Erich R. Gundlach, David S. Page, Jerry M. Neff, and Paul D. Boehm

11.1 Introduction

Coastal shorelines teem with life. The intersection of the land with the sea, combined with tidal fluctuations and coastal currents, creates an array of habitats that supports an amazing diversity of plants and animals – limpets, starfish, anemones, crabs, rockweed, eelgrass, snails, tubeworms, and the like – that live on the surface and in the sediments of the intertidal zone. When floating oil from a marine oil spill strikes a shoreline, the potential effects on these organisms (the shoreline biota) may be severe. Even species that are not directly affected by spill may suffer its effects if the shoreline prey on which they feed are diminished. Understanding how a spill affects the shoreline biota is therefore important to assessing the potential effects on the broader shoreline and coastal ecosystems.

During the *Exxon Valdez* spill, oil first spread over shorelines in Prince William Sound (PWS) and later extended outside of PWS to the Kenai Peninsula, Kodiak Island, and Alaska Peninsula (see Map 1, p. v). The effects of the spill and the need to respond rapidly were of enormous concern, particularly within PWS, where oil quantities and potential toxicity were greatest. In this chapter, we discuss three major programs undertaken to assess the effects of the *Exxon Valdez* oil spill on shoreline biota in PWS, including studies to determine the effects of intensive cleanup efforts.[1]

The three programs undertaken in PWS were:

1. *Shoreline Ecology Program (SEP)* – This program had two components: (a) a stratified random sampling (SRS) study involving random site selection that allowed results to represent the PWS spill zone as a whole; and (b) a nonrandom, fixed site (FS) study in which results applied only to the heavily oiled sites selected for field investigation. This program was sponsored by Exxon.
2. *Coastal Habitat Injury Assessment (CHIA)* – This study used a matched-pairs design to compare PWS habitats that were moderately or heavily oiled with unoiled sites. It was sponsored by the *Exxon Valdez* Oil Spill Trustee Council ("Trustees").
3. *Biological Monitoring Survey (BMS)* – This program used nonrandomly selected sites to determine effects and recovery on aggressively cleaned shorelines. It was sponsored by the US National Oceanic and Atmospheric Administration (NOAA).

---

[1] Studies undertaken outside of PWS were limited in scope and of shorter duration (Gilfillan *et al.*, 1995b; Stekoll *et al.*, 1996).

Oil in the Environment: Legacies and Lessons of the Exxon Valdez *Oil Spill*, ed. J. A. Wiens. Published by Cambridge University Press. © Cambridge University Press 2013.

The SEP–SRS and CHIA studies were undertaken in response to natural resource damage assessment (NRDA) federal regulations,[2] which require a determination of whether the resource (i.e., shoreline biota) has been injured. *Injury* is defined as a measurable adverse change (such as a reduction in abundance or viability) due to exposure to oil (Chapter 1, Box 1.3). Exposure may be direct, through contact with surface oil or oil in the water column, or indirect, as potentially cascading effects to fish, mammals, and birds that may contact and/or consume intertidal species.

The BMS program had different objectives than SEP and CHIA; it was designed to determine effects of the high-pressure, hot-water shoreline cleanup treatment applied in the summer of 1989. The Federal On Scene Coordinator (FOSC) approved this treatment with the support of NOAA after field tests in April 1989, assuming that shorelines could recover in one or two years rather than five or six. The FOSC then pressed Exxon to plan for that approach, with restrictions (e.g., no application on algal beds and soft-sediment areas).[3] The resultant boiler-based systems were mounted on barges that were able to provide heated (60°C) seawater to the oiled shoreline at 950–2000 L/min (Nauman, 1991).

Although successful at removing most of the surface oil, the immediate negative visual impact of this technique on shoreline biota created controversy. Most plants and animals were killed in place ("cooked"; Whitney, 1991, p. B1; Piper, 1993, p. 63) or physically removed because of the combination of high pressure and hot water. NOAA announced at a 1991 press conference that its approval of the technique was "on balance, a mistake" and "the indication is that the treatment process [caused] most of the problems" (Whitney, 1991, p. B1) and closed with "sometimes the best thing to do in an oil spill is nothing" (Piper, 1993, p. 64).

Impelled by this controversy, the BMS program was developed to answer questions related to the immediate effects on moderate and heavy oiled sites (Houghton *et al.*, 1993b). It continued to monitor long-term recovery until 1997. As BMS sites were not randomly chosen, the results cannot be extrapolated to other areas.

11.2 The PWS shoreline

PWS is an estuarine embayment of the Gulf of Alaska (GOA) with numerous rugged islands and a mountainous coastline dissected by deep fjords. The coastline is over 4800 km long and encompasses about 9000 km^2 of ocean. Much of the shoreline in the PWS spill zone is geologically young, a result of the 1–5-m elevation of shorelines during the 1964 earthquake that produced major changes in shoreline ecology (Plafker, 1965; Haven, 1971; Chapter 1).

The environment is subarctic and harsh, with wide temperature fluctuations (averages from −12°C in winter to 20°C in summer), a large tidal range (~5 m), high precipitation (up to 6 m per year), and frequent severe winter storms. Under these conditions, shoreline organisms and communities are far from fragile. Life in the intertidal zone is also exposed to long-term variations in temperature and salinity that may produce substantial ecological changes (Anderson and Piatt, 1999).

Approximately 99% of the shoreline in the PWS spill zone falls into one of four categories: exposed bedrock/rubble (~17%), sheltered bedrock/rubble (~56%),

---

[2] US Code of Federal Regulations 43 CFR §11
[3] Details of the government-approval process are in the FOSC's final report (Leschine *et al.*, 1993).

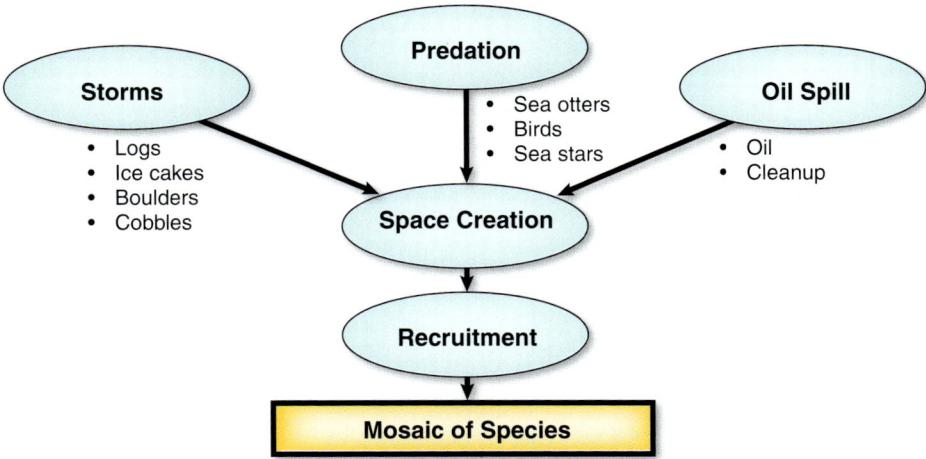

Figure 11.1 Factors influencing the process of space creation and recolonization (patch dynamics) on oiled shorelines in PWS.

boulder/cobble/gravel (~23%), or mixed pebble/gravel (~3.7%) (Chapters 1 and 6). Mud- and peat-dominated shorelines are rare, and there are no sand beaches (Page et al., 1995; Gilfillan et al., 1995a).

The ecology of the rocky shores greatly depends on the degree of wave exposure. Where wave exposure is high, wave-carried beach material, logs, and ice can remove biota from the surface, creating a patchy mosaic of species. Such "patch dynamics" (Fig. 11.1) (Sousa, 1985) strongly influence the shoreline ecology of PWS (Gilfillan et al., 1995a).

11.3 Spill-related studies of PWS shoreline biota

All three programs assessed the effects of the *Exxon Valdez* spill on shoreline organisms, but they did so in different ways (Table 11.1). Only the BMS program analyzed the effects of cleanup that used high-pressure, hot-water flushing. The SEP and CHIA programs considered that observed effects were due to oiling and cleanup together. No study attempted to separate out the potential effects of bioremediation that was applied on over 110 km of oiled shorelines in August–September 1989 and at more limited locations in 1990 (Chianelli et al., 1991; Chapter 9). All shoreline biota studies used the categories of light, moderate, and heavy oiling, based on the width of surface oiling in summer 1989 (Neff et al., 1995; Chapter 4).

11.3.1 Shoreline Ecology Program (SEP) method, 1990–99

Both SEP studies, SRS and FS, based sampling on the sediment quality triad approach (Long, 1989), in which sediment chemistry, toxicity, and ecology are all studied concurrently (Page et al., 1995).

11.3.1.1 SEP–stratified random sampling (SRS) method, 1990

The 64 sites for the SRS study included all three oiling categories (plus unoiled reference sites) and all four shoreline types. Because the ecological community was likely to change over the summer season, each shoreline type was sampled during the same 2-week period in 1990 (Gilfillan et al., 1995a).

Table 11.1 Summary of major *Exxon Valdez* shoreline biota assessment studies in PWS (references in text).

SEP–SRS	CHIA	SEP–FS	BMS
Study objective			
Assess injury and recovery of entire spill zone.	Quantify injuries to intertidal biota of the (moderate + heavy oiled) oil spill area.	Assess injury and recovery of selected heavy oiled and soft-sediment sites.	Assess recovery of intertidal resources and influence of high-pressure, hot-water treatment.
Duration of sampling program			
Summer 1990	Spring and summer 1990–91 and selected sites through 1994 for algae.	1990–91 (21 sites), 1998–99 (4 boulder/cobble + reference sites)	Annual sampling 1989–97. Early studies: ~31 sites. 9 long-term sites.
Additional sampling/analyses			
Wave exposure, sediment grain size, total organic carbon, sediments and tissue for hydrocarbon chemistry, sediments for toxicity bioassays (triad approach).	Hydrocarbon chemistry of selected sediment and tissue samples.	Same as SEP–SRS.	Varies by year and study, includes grain size, total organic carbon, nitrogen, hydrocarbon chemistry of sediments and tissues, and geomorphic profiling
Number of sites and category of shoreline oiling			
64 randomly selected sites: 46 oiled, 18 unoiled. Light (15), moderate (10), heavy (21), none (18).	37 sites: 19 randomly selected moderate + heavy and 18 nonrandom unoiled to form 19 matched pairs.	1990–91 = 21 sites: 11 selected oiled sites: 10 reference (includes 9 from SRS). Light (1), moderate (2), heavy (8), none (10).	1989–97: 9 selected sites: moderate + heavy (6) unoiled (3). 4 sites = geomorphology + chemistry + biology.
Shoreline types			
Boulder/cobble, Exposed bedrock, Pebble/gravel, Sheltered bedrock.	Exposed rocky, Coarse-textured, Sheltered rocky, Estuarine.	Boulder/cobble, Pebble/gravel, Soft sediment.	Rocky, Boulder/cobble, Mixed soft.
Number of sampling stations, transects per site			
12 stations per site. 3 transects perpendicular to shoreline, separated by 20–30 m, with 4 stations (upper, middle, lower intertidal and shallow subtidal).	18 stations per site. 6 transects with 3 stations (upper, middle, and lower intertidal).	Same as SEP-SRS.	Varies by study. 1989–97: 25 stations per site. 3 transects parallel to waterline, 5 or 10 stations per transect.

At each site, three transects were established 30 m apart and perpendicular to the waterline. Along each transect, four stations (upper, middle, and lower intertidal, and shallow subtidal (–3 m)) were sampled. On rocky substrates, all surface biota (epibiota) within a 12.5 × 25-cm quadrat were collected for analysis (Fig. 11.2). At the same location, 50 × 50-cm quadrats were photographed to document surface conditions and coverage by surface biota (Fig. 11.3). On sedimentary substrates, core samples (10-cm diameter, 10-cm deep) were taken whenever possible to assess fauna within the sediment (infauna). Both scrape and core samples were sieved in the field to collect organisms and other material greater than 1.0-mm mesh size. Algae were also sorted

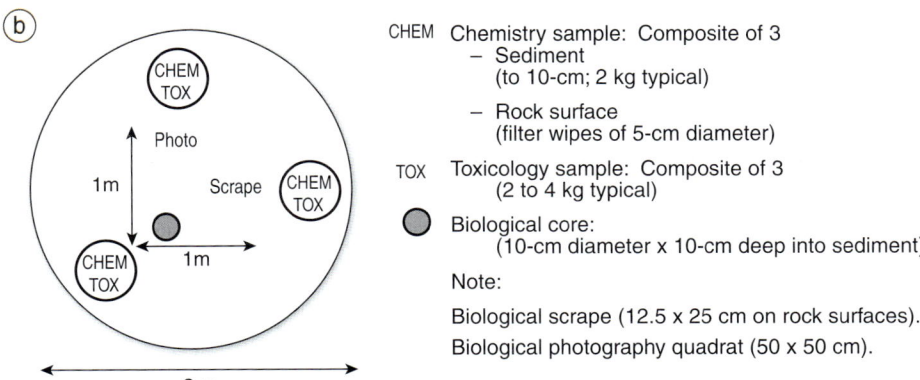

Figure 11.2 Site sampling layout for the SEP (from Page et al., 1995). (a) Plan view of shoreline transects and sampling stations by tidal zones. In successive years, sampling stations were shifted laterally to avoid resampling the same location. (b) Diagram of samples taken at each station using the sediment quality triad approach. Reprinted, with permission, from ASTM STP 1219, Exxon Valdez *Oil Spill: Fate and Effects in Alaskan Waters*, copyright ASTM International, 100 Bar Harbor Drive, West Conshohocken, PA 12428.

and measured by weight in the field. In the laboratory, organisms were sorted into eight taxonomic groups and then identified to the lowest practical taxonomic level.

Sediment samples for chemistry and sediment toxicity measurements were taken concurrently at all locations in accordance with sediment quality triad procedures. In addition, mussels (*Mytilus trossulus*) were taken from several locations at each site, mixed together, and analyzed to derive a single hydrocarbon concentration for the site. This approach enabled the determination of ecological recovery, sediment contamination, and petroleum hydrocarbon uptake by resident biota. Boehm et al. (1995) summarize the analytical chemistry methods and results.

Figure 11.3 Sampling of biota in the lower intertidal zone at the SEP–SRS exposed bedrock site on Little Smith Island in July 1990. (Photo: David S. Page)

At all SEP sites, nonspill-related differences between oiled and reference sites were controlled by measuring covariates (wave energy, total organic carbon, and grain size) (Page et al., 1995). Measuring covariates and using the three transects as independent samples at each study site increased the power of the SEP–SRS to detect an oiling effect (Gilfillan et al., 1999; see Chapter 10).

The data were analyzed at community and species levels (Gilfillan et al., 1995a). The community-level analyses included organism total abundance, species richness, Shannon diversity, and algal biomass. Potential oiling effects were determined using (1) univariate statistics that consider oiling as the only factor influencing the community, and (2) multivariate correspondence analyses that enable multiple dependent variables to be considered. At the species level, only univariate modeling was applied. Gilfillan et al. (1995a, 1999) discuss the statistical models applied to these data sets.

11.3.1.2 SEP–fixed site (FS) method, 1990–99

In contrast to the randomized selection of SRS sites, the FS study selected 11 sites that represented worst case, heavily oiled areas (six mixed pebble/gravel shorelines, four boulder/cobble/gravel beaches, and one soft-sediment tidal flat; Chapter 6). Reference sites were selected out of the SRS program.

The initial FS program included sampling in 1990 and 1991 (Gilfillan et al., 1995a). Field and statistical methods were the same as in the SEP–SRS program. Follow-up surveys were conducted in the summers of 1998 and 1999 at just the four boulder/cobble/gravel shorelines and their four reference sites (Page et al., 1999; Gilfillan et al., 2000). Boulder/cobble/gravel beaches were selected for follow-up because oil residues persisted longest on this shoreline type, so they would be most likely to show spill effects. The full set of multiyear data was analyzed using a two-way analysis of covariance (years and location) with data from each intertidal location analyzed separately (Gilfillan et al., 2001).

11.3.2 Coastal Habitat Assessment Program (CHIA) method, 1990–91

The CHIA study used 19 matched pairs of sites to assess spill effects on moderate to heavy oiled shores (~30% of the total oiled shoreline in PWS) across four shoreline types. The design assumes that differences between oiled and reference sites are due solely to oiling,

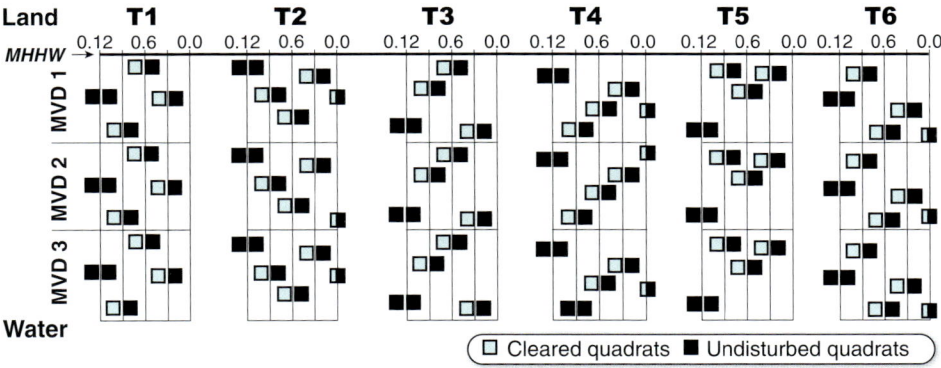

Figure 11.4 Plan view of CHIA field sampling layout at a typical station, illustrating sampling along six transects (T1–T6) at three intertidal levels (MVD = mean vertical drop below mean higher high water [MHHW]). The numbers 0.0, 0.3 0.6, 0.9, and 0.12 correspond to the locations of transects in 1989 (limited collection), spring 1990, summer 1990, spring 1991, and summer 1991, respectively. In summer 1991, only algae percentage cover was recorded and no quadrats were cleared. Modified from Highsmith et al., 1996, with permission of the American Fisheries Society.

so it was important to select reference sites with similar physical and biological characteristics (see Chapter 10). Lightly oiled shorelines showed little to no oiling when surveyed by CHIA, so they were not included as an oiled category but were matched as reference sites to some moderate or heavily oiled sites (Sundberg et al., 1996). Surveys were undertaken during four periods in spring and summer 1990 and 1991, although not all intertidal components and sites were measured during each survey.

At each survey site, six transects were placed perpendicular to the shore. Beginning 1 m below the mean higher high-water (MHHW) line, 0.1-m^2 quadrats were sampled at three locations every 1 m of vertical elevation, thereby capturing data for the upper, middle, and lower intertidal shoreline communities (Fig. 11.4). Algal cover was determined from photographs taken of a 0.4-m^2 area, while algal scrapes and infaunal cores (10-cm depth) were taken within a 0.1-m^2 area. All organisms retained on a 1-mm mesh sieve were sorted to the lowest possible taxonomic category (Highsmith et al., 1996).

The lowest part of the transect, at 3 m below MHHW, was approximately 0.5 m above the mean low water (Stekoll et al., 1996), about the same elevation as the SEP lower intertidal sites. Middle and upper intertidal sites of the SEP study were about 0.5 m above the comparable CHIA sites.

Various statistical tests were used to compare reference and oiled site abundance, biomass, and percentage of cover for ~144 taxonomic categories (species and species groups) during each sampling period, tidal level, region, and habitat (Highsmith et al., 1996). The statistical approach using the matched pairs for analysis is compared to SEP methods in Peterson et al. (2001, 2002) and Gilfillan et al. (1999, 2002).

11.3.3 Biological monitoring survey (BMS) method

BMS sites were selected on sheltered rocky, exposed boulder/cobble/gravel, and mixed soft-sediment (combined rocky and mixed sand/gravel) shorelines. Three categories of shoreline-oiling treatment were considered: (1) oiled–hot-washed (high-pressure, high-temperature flushing); (2) oiled not hot-washed, but may have been treated by other methods such as ambient-temperature water flushing and/or bioremediation; and (3) reference sites having no oiling or treatment. Sampling was within three zones: upper

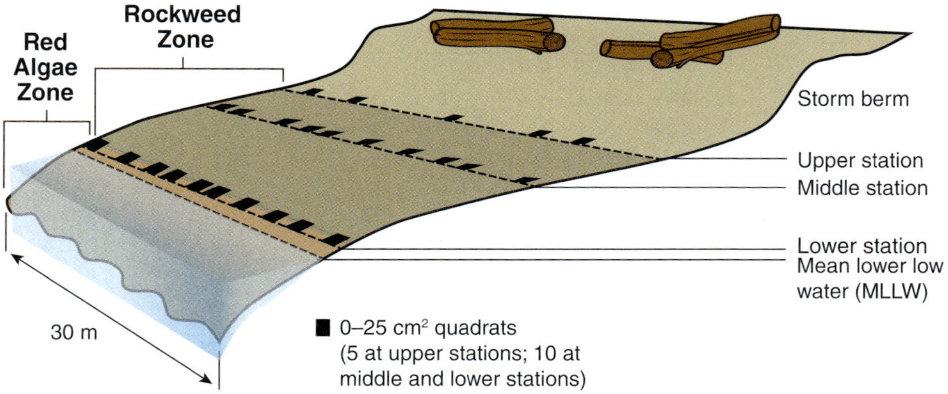

Figure 11.5 Typical site layout of sampling at BMS sites. From Houghton et al., 1993b.

intertidal (defined as near the upper limit of attached macrobiota), middle intertidal (upper portion of the *Fucus* rockweed zone), and lower intertidal (along the lower edge of the rockweed zone). Within each zone, 5 or 10 quadrats of 0.25-m² were analyzed on rocky shores and 5 infaunal core samples (10-cm diameter × 15-cm deep) were taken in soft sediments (Hoff and Shigenaka, 1999). Figure 11.5 shows a typical site layout for BMS field sampling.

The number of sites surveyed varied by year. In 1992, 72 stations at 31 study sites from 18 locations were sampled (Houghton et al., 1993b). Fewer sites and species were sampled through 1995 (Houghton et al., 1997; Hoff and Shigenaka, 1999). Nine sites (three within each shoreline category) were surveyed almost annually from 1989 to 1997 (Coats et al., 1999). Four sites also received geomorphic profiling and analysis of sedimentary and tissue hydrocarbons (Shigenaka et al., 1997). Studies continued at stations with hard-shell clams until 2007 (Lees and Driskell, 2007; Shigenaka et al., 2008; Coats and Fukuyama, 2008).

Three sites of the BMS program were purposely placed on special study areas called "set-asides" designed to monitor natural recovery without cleanup. Throughout PWS, a total of nine sections of shoreline (~2 km total) were approved as official set-asides (Leschine et al., 1993). Despite this designation, several set-asides were cleaned, such that by the end of 1991, only four of the original nine set-asides had not received any treatment. Houghton et al. (1993a) reviewed the cleanup history of each BMS site and confirmed that the three sites (two of the set-asides in Herring Bay, the other between two set-asides in Snug Harbor) used as oiled not hot-washed had indeed received no treatment. For reference, NOAA's geomorphology study (Hayes and Michel, 1998) and the previously discussed SEP–FS program also utilized set-aside areas for sampling (three and four set-aside sites, respectively).

11.3.4 SEP–SRS results

In 1990, SEP–SRS found that sites having the greatest quantity of remaining oil had the highest sediment toxicity and the most negative ecological effects (Table 11.2). These sites were primarily along the upper and middle intertidal zones of boulder/cobble/gravel beaches (Boehm et al., 1995; Gilfillan et al., 1995a). SEP defined *recovery* as the absence of significant differences (positive or negative) between biological communities at oiled and reference sites (see Chapter 1, Box 1.3). Recovery at SEP–SRS sites was as follows (Gilfillan et al., 1995a):

Table 11.2 Summary of results from shoreline biota studies, 1990–2007 (references in text).

SEP–SRS	CHIA	SEP-FS	BMS
Light, moderate, and heavy oiled shores	Moderate and heavy oiled shores	Heavy oiled shores	Moderate and heavy oiled shores
1990–1991 (1.25 to 2.25 years postspill)			
Shoreline recovery = 73% to 91%. Effects noted on rockweed, snails, limpets, and mussels depending on oiling category, habitat, and tidal zone.	5280 variable tests found ~86% with no statistical difference.	Significant differences for species abundance and diversity in soft-sediment habitat and in middle intertidal of boulder/cobble shores.	No significant differences for oiled, not hot-washed sites. Few differences for biota of oiled, hot-washed shores.
1992 (3 years postspill)			
	Fucus shows no differences in middle and lower intertidal zones.		Total infauna abundance, 12 algal species, and 20 invertebrates show equilibrium on oiled, hot-washed sites.
1993 (4 years postspill)			
			Intertidal populations not different for oiled, hot-washed sites, including infauna, algae, and epifaunal invertebrates.
1994 (5 years postspill)			
	Fucus has no differences in upper intertidal zone.		All taxa reach equilibrium.
1996 (7 years postspill)			
			Fucus age structure, dynamics, synchrony continue to differ.
1997–98 (8 to 9 years postspill)			
		No differences evident for community parameters and for 17 species on boulder/cobble shores.	Lower equilibrium clam population possibly linked to change in sediment grain size caused by hot-water flush or other (natural) causes.
2002 (13 years postspill)			
			Lower equilibrium clam population possibly linked to change in beach armor caused by hot-water flush.
2007 (18 years postspill)			
			Severe regional drop in clam population in ~2006 not related to spill.

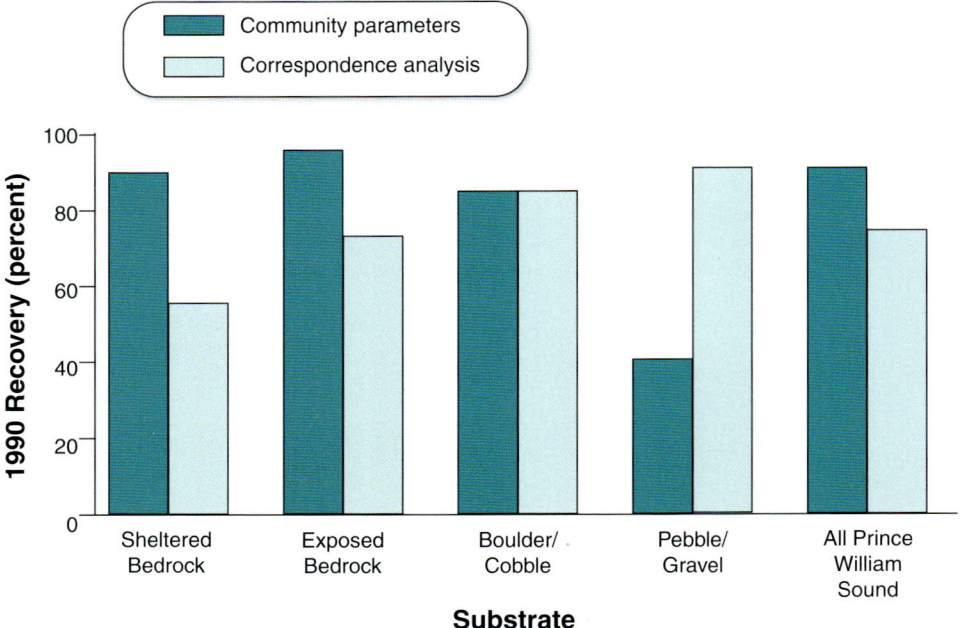

Figure 11.6 Summary of shoreline recovery found by the SEP–SRS survey on different shoreline substrates 15–18 months after the release of oil in PWS. (Data from Gilfillan et al., 1995a.)

- In 1990 (15–18 months after the spill), the overall estimated recovery of shoreline biota was between 73% and 91% for all of PWS. Recovery was between 40% and 97% for individual habitat types, depending on the statistical method applied (Fig. 11.6).
- Oiled and reference sites were statistically indistinguishable for about 81% of 443 species across all habitats and tidal zones.
- The effect of oiling was relatively small: about 11% using univariate community analysis and about 14% using multivariate correspondence analysis. Natural stress factors and covariates were responsible for much greater variation.
- For key intertidal species (rockweed, *Fucus gardneri*; littorine snails; limpets; and mussels), significant differences were present between oiled and reference sites, but these differences were not consistent with respect to habitat type, oiling level, or tidal zone.
- Both correspondence and community-structure analyses showed greatest recovery along the lightly oiled shorelines (about 75% and 94%, respectively) and least recovery in moderately oiled habitats (about 65% and 75%, respectively).
- Chemical analyses showed rapidly decreasing and degraded hydrocarbon concentrations, but many of the oiled SRS sites still had higher concentrations than reference sites in 1990. There was usually no sediment toxicity due to oil at SRS sites in 1991 (Boehm et al., 1995).

11.3.5 SEP–FS results

From the 11 heavily oiled sites in this study, only 11 of 177 samples (9.4%) showed a statistically significant difference between FS and unoiled reference sites in the summer of 1990: 3 were from boulder/cobble/gravel sites and 8 from mixed pebble/gravel sites.

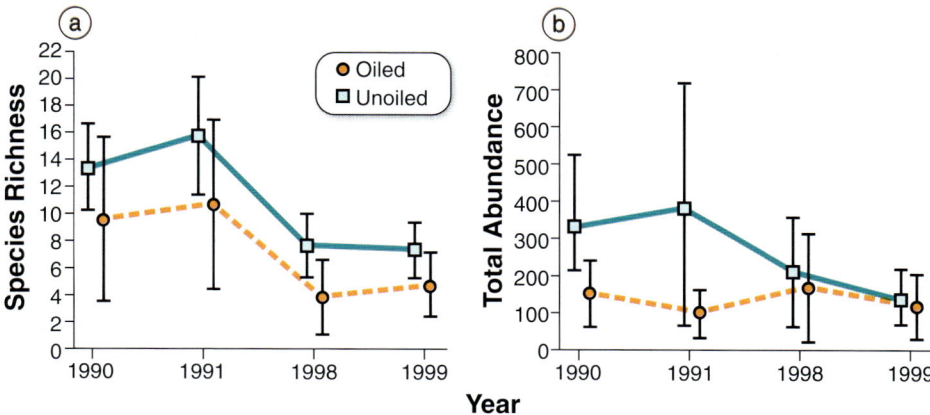

Figure 11.7 Interaction plots of middle tidal zone results from the SEP–FS survey, 1990–99. The error bars are 95% confidence limits. (a) Species richness showing no interaction but large interannual variability and consistent differences between oiled and unoiled samples. (b) Total abundance (number of individuals of all species) showing diverging lines in 1991, indicating a significant interaction between year and location (oiled versus unoiled). From Gilfillan *et al.*, 2001, courtesy International Oil Spill Conference.

Seven of the 11 differences were from set-aside areas where there was never any cleanup. In 1991, most differences were also located on set-asides even though they represented only 4 of the 11 shorelines sampled (Gilfillan *et al.*, 1995a).

The follow-up study in 1998 and 1999 at the four boulder/cobble/gravel sites and their reference sites found significant oiling effects in only two parameters: abundance and diversity of fauna in the middle intertidal zone. These differences were detectable in 1990–91, but not in 1998–99 (Gilfillan *et al.*, 2001). Species richness over time from the middle intertidal zone is plotted in Figure 11.7a (no interaction indicated) and species abundance in Figure 11.7b (interaction indicated by lines diverging in 1990–91). In analyzing abundances of 17 species in the same manner, only the snail *Alvania compacta* showed differences due to oiling of this habitat type in 1990–91.

Nearly all changes in intertidal communities between 1990 and 1999 were attributed to natural yearly variations or were related to long-term climate trends (Anderson and Piatt, 1999). The long-term SEP–FS program demonstrates the importance of a study design that separates oiling effects from natural factors affecting the biological community.

Mean petroleum hydrocarbon concentrations, particularly of total polycyclic aromatic hydrocarbon (TPAH), were higher in oiled FS sediments and mussel tissues than in their references in 1990 and 1991. Toxicity of sediments to amphipods was significantly higher at upper intertidal boulder/cobble/gravel beaches in 1990 and 1991 and at upper intertidal pebble/gravel beaches in 1990 – where initial oiling was greatest (Chapter 6). Differences for other intertidal zones and the shallow subtidal area were not significant (Boehm *et al.*, 1995; Chapter 6).

11.3.6 CHIA results

Spill effects were clearly present in 1990–91 at the CHIA sites in PWS; however, an overall pattern based on species, habitat, or tidal zone was not evident. Conclusions about the effects of oil and the recovery of algal and invertebrate populations (Highsmith *et al.*, 1996; Stekoll *et al.*, 1996) are:

- Most habitats were recovering but had not fully recovered by 1991, where recovery was defined as no statistical difference ($p \leq .05$) between oiled and unoiled sites for the parameter (biomass or abundance) being evaluated. Results of all species-abundance tests (5280) for 1990–1991 at all tidal elevations showed that about 11% were significantly positive (more organisms at the reference site than at the oiled) and about 3% were significantly negative (more organisms at the oiled site than at the reference) (Stekoll et al., 1996).
- The spill oiling and cleanup affected invertebrates more than algae, as shown by invertebrates having a greater percentage of tests showing significant differences for the same habitat type.
- The greatest effects (about 26% either negative or positive) on algae occurred in 1991 in the upper intertidal zone of sheltered rocky shorelines, followed closely by estuarine shores in the same intertidal zone and same year (Fig. 11.8). Algae on exposed rocky shores showed the fewest differences compared to reference sites. In many cases, the effects on algae were greater in 1991 than in 1990.
- Stekoll and Deysher (2000) monitored rockweed abundance annually until 1994 at the CHIA sheltered rocky sites. Significant differences between oiled and reference sites ended after summer 1991, except for the upper tidal area, which showed a lower percentage of algal cover in oiled sites until 1994.
- The greatest effects on invertebrates occurred on coarse-textured beaches (mixed sand and gravel beaches, gravel beaches, or exposed tidal flats), with about 30% of all tests showing a significant difference between oiled and reference sites. Invertebrates on sheltered rocky shores showed about 10–15% differences for all tests extending across all tide zones for both 1990 and 1991, frequently with more invertebrates on oiled, cleaned sheltered rocky shores than at reference sites.

11.3.7 BMS results

Early results from BMS sites (Houghton et al., 1993b) showed that:

- Oiled sites, including those that had been hot-washed, were well on the way to recovery by 1991. No significant differences remained between the biota of unoiled sites and that of oiled sites that were not hot-washed, and few significant differences were found between unoiled and oiled, hot-washed shorelines, although an altered age-class structure was apparent.
- Apparent effects on the infaunal community at mixed soft-sediment beaches (rocky and mixed sand and gravel) decreased significantly through 1992, but some effects of hot-washing could still be discerned.

Data through 1997 were analyzed to determine recovery based on "parallelism" (Coats et al., 1999; Hoff and Shigenaka, 1999; Skalski et al., 2001). Parallelism recognizes that there were likely natural ecological differences between the reference and spill-affected sites before the spill happened. Thus, *recovery* occurs when the parameters measured at oiled and unoiled sites vary with time in a parallel manner (this is different from the definition used by SEP). This approach takes into account inherent differences (e.g., exposure, adjacent ecological communities, substrate variability) between oiled and

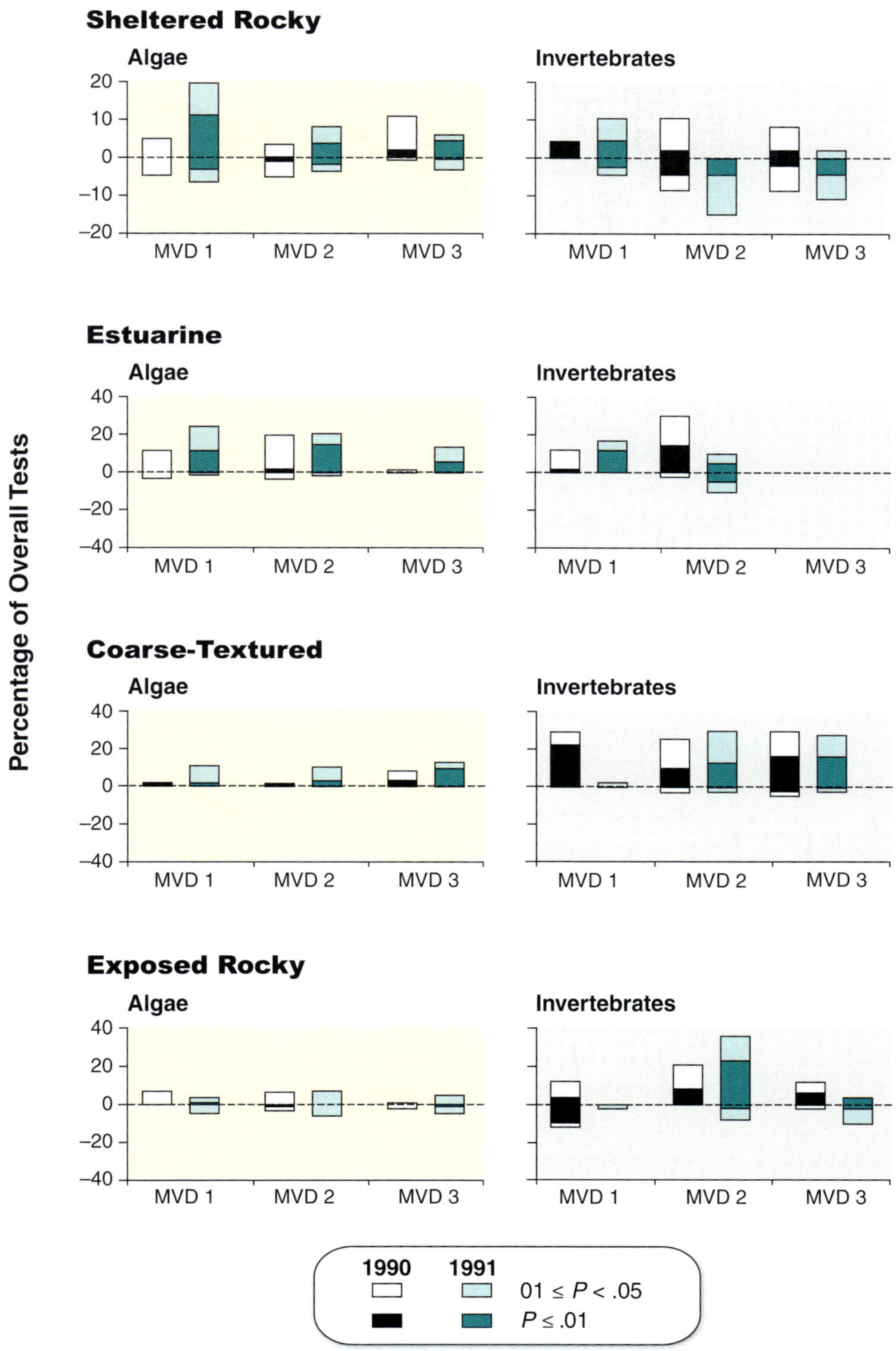

Figure 11.8 Percentage of overall tests for algae and invertebrates in PWS by habitat type and tidal zone that showed significant ($p \leq .05$) differences between the means of variables (species and species groups) measured at CHIA oiled/treated and reference sites in 1990 and 1991. Positive differences are those where the reference mean was greater than at oiled sites. Negative differences are cases where the mean at oiled sites was greater than at reference sites. MVD locations are shown in Figure 11.4. Modified from Stekoll et al. (1996) with permission of the American Fisheries Society.

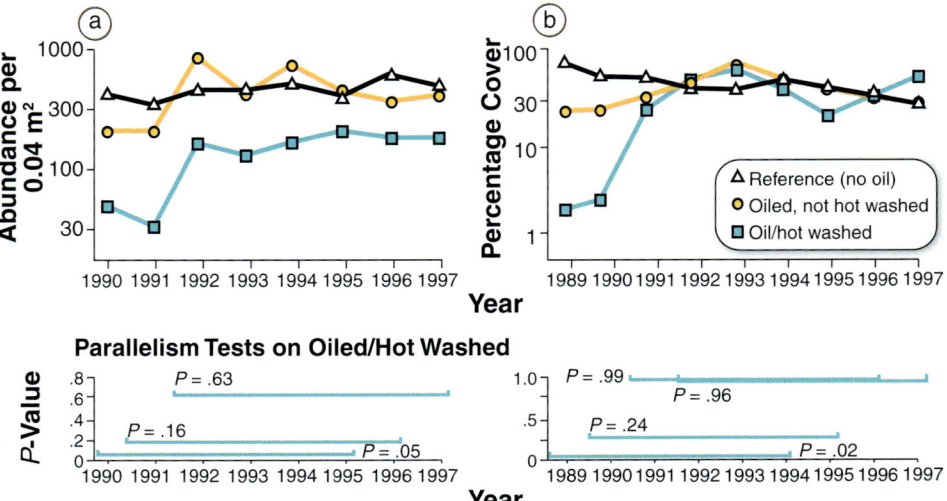

Figure 11.9 (a) Infaunal abundance (log scale) and (b) percentage cover (log scale) of middle intertidal rockweed (*Fucus gardneri*) for three sites in each of three oiling categories. At the bottom of each graph are results of parallelism tests for 6-year time spans with associated likelihood (*p*-value) of rejecting the null hypothesis that parallelism is not occurring. The lower the bar, the more likely that recovery is ongoing within the time window; the higher the bar, the greater the likelihood that parallelism is occurring. Modified from Coats et al., 1999.

reference sites that are not spill related (see Chapter 10). For the nine sites surveyed, parallelism studies showed that:

- Intertidal populations – including infauna, algae, and epifaunal invertebrates – showed significant recovery by 1993 at both oiled, hot-washed and oiled, not hot-washed sites.
- Population increases were greater for oiled, hot-washed sites because of their greater initial losses. However, there was no evidence that hot-washing delayed recovery because the timing and duration of repopulation were remarkably similar across all spill-affected sites, tidal zones, and intertidal species.
- The total mean infaunal abundance increased substantially from 1989 until 1992, when it reached parallel conditions at oiled sites that had been both hot-washed and not (Fig. 11.9a). Parallelism analysis showed that recovery had occurred on hot-washed sites, even though the population size was lower.
- Differences in *Fucus gardneri* coverage were significant between oiled and reference sites until 1990. Coverage remained slightly lower at oiled sites until 1992, and thereafter showed mixed patterns. High *p*-values for 6-year intervals from 1991 onward indicate that middle intertidal sites had achieved parallelism (Fig. 11.9b). Recolonization along the upper intertidal took longer, stabilizing in 1993.
- Motile epifauna (snails, crabs, limpets, etc.) in the middle intertidal zone had largely recovered by 1991. Natural patchiness made it difficult to assess parallelism for upper intertidal organisms, although growing populations of epifauna at oiled, hot-washed sites indicated that recovery was in process.

Using a similarity analysis of composition and abundance, Kimura and Steinbeck (1999) found the same results as the above studies for 12 algal and 20 invertebrate species from middle intertidal zones. By summer 1992, and continuing through 1997, both species groups within all spill-affected sites had recovered to the same range as at unoiled sites.

Mussel tissues from 23 oiled sites contained significantly higher mean TPAH concentrations than mussels from reference sites through 1992, but not after. TPAH concentrations were lower in tissues of mussels than in the sediments where they were collected, indicating that polycyclic aromatic hydrocarbons (PAH) in the weathered subsurface oil had a low bioavailability. The degree of oil weathering in intertidal sediments varied with how exposed and how deep the oil was buried (Hoff and Shigenaka, 1999; see Chapter 6).

Even though littleneck clams (*Protothaca staminea*) also showed a pattern of parallelism similar to total infaunal abundance (including lower values in the oiled, hot-washed sites), several researchers have questioned whether recovery had truly occurred. Shigenaka *et al.* (1999) thought that the lower population might have resulted from changes in sedimentary characteristics caused by the hot-wash cleanup, although other factors, such as inherent differences in grain sizes, might also be at work. After 2002 field surveys, Lees and Driskell (2007) linked the lower population to a possible but unmeasured change in beach surface characteristics caused by cleanup. The effect of increased sea otter (*Enhydra lutris*) populations was also a possible cause (Shigenaka, quoted in Raloff, 2009). This issue may never be resolved because littleneck clam populations showed a steep decline in 2007 that was unrelated to the spill at several oiled and unoiled areas in PWS (Coats and Fukuyama, 2008; Bodkin *et al.*, 2009) and in the northeastern Pacific Ocean region in general.

Driskell *et al.* (2001) also questioned the state of recovery, in this case based on analysis of the size-class structure, dynamics, and synchrony of *Fucus*, which remained different between oiled and reference sites until at least 1996. (Results were based on 1989–96 sampling from the middle tidal zone at seven BMS sites: three oiled, not hot-washed; two oiled, hot-washed; and two unoiled reference sites.) Driskell *et al.* (2001) equated the differences to the predominance of a single year-class of algae re-establishing after the spill, which then naturally aged and declined across the habitat several years later. Over time, algal year-class heterogeneity will be re-established and differences between oiled and unoiled sites are expected to diminish.

11.4 Discussion

All of these studies faced a daunting challenge: identifying oil-spill effects in an area of high natural disturbance and resulting biological variability. The distribution and persistence of oil along the shoreline was also highly variable, often changing with small differences in geomorphology and exposure to wind and waves. The standardized spill-concentration categories (light, moderate, heavy) used by almost all shoreline studies do not consider how long oil actually remained at that concentration at the site being studied, nor do they reflect small localized differences in quantity and thickness, as occurred particularly on rocky shorelines. In addition, cleanup is a major variable that, for the most part, is understood only in general terms. Cleanup varied – even across short sections of shoreline – by type and duration, depending on geomorphology, weather conditions, tidal stage, amount and duration of oil locally present at the time, and other factors (Houghton *et al.*, 1993a). All these factors altered the effects of the oil and cleanup on shoreline biota and increased the difficulty of finding statistically significant differences that could be attributed unambiguously to oiling or cleanup.

The studies dealt with these challenges differently based on their objectives. In responding to NRDA requirements, two studies attempted to represent all spill-affected environments in PWS. The SEP–SRS program successfully produced results applicable to all habitats, tidal zones, and oiling categories. Although samples were concurrently collected for chemical and ecotoxicity analyses, the study was limited to a single-year survey (1990) and was extremely labor intensive. The CHIA program, also very labor intensive, had 2 years of data collection (1990 and 1991) and considered three intertidal zones and all habitat types, but it was limited to only one oiling category (moderate and heavy combined). The other two studies, SEP–FS and BMS, did not attempt broad representation but instead selected worst case sites to follow over time (intermittently to 1998 for FS and nearly annually until 1997 for BMS). The BMS was the only survey to separate out the effects of aggressive hot-wash cleanup.

Despite different objectives and methods, all surveys consistently showed that most shoreline biota in oiled areas had recovered by 1990 or 1991:

- SEP–SRS found that 73% to 91% of the oiled PWS shoreline had recovered by 1990. Overall, 82% of all species tests showed no difference between oiled and reference sites (Gilfillan et al., 1995a).

- The extended SEP–FS program (1990–99) used interaction-plot analysis to show that oil effects were already quite limited by 1991 and that the changes in intertidal community parameters between 1990–91 and 1998–99 were related to natural interannual variation unrelated to the spill (Gilfillan et al., 2000, 2001).

- The CHIA survey found no statistical differences for ~86% of all tests performed on 1990–91 data for moderate and heavy oiled sites using matched-pair analysis (Stekoll et al., 1996). As moderate and heavy oiled sites represented only 30% of the total oiled shoreline, species-recovery values would be much higher had lightly oiled shorelines also been considered.

- The BMS found that, by 1991, no significant differences remained between biota on unoiled versus oiled, not hot-washed sites, and only a few species showed significant differences at unoiled and hot-washed shores. The community size–class structure, however, was altered (Houghton et al., 1993b). Differences did not persist past 1991 for algae and invertebrates in middle intertidal zones (Kimura and Steinbeck, 1999), and by 1993 all intertidal biota showed significant recovery and achieved stability with their environment (Coats et al., 1999). Recovery was equal in timing and scope for both oiled, hot-washed and oiled, not hot-washed sites.

As shown above, the recovery findings of the BMS, CHIA, and SEP studies are highly congruent and consistent.

Not all researchers agree, however, with the definition of "recovery" used in these studies. Some have argued that recovery occurs only when differences in the size-class structure (closely related to age-class) of organisms are statistically undetectable between affected and reference areas (Southward, 1982; Suchanek, 1993; Paine et al., 1996). Hawkins and Southward (1992) showed differences in the age-class structure of shoreline biota nearly 15 years after the application of dispersants to rocky shorelines at the *Torrey Canyon* oil-spill site. At the *Exxon Valdez* spill site, oiling and intensive cleanup sometimes removed most if not all algae and epifauna from parts of the intertidal zone. Because recolonization started from near zero, it is not surprising that the size-class distribution in rockweed was different between oiled and unoiled areas at least until 1996 (Driskell et al., 2001). These differences will decrease as the cycle of algal settlement and growth continues.

11.5 Lessons learned

These three major shoreline ecology programs provide guidance for future shoreline investigations after a large spill.

- Use a random site-selection methodology to ensure that the entire spill zone is represented. That being said, time and cost factors may preclude sampling of lightly oiled shorelines. If lightly oiled shores are excluded from the analysis, then results should clearly reference this fact and measure percentage of recovery accordingly.

- Ensure that site selection is based on an accurate characterization of shoreline types and oiling levels. Both habitat type and shoreline oiling level are needed to properly design a field sampling program representative of the spill area (Chapter 4).

- Collect sediment samples concurrently with biota samples. Having physical and chemical data about sediments is a key to establishing a causal relationship between oil exposure and biological effects. Concurrent analysis of sediment and biota is particularly important in later phases of the spill, when visible surface oil residues have disappeared, to determine whether persistent subsurface oil residues are still toxic or are no longer bioavailable to intertidal biota. Other influences – such as berm relocation, cleanup, and other human disturbances – should also be considered.

- Sample primary and indicator species or taxonomic groups. Focusing on a few dominant species enables a substantial reduction in the collection effort and permits additional field sites to be sampled, increasing the statistical power to discern spill-related effects. In PWS, 79% of the biomass on surface shores is represented by three species groups: rockweed, mussels, and barnacles (Stekoll *et al.*, 1996).

- Intertidal species that feed important wildlife species, such as sea otters and shorebirds, should be a focus of any wildlife-exposure assessment (Neff *et al.*, 2011). Important forage species on PWS shores include mussels, clams, snails, and crustaceans.

- Select an appropriate number of sites and samples. Using the BMS data set for 270 intertidal taxa, Coats and Shigenaka (2005) developed sample-size curves that indicate the number of replicate samples needed for abundant taxa at each site. For example, if sampling is planned for three or more sites within each treatment, then at least four but no more than eight replicate samples should be collected within each tidal zone–habitat-type combination. Some sites require more effort to collect samples, causing scientists to make tradeoffs between comprehensive sampling at fewer sites versus meager sampling at many sites. For the BMS study, oiled sites required twice the amount of effort to collect data as reference sites, yet the reference sites had the most biological variability (Hoff and Shigenaka, 1999). In this case, it is better to establish additional reference sites than to collect data from more oiled areas.

- Do not automatically prohibit shoreline cleanup techniques that might be too aggressive. In spite of initial warnings of dire consequences, the BMS found no evidence that high-pressure, hot-water flushing measurably delayed recovery. Hot-washing is not always appropriate, and must be avoided on soft-sediment shorelines and in the biologically productive lower intertidal zone. As shown at other spills (e.g., *Metula*, BIOS, and *Arrow*; Chapter 6), however, leaving large quantities of untreated oil on boulder/cobble or pebble/gravel shorelines would have resulted in long-term oil persistence, continued environmental degradation, and increased risk to wildlife.

- Cautiously select set-aside sites (oiled but no cleanup treatment). Following the *Exxon Valdez* spill, there was controversy over using aggressive cleanup. The set-asides were part of a net environmental benefit analysis for different shoreline cleanup techniques (including no treatment) that could be applied to the next spill. Although set-asides were not always completely left alone as intended, the sites were still essential to determining the effects of hot-washing. Early agreement among all stakeholders (the responsible party, governmental agencies, landowners) is required to demarcate set-asides before cleanup operations reach them.

REFERENCES

Anderson, P.J. and J.F. Piatt (1999). Community reorganization in the Gulf of Alaska following ocean climate regime shift. *Marine Ecology Progress Series* **189**: 117–123.

Bodkin, J.L., T.A. Dean, H.A. Coletti, and K.A. Kloecker (2009). *Nearshore Data Management and Monitoring.* Anchorage, AK, USA: US Geological Survey, Alaska Science Center; *Exxon Valdez* Oil Spill Restoration Project Final Report (Restoration Project 070 750). [http://www.evostc.state.ak.us/Files.cfm?doc=/Store/FinalReports/2007-070750.pdf&]

Boehm, P.D., D.S. Page, E.S. Gilfillan, W.A. Stubblefield, and E.J. Harner (1995). Shoreline Ecology Program for Prince William Sound, Alaska, following the *Exxon Valdez* oil spill: Part 2 – Chemistry and toxicology. In Exxon Valdez *Oil Spill: Fate and Effects in Alaskan Waters.* P.G. Wells, J.N. Butler, and J.S. Hughes, eds. Philadelphia, PA, USA: American Society for Testing and Materials; ASTM Special Technical Publication 1219; ISBN-10: 0803118961; pp. 347–397.

Chianelli, R.R., T. Aczel, R.E. Bare, G.N. George, M.W. Genowitz, M.J. Grossman, C.E. Haith, F.J. Kaiser, R.R. Lessard, R. Liotta, R.L. Mastracchio, V. Minak-Bernero, R.C. Prince, W.K. Robbins, E.I. Stiefel, J.B. Wilkinson, S.M. Hinton, J.R. Bragg, S.H. McMillen, and R.M. Atlas (1991). Bioremediation technology development and application to the Alaskan spill. In *Proceedings of the 1991 International Oil Spill Conference (Prevention, Behavior, Control, Cleanup), March 4–7, 1991, San Diego, California.* Washington DC, USA: American Petroleum Institute; Technical Publication 4529; pp. 549–558.

Coats, D.A. and A.K. Fukuyama (2008). Population recovery status of littleneck clams *(Leucoma staminea)* in Prince William Sound: an unexpected turn of events. In *Alaska Marine Science Symposium 2008 Book of Abstracts for Oral Presentations and Posters, January 20–23, 2008, Anchorage, Alaska.* Anchorage, AK, USA: Alaska Marine Science Symposium. [http://doc.nprb.org/web/symposium/2008/Abstract%20Book%202008.pdf]

Coats, D.A., E. Imamura, A.K. Fukuyama, J.R. Skalski, S. Kimura, and J. Steinbeck (1999). *Monitoring the Biological Recovery of Prince William Sound Intertidal Sites Impacted by the* Exxon Valdez *Oil Spill. 1997 Biological Monitoring Survey.* G. Shigenaka, R. Hoff, and A. Mearns, eds. Seattle, WA, USA: National Oceanic and Atmospheric Administration, Hazardous Materials Response Division, Office of Response and Restoration; NOAA Technical Memorandum NOS OR&R 1.

Coats, D.A. and G. Shigenaka (2005). Sampling needed to assess intertidal impacts: lessons learned from 11 years of monitoring in Prince William Sound. In *Proceedings of* the 2005 *International Oil Spill Conference (Prevention, Preparedness, Response and Restoration – Raising Global Standards), May 15–19, 2005, Miami, Florida, USA*. Washington DC, USA: American Petroleum Institute.

Driskell, W.B., J.L. Ruesink, D.C. Lees, J.P. Houghton, and S.C. Lindstrom (2001). Long-term signal of disturbance: *Fucus gardneri* after the *Exxon Valdez* oil spill. *Ecological Applications* **11**(3): 815–827.

Gilfillan, E.S., E.J. Harner, J.E. O'Reilly, D.S. Page, and W.A. Burns (1999). A comparison of shoreline assessment study designs used for the *Exxon Valdez* oil spill. *Marine Pollution Bulletin* **38**(5): 380–388.

Gilfillan, E.S., E.J. Harner, and D.S. Page (2002). Comment on Peterson *et al*. (2001): "Sampling design begets conclusions." *Marine Ecology Progress Series* **231**: 303–308.

Gilfillan, E.S., D.S. Page, E.J. Harner, and P.D. Boehm (1995a). Shoreline Ecology Program for Prince William Sound, Alaska, following the *Exxon Valdez* oil spill: Part 3 – Biology. In Exxon Valdez *Oil Spill: Fate and Effects in Alaskan Waters*. P.G. Wells, J.N. Butler, and J.S. Hughes, eds. Philadelphia, PA, USA: American Society for Testing and Materials; ASTM Special Technical Publication 1219; ISBN-10: 0803118961; pp. 398–443.

Gilfillan, E.S., D.S. Page, J.M. Neff, K.R. Parker, and P.D. Boehm (2000). 1999 shoreline conditions in the *Exxon Valdez* oil spill zone in Prince William Sound. In *Proceedings of the 23rd Arctic and Marine Oilspill Program (AMOP) Technical Seminar, June 14–16, 2000, Vancouver, British Columbia, Canada*. Ottawa, ON, Canada: Environment Canada; pp. 281–294.

Gilfillan, E.S., D.S. Page, J.M. Neff, K.R. Parker, and P.D. Boehm (2001). A 10-year study of shoreline conditions in the *Exxon Valdez* spill zone, Prince William Sound, Alaska. In *Proceedings of the 2001 International Oil Spill Conference (Global Strategies for Prevention, Preparedness, Response, and Restoration), March 26–29, 2001, Tampa, Florida*. Washington DC, USA: American Petroleum Institute; Special Technical Publication I4710A (CD), 14710B (paper); pp. 559–567. [http://ioscproceedings.org/doi/pdf/10.7901/2169-3358-2001-1-559]

Gilfillan, E.S., T.H. Suchanek, P.D. Boehm, E.J. Harner, D.S. Page, and N.A. Sloan (1995b). Shoreline impacts in the Gulf of Alaska region following the *Exxon Valdez* oil spill. In Exxon Valdez *Oil Spill: Fate and Effects in Alaskan Waters*. P.G. Wells, J.N. Butler, and J.S. Hughes, eds. Philadelphia, PA, USA: American Society for Testing and Materials; ASTM Special Technical Publication 1219; ISBN-10: 0803118961; pp. 444–481.

Haven, S.B. (1971). Effects of land-level changes on intertidal invertebrates, with discussion of post earthquake ecological succession. In *The Great Alaska Earthquake of 1964: Biology*. Washington DC, USA: National Academy of Science, National Academy Press; NAS Publication 1604; pp. 82–126.

Hawkins, S.J. and A.J. Southward (1992). The *Torrey Canyon* oil spill: Recovery of rocky shore communities. In *Restoring the Nation's Marine Environment*. G.W. Thayer, ed. College Park, MD, USA: University of Maryland Sea Grant; ISBN-10: 0943676576; pp. 583–631.

Hayes, M.O. and J. Michel (1998). *Evaluation of the Condition of Prince William Sound Shorelines Following the* Exxon Valdez *Oil Spill and Subsequent Shoreline Treatment: 1997 Geomorphological Monitoring Survey*. Seattle, WA, USA: US National Oceanic and Atmospheric Administration, National Ocean Service, Office of Ocean Resources Conservation and Assessment; NOAA Technical Memorandum NOS ORCA 126.

Highsmith, R.C., T.L. Rucker, M.S. Stekoll, S.M. Saupe, M.R. Lindeberg, R.N. Jenne, and W.P. Erickson (1996). Impact of the *Exxon Valdez* oil spill on intertidal biota. In *Proceedings of the* Exxon Valdez *Oil Spill Symposium*. S.D. Rice, R.B. Spies, D.A. Wolfe, and B.A. Wright, eds. Bethesda, MD, USA: American Fisheries Society; Symposium 18; ISBN-10: 0913235954; ISSN: 08922284; pp. 212–237.

Hoff, R.Z. and G. Shigenaka (1999). Lessons from ten years of post-*Exxon Valdez* monitoring on intertidal shorelines. In *Proceedings of the 1999 International Oil Spill Conference (Beyond 2000 – Balancing Perspective), March 8–11, 1999, Seattle, Washington*. Washington DC, USA: American Petroleum Institute; Special Technical Publication 4686B; pp. 111–117.

Houghton, J.P., A.K. Fukuyama, D.C. Lees, H. Teas, III, H.L. Cumberland, P.M. Harper, T.A. Ebert, and W.B. Driskell (1993a). *Evaluation of the 1991 Condition of Prince William Sound Shorelines Following the* Exxon Valdez *Oil Spill and Subsequent Shoreline Treatment. Vol. III. 1991 Biological Monitoring Survey*. Seattle, WA, USA: US National Oceanic and Atmospheric Administration, National Ocean Service, Office of Ocean Resources Conservation and Assessment. NOAA Technical Memorandum NOS ORCA 67. [http://docs.lib.noaa.gov/noaa_documents/NOS/ORCA/TM_NOS_ORCA/nos_orca_67v2.pdf]

Houghton, J.P., R.H. Gilmour, D.C. Lees, W.B. Driskell, and S.C. Lindstrom (1997). *Evaluation of the Condition of Prince William Sound Shorelines Following the* Exxon Valdez *Oil Spill and Subsequent Shoreline Treatment. Vol. I. 1995 Biological Monitoring Survey*. Seattle, WA, USA: US National Oceanic and Atmospheric Administration, National Ocean Service, Office of Ocean Resources Conservation and Assessment. NOAA Technical Memorandum NOS ORCA 110. [http://www.ccma.nos.noaa.gov/publications/tm110.pdf]

Houghton, J.P., D.C. Lees, and W.B. Driskell (1993b). *Evaluation of the Condition of Prince William Sound Shorelines Following the* Exxon Valdez *Oil Spill and Subsequent Shoreline Treatment. Volume II. 1992 Biological Monitoring Survey*. Seattle, WA, USA: US National Oceanic and Atmospheric Administration, National Ocean Service, Office of Ocean Resources Conservation and Assessment; NOAA Technical Memorandum NOS ORCA 73. [http://docs.lib.noaa.gov/noaa_documents/NOS/ORCA/TM_NOS_ORCA/nos_orca_73v3.pdf]

Kimura, S. and J. Steinbeck (1999). Can post-oil spill patterns of change be used to infer recovery? In *Proceedings of the 1999 International Oil Spill Conference (Beyond 2000: Balancing Perspective), March 8–11, 1999, Seattle, Washington*. Washington DC, USA: American Petroleum Institute Special Technical Publication 4686B; pp. 339–347.

Lees, D.C. and W.B. Driskell (2007). *Assessment of Bivalve Recovery on Treated Mixed-soft Beaches in Prince William Sound*. Juneau, AK, USA: US National Oceanic and Atmospheric Administration, National Marine Fisheries Service, Office of Oil Spill Damage & Restoration; *Exxon Valdez* Oil Spill Restoration Project 040574 Final Report. [http://www.evostc.state.ak.us/Files.cfm?doc=/Store/FinalReports/2004-040574-Final.pdf&]

Leschine, T.M., J. McGee, R. Gaunt, A. van Emmerik, D.M. McGuire, R. Travis, and R. McCready (1993). T/V Exxon Valdez *Oil Spill: Federal On Scene Coordinator's Report*. Washington DC, USA: United States Department of Transportation, United States Coast Guard; Report DOT-SRP-94-1; National Technical Information Service Order Number PB94–121845 (Volume 1).

Long, E. (1989). The use of the sediment quality triad classification of sediment contamination. In *Contaminated Marine Sediments: Assessment and Remediation.* Washington DC, USA: National Academy Press; ISBN-10: 030908671X; ISBN-13: 9780309086714; pp. 78–99.

Nauman, S.A. (1991). Shoreline cleanup, equipment and operations. In *Proceedings of the 1991 International Oil Spill Conference (Prevention, Behavior, Control, Cleanup), March 4–7, 1991, San Diego, California.* Washington DC, USA: American Petroleum Institute; Technical Publication 4529; pp. 141–147.

Neff, J.M., E.H. Owens, S.W. Stoker, and D.M. McCormick (1995). Shoreline oiling conditions in Prince William Sound following the *Exxon Valdez* oil spill. In Exxon Valdez *Oil Spill: Fate and Effects in Alaskan Waters.* P.G. Wells, J.N. Butler, and J.S. Hughes, eds. Philadelphia, PA, USA: American Society for Testing and Materials; ASTM Special Technical Publication 1219; ISBN-10: 0803118961; pp. 312–346.

Neff, J.M., D.S. Page, and P.D. Boehm (2011). Exposure of sea otters and harlequin ducks in Prince William Sound, Alaska, USA, to shoreline oil residues 20 years after the *Exxon Valdez* oil spill. *Environmental Toxicology and Chemistry* **30**: 659–672.

Page, D.S., E.S. Gilfillan, P.D. Boehm, and E.J. Harner (1995). Shoreline Ecology Program for Prince William Sound following the *Exxon Valdez* oil spill. Part 1: Study design and methods. In Exxon Valdez *Oil Spill: Fate and Effects in Alaskan Waters.* P.G. Wells, J.N. Butler, and J.S. Hughes, eds. Philadelphia, PA, USA: American Society for Testing and Materials; ASTM Special Technical Publication 1219; ISBN-10: 0803118961; pp. 263–295.

Page, D.S., E.S. Gilfillan, J.M. Neff, S.W. Stoker, and P.D. Boehm (1999). 1998 shoreline conditions in the *Exxon Valdez* oil spill zone in Prince William Sound. In *Proceedings of the 1999 International Oil Spill Conference (Beyond 2000: Balancing Perspective), March 8–11, 1999, Seattle, Washington.* Washington DC, USA: American Petroleum Institute; Special Technical Publication 4686B; pp. 119–126.

Paine, R.T., J.L. Ruesink, A. Sun, E.L. Soulanille, M.J. Wonham, C.D.G. Harley, D.R. Brumbaugh, and D.L. Secord (1996). Trouble on oiled waters: Lessons from the *Exxon Valdez* oil spill. *Annual Review of Ecology, Evolution, and Systematics* **27**: 197–235.

Peterson, C.H., L.L. McDonald, R.H. Green, and W.P. Erickson (2001). Sampling design begets conclusions: The statistical basis for detection of injury to and recovery of shoreline communities after the *Exxon Valdez* oil spill. *Marine Ecology Progress Series* **210**: 255–283.

Peterson, C.H., L.L. McDonald, R.H. Green, and W.P. Erickson (2002). Reply comment: the joint consequences of multiple components of statistical sampling designs. *Marine Ecology Progress Series* **231**: 309–314.

Piper, E., ed. (1993). *The* Exxon Valdez *Oil Spill: Final Report, State of Alaska Response.* Anchorage, AK, USA: Alaska Department of Environmental Conservation. [http://docs.lib.noaa.gov/noaa_documents/NOAA_related_docs/oil_spills/ExxonValdez_oil_spill_final_report_1993.pdf]

Plafker, G. (1965). Tectonic deformation associated with the 1964 Alaska earthquake. *Science* **148**(3678): 1675–1687.

Raloff, J. (2009). *Exxon Valdez*: tidal waters still troubled. *Science News*. Web edition; Monday, March 30, 2009. [http://www.sciencenews.org/view/generic/id/42311/title/Exxon_Valdez_Tidal_waters_still_troubled]

Shigenaka, G., D.A. Coats, and A.K. Fukuyama (2008). *Population Recovery Status of Littleneck Clams in Prince William Sound: an Unexpected Turn of Events*. Seattle, WA, USA: US National Oceanic and Atmospheric Administration; *Exxon Valley* Oil Spill Restoration Project 070829 Draft Report; unpublished. [not available at http://www.evostc.state.ak.us/]

Shigenaka, G., D.A. Coats, A.K. Fukuyama, and P.O. Roberts (1999). Effects and trends in littleneck clams (*Protothaca staminea*) impacted by the *Exxon Valdez* oil spill. In *Proceedings of the 1999 International Oil Spill Conference (Beyond 2000: Balancing Perspective), March 8–11, 1999, Seattle, Washington*. Washington DC, USA: American Petroleum Institute; Special Technical Publication 4686B; pp. 349–356.

Shigenaka, G., M.O. Hayes, J. Michel, C.B. Henry Jr., P. Roberts, J.P. Houghton, and D.C. Lees (1997). *Integrating Physical and Biological Studies of Recovery from the* Exxon Valdez *Oil Spill. Case Studies of Four Sites in Prince William Sound, 1989–1994*. Seattle, WA, USA: US National Oceanic and Atmospheric Administration, National Ocean Service; NOAA Technical Memorandum NOS ORCA 114. [http://archive.orr.noaa.gov/book_shelf/966_TM114.pdf]

Skalski, J.R., D.A. Coats, and A.K. Fukuyama (2001). Criteria for oil spill recovery: a case study of the intertidal community of Prince William Sound, Alaska, following the *Exxon Valdez* oil spill. *Environmental Management* **28**(1): 9–18.

Sousa, W.P. (1985). Disturbance and patch dynamics on rocky intertidal shores. In *The Ecology of Natural Disturbance and Patch Dynamics*. S.T.A. Pickett and P.S. White, eds. Orlando, FL, USA: Academic Press; ISBN-10: 0125545207; pp. 101–124.

Southward, A.J. (1982). An ecologist's view of the implications of the observed physiological and biochemical effects of petroleum compounds on marine organisms and ecosystems. *Philosophical Transactions of the Royal Society B* **297**(1087): 241–255.

Stekoll, M.S. and L. Deysher (2000). Response of the dominant alga *Fucus gardneri* (Silva) (Phaeophyceae) to the *Exxon Valdez* oil spill and cleanup. *Marine Pollution Bulletin* **40**(11): 1028–1041.

Stekoll, M.S., L. Deysher, R.C. Highsmith, S.M. Saupe, Z. Guo, W.P. Erickson, L. McDonald, and D. Strickland (1996). Coastal habitat injury assessment: intertidal communities and the *Exxon Valdez* oil spill. In *Proceedings of the* Exxon Valdez *Oil Spill Symposium*. S.D. Rice, R.B. Spies, D.A. Wolfe, and B.A. Wright, eds. Bethesda, MD, USA: American Fisheries Society; Symposium 18; ISBN-10: 0913235954; ISSN: 08922284; pp. 177–192.

Suchanek, T.H. (1993). Oil impacts on marine invertebrate populations and communities. *American Zoologist* **33**(6): 510–523.

Sundberg, K., L. Deysher, and L. McDonald (1996). Intertidal and supratidal site selection using a geographical information system. In *Proceedings of the* Exxon Valdez *Oil Spill Symposium*. S.D. Rice, R.B. Spies, D.A. Wolfe, and B.A. Wright, eds. Bethesda, MD, USA: American Fisheries Society; Symposium 18; ISBN-10: 0913235954; ISSN: 08922284; pp. 167–176.

Whitney, D. (1991). Hot washing oily beaches was a mistake. *Anchorage Daily News*, Metro Section, Final Edition April 10, 1991 p. B1.

CHAPTER TWELVE

Oiling effects on pink salmon

Ernest L. Brannon, Matthew A. Cronin, Alan W. Maki, Larry L. Moulton, and Keith R. Parker

12.1 Introduction

Alaskan salmon are of major sport and commercial importance, figure importantly in the traditions and livelihood of native cultures, and support food webs for an array of carnivores and scavengers. Of the five Pacific salmon species, pink salmon (*Oncorhynchus gorbuscha*) are the most abundant in Prince William Sound (PWS). Annual harvests yield 20–70 million adult pink salmon, with a value that averaged over $29 million annually between 2001 and 2010 (Fig. 12.1). The subsistence and commercial importance of the pink-salmon fishery, combined with the overlap of the 1989 *Exxon Valdez* oil spill with the early life stages of the salmon, make understanding the effects of the spill both critical and challenging.

Following the spill, the commercial pink-salmon fishery was closed. In addition, an Oil Spill Health Task Force was organized to ensure the safety of subsistence foods. The Task Force used analytical data on hydrocarbons in pink salmon (and other subsistence foods) (Field *et al.*, 1999) and determined that there were no *Exxon Valdez* polycyclic aromatic hydrocarbons (PAH) in sampled edible salmon tissues in 1989 and 1990.

In this chapter, we address the immediate and long-term effects of the *Exxon Valdez* spill on pink salmon in PWS. The focus is on what studies were conducted, how and why they were done, their main findings, and what lessons were learned that may help guide future investigations of the effects of environmental disruptions on fish.

12.2 Background: PWS pink salmon and the *Exxon Valdez* oil spill

12.2.1 Life cycle of pink salmon

Pink salmon are unique among the Pacific salmon in that they have a 2-year life cycle (Fig. 12.2). Adults return to their stream of origin in the fall to spawn. Females, which normally carry about 1700 eggs, excavate nests (called *redds*) 15–25 cm deep in the stream substrate. They deposit their eggs, which are externally fertilized by males, and

---

Oil in the Environment: Legacies and Lessons of the Exxon Valdez *Oil Spill*, ed. J. A. Wiens. Published by Cambridge University Press. © Cambridge University Press 2013.

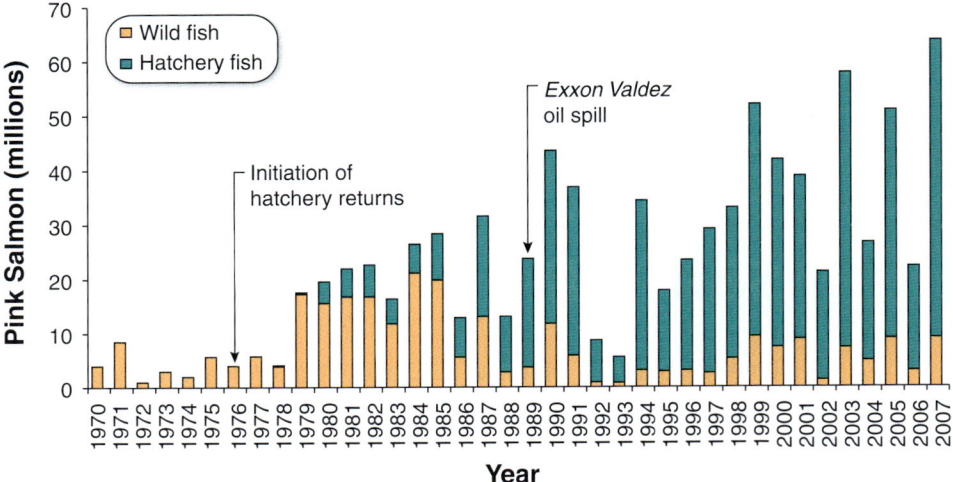

Figure 12.1 Adult pink salmon returns to PWS from 1960 to 2007, showing the record returns of adults in 1990 and 1991, which had experienced the highest oil concentrations of *Exxon Valdez* crude oil during incubation in 1989 and 1990, and the increasing proportion of hatchery-reared fish from 1988 onward.

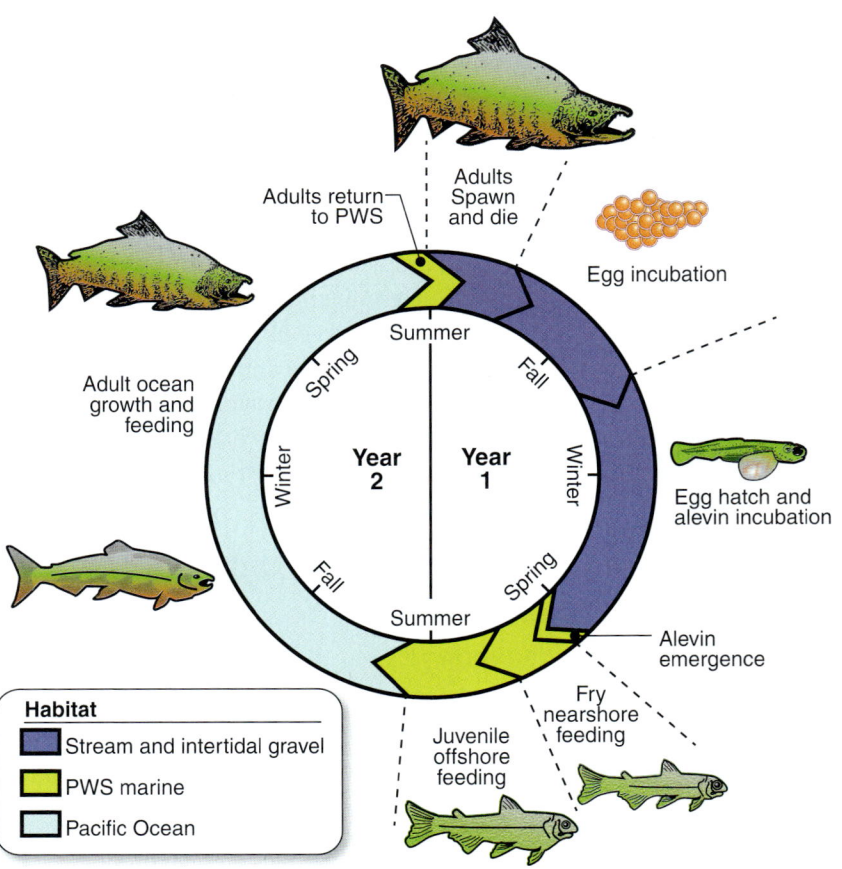

Figure 12.2 Two-year life cycle of pink salmon from spawning, incubation, emergence, and marine residence to return.

cover the redds with gravel from immediately upstream. One female may create five to eight redds, until she exhausts her egg supply and dies.

The eggs incubate through the fall and hatch in late fall to early winter. The hatchlings, called *alevins*, remain in the redds until spring, using their yolk-sac nutrient reserves. The emergent salmon, called *fry*, are pre-acclimated to salt water and migrate directly to nursery areas in nearshore marine waters to start feeding, where they grow into *fingerlings*. Shortly thereafter, they move offshore into deeper water as *juveniles*, entering the Gulf of Alaska (GOA) in late August through September. After a single, 1-year circuit around the North Pacific gyre, they return in the fall to their stream of origin as 2-year *adults*, and the cycle is repeated. Thus, there are two distinct year classes (or *brood years*): one spawning in odd years and one in even years.

12.2.2 Susceptibility of pink salmon to the *Exxon Valdez* oil spill

The life-cycle peculiarities of pink salmon could make them susceptible to a marine oil spill, especially in PWS where eggs and alevins could be exposed to oil in intertidal areas. Eggs (embryos) and alevins are particularly vulnerable to the effects of oil. Alevins are the most sensitive life stage because of their high ratio of body surface area per unit of mass (Rice *et al.*, 1975). Eggs and alevins incubate in stream gravel for several months, providing sufficient opportunity for chronic exposure to oil if it infiltrates the incubation environment. Thus, there was considerable concern that incubating pink salmon would suffer injuries from the spill.

When pink salmon fry emerge from their redds, they must begin feeding immediately upon moving downstream to the marine nursery areas. Even subtle effects of oil on feeding behavior, feeding opportunity, or prey could affect fry growth and condition during their first few weeks of marine residence.

When the *Exxon Valdez* grounded in March 1989, alevins in the 1988 brood year were ending their 8-month incubation. They were about to emerge as fry from their potentially oiled streams and enter nearshore waters. As time progressed, successive life stages, successive brood years, and numerous potential exposure pathways became the foci of concern and study (Fig. 12.3). Therefore, potential injury was also a concern during the early stages of their marine residence.

Interpreting the effects of the oil spill on pink-salmon populations is immediately confounded by the fact that PWS has a number of pink-salmon hatcheries. Hatchery production, which began in 1978, first accounted for a small portion of the harvest, but by 1988 well over 70% of pink salmon harvested originated from hatcheries. The hatchery harvest has remained near that level ever since (Fig. 12.1). In addition, *strays* (spawning salmon that return not to their hatcheries of origin but to nearby streams) can account for as much as 60% of spawning populations in streams near hatcheries (Sharr *et al.*, 1994; Joyce and Evans, 1999; Cronin and Maki, 2004; Collins *et al.*, 2009a).

The hatcheries in PWS were protected from spilled oil during incubation. The release of hatchery fry was delayed until late April and early May to avoid releasing fry into waters with possibly higher hydrocarbon concentrations immediately after the spill. Once released, however, hatchery fry followed the same pattern of distribution and migration through PWS and confronted the same conditions as the wild fry. The potential for damage to wild and hatchery salmon, therefore, was prominent in the minds of spill responders and researchers.

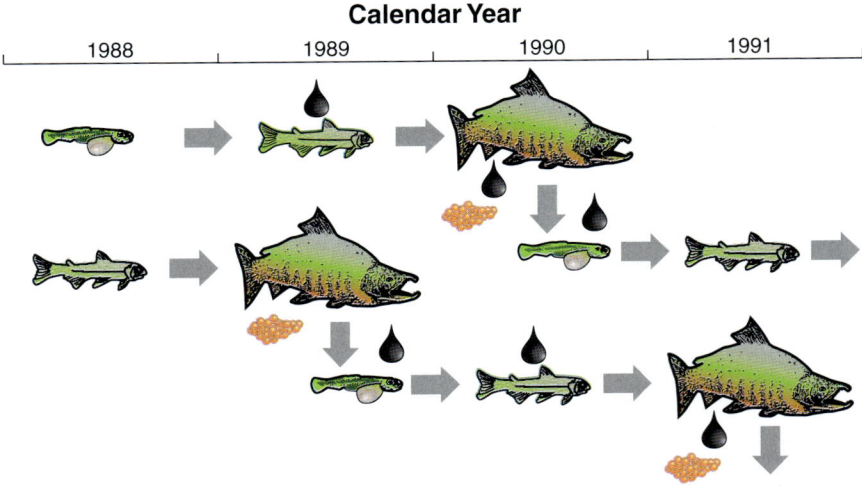

Figure 12.3 Year and life stage of PWS pink salmon that experienced potential risk of oil exposure.

12.3 Research protocols

Scientific concerns and uncertainties centered on the extent of exposure and what effects such exposure might have. The specific questions were:

- What effects did oil have on current and future generations of pink-salmon populations in PWS?
- What were the exposure pathways and mechanisms through which such effects could be realized?
- How can we assess those effects?

To answer these questions, government- and Exxon-sponsored scientists developed research protocols to determine the effects of oil exposure on pink salmon. The biological areas of concern were (1) effects from exposure during incubation and (2) effects from exposure during early life in nearshore marine waters. Research in these areas involved examining the survival and developmental success of stream-specific incubating populations and assessing the growth and condition of juveniles in nearshore marine nursery areas and offshore open waters. Ultimately, effects on survival during these life stages would influence the numbers of returning adults.

12.3.1 Effects from exposure during incubation

12.3.1.1 Analysis of tissue total polycyclic aromatic hydrocarbons

The most crucial exposure data are the concentrations of total PAH (TPAH) in the tissues of eggs and alevins (Heintz *et al.*, 1999). Tissue TPAH concentrations are direct measures of TPAH assimilation and are a function of water-column and sediment concentrations of TPAH to which individuals may be exposed. Tissue TPAH concentrations are contrasted with background concentrations in the incubation environment (i.e., water and sediments) and with concentrations demonstrated to be harmful as reflected in regulations and from laboratory assays.

12.3.1.2 Assessment of mortality

A potential indicator that a toxicity problem may exist is increased mortality of eggs and alevins above normal losses (~22%; Collins *et al.*, 2000) in an unoiled environment. (Incubation mortality varies extensively under normal conditions, owing to such factors as anoxia from reduced water flow through the redd or physical disturbance during embryo gastrulation.) Mortality can be assessed by examining live-to-dead ratios of incubating eggs, comparing alevin survival between oiled and reference streams, or by observing differences in survival success at emergence by counting fry leaving their redds.

12.3.1.3 Presence of developmental deformities

A high prevalence of developmental deformities in alevins can be an important indicator of toxic conditions. Deformities are relatively rare under normal stream conditions ($< 1\%$). Therefore, a comparison of field data with the results of laboratory bioassays, such as those for ascites (blue-sac disease), can help to confirm or reject whether oil contamination levels in the streams were toxic.

12.3.1.4 Laboratory oil exposure studies of eggs and alevins

Laboratory bioassays that expose eggs and alevins to a range of oil concentrations can reveal the toxic threshold level of petroleum hydrocarbons to incubating pink salmon. In the absence of field measurements of oil levels or tissue loads, the bioassay data would indicate the minimum oil concentration necessary under field conditions to account for an increase in mortality during incubation.

12.3.1.5 Identification of intergenerational or genetic effects

Identifying potential intergenerational or genetic effects is important. If toxicity is present, did it cause heritable genetic mutations in the germline or somatic mutations that result in gamete inviability? These effects do not follow inevitably from exposure to petroleum but, were such anomalies to occur, they could result in mortality of the developing embryos or reduction of fitness in subsequent generations.

12.3.2 Effects from exposure during marine residence

The principal effects that can be caused by the exposure of juveniles to oil when they migrate to marine nearshore and offshore waters are reduced growth rates, declines in the condition of fish, disruption of feeding or food sources, and failure to return to spawning streams as adults.

Determining these effects requires comparative analyses of samples from exposed and unexposed areas, which in turn require proper sampling and statistical methods. For valid comparisons to be made:

- The same population must be randomly distributed between oiled and unoiled areas;
- Samples from oiled and unoiled areas must represent the same age structure;
- The oiled and unoiled areas sampled must be environmentally and biologically similar; and
- Sampling gear must be unbiased (e.g., nets must not be size-selective).

When all of these conditions are satisfied, comparative analyses are justified. When these conditions for comparisons cannot be met, other empirical data can be used to resolve uncertainties about the effect in question in a weight-of-evidence approach. For example, where reduced growth rate is hypothesized to be due to exposure to oil, it is valid to look at other evidence that might support or reject the hypothesis. Reduced growth could occur from oil toxicity affecting juvenile growth physiology, toxicity reducing the food base, or nonlethal effects from consumption of oil-contaminated prey. Controlled laboratory tests can also help confirm whether exposure levels in the field are high enough to reduce growth.

12.3.2.1 Analysis of tissue total polycylic aromatic hydrocarbons

Just as for assessing oil's effect on eggs and alevins, tissue TPAH concentrations directly measure exposure in juveniles and adults. Tissue TPAH concentrations are generally measured in laboratory experiments (such as feeding studies), investigations of what levels of oil exposure lead to mortality, and examinations of other toxic effects (e.g., reduced growth).

12.3.2.2 Examination of reduced growth

Reduced growth may affect survival, because smaller juveniles are more susceptible to predation (Parker, 1971; Hargreaves and Lebrasseur, 1985; Heard, 1991; Moss *et al.*, 2005; Cross *et al.*, 2008). The sources and age structure of the juveniles in the samples, however, may not be determinable to the accuracy needed. Pink salmon juveniles can increase their weight more than 4% per day (Willette, 1996), so a difference of a day or two in mean age of the sample can account for significant differences in size. This can make it difficult to determine whether growth rate or age is responsible for size differences.

12.3.2.3 Examination of fish condition

The condition index ($100 \times weight/length^3$) of individual juveniles relates to the fitness of the individual. Low weight for a given length of fish indicates that nutrient uptake is less than optimum or that the fish is nutritionally stressed. However, the index does not indicate anything about the *rate* of growth, and the index value tends to decrease with age regardless of nutrition.

12.3.2.4 Identification of feeding disruptions

Feeding and availability of food resources are important factors affecting growth that can be measured directly. Feeding activity can be measured from comparisons of stomach contents and fullness between oiled and unoiled areas. Pink salmon juveniles feed primarily on zooplankton and epibenthic crustaceans, whose abundances can be measured in samples taken from oiled and unoiled nursery areas for comparative analyses of prey composition, biomass, and density.

12.3.2.5 Determination of adult returns to spawn

One of the most important indices of oil effects is the number of adult pink salmon that return to their stream of origin (and their spawning success). In the absence of measures to determine survival during incubation and early marine residence, counts of spawners at oiled and unoiled streams can provide comparative data indicating whether oil may have reduced survival during early life stages.

Table 12.1 Pink salmon studies undertaken in government- and Exxon-sponsored research on survival, growth, and adult return following the *Exxon Valdez* oil spill in Prince William Sound, Alaska.

Study	Agency[a]
Egg and alevin incubation in the field	
Alevin survival in oiled/non-oiled streams, 1989	ADFG
Survey of egg mortality in oiled/non-oiled streams, 1989 to 1997	ADFG
Survival of eggs in artificial redds in oiled/non-oiled streams, 1989	Exxon
Cytochrome P450 induction and histopathology in oiled streams, 1989	ADFG
Egg, alevin, and fry survival in oiled/non-oiled streams, 1990–91	Exxon
Effects of hydraulic sampling on egg survival in PWS streams, 1998	Exxon
Survey of incubating embryos in PWS streams for K-*ras* mutations, 1999	Exxon
Resistance to shock of naturally incubating eggs, 2000	NMFS
Assessment of buried egg survival in residual oil in PWS deltas, 2001	Exxon
Egg and alevin incubation in the laboratory	
Histology, survival, cell anomaly, and cytochrome P450 1A in 1992 lab oil bioassay	NMFS
Sensitivity of embryos to weathered crude oil in 1993 bioassay	NMFS
K-*ras* mutations in embryos exposed to *Exxon Valdez* crude oil, 1995	NYU
Cytochrome P450 1A induction in oil-exposed embryos and survival, 1999	NMFS
Classification of live, mechanically damaged, and dead embryos, 2000	NMFS
Toxicity of weathered *Exxon Valdez* crude oil to incubating pink salmon, 2001	U Idaho
Bioassay on mechanical sensitivity of developing embryos, 2001	U Idaho
Toxicity of North Slope crude oil to incubating pink salmon, 2005	OSU
Effect of oil exposure on prolonged sensitivity of embryos to shock, 2008	NMFS
Oil exposure and growth of fingerlings in PWS marine waters	
Hydrocarbon contamination of nearshore pink salmon, 1989	NMFS
Diets of pink salmon in oiled and non-oiled nearshore habitat, 1989–90	NMFS
Abundance/growth of pink salmon in oiled/non-oiled waters, 1989–90	NMFS
Distribution, size, and growth of pink salmon in PWS, 1989–90	NMFS
Migration/growth/survival of pink salmon in PWS, 1989–91	ADFG
Growth of pink salmon in oiled/non-oiled waters, 1990–91	Exxon
Percentage of hatchery strays in PWS streams by otolith marks, 1995–98	ADFG
Percentage of hatchery strays in PWS streams by otolith marks, 2001–03	Exxon
Oil-exposed PWS adult return survival and gamete viability	
Return success and egg viability of oil-exposed embryos, 1995, 1997, 2000	NMFS
Spill effects on adult return and distribution, 1989–91	Exxon
Assessment of adult returns in oiled/non-oiled streams 1989–2006	Exxon
Run reconstruction on the wild pink salmon fishery in PWS, 1990–91	ADFG
Habitat recovery	
Recovery of spawning habitat in PWS streams, 1995	NMFS
Assessment of habitat recovery of oiled streams, 1999	NMFS
Oil exposure and growth of fingerlings in laboratory bioassays	
Growth, survival, and total nucleic acids of fry fed oil-contaminated feed, 1991	NMFS

[a] ADFG = Alaska Department of Fish and Game; NMFS = National Marine Fisheries Service; NYU = New York University; U Idaho = University of Idaho; OSU = Oregon State University.

12.3.3 Execution of studies

Government- and Exxon-sponsored scientists with experience in salmon biology and life history conducted field studies in the incubation environment and marine waters of PWS, as well as laboratory research. They were supported by marine biologists, chemists, statisticians, toxicologists, hydrologists, and engineers. A multidisciplinary approach to assessing oil effects on pink salmon proved invaluable (Table 12.1).

Investigations continued well beyond the time of the spill and its cleanup (March 1989–June 1992) and followed a variety of complementary and independent (and sometimes divergent) paths (Table 12.1). In the following sections, we discuss those studies and the lessons learned that can be applied to future spills and environmental accidents.

12.4 Studies of pink salmon and the *Exxon Valdez* oil spill

12.4.1 Levels of hydrocarbon exposure

The literature shows that open-water spills generally do not cause extensive injury to fish or other water-column organisms (Longhurst, 1982; Wells et al., 1985; McAuliffe, 1987). The PWS pink salmon spawning areas vulnerable to the spill were 60–100 km down-current from Bligh Reef, the site of the *Exxon Valdez* grounding, and the oil slick did not reach those areas until several days after the spill. During that time, most of the acutely toxic and volatile benzene, toluene, ethylbenzene, and xylenes (BTEX) dissipated (Wolfe et al., 1994). When the oil stranded on shorelines, the principal toxic compounds of concern were PAH (Chapters 1, 6).

Most importantly for pink salmon, of the ~1300 spawning streams in PWS, only 30 (~2% of the total) were significantly oiled in 1989. All of these 30 oiled streams were among the 220 salmon streams located in the PWS Southwest, Montague, and Eshamy Alaska Department of Fish and Game (ADFG) fishery management districts (Maki et al., 1995).

The protocols and methodologies to assess levels of oil contamination in the water column and on shoreline substrates have been covered in Chapters 3, 4, 6, and 7. The standard protocols of the US Environmental Protection Agency (US Environmental Protection Agency, 1985, 1988) for determining aqueous toxicity levels in the affected areas were also used.

12.4.1.1 Water column

It is unlikely that wild pink salmon fry or juveniles were exposed to potentially harmful TPAH concentrations in the water column in 1989 or 1990. Boehm et al. (2007) synthesized all government and Exxon water-column PAH-concentration data for 1989–2007. They related the data to the exposure of pink salmon to dissolved and dispersed PAH in PWS. Fry or juveniles in nearshore waters generally encountered 0.1–1 µg/L TPAH from April to July 1989. Juveniles encountered < 0.1 µg/L TPAH when they migrated to offshore waters in July 1989. Although TPAH concentrations were much higher near heavily oiled shorelines (2–7 µg/L) from March 31 to April 4, 1989 (Short and Harris, 1996), by 1990 they had returned to prespill background levels (Boehm et al., 2007). Fry and juveniles are much less sensitive to oil than are the eggs and alevins for which Heintz et al. (1999) determined a no-observed-effect concentration of 7.8 µg/L TPAH and a lowest-observed-effect concentration of 18 µg/L TPAH.

These TPAH concentrations are consistent with the toxicological assessment of three indicator species in PWS marine waters. Using the US Environmental Protection Agency (EPA) standard protocols (US Environmental Protection Agency 1985, 1988), scientists observed no mortality in mysid shrimp (*Mysidopsis bahia*) or larval sheepshead minnow (*Cyprinodon variegatus*) upon exposure to water samples taken at the height of oil contamination of surface waters during April to July 1989 (Neff and Stubblefield, 1995). Marine diatom (*Skeletonema costatum*) cell growth and larval fish growth showed no negative effects from exposure to offshore PWS water.

12.4.1.2 Stream sediments

Measurements demonstrate that the potential for severe petroleum effects in intertidal areas was very limited. In 1989, the mean sediment TPAH concentrations in intertidal areas of oiled streams ranged from 0.5 to 267 ng/g (Brannon et al., 1995). These are well below the toxic TPAH levels reported by Long and Morgan (1991, Table 69): 870 ng/g lowest-observed-effect level; 4000 ng/g effects range-low; and 35 000 ng/g effects range-medium.

When stream sediment TPAH was shown to be below toxic levels, attention then turned to oil deposits on the stream deltas, but these levels were also well below toxic thresholds. The TPAH content of the intertidal reaches ranged from < 35 to ~500 ng/g during the first 3 years following the spill (Boehm et al., 1995; Short and Babcock, 1996; O'Clair et al., 1996; Wolfe et al., 1996).

It was suggested that PAH in subsurface oil residues in delta sediments were leaching into salmon stream beds (Heintz et al., 1999; Murphy et al., 1999; Carls et al., 2003; Rice et al., 2007). When Carls et al. (2003) injected water-soluble tracer dyes into shoreline sediments adjacent to two salmon streams during ebb tides, the dyes rapidly dispersed through the intertidal zone. The tracer entered the salmon streams through streambed gravel. This rapid transport indicates that the tracer study was conducted in highly permeable sands and gravels, where rapid tracer movement might erroneously suggest good connectivity between the subsurface oiled sediment and the streams. The tracer movement, however, is not a surrogate for PAH transport. Unlike PAH, the tracer is completely water soluble and flows preferentially through high-permeability paths. When tidal waters carrying tracer encountered low-permeability layers where oil was trapped, they always took the path of least resistance around the oil deposits. This inhibited flowing water from picking up PAH from oil residues and carrying them to salmon streambeds (see Chapter 7).

Another study that investigated the potential for PAH to leach into salmon stream beds examined pink salmon eggs buried in weathered oil deposits or contaminated sediments next to incubation streams (Brannon et al., 2007). These eggs were exposed to sediment TPAH concentrations as high as 27 500 ng/g. After 3 weeks of incubation, however, the tissue TPAH concentrations never exceeded 650 ng/g – well below toxic levels. This confirms that, even though the oil in the sediments can contain very high PAH concentrations, the concentrations dissolved in groundwater flowing through or near contaminated sediment may not be very high. This is because the solubility of most PAH in water is low, even if the water is in direct contact with the oiled sediment.

More importantly, in low-permeability sediments, groundwater flow and velocity are low. Dissolved PAH from sequestered oil residues are rapidly diluted as they reach adjacent, higher permeability sediments where groundwater flow and velocity are high. Therefore, dissolved PAH are quickly diluted and do not reach concentrations high enough to cause harm to the eggs.

Chapter 7 describes in detail the physical principles governing the flow and transport of PAH leaching from subsurface oil residues and why the observed concentrations in nearshore waters were at background levels. The residues in question were sequestered in low-permeability sediments that inhibit the flow of groundwater through them (Li and Boufadel, 2010; Pope et al., 2011a, b).

12.4.2 Effects from exposure during incubation
12.4.2.1 Analysis of tissue total polycylic aromatic hydrocarbons

Average tissue TPAH concentrations in alevins in oiled streams in 1990 and 1991 were < 95 ng/g (Brannon et al., 1995). This is well below the lethal threshold of > 7100 ng/g for eggs exposed in the laboratory to artificially weathered oil removed from the *Exxon Valdez* (Brannon et al., 2006a). The 1990 data were particularly important because, although those alevins were the first to have been exposed to oil during their entire incubation, they showed no elevated mortality. It is apparent that PAH levels at the measured concentrations in stream sediments or interstitial flows through oiled delta sediments were not a threat to any significant fraction of incubating eggs.

12.4.2.2 Assessment of mortality

In the spring of 1989, ADFG assessed the effects of exposure in incubating pink salmon by sampling alevins from several streams immediately before and 2 weeks after stream contamination (Sharr et al., 1994). ADFG hydraulically flushed material from redds and then captured the alevins in fyke nets placed immediately downstream. ADFG found no differences between prespill and postspill mortality of alevins. This was particularly significant because the comparison was made at the time when oil contamination in those streams was most severe and when alevins were very susceptible to oil toxicity (Rice et al., 1975; Moles et al., 1987).

In April 1990 and 1991, Exxon-supported scientists also evaluated incubation success using hydraulic flushing in a subset of 10 oiled and unoiled streams used in the ADFG stream surveys. Again, no differences in the survival of eggs or alevins were observed between oiled and unoiled streams (Brannon et al., 1995); in addition, no differences in the timing of fry emergence from their redds were observed. From these data, it was concluded that oil contamination caused no detectable difference in incubation success or in the timing of emergence (Brannon et al., 1995).

In September and October 1989 through 1997, ADFG surveyed egg mortality in oiled and unoiled streams using the same hydraulic flushing method as they had used with alevins. In contrast to their spring 1989 studies on alevins, they reported that eggs from oiled streams had a mortality as much as 50% higher than that from reference streams from 1989 to 1993 and again in 1997 (Bue et al., 1996, 1998). ADFG concluded that the higher mortality was initially from direct exposure to oil. (Higher mortality after 1991 was attributed to genetic mutations or physiological feedback from exposure as embryos – see section 12.4.2.5)

In late August and early September 1989, however, Moulton (1996) compared the survival of eggs in some of the same oiled and unoiled streams studied by ADFG. Eggs were artificially spawned from females in each stream and then put in incubation containers in the same streams from which the eggs were taken. The containers were recovered near the time of the eggs' hatching. Moulton found 3.5% mortality of eggs in oiled streams and 4.9% mortality in unoiled streams. The low mortality indicates that there were no negative survival effects from exposure to oil in the incubation environment of those intertidal streams.

The contrasting results of these studies present a conundrum – why did the findings differ, and which were right? The sampling techniques were similar and some sampled streams

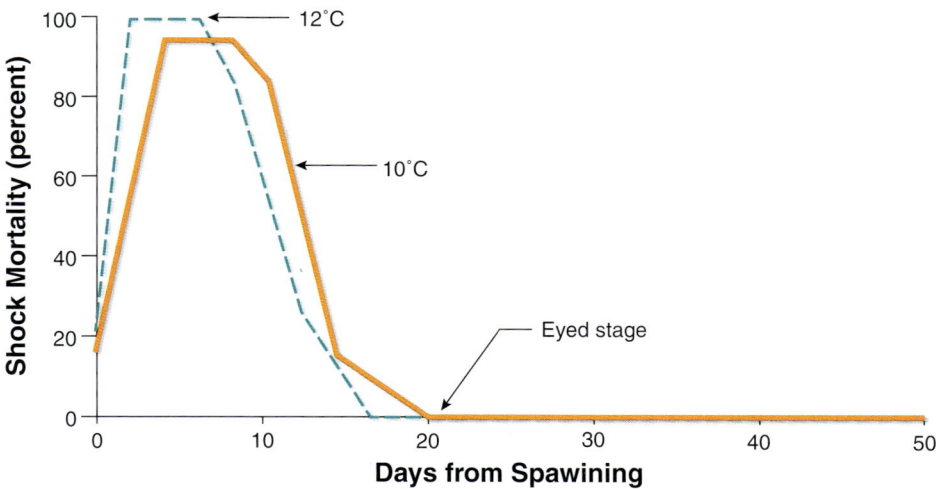

Figure 12.4 Sensitivity of eggs (expressed as percentage mortality) when exposed to mechanical shock by vertical drops into water from a 100-cm height at 10°C (orange solid line; Jensen and Alderdice, 1989) and by pouring the eggs submerged in water from one container to another from a 20-cm height at 12°C (blue dashed line; Jensen and Collins, 2003).

were the same. Misclassification of live and dead eggs was not the source of the difference: it is easy to differentiate between dead and live eggs, since mortality is clearly apparent from coagulated, cream-colored yolk material (compared to the near translucent orange yolk of live eggs). The explanation had to be something about the sampling process itself.

In comparing sampling protocols, it was apparent that eggs were sampled at different times relative to their spawning dates. In particular, ADFG sampled streams in the fall shortly after spawning was completed, and they sampled oiled streams earlier than unoiled streams. When sampling was early in the fall, close to spawning, mortalities were higher; when sampling was later in the fall, mortalities were lower (Brannon et al., 1995; Craig et al., 2002). This timing pattern was seen irrespective of whether the streams were oiled or not.

Salmon eggs are very sensitive to physical shock for up to 28 days after spawning, depending on temperature (Fig. 12.4). The delicate vitelline membrane that surrounds the yolk is very sensitive to rupture. Sensitivity continues until the membrane is replaced with epithelial tissue during gastrulation. If the eggs are disturbed before gastrulation, the membrane ruptures and water coagulates the yolk, killing the embryo (Jensen and Alderdice, 1989; Jensen and Collins, 2003). Sensitivity to severe shock continues (at a slightly lesser degree) through the rest of gastrulation to the pre-eyed stage of the embryo. Thus, the physical (e.g., hydraulic) sampling itself causes egg mortality when conducted too soon after fall spawning; later in the fall, after the embryos develop to the eyed stage, they are no longer vulnerable to sampling shock (Collins et al., 2000; Brannon et al., 2001, 2012; Thedinga et al., 2005; Carls and Thedinga, 2010).

ADFG's earliest samples overlapped with the period of embryo sensitivity, resulting in increased mortality. The effect is shown by comparing mortality with the mean number of postspawning days when egg samples were taken (Table 12.2): mortality decreased as the number of postspawning days before sampling increased. Sample timing and lethal shock effects are also shown in Figure 12.5, where egg mortality data (Craig et al., 2002) are plotted against spawn timing from 1989 to 1997.

Table 12.2 Percentage embryo mortality by mean number of days after the end of spawning when eggs were extracted from redds by ADFG, irrespective of year and oil contamination. Data are from 10 intertidal streams studied by both ADFG and Exxon from 1989 to 1998.

Mean days post-spawn	Percentage mortality
1.8	66.3
2.4	59.6
4.2	47.0
8.6	42.2
9.4	37.2
11.0	28.1
12.0	23.5
15.5	26.0
16.8	34.5
18.1	30.9
19.8	18.6
24.3	19.9
26.0	26.2
32.2	14.9

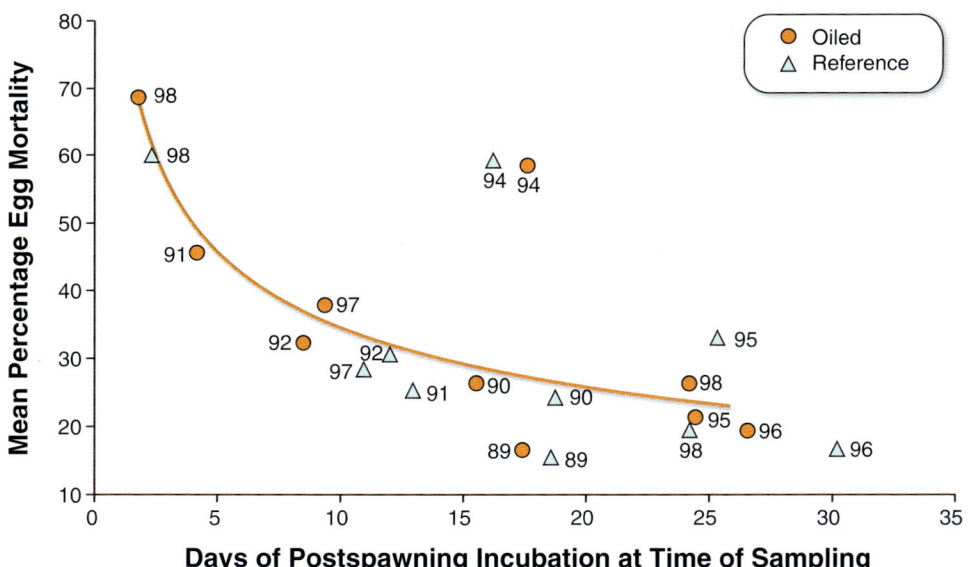

Figure 12.5 Mean percentage egg mortality in oiled and unoiled reference streams versus days of postspawning incubation at the time of sampling for the 10 study streams surveyed in PWS from 1989 to 1998 (line fit by eye). Mortality data from Craig et al. (2002); ADFG unpublished annual data reports (1996 and 1997); and Collins et al. (2000).

Oiled streams showed higher mortality than unoiled streams because they were sampled earlier than unoiled streams, as demonstrated in a separate study on the subset of 10 streams used in the ADFG surveys (Collins et al., 2000). In that study, hydraulic sampling in the 10 streams was scheduled to correspond with (1) the earlier ADFG survey dates, September 21–27, and (2) the later survey dates, October 16–19 (Craig et al., 2002). Irrespective of oiling, the September sample set had a 42.3% shock mortality, which dropped to 2.2% in the same streams in the later October sample set (Table 12.3).

Table 12.3 Total eggs, total mortality, percentage mortality, percentage dense opaque mortality (deaths prior to sampling), percentage shock mortality, and eyed eggs present at the time of sampling in 10 PWS study streams during the earlier September and later October ADFG sampling periods repeated in 1998 (percentage based on sum of all streams).

Sample date	Total eggs	Total mortality	% total mortality	% dense opaque	% shock mortality	% eyed eggs
Sept 21–27	27 755	17 654	63.6	21.3	42.3	26.8
Oct 16–19	34 202	7 807	23.1	22.8	2.2	77.2

Thus, when lethal shock is taken into account, there are no significant mortality differences between oiled and unoiled streams. The low mortality observed by Moulton (1996) among eggs manually buried in the oiled streams where Bue et al. (1996) reported high mortality further confirmed that oil was not the cause of the embryo deaths. The effects of lethal shock were also apparent in the ADFG data as well. The higher egg mortality observed in oiled streams sampled in the fall was absent in the samples taken in the following spring (Craig et al., 2002), when only natural mortality would be apparent.

12.4.2.3 Presence of developmental deformities

Advanced alevins approaching emergence were sampled and examined for deformities in the spring of 1990 and again in 1991 (Brannon et al., 1995). No physical anomalies were detected. There were also no observations of alevins with fluid-distended yolks (ascites or blue-sac disease) in any of the samples from oiled streams (Brannon et al., 1995).

These findings are in contrast to laboratory observations (see section 12.4.2.4), where deformities were observed (Marty et al., 1997; Heintz et al., 1999; Carls et al., 2005) and ascites was commonly reported (Marty et al., 1997). However, embryos and alevins in the laboratory were exposed to oil at much higher concentrations than were those in the field.

The lack of deformities in the field shows that the contamination levels tested in the laboratory were not representative of the incubation conditions present in spill-path streams. Thus, the laboratory results were not applicable to field conditions.

12.4.2.4 Laboratory oil exposure studies of eggs and alevins

Based on the laboratory studies of Heintz et al. (1999), it was assumed that similar concentrations of TPAH leaching from shoreline subsurface oil residues (Heintz et al., 1999; Murphy et al., 1999) would cause the level of mortality reported in oiled streams (Bue et al., 1996, 1998). This was before the role of lethal physical shock during egg sampling had been established.

To simulate the exposure of pink salmon eggs to weathered oil residues leaching from intertidal sediments, fresh water and seawater were alternately pumped upward through vertical columns. The columns had been packed with gravel coated with a single dose of very weathered oil (VWO) from Prudhoe Bay, or with different doses of artificially weathered oil (AWO), also from Prudhoe Bay. Eggs were placed directly on the gravel at the top of the column or on a perforated tray above the gravel.

From the results of these studies, Heintz *et al.* (1999) concluded that:

- PAH accumulated in eggs through aqueous transport, not by direct contact with oil (i.e., dissolved PAH were the sole toxic agents);
- Adverse effects from VWO exposure began at an aqueous TPAH concentration of 1.0 µg/L;
- Oil toxicity increased with weathering;
- Pink salmon eggs in PWS streams may have been exposed to lethal concentrations of PAH leaching from shorelines years after the spill; and
- Toxic effects from AWO exposure began at an aqueous TPAH concentration of 18 µg/L – a very conservative lowest-observed-effect concentration for crude oil, but not inconsistent with accepted values (McGrath and Di Toro, 2009).

Landrum *et al.* (2012) and Page *et al.* (2012a, b, c) reviewed the Heintz *et al.* (1999) experiments in detail (see also Heintz *et al.* 2012a, b). They showed that the Heintz *et al.* data do not support the study's own conclusions about VWO; instead, they concluded that:

- Effluent from the oiled column contained oil droplets;
- The adverse effect at 1.0 µg/L in the VWO treatment was not PAH-related, because the VWO treatment showed toxicity at water or embryo PAH concentrations 2.5 to 7.8 times lower than nontoxic water or embryo PAH concentrations for AWO treatments;
- Weathering did not increase the toxicity of oil because the absolute concentrations of alkylphenanthrenes in the VWO treatment, said by Heintz *et al.* (1999) to be the toxic agents in weathered oil, were much lower than those in nontoxic AWO treatments; and
- Water PAH concentrations measured in PWS after the spill (Chapter 3) also do not support the claim by Heintz *et al.* (1999) that salmon eggs may have been exposed to toxic concentrations of PAH from weathered oil.

Injury to fish embryos and larvae can occur from exposure to stressors other than PAH, such as the products of microbial degradation of petroleum and other organic matter (Landrum *et al.*, 2012). In the case of the Heintz *et al.* (1999) study, chemistry data for the VWO treatment demonstrated extensive microbial degradation during prior use and storage. Observations of microbial contamination were noted in the laboratory records from these oiled gravel-column studies (Page *et al.*, 2012b).

Brannon *et al.* (2006a) were unable to replicate the experiments of Heintz *et al.* (1999) using the Heintz *et al.* experimental design and naturally weathered *Exxon Valdez* oil collected in PWS. Higher mortality was not observed in eggs exposed to aqueous 8.27 µg/L TPAH from naturally weathered oil. Brannon *et al.* found microdroplets of oil in their test solutions, and preliminary data in a later study indicate that toxicity with microdroplets present is greater than that from dissolved PAH only (Brannon *et al.*, 2012). Furthermore, under test conditions where oil is in contact with the chorion of the egg, petroleum hydrocarbons directly cross the egg membrane without entering the aqueous solution (Collins *et al.*, 2009b). This creates higher TPAH exposures than represented in just the aqueous dissolved fraction (Redman *et al.*, 2012).

12.4.2.5 Identification of intergenerational or genetic effects

As described above, ADFG observed elevated egg mortality in oiled streams from 1989 to 1993 and in 1997 (Bue *et al.*, 1996, 1998; Craig *et al.*, 2002). Because mortality continued after 1991, when oil was generally gone from sediments, it was attributed to germline mutations or feedback from adults having experienced chronic oil exposure as embryos in 1989.

To investigate this proposed genotoxicity, in 1993 Bue et al. (1998) artificially spawned fish taken from oiled and unoiled streams and incubated the eggs in a hatchery. They found higher egg inviability in fish taken from oiled streams and concluded that this supported genotoxicity.

However, in 1991 (the parent year of the spawners used by Bue et al. in their 1993 study), Brannon et al. (1995) had conducted a viability study with the same objective and found no difference between eggs taken from spawners in oiled and unoiled streams of that particular brood. Brannon et al. took eggs from adults in two streams that had experienced high oil contamination in 1989 and for which ADFG stream surveys had reported a mean mortality of 61%. In contrast, when artificially incubated in a hatchery, eggs taken from adults in the same two highly contaminated streams had less than 4% mortality. This demonstrated that genetic or physiological problems in adults were not causing the egg mortality observed in those stream surveys. As discussed earlier, the true cause of the high egg mortality observed in the field was the shock from hydraulic sampling. The mortality that Bue et al. (1998) observed in 1993 must have occurred from handling and preparing the gametes for study, not from oil.

Roy et al. (1999) looked at the genotoxicity question by exposing pink salmon eggs to high doses of *Exxon Valdez* oil in a laboratory experiment, then detecting K-*ras*[1] mutations. (No salmon from the region affected by the spill were assayed.) They suggested that K-*ras* mutations could be causing the egg mortality in oiled streams – although they said it was not verified and that the heritability of the mutations was not established.

Cronin et al. (2002) used Roy et al.'s (1999) methods to assess K-*ras* mutations in fish from unoiled streams and streams affected by the *Exxon Valdez* spill and by oil-seep areas on the Alaska Peninsula. They did not detect the K-*ras* mutations reported by Roy et al. (1999), suggesting that the oil concentrations in areas affected by the *Exxon Valdez* spill or the seep areas (which were less than Roy et al.'s (1999) laboratory concentrations) were not high enough to induce K-*ras* mutations. Further, lethal germline mutations will decline rapidly following short-term exposure to a mutagen; the highly variable egg mortality observed from year to year in oiled streams is not consistent with this reality (Cronin and Bickham, 1998).

Heintz (2002) further investigated genotoxicity by studying the viability of eggs from adults that had experienced tissue TPAH levels greater than 5000 ng/g as embryos in laboratory incubation before being released as fry in 1998. Upon artificially spawning the returning adults, the eggs showed no viability impairment. The laboratory tissue TPAH concentrations were over 50 times the mean tissue TPAH levels of incubating pink salmon in oiled streams, yet egg viability was still unaffected in the Heintz study.

12.4.3 Effects from exposure during marine residence
12.4.3.1 Analysis of tissue total polycylic aromatic hydrocarbons

The tissue TPAH concentrations of pink salmon fry off oiled bays in 1989 ranged between 180 and 218 ng/g, and those within oiled corridors between islands ranged between 109 and 255 ng/g tissue TPAH (Carls et al., 1996b). These tissue concentrations were well below the 1900–3000 ng/g tissue TPAH levels to which embryos were exposed

---

[1] K-*ras* is a type of gene called an oncogene that can cause cancer when activated. It can be activated by mutations caused by exposure to chemicals such as those in crude oil.

in the laboratory with no subsequent reduction in growth (Heintz et al., 2000; Carls et al., 2005). This suggests that fry in open marine waters did not experience any reduction in growth from the low level of oil exposure experienced.

12.4.3.2 Examination of reduced growth

Brannon and Maki (1996) examined the growth of fingerlings from oiled areas and unoiled areas offshore of the major nursery basins in 1989 and 1990 and found no differences. However, recalling the four conditions for comparisons,[2] this lack of a difference does not mean that there were no growth effects because the source and age structure of the samples were unknown. Pink salmon are migratory, so the site of interception may not represent their previous experience in another part of PWS.

Wertheimer and Celewycz (1996) sampled marine resident juveniles in 1989 and 1990 in both oiled and unoiled bays and corridors between islands (Fig. 12.6). Unlike Brannon et al., they did not sample any offshore sites. The corridor samples were taken with beach seines in the same manner as in bays, so the only difference from bays was in location along the shoreline. Wertheimer and Celewycz measured length and weight of juveniles in their samples. They found no differences in the weight of pink salmon and the sympatric chum salmon (*Oncorhynchus keta*) between oiled and unoiled bays and corridors.

The lengths of pink salmon juveniles, however, were shorter in 1989 in oiled corridors than in unoiled corridors. Lengths were no different at any other time or location for either pink or chum salmon. Curiously, the differences in pink salmon juvenile length occurred in oiled corridors where the aqueous-TPAH concentrations were < 1 µg/L and where zooplankton and epibenthic crustaceans were more abundant than in unoiled corridors. If oil were the cause, then juveniles in oiled bays should have shown a size reduction compared to those from unoiled bays; however, length was not affected in oiled bays where the aqueous TPAH concentrations were 2–7 µg/L. Additionally, given that both weight and length were affected when growth was measured in laboratory oil-contaminated feeding studies (Carls et al., 1996a), it leaves doubt that differences in only one size parameter would be affected in the field studies. Nonetheless, Wertheimer and Celewycz (1996) concluded that the shorter length of juveniles in oiled corridors was related to reduced growth of pink salmon because of oil exposure in the marine environment.

Wertheimer and Celewycz (1996) were careful in reporting their methods and data, which allowed ready re-assessment of their study. Samples from the corridors were taken by beach seines close to shore, thereby excluding any larger fish that had moved offshore; consequently, their fish had a mean length in early June of 45.5 mm compared to a mean length of 51.8 mm in samples taken at the same time by Brannon et al. (1995) using surface trawls over deeper water (Fig. 12.7). It appears that the size disparity also occurred because different subpopulations were sampled: pre-fed hatchery fish that were released in a narrow time frame dominated unoiled corridor sites, whereas wild juveniles entering the migrating population over the whole emergence period dominated oiled corridor sites (Wertheimer and Celewycz, 1996). Consequently, the sampling favored larger fish in the unoiled corridors and most likely slightly lower ages in the oiled corridors (Brannon et al., 2012).

---

[2] The same population must be randomly distributed between oiled and unoiled areas; samples from oiled and unoiled areas must represent the same age structure; the oiled and unoiled areas sampled must be environmentally and biologically similar; and sampling gear must be unbiased.

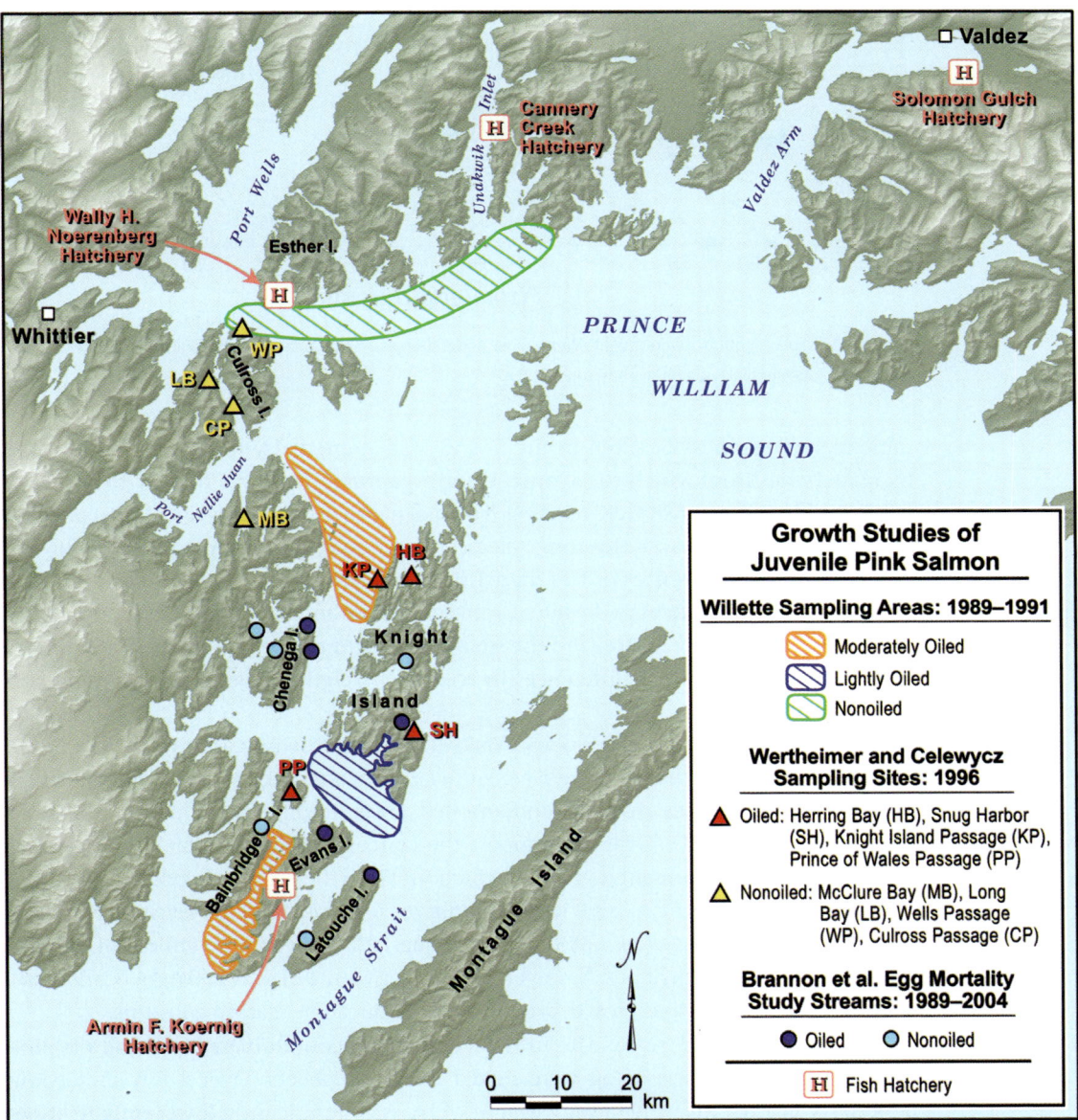

Figure 12.6 Map of PWS showing streams used by Brannon *et al*. (1995) for egg mortality studies, oiled and unoiled bays and corridors (passages) used by Wertheimer and Celewycz (1996), and hatchery release sites and oiled and reference recovery areas used by Willette (1996) for assessment of juvenile pink salmon growth.

Willette (1996) also looked at growth patterns in 1989–91, sampling some of the same areas as Wertheimer and Celewycz (Fig. 12.6), but using marked fish from two hatcheries for analysis. In one of three 1989 marked time-release test groups, samples recovered in oiled areas had a lower mean weight than members of the same marked group recovered in the reference area. Willette concluded that the weight differences were growth-related consequences of oil exposure. (Willette found no differences in length.)

Even with marked fry, however, there was uncertainty about the age composition of the samples. Further insight was possible because of Willette's care in recording methods and data. The marked groups that showed weight differences received the

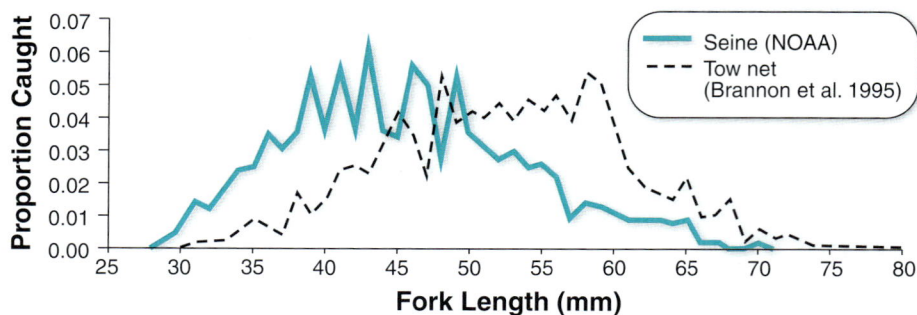

Figure 12.7 Fork length of seine-caught (Wertheimer and Celewycz, 1996) and tow-net-caught (Brannon et al., 1995) PWS juvenile pink salmon in early June, 1989.

same mark, regardless of when in the 5- and 8-day periods fry were released from the two hatcheries from which they came. Upon recapture, there was no way to know on which day individual fish were released. The mean growth rate in weight (4.7% per day) of the hatchery fish over the long release period could more than account for the observed weight differences of less than 1.0% per day. The unknown age composition of samples, the differences in distances between release and recapture sites, temperature variations in migratory routes, and the small number of recaptured fry create too many variables to conclude that differences in recapture weights represented differences in rates of growth.

Consequently, for the Brannon and Maki (1996), Wertheimer and Celewycz (1996), and Willette (1996) studies, there were sufficient uncertainties so as to prevent interpreting size differences in oiled and unoiled areas as differences in growth rates. Laboratory research also does not support the conclusion that smaller sized juveniles in marine samples represent a growth reduction. Growth was not reduced in juveniles reared in net-pens after exposure to oil as embryos until tissue TPAH levels were greater than 1900 ng/g (Heintz et al., 2000). Mean tissue TPAH levels in juvenile pink salmon from oiled bays and corridors, however, ranged between 175 and 200 ng/g (Carls et al., 1996b). This suggests that growth would not be reduced in marine corridors.

Wang et al. (1993) assessed the influence of a diet contaminated with oil on pink salmon fry. Three levels of oil were mixed into feed pellets. Over a 6-week feeding period, fry growth measured as *dry* weight of fry was significantly lower only in those fed the highest-oiled feed, 35 mg oil/g of feed.

Carls et al. (1996b) reworked the Wang et al. (1993) data in terms of *wet* weight. They concluded that ingested oil at the lowest concentration, 0.37 mg oil/g feed, affected wet weight by the end of the 6-week period. The 0.37 mg oil/g of feed represents 13 000 ng PAH/g of feed (Carls et al., 1996b). Tissue PAH levels resulting from consuming the contaminated feed were not reported. However, based on laboratory evidence that tissues will accumulate concentrations greater than the aqueous PAH levels present during exposure through water (Heintz et al., 1999; Brannon et al., 2006a), we expect that feed PAH doses of 13 000 ng/g ingested over a 6-week period would result in the accumulation of tissue PAH loads higher than 13 000 ng/g. That assumption would have to be verified, of course. If this is the case, we would not anticipate any growth reduction among fingerlings with tissue PAH levels as low as the ~175 to 200 ng/g measured in juveniles captured from oiled corridors and bays (Carls et al., 1996b).

12.4.3.3 Examination of fish condition

The fingerling condition indices ($100 \times weight/length^3$) for samples taken in 1989 and 1990 were all in the range considered healthy, with no reduction in condition in oiled areas as compared to unoiled areas (Brannon et al., 1995; Brannon and Maki, 1996; Wertheimer and Celewycz, 1996; Willette, 1996).

12.4.3.4 Identification of feeding disruptions

Zooplankton biomass and density were not reduced by oil in the marine environment and were greater in corridors away from shore than in the bays (Celewycz and Wertheimer, 1996). Epibethic crustaceans were more abundant in oiled than in unoiled areas (Wertheimer et al., 1996), as were other zooplankton in some cases. Calanoid copepods, an important food source for young salmon, dominated the biomass of zooplankton (Wertheimer et al., 1996).

The stomach contents and fullness of pink salmon fingerlings in oiled and unoiled areas indicated that feeding had not been reduced by oil exposure and that the preferred zooplankton and epibenthic crustaceans were represented as expected (Celewycz and Wertheimer, 1996; Wertheimer et al., 1996).

Based on stomach weight and fullness, there were no differences in the feeding performance of pink salmon juveniles that had been exposed to low levels of oil in their marine nursery and migratory areas (Sturdevant et al., 1996; Wertheimer and Celewycz, 1996; Carls et al., 2005).

12.4.3.5 Determination of adult returns to spawn

In 1990, 11.8 million wild and 31.8 million hatchery salmon returned to PWS. Geiger et al. (1996) estimated that, because of the spill, 1.9 million wild pink salmon failed to return, about 4% of the return (Fig. 12.1). This estimate was based primarily on higher egg mortality and the problematic conclusions of poor growth in fry and juveniles due to oil, as discussed above (Wertheimer and Celewycz, 1996; Willette, 1996). Consequently, the evidence shows that Geiger et al. (1996) most likely overestimated the loss in returns.

Research to assess the effect of substrate oil on mean distribution and density of adult spawners in 1989 to 1992 showed no differences between oiled and unoiled streams (Maki et al., 1995). In addition, comparisons of survival of wild adults returning to a subset of those streams over a 12-year period also revealed no differences in return and reproductive success (Brannon et al., 2006b).

Figure 12.1 summarizes adult returns, both wild and hatchery, from 1960 to 2007. Given the abundance of oil in the PWS environment following the spill, it seems improbable that oiling effects on the most numerous fish species in PWS would not be clearly evident if the smaller-scale field and laboratory assessments of damage were true. But pink salmon had record returns in 1990 and 1991 for the 1989 and 1990 brood-year adults, respectively, indicating no serious population-level effects in the years of greatest potential exposure to oil.

12.4.4 Discussion

There were major differences in what scientists concluded about the effects of the *Exxon Valdez* oil spill on pink salmon in PWS. The importance of having a null hypothesis as well as alternative hypotheses at the onset of investigations cannot be overemphasized

(Chapter 10). Because a null hypothesis and alternative hypotheses were not used, an entire oil toxicity narrative was developed to explain egg mortality, to describe how toxic oil entered the incubation redds, and to support the conclusion of long-term negative effects on survival and reproductive efficiency.

Understandably, the expectation that the oil spill would have major effects on pink salmon would tend to influence interpretation of results, and that turned out to be the initial problem. It was assumed that oil was the cause of the increased egg mortality observed in the ADFG stream surveys (Bue *et al.*, 1996). Although the lethal effect of sampling too soon after spawning was pointed out by Brannon and Maki (1996), it was not until much later that ADFG recognized that sampling date may have affected their results (Craig *et al.*, 2002), but by that time laboratory research on oil toxicity was well on its way.

Given the mistaken conclusion that oil was the cause of the egg mortality (Heintz *et al.*, 1999; Rice *et al.*, 2001, 2007), the problem was further confounded by the manner in which the laboratory work was interpreted. Laboratory research concentrated on determining the level of oil exposure that would induce the level of mortality thought to have been observed in ADFG stream surveys. It was then assumed that those were the levels of exposure that eggs experienced in oiled streams, without any government-supported field studies undertaken to verify that eggs in spill-path streams actually experienced toxic oil concentrations. In fact, the laboratory test concentrations far exceeded anything in spill-path streams, or in tissue loads of pink salmon alevins (Brannon *et al.*, 1995). The result was an oil toxicity narrative of unsupported conclusions about levels of exposure in oiled streams, mortality at later life stages, and speculation about reduced gamete viability of returning adults (Heintz *et al.*, 1999, 2000; Heintz, 2007; Rice *et al.*, 2001, 2007).

Evidence of egg mortality from sampling shock, careful examination of laboratory studies, and the absence of intergenerational genetic effects lead to the conclusion that this toxicity narrative was incorrect. There had been no attempt to verify the narrative in the field where the incubating eggs and alevins were actually exposed to the oil in those first three critical years following the spill, where follow-up would have alleviated much of the disagreement. Part of the problem was that litigation tended to limit exchange of data among government- and Exxon-sponsored scientists, but in the end it was those multiple approaches taken by all of the scientists that provided resolution to the toxicity issues.

Similarly, three studies on early juvenile pink salmon growth in marine waters differed in their conclusions and in the parameters affected (Brannon and Maki, 1996; Wertheimer and Celewycz, 1996; Willette, 1996). Here, too, it was anticipated that the spill would affect the early marine life history of the species. However, the actual level of water-column contamination was below the level that would induce negative physical effects on growth of sensitive indicator organisms (Neff and Stubblefield, 1995), and food production was actually enhanced in some of the oiled areas (Wertheimer *et al.*, 1996; Sturdevant *et al.*, 1996). The variations in population structure, age, and distribution did not satisfy the conditions necessary for interpreting size variation as differences in growth rates, and the empirical evidence on conditions in the marine waters of the Sound did not support the conclusions that growth rates were reduced in those oiled areas. In this case, the several indirect studies of growth involving oil contamination of the water column, juvenile tissue chemistry, responses in plankton production, temperature differences, and pen-rearing trials with juvenile pink salmon collectively provided compelling evidence that growth was not reduced by exposure to the levels of oil in marine waters in 1989.

In the final analysis, the lack of oil-spill effects on the pink salmon population is shown by the 1990 and 1991 record returns of adult pink salmon (Fig. 12.1). Those two brood years were exposed as embryos or fry to the greatest toxic risk from oil in 1989 and 1990, yet they returned the largest runs recorded up to that time. The multiple sources of evidence that culminated with the successful returns of those populations exposed to oil clearly show that pink salmon were not damaged at any detectable population level by the *Exxon Valdez* oil spill.

12.5 Lessons learned

- The null hypothesis and alternative hypotheses that identify alternative causes for an observed effect must be properly considered and incorporated in the analyses. This seems like a scientific homily, but in the chaotic environment that follows a spill, it can be overlooked.
- Multiple approaches provide checks and balances to investigations, and are often required to understand the disparities between the conclusions arising from different investigations.
- The life history and biology of the species must be taken into account when investigating environmental disturbances. A clear example is the sensitivity of pink salmon eggs to lethal shock from sampling during early embryo development that was mistaken for oil effects.
- Well-trained sampling teams are essential, especially when critical life stages or biological particularities of a species are involved.
- Logistics of sampling and field study must be carefully considered to avoid inducing biological effects. The life history of a species can present formidable logistical problems. For example, in an ideal paired treatment versus control study, oiled and unoiled streams would be sampled at the same time. However, given the size of PWS, the distances between oiled and unoiled streams, variation in sample timing, inclement weather, and limitations on personnel and transportation, ideal sample timing can rarely be achieved. Sampling timing issues, even the unintentional ones described in the text, must be carefully evaluated.
- Natural variations and environmental circumstances need to be taken into account along with the accident or perturbation under study. There can be other influencing factors, such as sampling locations, temperature differences, weather, ocean regime changes, changes in predator–prey relationships, and variation in timing of adult returns, that can affect study results.
- If laboratory studies are conducted to assess the effects of an event in the field, precautions must be taken to reflect the actual field conditions they are intended to represent.
- When assessing specific potential environmental damages and their causes, population-level effects are often more important in the end than are individual or subpopulation effects. However, subpopulations may require further study if they are particularly large, spatially isolated, or of special significance (e.g., commercially or for subsistence use).

REFERENCES

Boehm, P.D., J.M. Neff, and D.S. Page (2007). Assessment of polycyclic aromatic hydrocarbon exposure in the waters of Prince William Sound after the *Exxon Valdez* oil spill: 1989–2005. *Marine Pollution Bulletin* **54**(3): 339–367.

Boehm, P.D, D.S. Page, E.S. Gilfillan, W.A. Stubblefield, and E.J. Harner (1995). Shoreline Ecology Program for Prince William Sound, Alaska, following the *Exxon Valdez* oil spill. Part 2: chemistry and toxicology. In Exxon Valdez *Oil Spill: Fate and Effects in Alaskan Waters*. P.G. Wells, J.N. Butler, and J.S. Hughes, eds. Philadelphia, PA, USA: American Society for Testing and Materials; ASTM Special Technical Publication 1219; ISBN-10: 0803118961; pp. 347–356.

Brannon, E.L., K. Collins, M.A. Cronin, L.L. Moulton, A.W. Maki, and K.R. Parker (2012). Review of the *Exxon Valdez* oil spill effects on pink salmon in Prince William Sound, Alaska. *Reviews in Fisheries Science* **20**(1): 20–60.

Brannon, E.L., K.C.M. Collins, M.A. Cronin, L.L. Moulton, K.R. Parker, and W. Wilson (2007). Risk of weathered residual *Exxon Valdez* oil to pink salmon embryos in Prince William Sound. *Environmental Toxicology and Chemistry* **26**(4): 780–786.

Brannon, E.L., K.C.M. Collins, L.L. Moulton, and K.R. Parker (2001). Resolving allegations of oil damage to incubating pink salmon eggs in Prince William Sound. *Canadian Journal of Fisheries and Aquatic Sciences* **58**(6): 1070–1076.

Brannon E.L., K.M. Collins, J.S. Brown, J.M. Neff, K.R. Parker, and W.A. Stubblefield (2006a). Toxicity of weathered *Exxon Valdez* crude oil to pink salmon embryos. *Environmental Toxicology and Chemistry* **25**(4): 962–972.

Brannon, E.L. and A.W. Maki (1996). The *Exxon Valdez* oil spill: analysis of impacts on Prince William Sound pink salmon. *Reviews in Fishery Science* **4**(4): 289–337.

Brannon, E., A.W. Maki, L. Moulton, and K. Parker (2006b). Results from a sixteen year study on the effects of oiling from *Exxon Valdez* on adult pink salmon. *Marine Pollution Bulletin* **52**(8): 892–899.

Brannon, E.L., L.L. Moulton, L.G. Gilbertson, A.W. Maki, and J.R. Skalski (1995). An assessment of oil spill effects on pink salmon populations following the *Exxon Valdez* oil spill – Part 1: Early life history. In Exxon Valdez *Oil Spill: Fate and Effects in Alaskan Waters*. P.G. Wells, J.N. Butler, and J.S. Hughes, eds. Philadelphia, PA, USA: American Society for Testing and Materials; ASTM Special Technical Publication 1219; ISBN-10: 0803118961; pp. 548–584.

Bue, B.G, S. Sharr, S.D. Moffitt, and A.K. Craig (1996). Effects of the *Exxon Valdez* oil spill on pink salmon embryos and preemergent fry. In *Proceedings of the* Exxon Valdez *Oil Spill Symposium*. S.D. Rice, R.B. Spies, D.A. Wolfe, and B.A. Wright, eds. Bethesda, MD, USA: American Fisheries Society; Symposium 18; ISBN-10: 0913235954; ISSN: 08922284; pp. 619–627.

Bue, B.G., S. Sharr, and J.E. Seeb (1998). Evidence of damage to pink salmon inhabiting Prince William Sound, Alaska, two generations after the *Exxon Valdez* oil spill. *Transactions of the American Fisheries Society* **127**(1): 35–43.

Carls, M.G., R.A. Heintz, G.D. Marty, and S.D. Rice (2005). Cytochrome P450 1A induction in oil-exposed pink salmon (Oncorhynchus gorbuscha) embryos predicts reduced survival potential. *Marine Ecology Progress Series* **301**: 253–265.

Carls, M.G., L. Holland, M. Larsen, J.L. Lum, D.G. Mortensen, S.Y. Wang, and A.C. Wertheimer (1996a). Growth, feeding, and survival of pink salmon fry exposed to food contaminated with crude oil. In *Proceedings of the* Exxon Valde *Oil Spill Symposium*. S.D. Rice, R.B. Spies, D.A. Wolfe, and B.A. Wright, eds. Bethesda, MD, USA: American Fisheries Society; Symposium 18; ISBN-10: 0913235954; ISSN: 08922284; pp. 608–618.

Carls, M.G. and J.F. Thedinga (2010). Exposure of pink salmon embryos to dissolved polynuclear aromatic hydrocarbons delays development, prolonging vulnerability to mechanical damage. *Marine Environmental Research* **69**(5): 318–325.

Carls, M.G., R.E. Thomas, M.R. Lilly, and S.D. Rice (2003). Mechanism for transport of oil-contaminated groundwater into pink salmon redds. *Marine Ecology Progress Series* **248**: 245–255.

Carls, M.G., A.C. Wertheimer, J.W. Short, R.M. Smolowitz, and J.J. Stegeman (1996b). Contamination of juvenile pink and chum salmon by hydrocarbons in Prince William Sound. In *Proceedings of the* Exxon Valdez *Oil Spill Symposium*. S.D. Rice, R.B. Spies, D.A. Wolfe, and B.A. Wright, eds. Bethesda, MD, USA: American Fisheries Society; Symposium 18; ISBN-10: 0913235954; ISSN: 08922284; pp. 593–607.

Celewycz, A.G. and A.C. Wertheimer (1996). Prey availability to juvenile salmon after the *Exxon Valdez* oil spill. In *Proceedings of the* Exxon Valdez *Oil Spill Symposium*. S.D. Rice, R.B. Spies, D.A. Wolfe, and B.A. Wright, eds. Bethesda, MD, USA: American Fisheries Society; Symposium 18; ISBN-10: 0913235954; ISSN: 08922284; pp. 564–577.

Collins, K.M., E.L. Brannon, and M.A. Cronin (2009a). *Pink Salmon (*Oncorhynchus gorbuscha*) Spawning Adults of Hatchery and Wild Origin in Prince William Sound as Determined with Otoliths*. Moscow, ID, USA: University of Idaho, Center for Salmonid and Freshwater Species at Risk; Research Bulletin 09–1.

Collins, K., E.L. Brannon, and K. Parker (2009b). *Sensitivity of Steelhead Trout to* Exxon Valdez *Crude Oil*. Moscow, ID, USA: University of Idaho, Center for Salmonid and Freshwater Species at Risk; Research Bulletin 09–3.

Collins, K.M., E.L. Brannon, L.L. Moulton, M.A. Cronin, and K.R. Parker (2000). Hydraulic sampling protocol to estimate natural embryo mortality of pink salmon. *Transactions of the American Fisheries Society* **129**(3): 827–834.

Craig, A.K., T.M. Willette, D.G. Evans, and B.G. Bue (2002). *Injury to Pink Salmon Embryos in Prince William Sound: Field Monitoring*. Anchorage, Cordova, and Soldotna, AK, USA: Alaska Department of Fish and Game, Division of Commercial Fisheries; *Exxon Valdez* Oil Spill Restoration Project 98191A-1 Final Report. [http://www.evostc.state.ak.us/Files.cfm?doc=/Store/FinalReports/1998-98191A1-Final.pdf&]

Cronin, M.A. and J.W. Bickham (1998). A population genetic analysis of the potential for crude oil spill to induce heritable mutations and impact natural populations. *Ecotoxicology* **7**(5): 259–278.

Cronin, M.A. and A.W. Maki (2004). Assessment of the genetic toxicological impacts of the *Exxon Valdez* oil spill on pink salmon (*Oncorhynchus gorbuscha*) may be confounded by the influence of hatchery fish. *Ecotoxicology* **13**(6): 495–501.

Cronin, M.A., J.K. Wickliffe, Y. Dunina, and R.J. Baker (2002). K-*ras* oncogene sequences in pink salmon in streams impacted by the *Exxon Valdez* oil spill: no evidence of oil-induced heritable mutations. *Ecotoxicology* **11**(4): 233–241.

Cross, A.D., D.A. Beauchamp, K.W. Myers, and J.H. Moss (2008). Early marine growth of pink salmon in Prince William Sound and the coastal Gulf of Alaska during years of low and high survival. *Transactions of the American Fisheries Society* **137**(3): 927–939.

Field, L.J., J.A. Fall, T.S. Nighswander, N. Peacock, and U. Varanasi, eds (1999). *Evaluating and Communicating Subsistence Seafood Safety in a Cross-Cultural Context: Lessons Learned from the Exxon Valdez Spill*. Pensacola, FL, USA: Society of Environmental Toxicology and Chemistry; ISBN-10: 1880611376.

Geiger, H.J., B.G. Bue, S. Sharr, A.C. Wertheimer, and T.M. Willette (1996). A life history approach to estimating damage to Prince William Sound pink salmon caused by the *Exxon Valdez* oil spill. In *Proceedings of the* Exxon Valdez *Oil Spill Symposium*. S.D. Rice, R.B. Spies, D.A. Wolfe, and B.A. Wright, eds. Bethesda, MD, USA: American Fisheries Society; Symposium 18; ISBN-10: 0913235954; ISSN: 08922284; pp. 487–498.

Hargreaves, N.B. and R.J. Lebrasseur (1985). Species selective predation on juvenile pink (*Oncorhynchus gorbuscha*) and chum salmon (*O. keta*) by coho salmon (*O. kisutch*). *Canadian Journal of Fisheries and Aquatic Sciences* **42**(4): 659–668.

Heard, W.H. (1991). Life history of pink salmon. In *Pacific Salmon Life Histories*. C. Groot and L. Margolis, eds. Vancouver, BC, Canada: University of British Columbia Press; ISBN-10: 0774803592; ISBN-13: 9780774803595; pp. 119–230.

Heintz, R.A. (2002). *Effects of Oiled Incubation Substrate on Pink Salmon Reproduction*. Juneau, AK, USA: National Oceanic and Atmospheric Administration, National Marine Fisheries Service; *Exxon Valdez* Oil Spill Restoration Project 01476 Annual Report. [http://www.evostc.state.ak.us/Files.cfm?doc=/Store/AnnualReports/2001–01476-Annual.pdf&]

Heintz, R.A. (2007). Chronic exposure to polynuclear aromatic hydrocarbons in natal habitats leads to decreased equilibrium size, growth, and stability of pink salmon populations. *Integrated Environmental Assessment and Management* **3**(3): 351–363.

Heintz, R.A., S.D. Rice, M.G. Carls, and J.W. Short (2012a). The authors' reply [to Page *et al.*, 2012b]. *Environmental Toxicology and Chemistry* **31**(3): 472–473.

Heintz, R.A., S.D. Rice, M.G. Carls, and J.W. Short (2012b). The authors' second reply [to Page *et al.*, 2012c]. *Environmental Toxicology and Chemistry* **31**(3): 475–476.

Heintz, R.A., S.D. Rice, A.C. Wertheimer, R.F. Bradshaw, F.P. Thrower, J.E. Joyce, and J.W. Short (2000). Delayed effects on growth and marine survival of pink salmon *Oncorhynchus gorbuscha* of exposure to crude oil during embryonic development. *Marine Ecology Progress Series* **208**: 205–216.

Heintz, R.A., J.W. Short, and S.D. Rice (1999). Sensitivity of fish embryos to weathered crude oil: Part II. Increased mortality of pink salmon (*Oncorhynchus gorbuscha*) embryos to weathered *Exxon Valdez* crude oil. *Environmental Toxicology and Chemistry* **18**(3): 494–503.

Jensen, J.O.T. and D.F. Alderdice (1989). Comparison of mechanical shock sensitivity of eggs of five Pacific salmon (*Oncorhynchus*) species and steelhead trout (*Salmo gairdneri*). *Aquaculture* **78**(2): 163–181.

Jensen, N. and K. Collins (2003). Time required for yolk coagulation in pink salmon and steelhead eggs exposed to lethal shock prior to eyeing. *North American Journal of Aquaculture* **65**(4): 339–343.

Joyce, T.L. and D.G. Evans (1999). *Otolith Marking of Pink Salmon in Prince William Sound Hatcheries, 1995–1998.* Anchorage and Cordova, AK, USA: Alaska Department of Fish and Game, Division of Commercial Fisheries; *Exxon Valdez* Oil Spill Restoration Project 99188 Final Report. (Note: dated September 2000 on the title page.) [http://www.evostc.state.ak.us/Files.cfm?doc=/Store/FinalReports/1999-99188CLO-Final.pdf&]

Landrum, P.F., P.M. Chapman, J. Neff, and D.S. Page (2012). Evaluating the aquatic toxicity of complex organic chemical mixtures: lessons learned from polycyclic aromatic hydrocarbon and petroleum hydrocarbon case studies. *Integrated Environmental Assessment and Management.* **8**(2): 217–230.

Li, H. and M.C. Boufadel (2010). Long-term persistence of oil from the *Exxon Valdez* spill in two-layer beaches. *Nature Geoscience* **3**(2): 96–99.

Long, E.R. and L.G. Morgan (1991). *The Potential for Biological Effects of Sediment-sorbed Contaminants Tested in the National Status and Trends Programs.* Seattle, WA, USA: National Oceanic and Atmospheric Administration, National Ocean Service, Office of Oceanography and Marine Assessment, Ocean Assessments Division, Coastal and Estuarine Assessment Branch; NOAA Technical Memorandum NOS OMA 52. [http://docs.lib.noaa.gov/noaa_documents/NOS/OMA/TM_NOS_OMA/nos_oma_52.pdf]

Longhurst, A., ed. (1982). *Consultation on the Consequences of Offshore Oil Production on Offshore Fish Stocks and Fishing Operations, October 27–28, 1980, Bedford Institute of Oceanography, Dartmouth Nova Scotia.* Dartmouth, NS, Canada: Bedford Institute of Oceanography, Department of Fisheries and Oceans, Ocean Science and Surveys–Atlantic; Canada Department of Fisheries and Oceans, Canadian Technical Report of Fisheries and Aquatic Sciences 1096. [http://www.dfo-mpo.gc.ca/Library/31272.pdf]

Maki, A.W., E.L. Brannon, L.G. Gilbertson, L.L. Moulton, and J.R. Skalski (1995). An assessment of oil spill effects on pink salmon populations following the *Exxon Valdez* oil spill. Part 2: Adults and escapement. In Exxon Valdez *Oil Spill: Fate and Effects in Alaskan Waters*. P.G. Wells, J.N. Butler, and J.S. Hughes, eds. Philadelphia, PA, USA: American Society for Testing and Materials; ASTM Special Technical Publication 1219; ISBN-10: 0803118961; pp. 585–625.

Marty, G.D., D.E. Hinton, J.W. Short, R.A. Heintz, S.D. Rice, D.M. Dambach, N.H. Willits, and J.J. Stegeman (1997). Ascites, premature emergence, increased gonadal cell apoptosis, and cytochrome P450 1A induction into pink salmon larvae continuously exposed to oil-contaminated gravel during development. *Canadian Journal of Zoology* **75**(6): 989–1007.

McAuliffe, C.M. (1987). Organism exposure to volatile/soluble hydrocarbons from crude oil spills: a field and laboratory comparison. In *Proceedings of the 1987 Oil Spill Conference (Prevention, Behavior, Control, Cleanup), April 6–9, 1987, Baltimore, Maryland*. Washington DC, USA: American Petroleum Institute; Technical Publication 4452; pp. 275–288.

McGrath, J.A. and D.M. Di Toro (2009). Validation of the target lipid model for toxicity assessment of residual petroleum constituents: Monocyclic and polycyclic aromatic hydrocarbons. *Environmental Toxicology and Chemistry* **28**(6): 1130–1148.

Moles, A., M.M. Babcock, and S.D. Rice (1987). Effects of oil exposure on pink salmon, *Oncorhynchus gorbuscha*, alevins in a simulated intertidal environment. *Marine Environmental Research* **21**(1): 49–58.

Moss, J.H, D.A. Beauchamp, A.D. Cross, K.W. Myers, E.V. Farley, Jr., J.M. Murphy, and J.H. Helle (2005). Evidence for size-selective mortality after the first summer of ocean growth by pink salmon. *Transactions of the American Fisheries Society* **134**(5): 1313–1322.

Moulton, L.L. (1996). Effects of oil-contaminated sediments on early life stage and egg viability of pink salmon in Prince William Sound Alaska. In *Proceedings of the 17th Northeast Pacific Pink and Chum Salmon Workshop, March 1–3, 1995, Bellingham, Washington*. H. Fuss and G. Graves, eds. Olympia, WA, USA: Northwest Indian Fisheries Commission and Washington Department of Fish and Wildlife; pp. 147–155.

Murphy, M.L., R.A. Heintz, J.W. Short, M.L. Larsen, and S.D. Rice (1999). Recovery of pink salmon spawning areas after the *Exxon Valdez* oil spill. *Transactions of the American Fisheries Society* **128**(5): 909–918.

Neff, J. and W.A. Stubblefield (1995). Chemical and toxicological evaluation of water quality following the *Exxon Valdez* oil spill. In *Exxon Valdez Oil Spill: Fate and Effects in Alaskan Waters*. P.G. Wells, J.N. Butler, and J.S. Hughes, eds. Philadelphia, PA, USA: American Society for Testing and Materials; ASTM Special Technical Publication 1219; ISBN-10: 0803118961; pp. 141–177.

O'Clair, C.E., J.W. Short, and S.D. Rice (1996). Contamination of intertidal and subtidal sediments by oil from the *Exxon Valdez* oil spill in Prince William Sound. In *Proceedings of the* Exxon Valdez *Oil Spill Symposium*. S.D. Rice, R.B. Spies, D.A. Wolfe, and B.A. Wright, eds. Bethesda, MD, USA: American Fisheries Society; Symposium 18; ISBN-10: 0913235954; ISSN: 08922284; pp. 61–93.

Page, D.S., P.M. Chapman, P.F. Landrum, J.M. Neff, and R.A. Elston (2012a). A perspective on the toxicity of low concentrations of petroleum-derived polycyclic aromatic hydrocarbons to early life stages of herring and salmon. *Human and Ecological Risk Management* **18**(2): 229–260.

Page, D.S., J.M. Neff, P.F. Landrum, and P.M. Chapman (2012b). Letter to the editor: Sensitivity of pink salmon (*Oncorhynchus gorbuscha*) embryos to weathered crude oil. *Environmental Toxicology and Chemistry* **31**(3): 469–471.

Page, D.S., J.M. Neff, P.F. Landrum, and P.M. Chapman (2012c). Authors' reply to Heintz *et al.* [Heintz et al., 2012a]. *Environmental Toxicology and Chemistry* **31**(3): 473–475.

Parker, R.R. (1971). Size selective predation among juvenile salmonids fishes in a British Columbia inlet. *Journal of Fisheries Research Board of Canada* **28**(10): 1503–1510.

Pope, G.A., K.D. Gordon, and J.R. Bragg (2011a). Fundamental reservoir engineering principles explain lenses of shoreline oil residue twenty years after the *Exxon Valdez* oil spill. In *Proceedings of the Society of Petroleum Engineers' Americas E&P Health, Safety, Security, and Environmental Conference, March 21–23, 2011, Houston, Texas*. Houston, TX, USA: Society for Petroleum Engineers; SPE Paper 141809.

Pope, G.A., K.D. Gordon, and J.R. Bragg (2011b). Using fundamental practices to explain field observations twenty-one years after the *Exxon Valdez* oil spill. In *Proceedings of the 2011 International Oil Spill Conference (Promoting the Science of Spill Response), May 24–26, 2011, Portland, Oregon, USA*. Washington DC, USA: American Petroleum Institute.

Redman, A.D., J.M. McGrath, W. Stubblefield, A. Maki, and D.M. Di Toro (2012). Quantifying the concentration of crude oil microdroplets in oil-water preparations. *Environmental Toxicology and Chemistry* **31**(8): 1814–1822.

Rice, S.D., A. Moles, and J.W. Short (1975). The effects of Prudhoe Bay crude oil on survival and growth of eggs, alevins, and fry of pink salmon. In *Proceedings of the 1975 Conference on Prevention and Control of Oil Pollution, March 25–27, 1975, San Francisco, California*. Washington DC, USA: American Petroleum Institute; pp. 503–507.

Rice, S.D., J.W. Short, M.G. Carls, A. Moles, and R.B. Spies (2007). The *Exxon Valdez* oil spill. In *Long-term Ecological Change in the Northern Gulf of Alaska*. R.B. Spies, ed. Amsterdam, The Netherlands: Elsevier; ISBN-10 044452960; ISBN-13: 9780444529602; pp. 419–520.

Rice, S.D., R.E. Thomas, M.G. Carls, R.A. Heintz, A.C. Wertheimer, M.L. Murphy, J.W. Short, and D.A. Moles (2001). Impacts to pink salmon following the *Exxon Valdez* oil spill: persistence, toxicity, sensitivity, and controversy. *Reviews in Fisheries Science* **9**(3): 165–211.

Roy, N.K., J. Stabile, J.E. Seeb, C. Habicht, and I. Wirgin (1999). High frequency of K-*ras* mutations in pink salmon embryos experimentally exposed to *Exxon Valdez* oil. *Environmental Toxicology and Chemistry* **18**(7): 1521–1528.

Sharr, S., J.E. Seeb, B.G. Bue, S.D. Moffitt, A.K. Craig, and C.D. Miller (1994). *Injury to Salmon Eggs and Pre-emergent Fry in Prince William Sound*. Cordova, AK, USA: Alaska Department of Fish and Game, Commercial Fisheries Management and Development Division; *Exxon Valdez* Oil Spill Restoration Project 93003 Final Report. [http://www.evostc.state.ak.us/Files.cfm?doc=/Store/FinalReports/1993-93003-Final.pdf&]

Short, J.W. and M.M. Babcock (1996). Prespill and postspill concentrations of hydrocarbons in mussels and sediments in Prince William Sound. In *Proceedings of the* Exxon Valdez *Oil Spill Symposium*. S.D. Rice, R.B. Spies, D.A. Wolfe, and B.A. Wright, eds. Bethesda, MD, USA: American Fisheries Society; Symposium 18; ISBN-10: 0913235954; ISSN: 08922284; pp. 149–166.

Short, J.W. and P.M. Harris (1996). Chemical sampling and analysis of petroleum hydrocarbons in near-surface seawater of Prince William Sound after the *Exxon Valdez* oil spill. In *Proceedings of the* Exxon Valdez *Oil Spill Symposium*. S.D. Rice, R.B. Spies, D.A.

Wolfe, and B.A. Wright, eds. Bethesda, MD, USA: American Fisheries Society; Symposium 18; ISBN-10: 0913235954; ISSN: 08922284; pp. 17–28.

Sturdevant, M.V., A.C. Wertheimer, and J.L. Lum (1996). Diets of juvenile pink and chum salmon in oiled and non-oiled nearshore habitats in Prince William Sound, 1989 and 1990. In *Proceedings of the Exxon Valdez Oil Spill Symposium*. S.D. Rice, R.B. Spies, D.A. Wolfe, and B.A. Wright, eds. Bethesda, MD, USA: American Fisheries Society; Symposium 18; ISBN-10: 0913235954; ISSN: 08922284; pp. 564–578.

Thedinga, J.F., M.G. Carls, J.M. Maselko, R.A. Heintz, and S.D. Rice (2005). Resistance of naturally spawned pink salmon eggs to mechanical shock. *Alaska Fishery Research Bulletin* **11**(1): 37–43. [http://www.adfg.alaska.gov/static/home/library/PDFs/afrb/thedv11n1.pdf]

US Environmental Protection Agency (1985). *Methods for Measuring the Acute Toxicity of Effluents to Freshwater and Marine Organisms*, 3rd edn. Cincinnati, OH, USA: US Environmental Protection Agency, Environmental Monitoring and Support Laboratory; National Service Center for Environmental Publications; EPA/600/4-84/013. [http://nepis.epa.gov (search for the title)]

US Environmental Protection Agency (1988). *Short-Term Methods for Estimating the Chronic Toxicity of Effluents and Receiving Waters to Freshwater and Marine Organisms*, 1st edn. Cincinnati, OH, USA: US Environmental Protection Agency, Environmental Monitoring and Support Laboratory; National Service Center for Environmental Publications; EPA 600/4-86/028. [http://nepis.epa.gov (search for the title)]

Wang, S.Y., J.L. Lum, M.G. Carls, and S.D. Rice (1993). The relationship between growth and total nucleic acids in juvenile pink salmon, *Oncorhynchus gorbuscha*, fed crude oil-contaminated feed. *Canadian Journal of Fisheries and Aquatic Sciences* **50**(5): 996–1001.

Wells, P.G., J.A. Percy, and F.R. Engelhardt (1985). Effects of oil on arctic invertebrates. In *Petroleum Effects in the Arctic Environment*. F.R. Engelhardt, ed. New York, NY, USA, and London, UK: Elsevier Applied Science Publishers, Ltd.; ISBN-13: 9780853343561; ISBN-10: 085334356X; pp. 101–156.

Wertheimer, A.C., N.J. Bax, A.D. Celewycz, M.G. Carls, and J.N. Landingham (1996). Harpacticoid copepod abundance and population structure in Prince William Sound, one year after the *Exxon Valdez* oil spill. In *Proceedings of the Exxon Valdez Oil Spill Symposium*. S.D. Rice, R.B. Spies, D.A. Wolfe, and B.A. Wright, eds. Bethesda, MD, USA: American Fisheries Society; Symposium 18; ISBN-10: 0913235954; ISSN: 08922284; pp. 551–563.

Wertheimer, A.C. and A.G. Celewycz (1996). Abundance and growth of juvenile pink salmon in oiled and non-oiled locations of western Prince William Sound after the *Exxon Valdez* oil spill. In *Proceedings of the Exxon Valdez Oil Spill Symposium*. S.D. Rice, R.B. Spies, D.A. Wolfe, and B.A. Wright, eds. Bethesda, MD, USA: American Fisheries Society; Symposium 18; ISBN-10: 0913235954; ISSN: 08922284; pp. 518–532.

Willette, M. (1996). Impacts of the *Exxon Valdez* oil spill on the migration, growth, and survival of juvenile pink salmon in Prince William Sound. In *Proceedings of the Exxon Valdez Oil Spill Symposium*. S.D. Rice, R.B. Spies, D.A. Wolfe, and B.A. Wright, eds.

Bethesda, MD, USA: American Fisheries Society; Symposium 18; ISBN-10: 0913235954; ISSN: 08922284; pp. 533–550.

Wolfe, D.A., M.J. Hameedi, J.A. Galt, G. Watabayashi, J. Short, C. O'Claire, S. Rice, J. Michel, J.R. Payne, J., Braddock, S. Hanna, and D. Sale (1994). The fate of the oil spilled from the *Exxon Valdez*. *Environmental Science & Technology* **28**(13): 561A–568A.

Wolfe, D.A., M.M. Krahn, E. Casillas, S. Sol, T.A. Thompson, J. Lunz, and K.J. Scott (1996). Toxicity of intertidal and subtidal sediments contaminated by the *Exxon Valdez* oil spill. In *Proceedings of the Exxon Valdez Oil Spill Symposium*. S.D. Rice, R.B. Spies, D.A. Wolfe, and B.A. Wright, eds. Bethesda, MD, USA: American Fisheries Society; Symposium 18; ISBN-10: 0913235954; ISSN: 08922284; pp. 121–139.

CHAPTER THIRTEEN

Pacific herring

Walter H. Pearson, Ralph A. Elston, Karen Humphrey, and Richard B. Deriso

13.1 Introduction

Following the *Exxon Valdez* oil spill in 1989, concern quickly arose about potential effects on Pacific herring (*Clupea pallasii*). As the most abundant forage fish in Prince William Sound (PWS), Pacific herring is a keystone species, consuming zooplankton and providing high-quality prey for birds, marine mammals, and other fish (Spies, 2007). Pacific herring are also commercially important, supporting five fisheries in PWS: two spring fisheries for roe (eggs as food), two spring fisheries for eggs on kelp (a delicacy), and a fall "fish" fishery for bait and food. The four spring fisheries are the first opportunity to conduct commercial fishing after the winter. These five fisheries provided landings (i.e., landed catches) worth about $12 million in the year before the spill (Brady *et al.*, 1991a). And Pacific herring are an important subsistence food for people living in the area.

Herring life history is complex (Fig. 13.1; Blaxter, 1985; Hay, 1985; McQuinn, 1997), and multiple factors – such as ocean conditions, prey availability, predation, and competition – structure herring population dynamics in PWS and the Gulf of Alaska (GOA). Establishing whether and how the spill affected herring required separating spill-related injuries from changes caused by the complex of other factors affecting herring. Moreover, little information about several aspects of herring life history and population dynamics in PWS was available at the time of the spill. Thus, although the overall goal of scientific investigations of herring was to assess the effects of the oil spill, it was also necessary to undertake basic research to fill in the information gaps necessary to assess the spill's effects.

In this chapter, we provide background on herring and the *Exxon Valdez* oil spill, and then consider both effective and problematic aspects of study design and implementation. The herring studies became a quest to separate oil-spill effects from natural and other anthropogenic effects. Active measures to assess oil exposure and to weigh alternative hypotheses were instrumental in the success of that quest. We conclude by highlighting some lessons learned that might be applied to other studies of such keystone species.

13.1.1 Background on Pacific herring and the *Exxon Valdez* oil spill

Studies before the *Exxon Valdez* oil spill had already shown the potential for oil exposure to injure herring eggs (Lindén, 1978; Smith and Cameron, 1979), particularly from direct contact with oil (Pearson *et al.*, 1985). During spawning, Pacific herring deposit

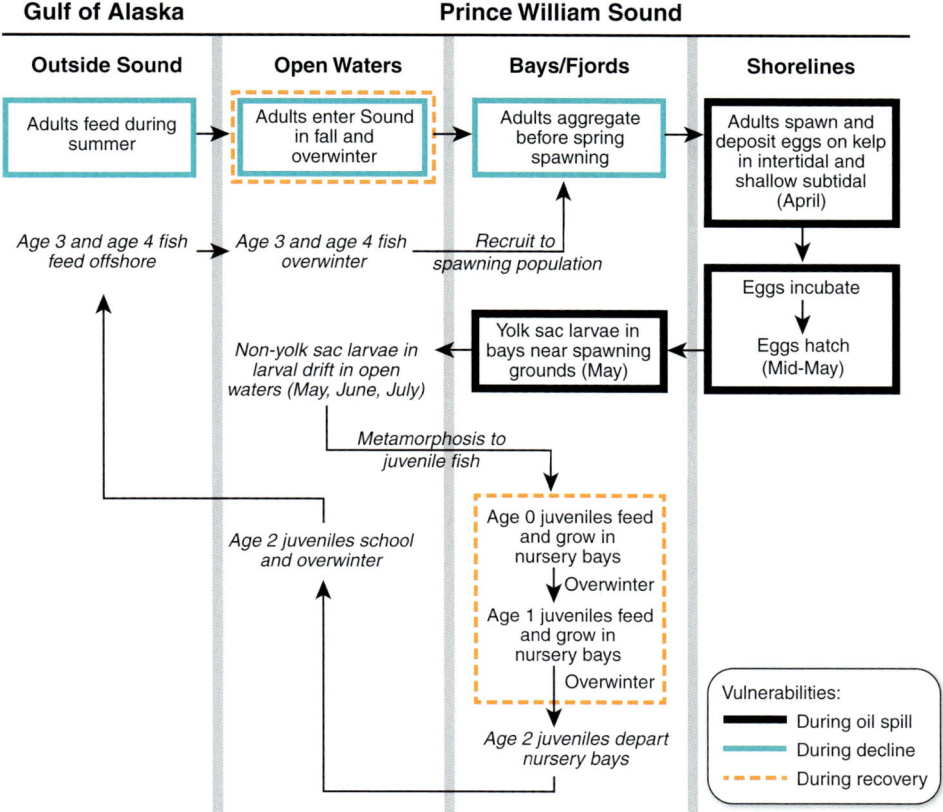

Figure 13.1 Life history of Pacific herring in PWS with vulnerabilities during the oil spill and during the decline and recovery periods. Compiled from Hay (1985), Norcross et al. (2001), Brown et al. (2002), and Pearson et al. (1999, 2012).

their eggs on vegetation in the intertidal and shallow subtidal zones. This behavior, coupled with the potential effects from direct contact with oil and the grounding of oil along beaches, makes the egg stage especially vulnerable in oil spills (Fig. 13.1). Eggs suffer substantial natural mortality from a number of other physical and biological factors (Alderdice and Hourston, 1985; Gunderson and Dygert, 1988), including bird predation (Outram, 1958; Bishop and Green, 2001), wave action (Rooper et al., 1999), temperature (Alderdice and Velsen, 1971; Purcell et al., 1990), salinity (Alderdice and Velsen, 1971), water depth (Taylor, 1971), and thickness of egg deposition (Taylor, 1971; Hourston et al., 1984).

Firm understanding of the distribution and timing of larval drift, metamorphosis, and the first 2 years of juvenile life, however, did not exist for PWS until the early 2000s. Back in 1989, scientists had little knowledge about the distribution and abundance of larval and juvenile herring in PWS. This lack hampered their ability to assess exposure and vulnerability of the larval and juvenile stages.

The timing of the spill in late March, just before spring spawning and the spring fisheries (late March and April), drove the decision to mount immediate studies of the potential effects of the oil spill on herring eggs and larvae. At the same time, the Alaska Department of Fish and Game (ADFG) closed all commercial fisheries in PWS. The herring studies took place during all of the three phases of oil spills described in Chapter 2:

Phase 1 includes the oil release and immediate response, which lasted several months in the case of the *Exxon Valdez* spill. In this phase, herring studies were planned and begun.

Phase 2, the cleanup, lasted about 3 years following the *Exxon Valdez* spill. Studies on the effects of the spill on vulnerable life stages of herring were primarily implemented during this period.

Phase 3 is the recovery period, the years after the cleanup period, which lasted almost 20 years for some studies of the *Exxon Valdez* spill. Normally, studies in the recovery phase emphasize data collection related to scaling and implementing environmental restoration projects. Phase 3 herring studies took a different focus because of two events: the decline of the PWS herring fisheries in 1993, and the subsequent poor recovery from that decline. Initial Phase 3 studies focused on determining the cause of the decline, especially any connection to the oil spill.

13.2 Phase 1: oil release and immediate response

Assessing potential pathways for herring to suffer oil exposure readily identified the eggs as being at high risk. Eggs are spawned in late March and early April and incubate through early to mid-May. Newly hatched herring larvae ("yolk sac" larvae) were thought to aggregate in the shallow subtidal areas of spawning grounds before entering larval drift. Given that the timing and distribution of larval drift was not known in 1989 and that larvae are less exposed to oil in open water, larval sampling focused on the bays where spawning had occurred. Later research would show that after larvae have absorbed their yolk sac (about 1 week post-hatch), the larvae drift in a counter-clockwise nearshore current in PWS (Norcross *et al.*, 2001). Therefore, the source of any larvae captured away from the spawning grounds could not have been known with any certainty.

The distribution of juveniles in PWS was not known at the time of the spill, and no sampling of juveniles was done during the spill or cleanup phases. Research in later years revealed that juvenile herring spend the first 2 years of life in nursery bays (Fig. 13.1; Stokesbury *et al.*, 1999, 2000; Norcross *et al.*, 2001; Brown *et al.*, 2002).

After spawning in late March and April, adult herring depart PWS for summer feeding areas in the GOA and return in the fall to overwinter (Fig. 13.1). Modest sampling of adult herring on and near spawning grounds was undertaken in late April and early May 1989.

The information available in 1989 enabled the herring studies to correctly target egg and larval success. Had more life-history information been available in the first few weeks of the spill, sampling for herring larvae could have been better timed and focused on the pathway for larval drift. Studies in the mid-1990s showed that most larvae drift near the shore in the northern and northeastern regions of PWS (Norcross *et al.*, 2001), not in the open waters that were sampled to a limited extent in 1989 (Norcross and Frandsen, 1996).

Also, information on nursery bays and relative juvenile abundance, which were not known until the late 1990s, would have been beneficial in assessing the relative exposure of juvenile herring in 1989. Of the 143 juvenile herring schools detected by aerial surveys in summer 1996, 34% were in the oiled area scientists assessed in 1989 (Norcross *et al.*, 2001). Ten percent of the 143 schools were along or near Green Island and the Knight Island Archipelago shores that had been moderately to heavily oiled. The remaining 66% of the schools were in the unoiled reference area.

13.3 Phase 2: cleanup

Phase 2 studies emphasized assessing potential injury to herring eggs. Based on studies in 1989 and 1990, *Exxon Valdez* Oil Spill Trustee (hereafter "Trustee") and Exxon studies agreed that any effects on herring eggs from the oil spill were limited to 1989 (Pearson *et al.*, 1995; Brown *et al.*, 1996). Some disagreement remains about how comparable the oiled and reference areas were, how much overlap there was between oiling and spawning, and how specific the criteria were for assessing injury. By using concomitant variables, scientists were able to establish that the 1989 effects on eggs attributable to the spill were localized rather than widespread (Pearson *et al.*, 1995). Effects on herring eggs did not occur beyond 1989, and the localized effects in 1989 did not lead to effects at the population level.

13.3.1 Oiled and reference areas differed in important aspects

Regulations require that injury studies provide a comparison of an assessed area (presumably oiled) and a reference area (presumably unoiled). Study designs in 1989 identified assessed and reference areas to study, based on the trajectory of spilled oil slicks. The trajectory was generally southwest from the *Exxon Valdez*'s grounding point at Bligh Reef (see Map 1, p. v; Chapter 1; Wolfe *et al.*, 1994). The assessed area was south of the northern side of the Naked Island archipelago and west of Rocky Bay on Montague Island, while the reference area was in northern and northeastern PWS (Fig. 13.2).

The selection and use of assessed and reference areas proved problematic. Assessed areas can differ from reference areas in physical characteristics, temperature regimes, and biological influences on herring growth and survival. For example, McGurk and Brown (1996) reported that their findings of smaller larval size and higher mortality from eggs to larvae at assessed sites (versus reference sites) were confounded by temperature differences between the two types of sites. Waves are known to cause high losses of herring eggs (Rooper *et al.*, 1999), and spawn at oiled sites along Montague Point and Graveyard Point on Montague Island were along open coastline, where waves undoubtedly lead to high losses of eggs. Lack of sheltered waters on the open coast of Montague Island probably subjected newly hatched larvae to greater rates of transport away from the spawning location.

13.3.2 Overlap of oiling and spawning was low

Areas where spawn distribution and shoreline oiling overlapped were infrequent (Fig. 13.2). This overlap was identified by annual aerial surveys by ADFG, which determined the distribution of spawning within PWS to support herring stock assessment, and by beach surveys by Shoreline Cleanup Assessment Technique (SCAT) teams, which assessed and mapped the distribution of oil (Chapter 4).

In reviewing these data for 1989–90, Pearson *et al.* (1999) reported that, at the most, 9–10% of the length of shorelines where herring were spawned was also oiled. Only 0.8% of the spawn length (1.3 km of the 158 km of spawn length observed by ADFG) overlapped with heavily oiled shorelines (Pearson *et al.*, 1995). In contrast, Brown *et al.* (1996) used spill trajectory and the accumulation of a toxic component of crude oil, polycyclic aromatic hydrocarbons (PAH), in mussels (Chapter 3) to estimate egg exposure to oil. They used diver surveys to estimate egg biomass and reported that 40–50% of the egg biomass was within the oil spill trajectory and therefore exposed in 1989–90.

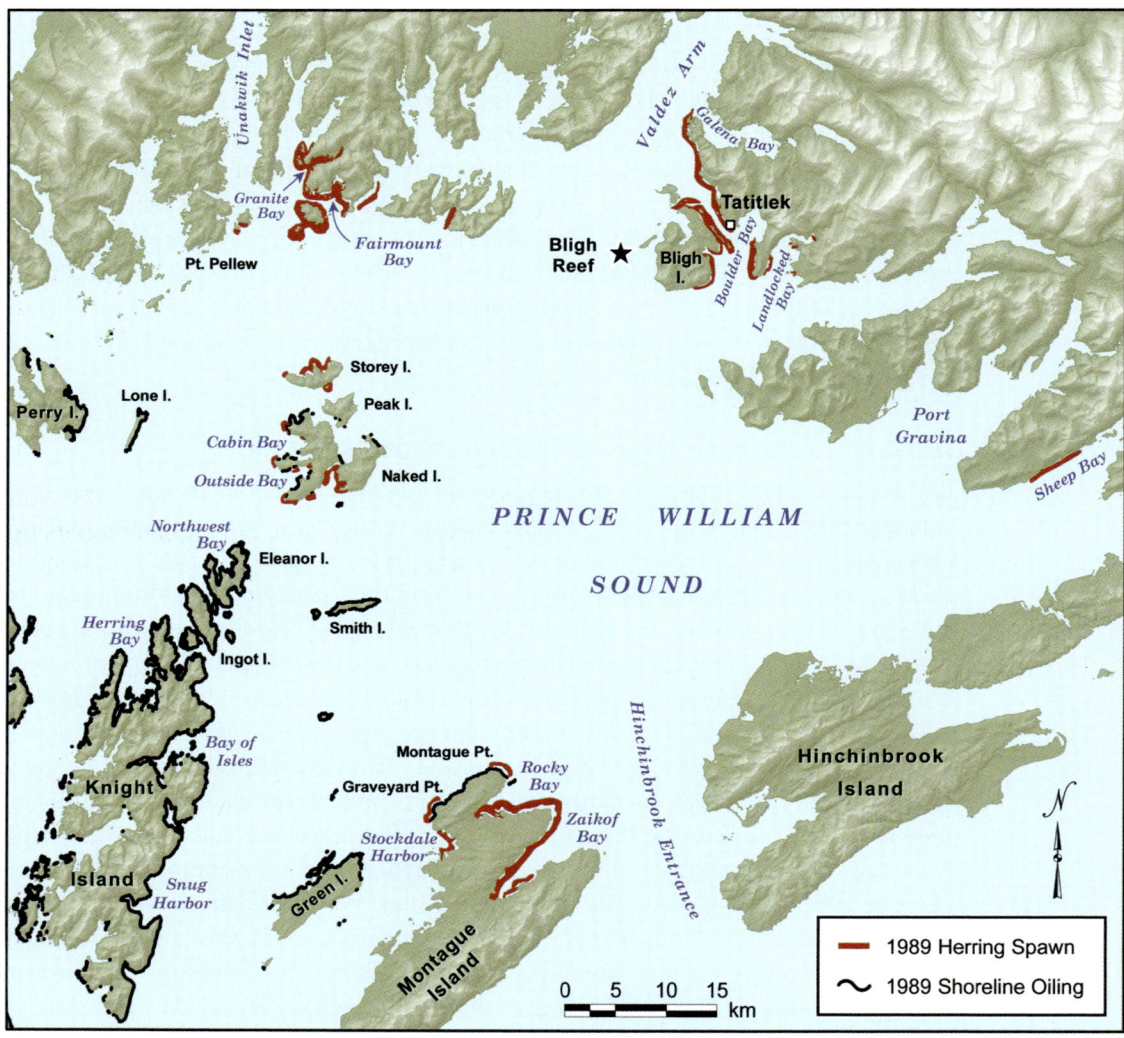

Figure 13.2 Geographical overlap of PWS herring spawn and shoreline oiling in 1989. Note overlap areas on the north end of Montague Island and on Naked Island. Data: spawn, Brady *et al*. (1991a); oiling, Galt *et al*. (1991), Pearson *et al*. (1999).

Both Pearson *et al*. and Brown *et al*. agreed that eggs from Cabin Bay on Naked Island were visibly coated with oil and had accumulated oil concentrations that could be traced to the *Exxon Valdez*. For the 1989 samples of Pearson *et al*. (1995), the highest mean total PAH (TPAH) concentration (~350 ng/g dry weight) occurred in Cabin Bay eggs from the intermediate tidal zone.

The poor agreement between the overlap estimates of Pearson *et al*. (1995) and Brown *et al*. (1996) reflects the use of different scales. Using the broad scale of the oil spill trajectory was appropriate for establishing the boundaries of the reference and assessed areas but did not take into account the patchiness of shoreline oiling and its observed low overlap with herring spawn (Fig. 13.2). Using the finer scales of shoreline oiling and aerial spawn-length maps verified by sampling eggs for PAH analysis, Pearson *et al*. addressed the patchy nature of shoreline oiling and enabled the use of the PAH results as a direct measure of oil exposure in subsequent statistical analyses.

As described below, whatever the precise extent and nature of exposure of herring eggs, such exposure did not produce effects at the population level. The small overlap of oiling and spawn greatly lessened exposure and the extent of potential effects (Rice and Carls, 2007).

13.3.3 Many selected endpoints lacked specificity

The ideal criteria (or endpoints) for determining injury would be morphologic, metabolic, or other abnormalities that are caused almost exclusively by PAH exposure and also can be measured with standardized techniques. However, specific criteria and standardized measurement techniques had not yet been established in 1989, although oil exposure was known to cause abnormalities in larvae hatching from exposed herring eggs (Smith and Cameron, 1979; Pearson et al., 1985). Lacking better methods, scientists used yolk-sac edema, pericardial edema, jaw or craniofacial abnormalities, finfold abnormalities, and spinal defects to assess injury. Unfortunately, many natural stressors induce larval abnormalities (Galkina, 1968; Alderdice and Velsen, 1971; Taylor, 1971; von Westernhagen, 1988; Purcell et al., 1990; Ojaveer, 2006). Lack of endpoint specificity for PAH exposure may lead to erroneous attribution of observed effects to oil exposure.

Methods for evaluating the prevalence and extent of injury were also not standardized in 1989 (Smith and Cameron, 1979; Pearson et al., 1985), so there was considerable potential for subjectivity in assigning categories of severity. Pearson et al. (1995) used the observed frequency of individual abnormalities as well as frequencies of developed eggs, hatched eggs, and live normal larvae (Table 13.1). Hose et al. (1996) used a Graduated Severity Index (GSI) to assign scores for combinations of abnormalities in individual larvae, based on their frequency and severity. GSI scores showed substantial variability that did not align with oiling history. Scores were significantly higher in assessed bays compared to reference bays in 1989 (Hose et al., 1996). In 1991, however, larvae from a reference bay had significantly higher GSI scores than two other bays (one reference and one previously oiled).

Recent studies have investigated a correlation between PAH exposure, early cardiac dysfunction, and pericardial edema in early-life-stage fish (Incardona et al., 2009). Pericardial edema has been identified consistently in a number of laboratory studies of the effects of oil exposure on herring eggs (Lindén, 1978; Smith and Cameron, 1979; Pearson et al., 1985; Carls et al., 1999). However, the specificity of pericardial edema for PAH exposure and standardized measurement techniques need to be established before it can be considered an unambiguous indicator of oil exposure and effects.

Without standardized methods of measurement for highly specific endpoints, conclusions about what caused morphologic abnormalities in herring should be approached conservatively. Where factors other than oil exposure are known to cause abnormalities, directly measuring oil exposure and using known and measureable concomitant variables becomes a requirement rather than an option.

13.3.4 Use of covariants enabled discernment of effects

The studies most successful at distinguishing spill effects were those that directly measured PAH accumulation in herring egg tissue and used such measurements along with covariants in statistical analyses (Table 13.1; Pearson et al., 1995). Localized effects on herring eggs were found in 1989, but no effects were found in 1990 or 1995 (Pearson et al., 1995; Brown et al., 1996; Johnson et al., 1997). In 1989, Pearson et al. (1995) found

Table 13.1 Endpoint response variables, main effects, and covariates used in the statistical analyses of Pearson et al. (1995)

Endpoints as response variables	Main effects	Covariants
1989		
Developed eggs	Reference versus assessed	Tidal zone
Empty egg cases	PAH concentration in eggs-on-kelp	Egg density
Dead larvae		
Abnormal larvae		
Larvae with scoliosis		
Larvae with pericardial edema		
Larvae with abnormal yolk sacs		
Larvae with depleted yolk sacs		
Viable larvae		
1990		
Developed eggs	Reference PWS versus assessed PWS versus Sitka Sound	Tidal zone
Empty egg cases	PAH concentration in eggs-on-kelp	Egg density
Dead larvae		Number of egg layers deposited
Larvae with pericardial edema		Minimum dissolved oxygen in sample upon arrival in laboratory
Viable larvae		Minimum temperature of samples upon arrival
		Difference between temperature at arrival and at collection

that eggs showed decreased development with increased PAH accumulation. One location with decreased egg development was Cabin Bay, where both Brown et al. (1996) and Pearson et al. (1995) detected PAH contamination traceable to *Exxon Valdez* oil. By adding Sitka Sound as another reference area in the 1990 studies, Pearson et al. (1995) showed that significant variation in hatching rates and larval abnormalities can occur in eggs deposited at different tidal elevations in an Alaskan locale completely free of *Exxon Valdez* oil.

13.3.5 No effects at the population level were evident

The effects observed in 1989 at the individual level (Pearson et al. 1995; Brown et al., 1996; Hose et al., 1996; McGurk and Brown, 1996) were confined to portions of PWS and did not translate to discernible effects at the population level (Pearson et al., 1999, 2012). Natural mortality in herring eggs is known to be substantial and to derive from factors other than oil exposure. Estimates of mortality from bird predation range from 30% (Bishop and Green, 2001) to 60% (Outram, 1958). Thick egg deposition can lead to egg mortality as high as 92% (Alderdice and Hourston, 1985) and abnormalities in the hatching larvae can lead to more than 90% being nonviable (Galkina, 1968). Egg loss associated with commercial harvesting is limited because the fisheries are managed by a quota to be no more than 20% of spawning biomass per year (Botz et al., 2010).

Because herring are adapted to high egg loss, population levels can be sustained despite high egg loss. Neither egg biomass nor egg mortality is directly correlated with recruitment or adult spawning biomass (Funk and Sandone, 1990; O'Farrell and Larson,

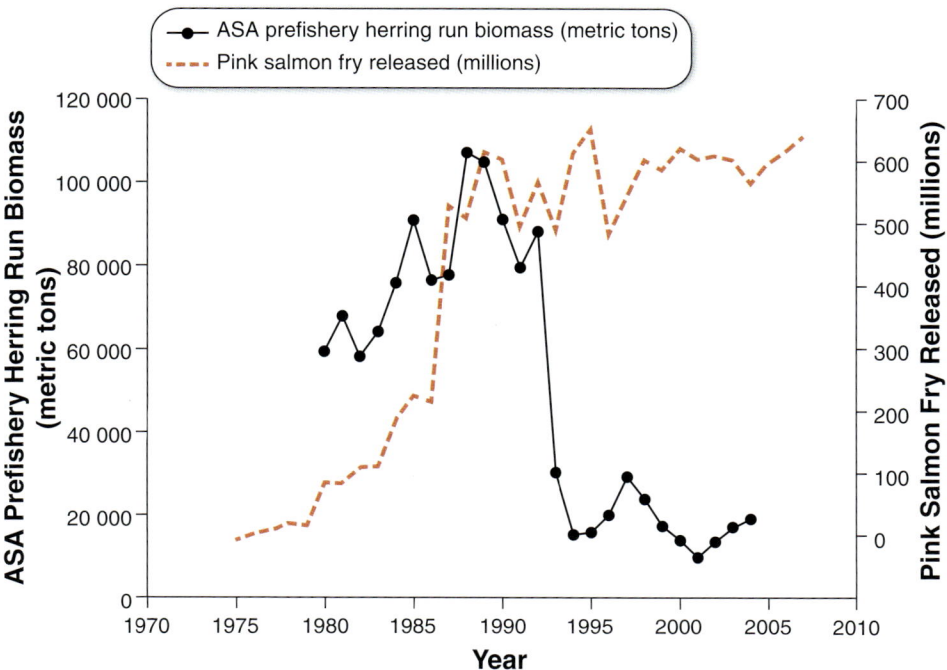

Figure 13.3 Spawning biomass of Pacific herring in PWS and number of juvenile pink salmon released from PWS hatcheries by year in PWS. ASA = age-structured assessment model. Data: Sandone (1988); Brady et al. (1988, 1990, 1991a, b); Biggs et al. (1992); Donaldson et al. (1992, 1993); Botz et al. (2010).

2005). Similarly, high prevalence of larval abnormalities (e.g., skeletal abnormalities in 55% of larvae) has been found to have no correlation with recruitment or herring spawning biomass (Hershberger et al., 2005).

Three observations at the population level provided site-specific evidence that the PWS herring population was not affected by the spill:

1. High biomasses and record high herring landings followed the spill in 1991 and 1992 (Fig. 13.3);
2. The postspill herring year-class composition (Fig. 13.4; Pearson et al., 1995, 1999) showed year classes recruiting as expected. In GOA herring populations, abundant year classes are expected every four years (Williams and Quinn, 2000a, b). Herring spawned in 1984 recruited abundantly before the spill and remained abundant, showing normal mortality in 1989. The 1988 year class was 1 year old at the time of the spill and living in nursery bays but recruited in high numbers as expected in 1991 and 1992 (Pearson et al., 2012). The 1989 year class was in the egg stage at the time of the spill. As the year class following an abundant year class, the abundance of the 1989 year class was expected to be low, and it was (Pearson et al., 1995, 1999); and
3. Subsequent age structured assessment (ASA) modeling did not show elevated mortality in adult herring in 1989 (Deriso et al., 2008; Hulson et al., 2008).

Comparing study results at the individual versus population levels indicates the need for population-level studies to correctly interpret the observations at the individual level.

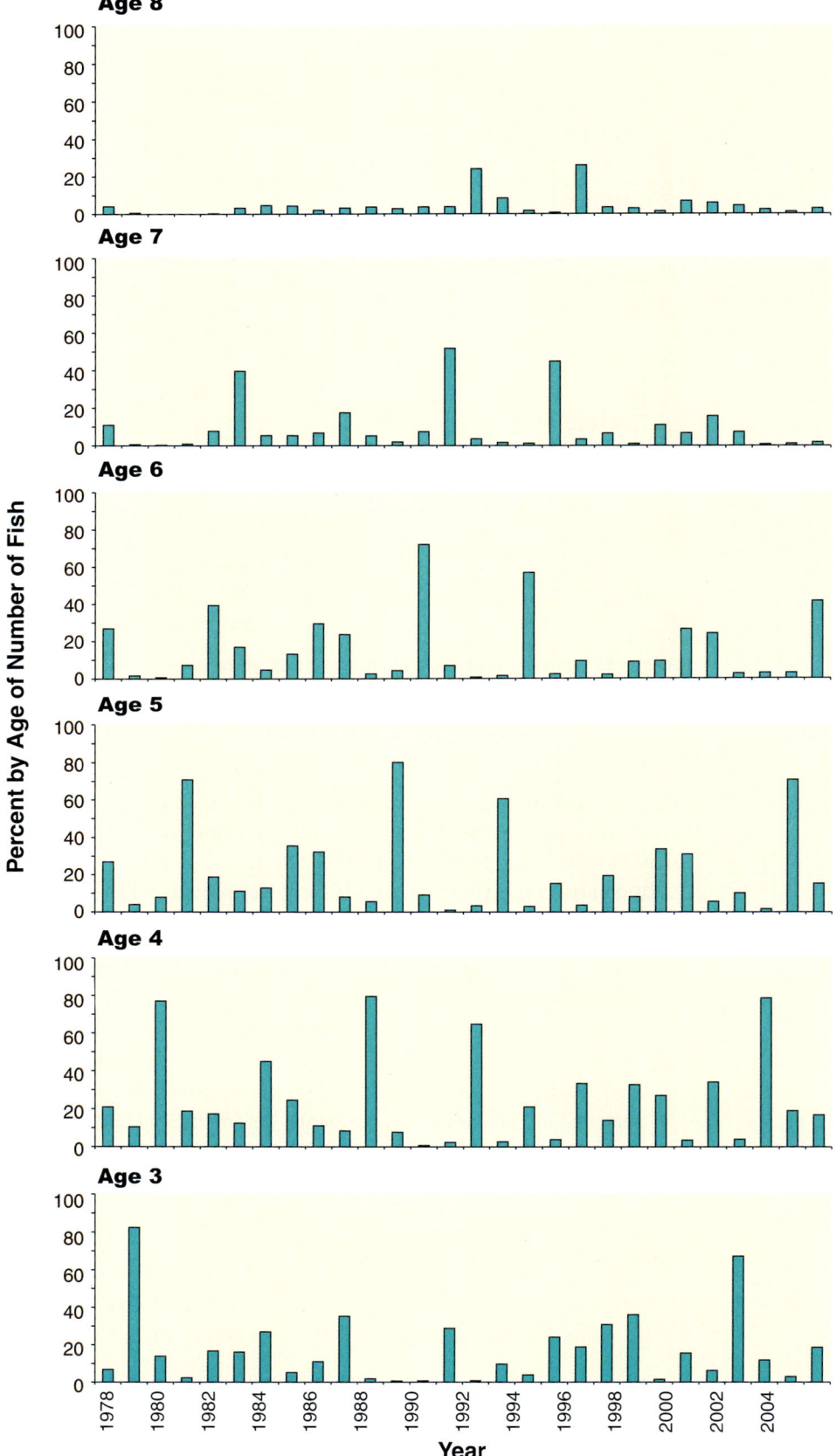

Figure 13.4 Age composition of PWS herring by year. Data: 1978–81, Biggs *et al*. (1992); 1982–2005, Moffitt (2006).

13.4 Phase 3: recovery

During the recovery phase, studies usually focus on activities related to environmental restoration. In the case of herring, however, the focus shifted. In 1993, the PWS herring fisheries declined and did not recover well. This decline and poor recovery led Phase 3 herring studies to emphasize determining the cause of the 1993 decline and, especially, whether there was any connection to the oil spill. After years of studies, it was finally concluded that there was no connection between the spill and the herring decline (Pearson *et al.*, 1999, 2012; Carls *et al.*, 2002; Rice and Carls, 2007; Deriso *et al.*, 2008; Hulson *et al.*, 2008; Elston and Meyers, 2009; Marty *et al.*, 2010), although some public perception to the contrary persists. As the poor recovery of the herring became evident, Phase 3 studies turned to determining the factors involved in the poor recovery (Marty *et al.*, 2010; Pearson *et al.*, 2012). The Phase 3 studies led to the broad recognition that natural and other anthropogenic factors overwhelmed any effects of the *Exxon Valdez* oil spill on the population dynamics of PWS herring.

13.4.1 The decline of herring in Prince William Sound

In the spring of 1993, the PWS spawning biomass was 20% of that expected (Fig. 13.3, Pearson *et al.*, 2012). The decline resulted from an acute adult mortality event between spring 1992 and spring 1993 rather than a recruitment failure (Quinn *et al.*, 2001; Deriso *et al.*, 2008; Hulson *et al.*, 2008).

Pearson *et al.* (1999) examined 14 hypotheses for the decline and found that poor nutritional status, either alone or in combination with disease or other natural factors, was most likely responsible for the 1993 collapse. The high herring population levels following the oil spill (Fig. 13.3), the lack of deviation from the expected age-class composition (Fig. 13.4), and the low levels of oil exposure in the years following the spill indicated that the 1989 oil spill was not a cause of or contributing factor to the decline. (Chapters 3 and 6 detail the rapid decline in water-column PAH concentrations and in shoreline oiling, respectively.)

Using more recent information and a more formal set of criteria, Pearson *et al.* (2012) again examined decline hypotheses (Table 13.2). The recent information included a disease analysis (Elston and Meyers, 2009) and results of three ASA modeling analyses (Deriso *et al.*, 2008; Hulson *et al.*, 2008; Marty *et al.*, 2010). The evaluation criteria were modified from Platt's classic (1964) paper on strong inference and included presence of a significant correlation or constant association, significance in the ASA model, presence of a plausible mechanism, and supporting or contrary evidence. Pearson *et al.* (2012) reaffirmed poor herring condition resulting from poor nutrition as the major factor in the 1993 decline. Disease was identified as a secondary response. There was no evidence that oil exposure, overfishing, foregone harvest, or loss of spawning habitat caused or contributed to the decline. Continued research now supports this widely held view that poor nutrition, not the oil spill, is the best explanation for the PWS herring decline.

13.4.1.1 Further studies changed conclusions about disease as a cause of the decline

Other investigators (Carls *et al.*, 2002; Marty *et al.*, 2003; Rice and Carls, 2007; Hulson *et al.*, 2008; Elston and Meyers, 2009) have reported that the decline most likely derived from poor nutrition, but the extent to which disease is seen as a causal or contributing

Table 13.2 Summary of hypotheses concerning the decline and subsequent poor recovery of PWS herring. Condensed from Pearson et al. (2012)

Decline	
Hypothesis	
Harvest, overfishing	Not a probable cause. Insufficient overfishing.
Harvest, foregone harvest	Not a probable cause. Timing of closed fisheries and decline do not align.
Spawning habitat loss	Not a probable cause. No spawning habitat was lost. Herring have varying spawning-site fidelity. Spawning site changes unrelated to oiled/unoiled conditions.
Disease, *Ichthyophonus hoferi*	Not a probable cause. ASA modeling provides support only when combined with an additional mortality factor. Loss of older age classes was not evident; observed age structure does not support causality. Prevalence pattern is contrary to an epizootic.
Disease, viral hemorrhagic septicemia	Not a cause but a probable secondary response after nutritional or other stress. Unsupported by modeling. VHS mortality primarily for younger fish; 1992–93 mortality was not primarily younger fish, rather age classes 3 to 9+.
Viral hemorrhagic septicemia and the spawn-on-kelp pound fishery	Not a probable cause. Confining herring in pounds (pens) increases prevalence of VHS. Unsupported by modeling. VHS mortality primarily on younger fish; 1992–93 mortality was not primarily younger fish, rather age classes 3 to 9+.
Oil exposure and viral hemorrhagic septicemia	Not a probable cause. Unsupported by modeling. Water-column PAH concentrations insufficient to cause adult mortality. Laboratory experiments failed to link oil exposure and VHS.
Oil exposure	Not a probable cause. Unsupported by modeling. Water-column PAH concentrations insufficient to cause mortality. Timing of exposure and decline not aligned. Exposure during surface air gulping too intermittent in time and space to induce mortality; mechanism not demonstrated.
Competition or predation by pink salmon	Not a probable cause. Modeling supports active role for pink salmon releases in population dynamics, but there is no mechanism for adult mortality.
Predation by sea lions	Not a probable cause. Unsupported by modeling. Sea lion distribution appears to be driven by herring biomass rather than vice versa. Sea lion abundance too small to be a significant source of mortality by predation.
Poor nutrition	A probable cause. Supported by modeling, particularly spring condition. Decreases in weight and condition preceded the decline. Condition in PWS poorer than at Sitka. Source of mortality still not directly demonstrated.
Ocean factors, stochastic population dynamics	A probable cause. Supported by modeling, particularly GOA winter water temperature. Parallel changes in biomass and year-class strength at PWS and Sitka indicate response to GOA-wide environmental changes. Source of mortality still not directly demonstrated.
Recovery	
Hypothesis	
Harvest, overfishing	Not a probable cause. PWS fisheries closed most years since 1993.
Harvesting, foregone harvest	Not a probable cause. No mechanism for lack of strong year classes. Foregone harvest would have led to increased biomass rather than decreased biomass.
Spawning habitat loss	Not a probable cause. Spawning habitat in good condition; no habitat lost. Herring have varying spawning-site fidelity. Spawning site changes unrelated to oiled/unoiled conditions.
Behavioral conservatism	Not a probable cause. If any conservatism occurred, too short-lived to explain time course of lack of strong year classes over 10 years.

Table 13.2 (cont.)

Recovery Hypothesis	
Disease, *Ichthyophonus hoferi*	Probable influence on older fish, adult biomass, but not a cause of poor recruitment; potential secondary response to poor nutrition. Supported by some modeling, not by all. Can cause mortality in older fish; some disruption of age class structure in late 1990s. However, loss of older age classes not consistent postdecline; age structure is as expected in 2000s; loss of older fish not a direct effect on recruitment.
Disease, viral hemorrhagic septicemia	Not a probable cause for lack of strong year classes. Weak to nonsupport by modeling. Affects younger fish; can lead to mortality when fish are stressed. However, not supported by disease analysis of Elston and Meyers (2009).
Viral hemorrhagic septicemia and the spawn-on-kelp pound fishery	Not a probable cause. No pound fishery since the decline.
Oil exposure and viral hemorrhagic septicemia	Not a probable cause. Unsupported by modeling. Water-column PAH concentrations insufficient to cause mortality. Laboratory experiments failed to link oil exposure and VHS.
Predation by humpback whales (predator pit)	Probable contributing factor for reducing adult biomass; improbable influence on recruitment. Whale abundance increasing; overwintering diet appears to be primarily herring. Data are supportive but estimates of consumption have uncertainties; more study is required.
Predation by adult coho salmon	Not a probable cause. Unsupported by modeling. No available evidence that increases in coho salmon have led to increased predation.
Predation by adult pink salmon	Not a probable cause. Adult pink salmon do not feed when returning to PWS.
Predation by juvenile pink salmon	Possible contributing factor but not empirically demonstrated. Supported by modeling. Predation on age-0 herring in July and August is likely.
Competition with juvenile pink salmon	Probable factor, more probable and substantial than ocean factors, but not empirically demonstrated. Strongly supported by modeling. Hatchery juvenile pink salmon releases increased dramatically before the decline and have remained high. Juvenile pink salmon and age-1 herring overlap in space and time and share prey. Field observations show significant disruption of herring feeding when juvenile pink salmon co-occur.
Poor nutrition	Probable factor, in association with ocean factors and juvenile pink salmon competition, but not empirically demonstrated. Supported by modeling, particularly spring condition. Continued low zooplankton abundance. Condition in PWS poorer than at Sitka. Nutrition for overwintering age-0 and age-1 herring is key to year class strength.
Ocean factors, stochastic population dynamics	Probable factor for lack of strong year classes, in association with nutrition and juvenile pink salmon competition, but not empirically demonstrated. Supported by modeling, particularly GOA winter water temperature. No strong year classes at either PWS or Sitka after 1993. Parallel changes in PWS and Sitka indicate response to GOA-wide environmental changes.

factor has continued to evolve. When moribund fish were found in spring 1993, the ADFG Pathology Laboratory identified viral hemorrhagic septicemia (VHS) disease in some fish. Their report cautioned, however, that the decline could not be attributed to this disease agent (Meyers *et al.*, 1994). Additional studies ensued, and other authors (Marty *et al.*, 1998) did attribute the decline to an epizootic of VHS.

Whether or not VHS was actually a cause of the decline became controversial, and this led to Elston and Meyers' (2009) comprehensive review of the role of VHS in PWS herring from 1989 to 2005. They found no support for a VHS effect on PWS herring spawning biomass or recruitment. VHS was not a primary cause of the 1993 decline because:

1. VHS is primarily a disease of juvenile herring, and the 1993 losses were distributed across all age classes – including older fish typically resistant to VHS;
2. There was not enough observed fish mortality to explain the loss. Most herring simply failed to appear at spawning grounds in 1993, so they could not be examined for cause of death; and
3. The poor condition of the herring over the winter of 1992–93 was sufficient to cause the decline.

Acquiring sufficient energy reserves during summer feeding is critical to herring surviving the winter, a season with poor feeding opportunities. Hulson *et al.* (2008) point to low PWS zooplankton abundance in the summer of 1992 as the precursor to poor nutrition, and this low 1992 abundance followed years of generally low zooplankton abundance since 1986 (Pearson *et al.*, 1999, 2012). The designation of the 1993 herring population loss as an "epidemic," which implied a primary infectious cause (e.g., Marty *et al.*, 2003; Marty, 2007), is incorrect.

Another herring pathogen, *Ichthyophonus hoferi*, was also considered as a potential factor in the 1993 herring decline. *I. hoferi* is a protistan pathogen known to be associated with mass mortality and population declines in Atlantic herring (*Clupea harengus*) (Sindermann, 1958; Tibbo and Graham, 1963; Patterson, 1996; Kramer-Schadt *et al.*, 2010). However, this pathogen was eliminated as a cause of the decline because its prevalence in PWS herring fluctuated in a different pattern than that expected if the pathogen were the cause (Pearson *et al.*, 2012). The pathogen had a very low prevalence in 1993, but increased in prevalence after the decline.

13.4.1.2 ASA models with covariates led to understanding which factors were influential

ASA modeling by several investigators was particularly effective in screening hypotheses. All the models showed that the decline was a one-time event, in which mortality of all adult age classes abruptly increased for a year (Quinn *et al.*, 2001; Deriso *et al.*, 2008; Hulson *et al.*, 2008). ASA modeling revealed that interactions with juvenile pink salmon (*Oncorhynchus gorbuscha*) released from PWS hatcheries influenced the population dynamics of PWS herring but did not cause the decline (Deriso *et al.*, 2008; Pearson *et al.*, 2012). Interactions with juvenile pink salmon did prove to be the dominant factor impeding recovery.

ASA modeling did not support the hypothesis proposed by Thomas and Thorne (2003) and Thorne and Thomas (2008) that the decline began in 1989 with the oil spill and was then accelerated by overharvesting. First, the acute mortality event was clearly between 1992 and 1993 in the three ASA models (Quinn *et al.*, 2001; Deriso *et al.*, 2008; Hulson *et al.*, 2008). Adult mortality in 1989 was normal. Second, Hulson *et al.* (2008) examined several scenarios to detect an overharvesting effect and found none.

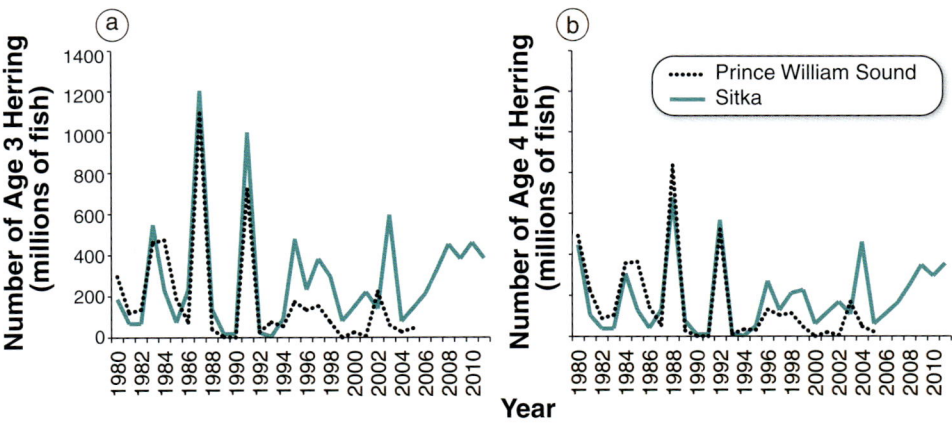

Figure 13.5 Numbers of (a) age-3, and (b) age-4 herring entering the total population at PWS and Sitka Sound by year. Data: PWS, Moffitt (2006); Sitka, Dressel (2011). Since the 2005 Sitka ASA model (Dressel, 2006) was run, ADFG has corrected the underlying age data. The estimated recruitment numbers for Sitka Sound here were drawn from the most recent (2011) ASA model and provided by ADFG (Dressel, 2011). Using the most recent age data influenced the estimated recruitment primarily for the years following 1999. Autocorrelation analysis with the most recent Sitka data shows that the significant pattern of strong year classes emerging every 4 years is evident at Sitka before but not after 1993.

13.4.2 Poor recovery of herring from the decline

Following the 1993 decline, PWS herring fisheries remained closed because of insufficient spawning biomass until 1997 and 1998, when modest harvests were allowed. In 1999, the fisheries were again closed because of insufficient biomass and remain closed as of this writing.

Poor recovery has two components. The first is a lack of the strong year classes that characterized the herring stocks of PWS and the GOA in the 1980s and early 1990s. Strong year classes generally recruit every 4 years and sustain the population and fishery during the intervening 3 years of lower recruitment. Strong year classes have not emerged in PWS (or anywhere else in the GOA) since 1992 (Fig. 13.5; Pearson et al., 2012). Autocorrelation analysis of the numbers of age-3 and age-4 herring showed a significant 4-year pattern in PWS and Sitka Sound before but not after 1993 (Pearson et al., 2012). Similar analysis with the recently updated Sitka data (Dressel, 2011) provides the same conclusion as that in Pearson et al. (2012). For a 4-year lag with age-4 herring, the autocorrelation coefficient (R) is $+.80$ ($P = .004$) before 1993 and $-.51$ ($P = .164$) after. Also, the correlation between PWS and Sitka recruitment ($R = .94$ for age-4, $P < .001$) is significant before 1993 but not after 1993 ($R = .28$, $P = .35$). Full recovery is unlikely without the re-emergence of strong year classes.

The second component of poor recovery is low adult biomass. For example, Sitka Sound also lacks strong year classes after 1992 (Fig. 13.5), but has enough biomass to sustain harvest (Pearson et al., 2012). Hypotheses to explain the poor recovery have to provide plausible mechanisms for one or both of these components.

Pearson et al. (2012) examined 16 factors potentially impeding recovery (Table 13.2). Overfishing, forgone harvest, the eggs-on-kelp fishery, loss of spawning habitat, VHS, predation by adult pink and coho salmon (*Oncorhynchus kisutch*), and direct or indirect oil exposure were eliminated. Disease from *I. hoferi* and predation by humpback whales

(*Megaptera novaengliae*) were judged to be potentially influential. Regional ocean factors and interactions with hatchery juvenile pink salmon were judged to be causal.

13.4.2.1 Oil exposure does not impede recovery

Exposure assessment for the recovery period indicates that water-column PAH concentrations had decreased exponentially from peak levels in 1989 to low values in 1990 and to background levels by 1995 (Boehm *et al.*, 2007; Chapter 3). Therefore, levels were too low to be toxic (Pearson *et al.*, 2012). The ASA modeling specifically tested the influence of oil exposure on herring recruitment: the results do not support any effect (Deriso *et al.*, 2008).

13.4.2.2 Disease from *Ichthyophonus hoferi* may partially impede recovery

Because of increasing prevalence of *I. hoferi* in PWS and the pathogen's known association with herring declines elsewhere, Marty *et al.* (2010) hypothesized that this pathogen could be impeding recovery. Considering the differential distribution of age classes of herring populations, their broad geographic distribution, and the clustering of *I. hoferi* infected fish, very extensive sampling and analysis effort is required to estimate the prevalence and effects of this disease (Holst *et al.*, 1997; Kramer-Schadt *et al.*, 2010). Because of these requirements, the PWS prevalence data need to be assessed with caution. Pearson *et al.* (2012) concluded that *I. hoferi* could be causing adult herring mortality in PWS, but not to the extent likely to affect herring recruitment.

13.4.2.3 Predation by overwintering humpback whales may impede recovery

Humpback whales preying on adult Pacific herring may be reducing PWS adult biomass (Rice, 2008, 2009); effects on recruitment are unlikely, however, because the whales do not prey heavily on juvenile herring (Pearson *et al.*, 2012). The number of whales spending the winter in PWS has increased in recent years, and studies indicate that their overwintering diet does include herring (Rice, 2008, 2009). Two separate studies provide preliminary estimates that overwintering whales may consume 20–30% of the overwintering adult herring in PWS (Rice, 2008; Pearson *et al.*, 2012). Ongoing studies still need to confirm the precise nature and extent of such consumption.

13.4.2.4 Regional ocean factors impede recovery

The absence of strong year classes at PWS, Sitka Sound, and other GOA locations indicates that large-scale ocean processes are affecting herring. The GOA herring stocks have synchronous interannual patterns in recruitment and weight at age that set them apart from stocks in the Bering Sea and British Columbia, which have different interannual patterns (Williams and Quinn, 2000a, b). Ocean processes appear to influence these patterns (Zebdi and Collie, 1995). Poor feeding conditions in the ocean have been linked to poor body condition and decreased adult biomass of herring in San Francisco Bay (California Department of Fish and Game, 2007) and the North Sea (Payne *et al.*, 2009). The continued poor condition of adult herring in PWS suggests that ocean processes over the entire GOA region may be changing the quantity and quality of herring's zooplankton prey.

13.4.2.5 Interactions with hatchery juvenile pink salmon impede recovery

Clearly, herring in PWS are responding to something besides ocean factors. Although both PWS and Sitka Sound lack strong year classes after 1993 (Fig. 13.5), the Sitka Sound herring population has fared better. The biomass at Sitka Sound decreased in 1993, but not abruptly, and had record levels and harvests after 1993. Compared to Sitka Sound,

PWS has more enclosed waters, colder temperatures, a longer overwintering period, and a 600-fold higher level of juvenile pink salmon releases from hatcheries.

In examining factors specific to PWS, Pearson *et al.* (2012) concluded that competitive and predatory interactions between herring ages 0–1 and juvenile pink salmon from hatcheries are likely to be impeding recovery. Competitive and predatory interactions were first raised as an influence on herring population dynamics during the decline studies (Pearson *et al.*, 1999), but they have received more attention in more recent studies (Rice, 2008, 2009; Pearson *et al.*, 2012).

Juvenile pink salmon releases were the most consistent and most statistically significant factor influencing population dynamics in the ASA model of PWS herring (Deriso *et al.*, 2008). Pink salmon releases into PWS rose abruptly in the mid-1980s, reaching over 500 million annually before the sudden decline of herring (Fig. 13.3; see Chapter 12, Fig. 12.1). In 2010, four pink salmon hatcheries in PWS released 647 million juvenile pink salmon; the two hatcheries near Sitka released 1.3 million juvenile pink salmon (White, 2011).

Because the ASA modeling identified the importance of the pink salmon releases but not whether the mechanism was competition or predation, Pearson *et al.* (2012) assessed the potential mechanisms for competition and predation. Some potential for the faster growing juvenile pink salmon to prey upon age-0 herring was found, but the main interaction appeared to be disruption of age-1 herring feeding by juvenile pink salmon. To affirm a competition hypothesis, four conditions need to be fulfilled:

1. *Juvenile salmon and herring have to co-occur*: juvenile pink salmon and age-1 herring co-occur in nearshore areas of bays from late spring through summer (Pearson *et al.*, 2012);
2. *Juvenile salmon and herring have to have dietary overlap*: the observed dietary overlap is sufficient for competition to occur (Sturdevant, 1999);
3. *Food intake by juvenile herring must decrease in the presence of juvenile pink salmon*: field studies have demonstrated substantial reduction in food intake by juvenile herring in the presence of juvenile pink salmon (Sturdevant, 1999); and
4. *Reduced food intake must affect juvenile herring survival*: adequate nutrition is necessary for growth and overwintering survival of juvenile herring and is influential in determining recruitment and year-class strength (Paul and Paul, 1998; Foy and Norcross, 1999, 2001; Brown and Norcross, 2001; Norcross *et al.*, 2001; Norcross and Brown, 2001).

13.4.2.6 Several factors acting together could impede recovery

The factors impeding herring recovery in PWS are not mutually exclusive and probably act together (Pearson *et al.*, 2012).

1. Regional-scale oceanic factors determine the prey field and, in turn, the growth and survival of both juvenile and adult PWS herring;
2. Interactions with pink salmon juveniles released from the hatcheries reduce age-0 herring survival through predation and reduce age-1 growth and survival through food competition and feeding disruption. Effects of restricted feeding on overwintering survival would be particularly detrimental to recruitment of strong herring year classes;
3. Disease as a secondary response following poor nutrition could reduce adult biomass levels; and
4. Overwintering whales could remove large portions of adult herring biomass through predation.

13.5 Lessons learned

- Use life history to identify at-risk resources and vulnerable life stages. Knowledge of life history and potential exposure pathways is necessary to readily identify resources at risk. Conceptual ecosystem models are an important component in risk assessment (Suter, 1999a, b; Chapter 16), but a conceptual ecosystem model specific to PWS was not available until the mid-2000s (Harwell and Gentile, 2006). Such models offer a broad framework within which to identify resource vulnerabilities and the dominant factors affecting exposure and resource condition.

- Select an appropriate scale of resolution for the exposure assessment. Addressing the exposure pathway is mandated by regulations, but broad assessments of exposure can be problematic. Spill trajectory is an important element in study design and sampling plans but is of insufficient resolution, by itself, to assess exposure. Because oil spills produce very patchy distributions of oiling, exposure assessments must also include fine-scale sampling of the water column and sediments. An area-specific conceptual ecosystem model can help focus such fine-scale sampling.

- Assess exposure levels with focused direct measurements. Sampling and analyses for exposure assessment need to address fingerprinting of the oil and other PAH sources and the establishment of background levels. Most importantly, direct measurements of contaminant levels in the tissues of the organisms are needed to establish a path to the contaminant source and tie any adverse changes observed to oil exposure.

- Use exposure assessment in all three phases of oil spill studies. The 1993 decline and poor recovery of the PWS herring populations point out sharply how external factors can influence the condition of a resource in the later phases of an oil spill. Understanding the magnitude and duration of exposure was particularly important in the assessments concerning the decline and poor recovery. Exposure assessment is needed throughout all three phases of oil-spill studies.

- Select appropriate injury criteria, considering life history, potential exposure, and state of the science. The more specific an indicator of toxic effects from oil exposure can be, the higher the confidence that an adverse change in that criterion indicates an injury related to the exposure. Injury criteria need to be appropriate and practical given the life history of the at-risk resource, and need to be used with an understanding of how natural factors can be influential.

- Use covariates to distinguish the influence of natural factors from the influence of the oil spill. The use of covariates as indicators of the influence of natural factors is critical to distinguish localized effects on herring embryos when used with direct measurements of PAH burdens in eggs on kelp. Greater confidence can be placed in an analysis that has taken into account known confounding factors.

- Actively assess alternative hypotheses. The appropriate null hypothesis is no spill effects (Chapter 10). To assume causation just because a spill and a resource change have occurred at the same time is to fall prey to several common cognitive biases in environmental assessments. Miller (1985) describes how attribution errors and selective attention can bias environmental judgments. Active assessment of alternative hypotheses is the remedy for such biases.

- Use ASA and other appropriate models to assess alternatives. With complex problems concerning fish species for which age and catch data are available, ASA models with covariates provide a statistically rigorous approach to testing which factors are influential (Quinn and Deriso, 1999). Other models based on age, size, or life-stage structure can be used with other resource species. Maunder and Deriso (2011) recently applied a stage-structured model with environmental covariates to separate the factors significantly affecting delta smelt (*Hyposmesus transpacifica*) from a complex of natural and anthropogenic candidate factors. Hoyle and Maunder (2004) and Baillie *et al*. (2009) used models to address the causes of population declines in a marine mammal and a bird, respectively.

- Seek biologically plausible mechanisms. ASA models helped explain the influence of hatchery releases of juvenile pink salmon on herring, but concluding that the hatchery releases were causal required examining the mechanisms, their plausibility, and the positive and negative evidence for the mechanisms. Scientists must challenge assumptions in models and seek to fully understand the modeled processes.

- Avoid a single-species perspective. A break-through to understanding resource population dynamics only occurs when the perspective shifts from a single species in isolation to that of a species having complex interactions with other ecosystem components (e.g., herring in combination with pink salmon, humpback whales, and ocean factors). Improved risk and injury assessments, as well as more effective fisheries management, are being called to move away from the single-species perspective to examine resource species and stressors in their ecosystem context (Suter, 1999a, b; Dickey-Collas *et al*., 2010). The ecosystem perspective is especially needed during the recovery phase, when natural and other anthropogenic factors intrude on the recovery process.

REFERENCES

Alderdice, D.F. and A.S. Hourston (1985). Factors influencing development and survival of Pacific herring (*Clupea harengus pallasi*) eggs and larvae to beginning of exogenous feeding. *Canadian Journal of Fisheries and Aquatic Sciences* **42**(1): 56–58.

Alderdice, D.F. and F.P.J. Velsen (1971). Some effects of salinity and temperature on early development of Pacific herring (*Clupea pallasi*). *Journal of the Fisheries Research Board of Canada* **28**(10): 1545–1562.

Baillie, S.R., S.P. Brooks, R. King, and L. Thomas (2009). Using a state-space model of the British song thrush *Turdus philomelos* population to diagnose the causes of population decline. In *Modeling Demographic Processes in Marked Populations. Environmental and Ecological Statistics, Volume 3*. D.L. Thomson, E.G. Cooch, and M.J. Conroy, eds. New York, NY, USA: Springer; ISBN-10: 0387781501; ISBN-13: 9780387781501; pp. 541–562.

Biggs, E.D., B.E. Haley, and J.M. Gilman, eds (1992). *Historic Database for Pacific Herring in Prince William Sound, Alaska, 1973–1991*. Anchorage, AK, USA: Alaska Department of Fish and Game; Regional Information Report 2C91–11. [http://www.sf.adfg.state.ak.us/fedaidpdfs/RIR.2C.1991.11.pdf]

Bishop, M.A. and S.P. Green (2001). Predation on Pacific herring (*Clupea pallasi*) spawn by birds in Prince William Sound, Alaska. *Fisheries Oceanography* **10**(Supplement s1): 149–158.

Blaxter, J.H.S. (1985). The herring: a successful species? *Canadian Journal of Fisheries and Aquatic Sciences* **42**(S1): s21–s30.

Boehm, P.D., J.M. Neff, and D.S. Page (2007). Assessment of polycyclic aromatic hydrocarbon exposure in the waters of Prince William Sound after the *Exxon Valdez* oil spill: 1989–2005. *Marine Pollution Bulletin* **54**(3): 339–356.

Botz, J., G. Hollowell, J. Bell, R. Brenner, and S. Moffitt (2010). *2009 Prince William Sound Area Finfish Management Report.* Anchorage, AK, USA: Alaska Department of Fish and Game, Divisions of Sport Fish and Commercial Fisheries; Fishery Management Report 10–55. [http://www.adfg.alaska.gov/FedAidPDFs/FMR10-55.pdf]

Brady, J., S. Morstad, E. Simpson, and E. Biggs (1991a). *Prince William Sound Management Area Annual Finfish Management Report 1989.* Cordova, AK, USA: Alaska Department of Fish and Game; Regional Information Report 2C90–07. [http://www.adfg.alaska.gov/FedAidPDFs/RIR.2C.1990.07.pdf]

Brady, J., S. Morstad, E. Simpson, and E. Biggs (1991b). *Prince William Sound Area 1990 Annual Finfish Management Report.* Anchorage, AK, USA: Alaska Department of Fish and Game; Regional Information Report 2C91–14. [http://www.adfg.alaska.gov/FedAidPDFs/RIR.2C.1991.14.pdf]

Brady, J.A., K. Schultz, E. Simpson, E. Biggs, S. Sharr, and K. Robertson (1990). *Prince William Sound Area Annual Finfish Management Report 1988.* Cordova, AK, USA: Alaska Department of Fish and Game; Regional Information Report 2C90–02. [http://www.adfg.alaska.gov/FedAidPDFs/RIR.2C.1990.02.pdf]

Brady, J.A., S. Sharr, K. Roberson, F.M. Thompson, and D. Crawford (1988). *Prince William Sound Area Annual Finfish Management Report 1987.* Cordova, AK, USA: Alaska Department of Fish and Game. [http://www.sf.adfg.state.ak.us/FedAidPDFs/AMR.CF.PWS.1988.pdf]

Brown, E.D. and B.L. Norcross (2001). Effect of herring egg distribution and environmental factors on year-class strength and adult distribution: Preliminary results from Prince William Sound, Alaska. In *Proceedings of the 18th Lowell Wakefield Fisheries Symposium, Herring 2000: Expectations for a New Millennium, February 23–26, 2000, Anchorage, Alaska.* F. Funk, J. Blackburn, D. Hay, A.J. Paul, R. Stephenson, R. Toresen, and D. Witherell, eds. Fairbanks, AK, USA: University of Alaska Sea Grant; Publication AK-SG-01–04; ISBN-10: 1566120705; pp. 335–345.

Brown, E.D., B.L. Norcross, and J.W. Short (1996). Introduction to studies on the effects of the (*Exxon Valdez*) oil spill on early life history stages of Pacific herring, *Clupea pallasi*, in Prince William Sound, Alaska. *Canadian Journal of Fisheries and Aquatic Sciences* **53**(10): 2337–2342.

Brown, E.D., J. Seitz, B.L. Norcross, and H.P. Huntington (2002). Ecology of herring and other forage fish as recorded by resource users of Prince William Sound and the Outer Kenai Peninsula, Alaska. *Alaska Fishery Research Bulletin* **9**(2): 75–101.

California Department of Fish and Game (2007). Pacific herring commercial fishing regulations, 2007. In *California Code of Regulations.* Sacramento, CA, USA: California

Department of Fish and Game; Schedule No. 98052052; Sections 163, 163.1, 163.5, and 164; Title 14.

Carls, M.G., G.D. Marty, and J.E. Hose (2002). Synthesis of the toxicological impacts of the *Exxon Valdez* oil spill on Pacific herring (*Clupea pallasi*) in Prince William Sound, Alaska, USA. *Canadian Journal of Fisheries and Aquatic Sciences* **59**(1): 153–172.

Carls, M.G., S.D. Rice, and J.E. Hose (1999). Sensitivity of fish embryos to weathered crude oil. Part I: Low-level exposure during incubation causes malformations, genetic damage, and mortality in larval Pacific herring (*Clupea pallasi*). *Environmental Toxicology and Chemistry* **18**(3): 491–493.

Deriso, R.B., M.N. Maunder, and W.H. Pearson (2008). Incorporating covariates into fisheries stock assessment models with application to Pacific herring. *Ecological Applications* **18**(5): 1270–1286.

Dickey-Collas, M., R.D.M. Nash, T. Brunel, C.J.G. van Damme, C.T. Marshall, M.R. Payne, A. Corten, A.J. Geffen, M.A. Peck, E.M.C. Hatfield, N.T. Hintzen, K. Enberg, L.T. Kell, and E.J. Simmonds (2010). Lessons learned from stock collapse and recovery of North Sea herring: a review. *ICES Journal of Marine Science: Journal du Conseil* **67**(9): 1875–1886.

Donaldson, W., S. Morstad, E. Simpson, and E. Biggs (1992). *Prince William Sound Management Area 1991 Annual Finfish Management Report.* Anchorage, AK, USA: Alaska Department of Fish and Game; Regional Information Report 2A92–09. [http://www.sf.adfg.state.ak.us/fedaidpdfs/RIR.2A.1992.09.pdf]

Donaldson, W., S. Morstad, E. Simpson, J. Wilcock, and S. Sharr (1993). *Prince William Sound Management Area 1992 Annual Finfish Management Report.* Anchorage, AK, USA: Alaska Department of Fish and Game; Regional Information Report 2A93–12. [http://www.sf.adfg.state.ak.us/fedaidpdfs/RIR.2A.1993.12.pdf]

Dressel, S. (2006). *ADF&G Age-Structured Assessment Excel Spreadsheet Program for Sitka Sound Herring.* Douglas, AK, USA: Alaska Department of Fish and Game, Division of Commercial Fisheries.

Dressel, S. (2011). *ADF&G Data on Estimated Numbers of Age 3 and Age 4 Herring from the 2011 Age-Structured Assessment Model for Sitka Sound Herring.* Douglas, AK, USA: Alaska Department of Fish and Game, Division of Commercial Fisheries.

Elston, R.A. and T.R. Meyers (2009). Effect of viral hemorrhagic septicemia virus on Pacific herring in Prince William Sound, Alaska, from 1989 to 2005. *Diseases of Aquatic Organisms* **83**(3): 223–246.

Foy, R.J. and B.L. Norcross (1999). Spatial and temporal variability in the diet of juvenile Pacific herring (*Clupea pallasi*) in Prince William Sound, Alaska. *Canadian Journal of Zoology* **77**(5): 697–706.

Foy, R.J. and B.L. Norcross (2001). Temperature effects on zooplankton assemblages and juvenile herring feeding in Prince William Sound, Alaska. In *Proceedings of the 18th Lowell Wakefield Fisheries Symposium, Herring 2000: Expectations for a New Millennium, February 23–26, 2000, Anchorage, Alaska.* F. Funk, J. Blackburn, D. Hay, A.J. Paul, R. Stephenson, R. Toresen, and D. Witherell, eds. Fairbanks, AK, USA: University of Alaska Sea Grant; Publication AK-SG-01-04; ISBN-10: 1566120705; pp. 21–35.

Funk, F. and G. Sandone (1990). *Catch-age Analysis of Prince William Sound, Alaska, Herring, 1973–1998*. Juneau, AK, USA: Alaska Department of Fish and Game; Fishery Research Bulletin 90–01. [http://www.sf.adfg.state.ak.us/fedaidpdfs/frb.1990.01.pdf]

Galkina, L.A. (1968). Survival of herring eggs and larvae in the White Sea spawning grounds during a period of massive spawning. *Voprosy Ikhtiologii* **8**: 544–551.

Galt, J.A., W.J. Lehr, and D.L. Payton (1991). Fate and transport of the *Exxon Valdez* oil spill. Part 4. *Environmental Science & Technology* **25**(2): 202–209.

Gunderson, G.R. and P.H. Dygert (1988). Reproductive effort as a predictor of natural mortality rate. *ICES Journal of Marine Science: Journal du Conseil* **44**(2): 200–209.

Harwell, M.A. and J.H. Gentile (2006). Ecological significance of residual exposures and effects from the *Exxon Valdez* oil spill. *Integrated Environmental Assessment and Management* **2**(3): 204–246.

Hay, D.E. (1985). Reproductive biology of Pacific herring (*Clupea harengus pallasi*). *Canadian Journal of Fisheries and Aquatic Sciences* **42**(S1): s111–s126.

Hershberger, P.K., N.E. Elder, J. Wittouck, K. Stick, and R.M. Kocan (2005). Abnormalities in larvae from the once-largest Pacific herring population in Washington State result primarily from factors independent of spawning location. *Transactions of the American Fisheries Society* **134**(2): 326–337.

Holst, J.C., A.G.V. Salvanes, and T. Johansen (1997). Feeding, *Ichthyophonus* sp. infection, distribution and growth history of Norwegian spring-spawning herring in summer. *Journal of Fish Biology* **50**(3): 652–664.

Hose, J.E., M.D. McGurk, G.D. Marty, D.E. Hinton, E.D. Brown, and T. Baker (1996). Sublethal effects of the *Exxon Valdez* oil spill on herring embryos and larvae: morphological, cytogenetic, and histopathological assessments, 1989–1991. *Canadian Journal of Fisheries and Aquatic Sciences* **53**(10): 2355–2365.

Hourston, A.S., H. Rosenthal, and H. Von Westernhagen (1984). *Viable Hatch from Eggs of Pacific Herring (*Clupea harengus pallasi*) Deposited at Different Intensities in a Variety of Substrates*. Nanaimo, BC, Canada: Department of Fisheries and Oceans; Canadian Technical Report of Fisheries and Aquatic Sciences No. 1274. [http://www.dfo-mpo.gc.ca/Library/69710.pdf]

Hoyle, S.D. and M.N. Maunder (2004). A Bayesian integrated population dynamics model to analyze data for protected species. *Animal Biodiversity and Conservation* **27**(1): 247–266.

Hulson, P.-J.F., S.E. Miller, T.J. Quinn, G.D. Marty, S.D. Moffitt, and F. Funk (2008). Data conflicts in fishery models: incorporating hydroacoustic data into the Prince William Sound Pacific herring assessment model. *ICES Journal of Marine Science: Journal du Conseil* **65**(1): 25–43.

Incardona, J.P., M.G. Carls, H.L. Day, C.A. Sloan, J.L. Bolton, T.K. Collier, and N.L. Scholz (2009). Cardiac arrhythmia is the primary response of embryonic Pacific herring (*Clupea pallasi*) exposed to crude oil during weathering. *Environmental Science & Technology* **43**(1): 201–207.

Johnson, S.W., M.G. Carls, R.P. Stone, C.C. Brodersen, and S.D. Rice (1997). Reproductive success of Pacific herring, *Clupea pallasi*, in Prince William Sound, Alaska, six years after the *Exxon Valdez* oil spill. *Fishery Bulletin* **95**(4): 748–761.

Kramer-Schadt, S., J. Christian, and D. Skagen (2010). Analysis of variables associated with the *Ichthyophonus hoferi* epizootics in Norwegian spring spawning herring, 1992. *Canadian Journal of Fisheries and Aquatic Sciences* **67**(11): 1862–1873.

Lindén, O. (1978). Biological effects of oil on early development of the Baltic herring *Clupea harengus membras*. *Marine Biology* **45**(3): 273–283.

Marty, G.D. (2007). Evidence that the *Exxon Valdez* oil spill did not cause the 1993 disease epidemic in the Pacific herring population of Prince William Sound, Alaska. In *Prince William Sound Herring: An Updated Synthesis of Population Declines and Lack of Recovery*. S.D. Rice and M.G. Carls, eds. Juneau, AK, USA: National Oceanic and Atmospheric Administration, National Marine Fisheries Service, Auke Bay Laboratory; *Exxon Valdez* Oil Spill Restoration Project 050794 Final Report; pp. 2.1–2.10. [http://www.evostc.state.ak.us/Files.cfm?doc=/Store/FinalReports/2005-050794-Final.pdf&]

Marty, G.D., E.F. Freiberg, T.R. Meyers, J. Wilcock, T.B. Farver, and D.E. Hinton (1998). Viral hemorrhagic septicemia virus, *Ichthyophonus hoferi*, and other causes of morbidity in Pacific herring *Clupea pallasi* spawning in Prince William Sound, Alaska, USA. *Diseases of Aquatic Organisms* **32**(1): 15–40.

Marty, G.D., P.-J.F. Hulson, S.E. Miller, T.J. Quinn, II, S. Moffitt, and R. Merizon (2010). Failure of population recovery in relation to disease in Pacific herring. *Diseases of Aquatic Organisms* **90**(1): 1–14.

Marty, G.D., T.J. Quinn, II, G. Carpenter, T.R. Meyers, and N.H. Willits (2003). Role of disease in abundance of a Pacific herring (*Clupea pallasi*) population. *Canadian Journal of Fisheries and Aquatic Sciences* **60**(10): 1258–1265.

Maunder, M.N. and R.B. Deriso (2011). A state-space multistage life cycle model to evaluate population impacts in the presence of density dependence: Illustrated with application to delta smelt (*Hyposmesus transpacifica*). *Canadian Journal of Fisheries and Aquatic Sciences* **68**(7): 1285–1306.

McGurk, M.D. and E.D. Brown (1996). Egg-larval mortality of Pacific herring in Prince William Sound, Alaska, after the *Exxon Valdez* oil spill. *Canadian Journal of Fisheries and Aquatic Sciences* **53**(10): 2343–2354.

McQuinn, I.H. (1997). Metapopulations and the Atlantic herring. *Reviews in Fish Biology and Fisheries* **7**(3): 297–329.

Meyers, T.R., S. Short, K. Lipson, W.N. Batts, J.R. Winton, J. Wilcock, and E.D. Brown (1994). Association of viral hemorrhagic septicemia virus with epizootic hemorrhages of the skin in Pacific herring *Clupea harengus pallasi* from Prince William Sound and Kodiak Island, Alaska, USA. *Diseases of Aquatic Organisms* **19**(1): 27–37.

Miller, A. (1985). Psychological biases in environmental judgments. *Journal of Environmental Management* **20**(3): 231–243.

Moffitt, S. (2006). *ADF&G Age-structured Assessment Excel Spreadsheet Program and Data for Prince William Sound Herring*. Cordova, AK, USA: Alaska Department of Fish and Game, Division of Commercial Fisheries.

Norcross, B.L. and E.D. Brown (2001). Estimation of first-year survival of Pacific herring from a review of recent stage-specific studies. In *Proceedings of the 18th Lowell Wakefield Fisheries Symposium, Herring 2000: Expectations for a New Millennium, February 23–26, 2000, Anchorage, Alaska*. F. Funk, J. Blackburn, D. Hay, A.J. Paul, R. Stephenson,

R. Toresen, and D. Witherell, eds. Fairbanks, AK, USA: University of Alaska Sea Grant; Publication AK-SG-01–04; ISBN-10: 1566120705; pp. 535–558.

Norcross, B.L., E.D. Brown, R.J. Foy, M. Frandsen, S.M. Gay, T.C. Kline, D.M. Mason, E.V. Patrick, A.J. Paul, and K.D.E. Stokesbury (2001). A synthesis of the life history and ecology of juvenile Pacific herring in Prince William Sound, Alaska. *Fisheries Oceanography* **10**(Supplement s1): 42 –57.

Norcross, B.L. and M. Frandsen (1996). Distribution and abundance of larval fishes in Prince William Sound, Alaska, during 1989 after the *Exxon Valdez* oil spill. In *Proceedings of the* Exxon Valdez *Oil Spill Symposium*. S.D. Rice, R.B. Spies, D.A. Wolfe, and B.A. Wright, eds. Bethesda, MD, USA: American Fisheries Society; Symposium 18; ISBN-10: 0913235954; ISSN: 08922284; pp. 463–486.

O'Farrell, M.R. and R.J. Larson (2005). Year-class formation in Pacific herring (*Clupea pallasi*) estimated from spawning-date distributions of juveniles in San Francisco Bay, California. *Fishery Bulletin* **103**(1): 130–141.

Ojaveer, E. (2006). On the external and parental effects in early development of herring (*Clupea pallasi*) at the NE Kamchatka. *Fisheries Research* **81**(1): 1–8.

Outram, D.N. (1958). *The Magnitude of Herring Spawn Losses due to Bird Predation on the West Coast of Vancouver Island*. Nanaimo, BC, Canada: Fisheries Research Board of Canada, Pacific Biological Station Progress Report No.111: pp. 9–13.

Patterson, K.R. (1996). Modeling the impact of disease-induced mortality in an exploited population: The outbreak of the fungal parasite (*Ichthyophonus hoferi*) in the North Sea herring (*Clupea harengus*). *Canadian Journal of Fisheries and Aquatic Sciences* **53**(12): 2870–2887.

Paul, A.J. and J.M. Paul (1998). Comparisons of whole body energy content of captive fasting age zero Alaskan Pacific herring (*Clupea pallasi* Valenciennes) and cohorts overwintering in nature. *Journal of Experimental Marine Biology and Ecology* **226**(1): 75–86.

Payne, M.R., E.M.C. Hatfield, M. Dickey-Collas, T. Falkenhaug, A. Gallego, J. Gröger, P. Licandro, M. Llope, P. Munk, C. Röckmann, J.O. Schmidt, and R.D.M. Nash (2009). Recruitment in a changing environment: the 2000s North Sea herring recruitment failure. *ICES Journal of Marine Science: Journal du Conseil* **66**(2): 272–277.

Pearson, W.H., R.B. Deriso, R.A. Elston, S. Hook, K. Parker, and J. Anderson (2012). Hypotheses concerning the decline and poor recovery of Pacific herring in Prince William Sound, Alaska. *Reviews in Fish Biology and Fisheries* **22**(1): 95–135.

Pearson, W.H., R.A. Elston, R.W. Bienert, A.S. Drum, and L.D. Antrim (1999). Why did the Prince William Sound, Alaska, Pacific herring (*Clupea pallasi*) fisheries collapse in 1993 and 1994? Review of hypotheses. *Canadian Journal of Fisheries and Aquatic Sciences* **56**(4): 711–737.

Pearson, W.H., E. Moksness, and J.R. Skalski (1995). A field and laboratory assessment of oil spill effects on survival and reproduction of Pacific herring following the *Exxon Valdez* spill. In Exxon Valdez *Oil Spill: Fate and Effects in Alaskan Waters*. P.G. Wells, J.N. Butler, and J.S. Hughes, eds. Philadelphia, PA, USA: American Society for Testing and Materials; ASTM Special Technical Publication 1219; ISBN-10: 0803118961; pp. 626–661.

Pearson, W.H., D.L. Woodruff, S.L. Kiesser, G.W. Fellingham, and R.A. Elston (1985). *Oil Effects on Spawning Behavior and Reproduction in Pacific Herring (Clupea harengus pallasi)*. Washington DC, USA: American Petroleum Institute, API Publication 4412; ISBN-10: 9996949974; ISBN-13: 9789996949975.

Platt, J.R. (1964). Strong inference. *Science* 146(3642): 347–353.

Purcell, J.E., D. Grosse, and J.J. Grover (1990). Mass abundance of abnormal Pacific herring larvae at a spawning ground in British Columbia. *Transactions of the American Fisheries Society* 119(3): 463–469.

Quinn, T.J., II, and R.B. Deriso (1999). *Quantitative Fish Dynamics*. New York, NY, USA: Oxford University Press; ISBN-10: 0195076311.

Quinn, T.J., II, G.D. Marty, J. Wilcock, and M. Willette (2001). Disease and population assessment of Pacific herring in Prince William Sound, Alaska. In *Proceedings of the 18th Lowell Wakefield Fisheries Symposium, Herring 2000: Expectations for a New Millennium, February 23–26, 2000, Anchorage, Alaska*. F. Funk, J. Blackburn, D. Hay, A.J. Paul, R. Stephenson, R. Toresen, and D. Witherell, eds. Fairbanks, AK, USA: University of Alaska Sea Grant; Publication AK-SG-01–04; ISBN-10: 1566120705; pp. 363–379.

Rice, S. (2008). *Significance of Whale Predation on Natural Mortality Rate of Pacific Herring in Prince William Sound*. Juneau, AK, USA: National Oceanic and Atmospheric Administration, National Marine Fisheries Service, Auke Bay Laboratory; *Exxon Valdez* Trustee Council Restoration Project 080804 Annual Report. [http://www.evostc.state.ak.us/Files.cfm?doc=/Store/AnnualReports/2008–080804-Annual.pdf&]

Rice, S. (2009). *Significance of Whale Predation on Natural Mortality Rate of Pacific Herring in Prince William Sound*. Anchorage, AK, USA: National Oceanic and Atmospheric Administration, National Marine Fisheries Service, Auke Bay Laboratory; *Exxon Valdez* Oil Spill Trustee Council Restoration Project 090804 Annual Report. [http://www.evostc.state.ak.us/Files.cfm?doc=/Store/AnnualReports/2009–090804-Annual.pdf&]

Rice, S. and M.G. Carls (2007). Executive summary. In *Prince William Sound Herring: An Updated Synthesis of Population Declines and Lack of Recovery*. S. Rice and M.G. Carls, eds. Juneau, AK, USA: National Oceanic and Atmospheric Administration, National Marine Fisheries Service, Auke Bay Laboratory; *Exxon Valdez* Trustee Council Restoration Project 050794 Final Report; pp. 9–22. [http://www.evostc.state.ak.us/Files.cfm?doc=/Store/FinalReports/2005–050794-Final.pdf&]

Rooper, C.N., L.J. Haldorson, and T.J. Quinn, II (1999). Habitat factors controlling Pacific herring (*Clupea pallasi*) egg loss in Prince William Sound, Alaska. *Canadian Journal of Fisheries and Aquatic Sciences* 56(6): 1133–1142.

Sandone, G. (1988). *Age, Sex, and Size Composition of Pacific Herring Sampled from the Prince William Sound Management Area, 1984–1987*. Anchorage, AK, USA: Alaska Department of Fish and Game; Regional Information Report 2A88–08. [http://www.adfg.alaska.gov/FedAidPDFs/RIR.2A.1988.08.pdf]

Sindermann, C.J. (1958). An epizootic in Gulf of Saint Lawrence fishes. In *Transactions of the 23rd North American Wildlife Conference, March 3–5, 1958, St. Louis, Missouri*. J.B. Trefethen, ed. Washington DC, USA: Wildlife Management Institute; pp. 349–360.

Smith, R.L. and J.A. Cameron (1979). Effect of water soluble fraction of Prudhoe Bay crude oil on embryonic development of Pacific herring. *Transactions of the American Fisheries Society* **108**(1): 70–75.

Spies, R.B., ed. (2007). *Long-Term Ecological Changes in the Northern Gulf of Alaska.* Amsterdam, The Netherlands: Elsevier; ISBN-10: 0444529608; ISBN-13: 9780444529602.

Stokesbury, K.D.E., R.J. Foy, and B.L. Norcross (1999). Spatial and temporal variability in juvenile Pacific herring, *Clupea pallasi*, growth in Prince William Sound, Alaska. *Environmental Biology of Fishes* **56**(4): 409–418.

Stokesbury, K.D.E., J. Kirsh, E.D. Brown, G.L. Thomas, and B.L. Norcross (2000). Spatial distributions of Pacific herring, *Clupea pallasi*, and walleye pollock, *Theragra chalcogramma*, in Prince William Sound, Alaska. *Fisheries Bulletin* **98**(2): 400–409.

Sturdevant, M.V. (1999). *Forage Fish Diet Overlap, 1994–1996.* Juneau, AK, USA: National Oceanic and Atmospheric Administration, National Marine Fisheries Service, Auke Bay Laboratory; *Exxon Valdez* Oil Spill Trustee Council Restoration Project 97163C Final Report. [http://www.evostc.state.ak.us/Files.cfm?doc=/Store/FinalReports/1997-97163C-Final.pdf&]

Suter, G.W., II (1999a). Developing conceptual models for complex ecological risk assessments. *Human and Ecological Risk Assessment* **5**(2): 375–396.

Suter, G.W., II (1999b). A framework for the assessment of ecological risks from multiple activities. *Human and Ecological Risk Assessment.* **5**(2): 397–413.

Taylor, F.H.C (1971). Variation in hatching success in Pacific herring (*Clupea pallasi*) eggs with water depth, temperature, salinity, and egg mass thickness. *Rapport et Procès-Verbaux des Réunions du Conseil International pour l'Exploration de la Mer* **160**: 34–41.

Thomas, G.L. and R.E. Thorne (2003). Acoustical-optical assessment of Pacific herring and their predator assemblage in Prince William Sound, Alaska. *Aquatic Living Resources* **16**(03): 247–253.

Thorne, R.E. and G.L. Thomas (2008). Herring and the "Exxon Valdez" oil spill: An investigation into historical data conflicts. *ICES Journal of Marine Science: Journal du Conseil* **65**(1): 44–50.

Tibbo, S.N. and T.R. Graham (1963). Biological changes in herring stocks following an epizootic. *Journal of Fish Research Board of Canada* **20**(2): 435–449.

von Westernhagen, H. (1988). Sublethal effects of pollutants on fish eggs and larvae. In *Fish Physiology. The Physiology of Developing Fish: Eggs and Larvae.* W.S. Hoar and D.J. Randall, eds. San Diego, CA, USA: Academic Press; ISBN-10: 0123504333; Volume 11, Part A; pp. 253–346.

White, B. (2011). *Alaska Salmon Fisheries Enhancement Program 2010 Annual Report.* Juneau, AK, USA: Alaska Department of Fish and Game; Fishery Management Report 11-04. [http://www.adfg.alaska.gov/FedAidPDFs/FMR11-04.pdf]

Williams, E.H. and T.J. Quinn, II (2000a). Pacific herring, *Clupea pallasi*, recruitment in the Bering Sea and North-East Pacific Ocean, I: Relationships among different populations. *Fisheries Oceanography* **9**(4): 285–299.

Williams, E.H. and T.J. Quinn, II (2000b). Pacific herring, *Clupea pallasi*, recruitment in the Bering Sea and North-East Pacific Ocean, II: Relationships to environmental variables and implications for forecasting. *Fisheries Oceanography* **9**(4): 300–315.

Wolfe, D.A., J.J. Hammedi, J.A. Galt, G. Watabayashi, J. Short, C. O'Claire, S. Rice, J. Michel, J.R. Payne, J. Braddock, S. Hanna, and D. Sale (1994). The fate of the oil spilled from the *Exxon Valdez*. *Environmental Science & Technology* **28**(13): 560A–568A.

Zebdi, A. and J.S. Collie (1995). Effect of climate on herring (*Clupea pallasi*) population dynamics in the Northeast Pacific Ocean. In *Climate Change and Northern Fish Populations, Conference, October 19–24, 1992, Victoria, British Columbia*. R.J. Beamish, ed. Ottawa, ON, Canada: NRC Research Press; Canadian Special Publication in Fisheries and Aquatic Sciences 121; ISBN-10: 0660157802; pp. 277–290.

CHAPTER FOURTEEN

Oil and marine birds in a variable environment

John A. Wiens, Robert H. Day, and Stephen M. Murphy

14.1 Introduction

Marine birds are among the most conspicuous elements of coastal ecosystems. Their variety, abundance, visibility, and behavior resonate with the public. They are also particularly vulnerable to oil spills. Most marine-bird species[1] forage at or beneath the water's surface, where they are at risk of exposure to floating oil. Additionally, many feed and nest along shorelines where floating oil can accumulate, and some form large breeding colonies that are susceptible to oil spills. Over 200 000 marine birds died as a direct consequence of the *Exxon Valdez* oil spill, and images of oiled birds dominated media coverage. Federal regulations requiring documentation of injuries to natural resources from oil spills (see Chapter 1) prompted multiple studies of marine birds, fueling debates about short- and long-term effects of the spill and subsequent recovery.

Determining how an oil spill (or any large environmental disturbance) affects marine birds is complicated by variation in the environment and in how birds respond to that environment. Coastal environments vary substantially from place to place and time to time. At high-latitude locations such as Prince William Sound (PWS), diurnal and monthly changes in tides; seasonal variations in storms, water temperatures, and productivity; and multiyear changes in ocean temperature and circulation such as El Niño events and the Pacific Decadal Oscillation can have dramatic effects. At any time, marine birds may be responding to a host of environmental factors over a range of scales, from the availability of suitable nesting sites along a small stretch of shoreline to the abundance of prey over a broad region or to variation in hemispheric-scale forces affecting the ocean. Because these influences do not disappear when an oil spill occurs, they and any associated environmental variations must be accounted for when assessing spill effects. Population changes attributed to an oil spill must be rigorously tested and documented, not just inferred.

Our objective in this chapter is to synthesize the results of scientific studies of the effects of the *Exxon Valdez* oil spill on marine birds that occur in and use the marine

---

[1] Seaducks, loons, grebes, tubenoses, cormorants, eagles and coastal raptors, shorebirds, gulls and terns, jaegers, alcids, kingfishers, and corvids.

Oil in the Environment: Legacies and Lessons of the Exxon Valdez *Oil Spill*, ed. J. A. Wiens. Published by Cambridge University Press. © Cambridge University Press 2013.

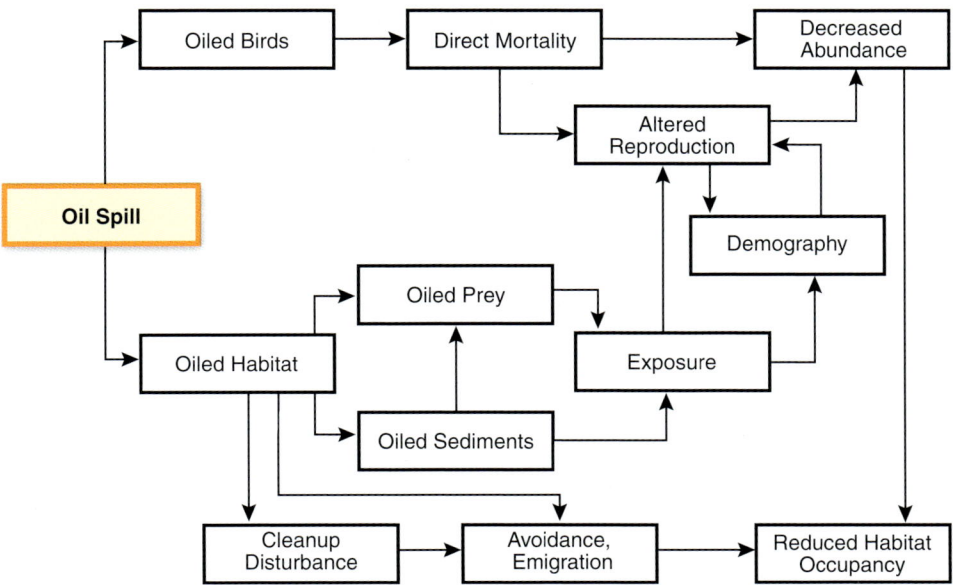

Figure 14.1 Major pathways of direct and indirect effects of a marine oil spill on marine birds.

and coastal environment in PWS and the Gulf of Alaska (GOA), particularly in the context of environmental variation. We do not review the scores of studies conducted on marine birds after the *Exxon Valdez* spill (see online bibliography). Instead, we summarize and evaluate major findings and conclusions, particularly as they bear on the following questions:

- What were the immediate, acute effects[2] of the oil spill on marine birds?
- When did recovery occur, and were there any long-term, chronic effects?
- How does variation among species affect analyses?
- How can we evaluate cause–effect relationships in a variable marine environment?
- What have we learned that can inform future assessments of large environmental accidents?

14.2 Acute effects

Scientific studies of the effects of the *Exxon Valdez* spill on marine birds began within weeks of the spill, one group supported by Exxon and another by the Alaska and federal governments (and, after its formation in 1989, the *Exxon Valdez* Oil Spill Trustee Council; hereafter, "Trustees").

The initial focus of both the Exxon and governmental studies was on establishing the magnitude and scope of immediate, acute spill effects. Spilled oil can get directly onto birds or into their habitat and food, and cleanup-related disturbances can adversely affect birds (Fig. 14.1). These oiling mechanisms largely operate on individuals

---

[2] "Acute" effects are short in duration but often severe, whereas "chronic" effects are persistent but often occur at a low level. In the context of an oil spill, acute effects occur immediately and over a few months or a year or two (or sometimes more) following the spill; chronic effects may continue for a decade or more or become evident only after some time (see Chapter 2).

Figure 14.2 Common and thick-billed murres accounted for nearly three-fourths of the immediate mortality of marine birds from the *Exxon Valdez* oil spill. (Photo: Robert H. Day)

(e.g., ingestion of oil by preening, changes in foraging behavior, increased susceptibility to hypothermia) but are manifested at the population level if sufficient numbers of individuals are affected. Rather than address these mechanisms, however, most studies instead focused on the consequences at the population level. Some of these consequences might be detected immediately, whereas others could take some time to appear.

Acute effects can be evaluated in terms of three general consequences: direct mortality and changes in abundance, reduced reproductive success or productivity, and changes in habitat occupancy (Fig. 14.1).

14.2.1 Mortality and abundance

The most obvious acute effect of the *Exxon Valdez* oil spill was thousands of dead birds. To determine the magnitude of this direct mortality, a major effort was made to collect oiled bird carcasses. Over 41 000 dead birds were recovered, of which 36 000 were identified and cataloged (Piatt and Ford, 1996). Although this tally may have included some birds that died of natural causes, Piatt *et al.* (1990) estimated that ~30 000 died directly from oiling. The majority (72%) of the carcasses were murres[3] (Ford *et al.*, 1991; Fig. 14.2). Most murre fatalities occurred along the Kenai Peninsula and near Kodiak Island, where birds were assembling before breeding (Boersma *et al.*, 1995). Most of the other carcasses were of other alcids (puffins, guillemots, and murrelets), shearwaters, seaducks, and larids (gulls, terns, and jaegers). Although PWS was the epicenter of the spill, the carcass count in PWS was ~3400 birds (Piatt *et al.*, 1990) – just 10% of that in the GOA.

---

[3] Scientific names of species are given in Table 14.1.

Converting from carcass counts to overall spill-mortality estimates is tricky and full of uncertainties. Carcasses at sea may be lost to scavengers, sink before striking land, or be carried out to sea. Even if a carcass is beached, it may be swept out to sea again, scavenged, buried, deposited on a beach that is never searched, or simply missed by an observer searching the beach. Models have been developed to incorporate these factors (Ford et al., 1991, 1996), and "drift experiments" that involve releasing tagged carcasses and recording their retrieval rates are often used to quantify model parameters (e.g., Bibby and Lloyd, 1977; Ford et al., 1996; Byrd et al., 2009).

Based on such models and experiments, the best estimates of total direct mortality from the spill were 250 000–375 000 birds, with extremes of 100 000–690 000 birds (Ford et al., 1996; Piatt and Ford, 1996). These estimates support the conclusion that the *Exxon Valdez* spill resulted in an unprecedented mortality of marine birds from an oil spill (Piatt and Lensink, 1989).

Different models, with different assumptions and using different parameters, produce different mortality estimates. For example, Patten et al. (2000) used carcass counts and the modeling approach of Ford et al. (1996) to estimate that 423 harlequin ducks died in PWS from direct exposure to oil. Trustee researchers killed another 132 harlequins in PWS for bioassays, so perhaps 555 harlequin ducks died in PWS as a consequence of the spill and associated research. Estimates based on a different modeling approach that incorporated projections of survival, reproduction, and movement, but considered just females, suggested that ~400 female harlequin ducks might have died in PWS as a consequence of the spill (Iverson and Esler, 2010).

Placing such mortality estimates in context requires relating them to the size of the total population. This is not easily done, however, for little prespill information about populations in the spill area was available. Estimates of the effects of immediate mortality on populations are most accurate for murres, which suffered the greatest mortality and which had been censused at breeding colonies for years (Sowls et al., 1978; US Fish and Wildlife Service, 2008). The ~185 000 murres estimated to have been killed by the spill represented ~41% of the prespill population in the spill area (Piatt and Anderson, 1996). Piatt et al. (1990) estimated that the largest murre colony in the area, on East Amatuli Island in the Barren Islands (see Map 2, p. vi), lost at least 50% of the breeding adults to the spill.

Although many murres died directly from the spill, several studies showed that numbers at breeding colonies in the spill path were either unaffected or were only somewhat reduced. Reasoning that such substantial mortality should be readily detectable, Boersma et al. (1995) surveyed in 1990–92 the same locations on East Amatuli that they had studied during the 1970s and 1980s. Overall counts of murres were similar after the spill to counts before it, with an increase of ~15% from 1990 to 1991 and then remaining unchanged in 1992. Erikson (1995) similarly found that murre numbers in 1991 at 32 colonies in the GOA were generally within historical ranges, with no significant differences associated with proximity to the spill path. Finally, Piatt and Anderson (1996) concluded that environmental changes in the 1970s to early 1990s had a greater effect on murre numbers than did the spill. They surveyed 16 murre colonies in the northern GOA and saw population declines between pre- and postspill periods, in contrast to the surveys of Boersma et al. (1995) and Erikson (1995). The greatest declines, however, occurred outside of the spill area and appeared to be regional, rather than specific to oiled areas.

What might explain the contradiction between high observed mortality yet little or no apparent population effect? The spill occurred when murres were aggregating on the water before moving to breeding colonies, so some of the birds that died may have been headed for colonies beyond the spill area. Alternatively, missing birds in affected populations may have been replaced by individuals from the large at-sea pool of nonbreeding and prebreeding individuals or by birds emigrating from other, larger colonies beyond the spill area (Boersma et al., 1995). Colonies in the Semidi Islands immediately southwest of Kodiak Island, for example, contain over a million murres (Hatch and Hatch, 1983).

Studies that evaluated immediate spill effects on the abundance of multiple marine-bird species found population effects in some cases but not in others and were often inconsistent in their conclusions about the same species (Table 14.1). Given the magnitude of the *Exxon Valdez* spill, one might expect most species to exhibit negative effects and most studies to agree in their assessments. Although 30 of the 52 species in Table 14.1 were reported to have been negatively affected in at least one study, only pelagic cormorants/"cormorants" and black oystercatchers were categorized as negatively affected in all four studies. Further, seven taxa with multiple studies were considered negatively affected in only a single study.

Two before-after-control-impact (BACI; Skalski and McKenzie, 1982) studies found that most bird species did not suffer significant effects. Klosiewski and Laing (1994) compared nearshore surveys in PWS from March, July, and August 1989–91 with prespill surveys conducted in 1972–73 (Dwyer et al., 1976) and 1984–85 (Irons et al., 1988). Of 28 species or species-groups (e.g., loons, cormorants, scoters) evaluated,[4] nine (32%) declined significantly more in oiled than in unoiled areas (Table 14.1). Murphy et al. (1997) compared midsummer 1989–91 surveys from 10 bays in PWS with the prespill surveys of Irons et al. (1988); of the 11 taxa evaluated, three (27%) showed negative effects (Table 14.1).

In contrast to these BACI analyses, Day et al. (1995, 1997b) and McKnight et al. (2008) compared postspill abundances of species among areas with different oiling levels (see Wiens and Parker, 1995, and Parker and Wiens, 2005, for discussion of study designs). Day and his coworkers used data from year-round surveys in 1989–91 to assess abundance patterns of 42 species in the same 10 study bays considered by Murphy et al. (1997); 19 species (45%) showed negative effects attributable to the spill (Table 14.1). A subsequent analysis, based on a longer time-series for the same study bays but restricted to midsummer, assessed 24 species and recorded negative effects on 10 (42%) (Wiens et al., 2004). McKnight et al. (2008) restricted their analysis to 20 species that were judged by previous studies (e.g., Klosiewski and Laing, 1994; Day et al., 1997; Murphy et al., 1997; Irons et al., 2000) as having been negatively affected by the spill; species and species groups that were considered not affected or recovered were not re-evaluated. On this basis, McKnight et al. (2008) reported that all 20 species continued to show negative spill effects through 2007.

Although these studies all used boat-based surveys, they differed in locations, seasons, numbers of years, analytical procedures, levels of statistical significance, and focus. Little wonder, then, that agreement among the studies was so low. The studies of

---

[4] Hereafter, "species."

Table 14.1 Comparison of the results of several studies that assessed the effects of the *Exxon Valdez* oil spill on marine birds in Prince William Sound, Alaska, 1989–2007. "Historical comparisons" involve comparisons of prespill surveys with postspill surveys; "postspill surveys" are based on comparisons of areas with different levels of oiling (Day et al., 1995; Wiens et al., 2004) or of oiled with unoiled reference areas (McKnight et al., 2008). * = species not included in study; none = no evidence of an effect; negative = statistically significant negative effect; positive = statistically significant positive effect; effects for which there was no evidence of recovery at the end of the survey period are denoted in bold.

Species/taxon	Scientific name	Historical comparisons		Postspill comparisons	
		Murphy et al. (1997)[a]	Klosiewski and Laing (1994)[b]	Day et al. (1995); Wiens et al. (2004)[c]	McKnight et al. (2008)[d]
Canada goose	*Branta canadensis*	*	*	None	*
American wigeon	*Anas americana*	*	*	None	*
Mallard	*Anas platyrhynchos*	*	*	Negative[e]	*
Green-winged teal	*Anas crecca*	*	*	None	*
Harlequin duck	*Histrionicus histrionicus*	None	Negative[e]	Negative[e]	**Negative**[e, f]
Surf scoter	*Melanitta perspicillata*	*	*	None	*
White-winged scoter	*Melanitta fusca*	*	*	Positive[e, f]	*
Black scoter	*Melanitta americana*	*	*	Negative[e]	*
"Scoters"	*Melanitta* spp.	*	**Negative**[f]	*	**Negative**[f, g]
Long-tailed duck	*Clangula hyemalis*	*	None	None	*
Bufflehead	*Bucephala albeola*	*	None	Negative[e, f]	Negative[e]
Common goldeneye	*Bucephala clangula*	*	*	None	*
Barrow's goldeneye	*Bucephala islandica*	*	*	Negative[e, f]	*
"Goldeneyes"	*Bucephala* spp.	*	None	*	**Negative**[e, f]
Common merganser	*Mergus merganser*	None	*	Negative	*
Red-breasted merganser	*Mergus serrator*	*	*	Negative[e, f]	*
"Mergansers"	*Mergus* spp.	*	None	*	**Negative**[e, f]
Common loon	*Gavia immer*	*	*	None	**Negative**[f, g]
"Loons"	*Gavia* spp.	*[h]	**Negative**[f]	*	**Negative**[f, g]
Horned grebe	*Podiceps auritus*	*	*	Negative[e, f]	*
Red-necked grebe	*Podiceps grisegena*	*	*	Negative[e, f]	*
"Grebes"	*Podiceps* spp.	None	None	*	Negative[e]
Fork-tailed storm-petrel	*Oceanodroma furcata*	*	None	None	*
Double-crested cormorant	*Phalacrocorax auritus*	*	*	None	*
Red-faced cormorant	*Phalacrocorax urile*	*	*	None	*

Table 14.1 (cont.)

Species/taxon	Scientific name	Historical comparisons		Postspill comparisons	
		Murphy et al. (1997)[a]	Klosiewski and Laing (1994)[b]	Day et al. (1995); Wiens et al. (2004)[c]	McKnight et al. (2008)[d]
Pelagic cormorant	*Phalacrocorax pelagicus*	**Negative**[f]	*	Negative[e, f]	*
"Cormorants"	*Phalacrocorax* spp.	*	**Negative**[e]	*	**Negative**[e, g]
Great blue heron	*Ardea herodias*	*	*	Negative[e, f]	*
Bald eagle	*Haliaeetus leucocephalus*	None	None	Negative[e, f]	**Negative**[e, f]
Black oystercatcher	*Haematopus bachmani*	**Negative**[f]	**Negative**[e, f]	Negative[e, f]	**Negative**[f]
Spotted sandpiper	*Actitis macularius*	*	*	None	*
Wandering tattler	*Tringa incana*	*	*	Negative[f]	*
Red-necked phalarope	*Phalaropus lobatus*	*	None	None	*
"Shorebirds"	Charadriidae or Scolopacidae	*	None	*	*
Pomarine jaeger	*Stercorarius pomarinus*	*	*	None	*
"Jaegers"	*Stercorarius* spp.	*	None	*	*
Black-legged kittiwake	*Rissa tridactyla*	**Positive**[f]	None	Positive[e, f]	**Negative**[e, f]
Bonaparte's gull	*Chroicocephalus philadelphia*	*	None	*	*
Mew gull	*Larus canus*	None	**Negative**[f]	Negative[f]	**Negative**[e, f]
Herring gull	*Larus argentatus*	*	None	None	*
Glaucous-winged gull	*Larus glaucescens*	None	None	Negative[e, f]	**Negative**[e, g]
"Gulls"	*Rissa* spp., *Chroicocephalus* spp., or *Larus* spp.	*	None	*	*
Arctic tern	*Sterna paradisaea*	*	**Negative**[f]	None	*
"Terns"	*Onychoprion* spp. or *Sterna* spp.	*	*	*	**Negative**[f]
Common murre	*Uria aalge*	*	*	Negative[e, f]	**Negative**[e, f]
"Murres"	*Uria* spp.	*	None	*	*
Pigeon guillemot	*Cepphus columba*	**Negative**[f]	**Negative**[e]	Positive	**Negative**[e, f]
Marbled murrelet	*Brachyramphus marmoratus*	None	None	Negative[e]	**Negative**[e, f]
Kittlitz's murrelet	*Brachyramphus brevirostris*	*	*	*	**Negative**[f]
"Murrelets"	*Brachyramphus* spp.	*	None	*	*
Parakeet auklet	*Aethia psittacula*	*	None	*	*
Horned puffin	*Fratercula corniculata*	*	None	*	*

Table 14.1 (cont.)

Species/taxon	Scientific name	Historical comparisons		Postspill comparisons	
		Murphy et al. (1997)[a]	Klosiewski and Laing (1994)[b]	Day et al. (1995); Wiens et al. (2004)[c]	McKnight et al. (2008)[d]
Tufted puffin	*Fratercula cirrhata*	*	None	Positive[e, f]	*
Belted kingfisher	*Megaceryle alcyon*	*	*	None	*
Steller's jay	*Cyanocitta stelleri*	*	*	None	*
Black-billed magpie	*Pica hudsonia*	*	*	None	*
Northwestern crow	*Corvus caurinus*	*	Negative[f]	Negative[e]	Negative[e, g]
Common raven	*Corvus corax*	*	*	None	*

[a] Murphy et al. compared 1984–85 prespill surveys during July/August (Irons et al., 1988) with 1989–91 postspill surveys.
[b] Klosiewski and Laing compared 1972–73 prespill surveys during March, July, and August (Dwyer et al., 1976) with 1989–91 postspill surveys.
[c] Day et al. and Wiens et al. analyzed 1989–2001 postspill, multiseason surveys.
[d] McKnight et al. analyzed 1989–1991, 1993, 1994, 1996, 1998, 2000, 2005, and 2007 postspill surveys during March and July.
[e] Effect detected in winter (includes fall for Day et al.).
[f] Effect detected in summer (includes spring for Day et al.).
[g] This taxon also was determined to have a positive effect in another season.

Day et al. (1995, 1997b) and Klosiewski and Laing (1994), for example, differed in how oiling levels were defined (quantitatively versus qualitatively) and in the selection of study areas (e.g., exclusion versus inclusion of glaciated fjords in the reference area).

These differences indicate how difficult it is to derive consistent conclusions about the effects of an oil spill on birds, even in a massive spill with so much mortality. They also illustrate the importance of using an appropriate study design (see Chapter 10).

14.2.2 Reproductive performance

Reproductive performance (breeding phenology and reproductive success) can be influenced by a variety of spill-associated mechanisms (Fig. 14.1).

Reproductive studies initially focused on murres because of the magnitude of their mortality. Boersma et al. (1995) had studied murre reproductive success and timing on East Amatuli during the 1970s in a 25-m^2 plot (an area that might include up to 275 breeding pairs). They returned to survey the same plot after the spill in 1990–92. Reproductive success was high in 1978 and 1991, intermediate in 1979 and 1990, and low in 1992. Time-lapse photographs of other areas in this breeding colony taken in 1991 indicated that reproductive success was well within the historical range (Boersma et al., 1995). The photographs also indicated that the timing of breeding might differ by as much as 10 days between plots less than 20 m apart. Thus, murre nesting phenology and success varied substantially among years and even within a colony, so any oil-spill effects would be superimposed on variations in a dynamic marine environment (Piatt and Anderson, 1996).

Another study that observed negative effects on murre reproduction after the spill did not include comparisons with good prespill data. After observing low reproductive success and delayed breeding at some murre colonies in 1989–1990, Nysewander et al.

(1993) speculated that, because the spill occurred early in the breeding season, adult breeders suffered the most mortality because they return to colonies earlier than do subadults. After they died, the open breeding sites would then be occupied by inexperienced birds, which typically breed later and are less productive than adults. Because breeding was delayed, Fry (1993, p. 32) asserted that winter storms in 1990–92 "swept more than 100 000 young chicks off the cliffs to their deaths." These speculations lacked empirical support (Piatt and Anderson, 1996).

Other studies investigated acute effects of the spill on reproduction in several other species: black-legged kittiwakes (Irons, 1996), black oystercatchers (Sharp *et al.*, 1996; Andres, 1997, 1999; Murphy and Mabee, 2000), bald eagles (White *et al.*, 1995; Bernatowicz *et al.*, 1996), harlequin ducks (Patten *et al.*, 2000), pigeon guillemots (Oakley and Kuletz, 1996), and marbled murrelets (Kuletz, 1996). These studies suggested that measures of reproductive performance were lower in oiled areas than in unoiled reference areas in 1989 for kittiwakes, oystercatchers, eagles, and harlequins (but see Wiens *et al.*, 2010) but were similar from 1990 onward. Kittiwake productivity remained low in oiled areas during 1990–94, although it also decreased by half in reference areas in 1990 and (as with Boersma *et al.*'s (1995) murres) was abnormally low in 1992 (Irons, 1996). These changes probably resulted from changes in prey availability (Irons, 1996).

Guillemots and murrelets were an exception to the lower postspill reproduction. This may be related to the location of the studies relative to the spill trajectory and to the birds' breeding phenology, however. Guillemots and murrelets were studied in the Naked Island group (see Map 3, p. vii), which was one of the first areas oiled. Most breeding guillemots had not yet arrived at the time of the spill, and much of the oil in that area had dissipated by the time they did arrive (Oakley and Kuletz, 1996; see also Chapters 3 and 6).

Taken together, the conclusions that emerge from these studies are that acute effects of the spill on marine-bird reproduction were variable, restricted in time and space, and largely gone after a few years. Most of these species are long-lived and have low reproductive potential, and annual variations in productivity are a fact of life (Zabala *et al.*, 2011). Because defining what is "normal" is problematic, it is difficult to establish a baseline for determining spill effects. The prespill baseline provided by Boersma *et al.*'s (1995) 25-m² plot is rarely available. In the absence of good prespill information, comparisons of oiled and reference areas often assume that they are similar in all respects except oiling history and that birds do not move between the two areas (see Chapter 10). The intra- and intercolony variation in reproductive timing and success of murres at East Amatuli casts doubt on the validity of the first assumption. Telemetry studies indicate that individuals of some bird species move back and forth between oiled and unoiled areas (e.g., kittiwakes; Ainley *et al.*, 2003), challenging the second assumption.

14.2.3 Habitat occupancy

The effects of oiling on habitat quality and availability were obvious immediately after the spill (see, for example, Chapter 1, Fig. 1.5). Determining how or whether oiling of habitats affected bird populations, however, proved to be difficult. In two studies, Day *et al.* (1995, 1997a, b) reasoned that reduced abundance of a species in a section of shoreline could be due either to oiling of the habitat or to the suitability of the habitat itself, independent of oil. To separate these effects, they asked whether there was a statistically significant relationship between the density of a species (individuals per kilometer of shoreline) and a quantitative measure of oiling intensity on sections of shoreline in PWS. If there was a statistically significant relationship, did that relationship change when quantitative measures of habitat features were included as covariates in the analyses? If the relationship changed, they concluded that variation in density among shoreline sections was primarily due to habitat suitability, rather than to oil. Including habitat measures changed the categorizations of four species from negatively affected to unaffected and of three species from unaffected to negatively affected (Day *et al.*, 1995).

Other studies (Esler *et al.*, 2000a, b) also considered habitat, focusing on single species (harlequin duck, Barrow's goldeneye). Both studies compared midwinter densities in 1995–96 and 1996–97 between Knight Island (oiled) and Montague Island (unoiled) (see Map 3, p. vii) with respect to habitat features (independent of oiling effects), prey biomass, and oiling history. Densities of both species were significantly related to habitat measures but not to prey biomass. After accounting for habitat effects, Barrow's goldeneye densities were determined to be unrelated to oil. Harlequin duck densities, however, were negatively related, suggesting that acute effects from the spill had not yet disappeared by 1997.

14.3 Recovery

The critical question in assessing the effects of an oil spill is, "How long do the effects last?" or, conversely, "When does recovery occur?" Acute effects end when exposure to oil or its residues no longer causes negative effects. The recovery process becomes apparent when the initial effects diminish over time (Chapter 1, Box 1.3). There are multiple pathways of recovery (Fig. 14.3) and multiple criteria for assessing recovery. For marine birds, recovery from effects of the *Exxon Valdez* spill was assessed by reproductive performance (as discussed above) or abundance and habitat occupancy (Wiens, 1995; Peterson, 2001; Integral Consulting, Inc., 2006; Rice *et al.*, 2007; *Exxon Valdez* Oil Spill Trustee Council, 2009).

These studies often yielded wildly differing conclusions (Table 14.2). For example, black oystercatchers were considered recovered in 1991 (two studies) and 1998 (one

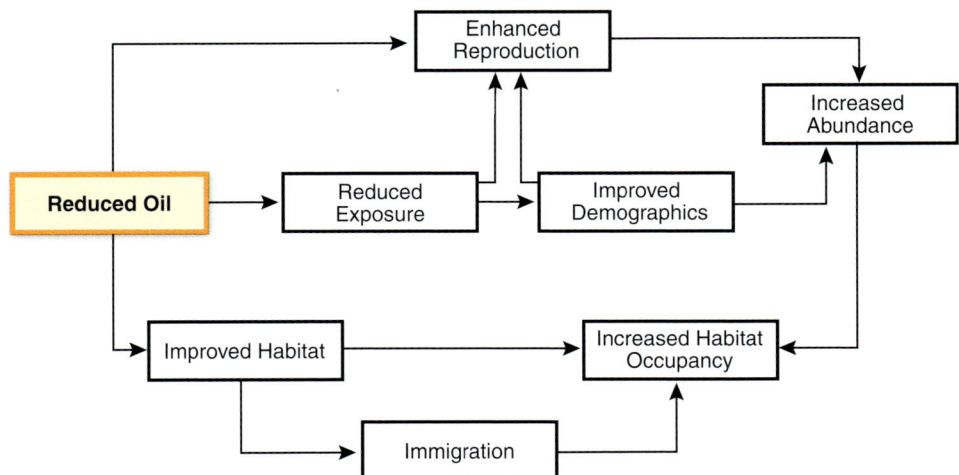

Figure 14.3 Major pathways of recovery from effects of a marine oil spill as the amount and distribution of oil remaining in the environment diminish over time.

Table 14.2 Recovery status and year for bird species listed as negatively affected by the *Exxon Valdez* oil spill in PWS. The studies differed in the time periods considered, in the seasons considered, and in the analytical criteria used to assess recovery (see text). NE = no effect or positive response; UN = information insufficient to assign a recovery status; NR = not recovered; RG = recovering; RD = recovered; W = winter; S = summer.

Species/taxon	Study						
	Day et al. (1997b); Murphy et al. (1997)	Day et al. (2003); Wiens et al. (2004)	Irons et al. (2000)[a]	Lance et al. (2001)[a]	McKnight et al. (2008)[a]	Integral Consulting, Inc. (2006)[a]	Exxon Valdez Oil Spill Trustee Council (2010)[a]
Time period	1989–1991	1984–2001	1984–1998	1989–1998	1989–2007	1989–2006	1989–2009
Analysis used (see text)	Oiling gradient + habitat; BACI	Oiling gradient + habitat; BACI	BACI	Homogeneity of slopes	Homogeneity of slopes	Synthesis of studies	Synthesis of studies
Mallard	RD 91	NE	na[b]	na	na	na	na
Harlequin duck	RG 91	NE	RG 98	RG 98	NR (W, S)	RG 06	NR 96–02; RG 06–10
Black scoter	RD 90	na	na	NR 98	RG (W), NR (S)	na	na
Barrow's goldeneye	NR 91	NE	NR 98	NR 98	NR (W, S)	na	RG 10[c]
Bufflehead	NR 91	na	na	RG 98	NR (W)	na	na
Common merganser	RD 90	RD 96	NR 98	NR 98	NR (W, S)	na	na
Red-breasted merganser	RD 91	na	na	na		na	na
Common loon	NE	na	na	na	RG (W), NR (S)	RD 06	UN 96[d]; NR 99–02; RD 06
Horned grebe	NR 91	na	na	na		na	na
Red-necked grebe	NR 91	na	na	NR 98	NR (W)	na	na
Double-crested cormorant	na	na	NR 98	na		RD 06	NR 96–02; RD 06[e]
Pelagic cormorant	RG 90	RD 91		NR 98	NR (W), RG (S)	na	
Red-faced cormorant	NE	na		na		na	
Great blue heron	RD 90	NE	na	na	na	na	na
Bald eagle	RD 90	RD 90	na	na	NR (W, S)	na	RG 90–95; RD 96
Black oystercatcher	RD 91	RD 91	RD 98	NR 98	NR (S)	na	RG 94–95; UN 96; RG 99; RD 02; RG 06–10
Wandering tattler	RD 90	na	na	NE	na	na	na
Spotted sandpiper	NE	RD 90	na	na	na	na	na

Table 14.2 (cont.)

Species/taxon	Study						
	Day et al. (1997b); Murphy et al. (1997)	Day et al. (2003); Wiens et al. (2004)	Irons et al. (2000)[a]	Lance et al. (2001)[a]	McKnight et al. (2008)[a]	Integral Consulting, Inc. (2006)[a]	Exxon Valdez Oil Spill Trustee Council (2010)[a]
Mew gull	NR 91	RD 91	na	NR 98	NR (W, S)	na	na
Glaucous-winged gull	RD 90	NE	na	NR 98	NR (W); RG (S)	na	na
Black-legged kittiwake	NE	na	RD 98	NR 98	NR (W, S)	na	na
Arctic tern	NE	na	na	NR 98	NR (S)	na	na
Common murre	RD 90	NE	NR 98	NR 98	NR (W, S)	na	RG 96–99; RD 02
Pigeon guillemot	RG 91[f]	RD 91	NR 98	NR 98	NR (W, S)	UN 06	NR 10
Marbled murrelet	RD 89	NE	na	NR 98	NR (W, S)	UN 06	NR 96; RG 99–02; UN 06–10
Kittlitz's murrelet	na	na	na	na	NR (S)	UN 06	UN 96–10
Belted kingfisher	NE	RD 96	na	na	na	na	na
Steller's jay	NE	RD 91	na	na	na	na	na
Northwestern crow	NR 91	RD 90	na	RG 98	NR (W), RG (S)	na	na
Total taxa not recovered	6	0	5	12	14 (W) 15 (S)	0	1
Total taxa recovering	2	0	1	3	2 (W) 3 (S)	1	3
Total taxa recovered	11	10	2	0	0 (W) 0 (S)	2 (3 unknown)	4 (2 unknown)

[a] Lance et al., Irons et al., Integral Consulting, Inc., McKnight et al., and the Trustees combined species of grebes, cormorants, scoters, mergansers, and murrelets into single species-group categories; in this table they were assigned to the most common species in the group.
[b] na = not analyzed.
[c] The Trustees added Barrow's goldeneyes to the list of injured resources in 2008.
[d] The Trustees added common loons to the list of injured resources in 1995.
[e] The Trustees added cormorants to the list of injured resources in 1996.
[f] Although Murphy et al. did not explicitly say so, the BACI analysis indicated a substantially smaller negative effect in 1991 than had been seen in 1989 and 1990, suggesting that recovery was beginning.

study), but not recovered in 1998 and 2006 (two studies). Based on several Trustee-sponsored studies, the Trustees categorized oystercatchers as recovering in 1995, unknown in 1996, recovering in 2000, recovered in 2002, and then recovering in 2009 (Table 14.2). Glaucous-winged gulls were considered not recovered in two studies/seasons (within a study), recovering in another season, recovered in one study, unaffected in one study, and positively affected in one study (Table 14.2). (The Trustees did not include glaucous-winged gulls in their assessments of affected bird species.)

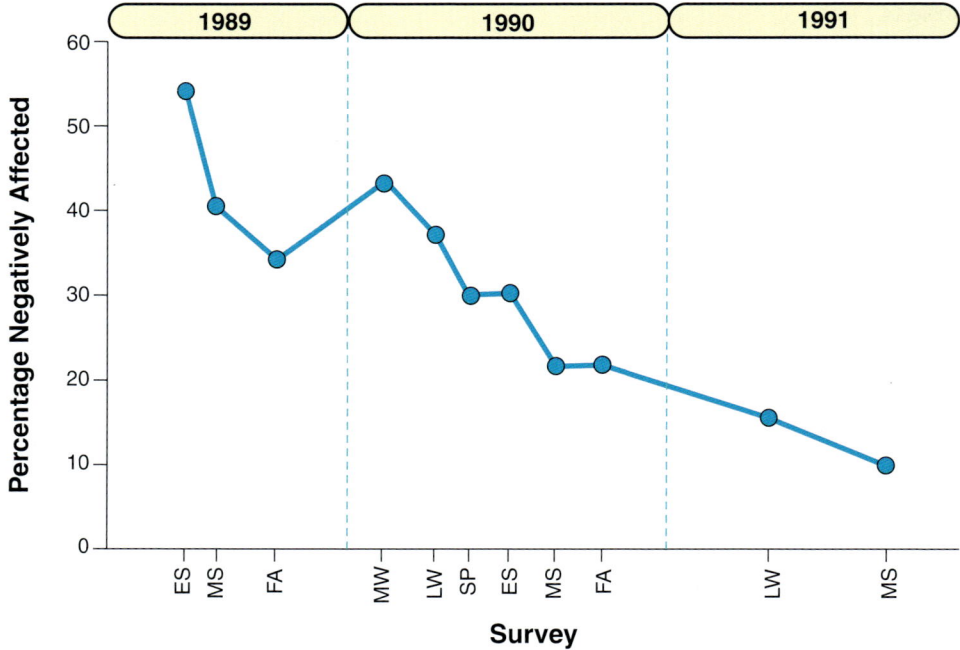

Figure 14.4 Percentage of marine bird species exhibiting negative effects of oiling from the *Exxon Valdez* on habitat occupancy and use of shorelines in PWS, based on surveys conducted in 1989–91. ES = early summer; MS = midsummer; LS = late summer; FA = fall; MW = midwinter; LW = late winter; SP = spring. From Day *et al.*, 1997b.

Both Day *et al.* (1997b) and Irons *et al.* (2000) conducted surveys of multiple species over several years, providing information on overall recovery of marine birds over time. In their initial survey 3 months after the spill, Day *et al.* (1997b) found that 54% of the species analyzed showed statistically significant negative effects on habitat occupancy. The proportion exhibiting negative effects decreased steadily thereafter, dropping to 10% by midsummer 1991 (Fig. 14.4). By 1996, Day and his colleagues concluded that all of the negatively affected species had recovered (Wiens *et al.*, 2004). In contrast, Irons *et al.* (2000) reported that 5 of 14 species or species-groups had not recovered by 1998. McKnight *et al.* (2008) used some of these same data, but with additional years and types of surveys and different analytical techniques. They concluded that 14 species showed no evidence of recovery by 2007.

Such contradictions, and others in Table 14.2, stem from differences among studies in the species considered, study design and analytical methods, definitions of recovery, and the null hypothesis being tested. For example, Day *et al.* (1995), Murphy *et al.* (1997), and Wiens *et al.* (2004) included all species with enough observations to permit statistical analyses (25–42 species) and tested the null hypothesis of no initial effects, whereas Lance *et al.* (2001) and McKnight *et al.* (2008) considered only species that had been previously identified by the Trustees as negatively affected (17 and 20 species, respectively). Although all of these studies used regression analyses, the Day *et al.* (1995), Murphy *et al.* (1997), and Wiens *et al.* (2004) studies considered relationships between densities and a quantitative index of shoreline oiling, with and without habitat covariates. The Lance *et al.* (2001) and McKnight *et al.* (2008) studies instead compared slopes of regressions over time for qualitatively defined oiled versus unoiled reference areas (i.e., a homogeneity-of-slopes or parallelism test; Skalski et al., 2001) and did not consider the effects of habitat.

The Day et al. (1995) and Wiens et al. (2004) studies defined recovery as the disappearance over time of a previously documented, statistically significant relationship between density and oiling level after habitat covariates were considered (i.e., the species had recovered).[5] In contrast, the Lance et al. (2001) and McKnight et al. (2008) studies defined recovery in terms of trends over time in previously oiled areas relative to those in reference areas (i.e., the species was recover*ing* but not recover*ed*). Hence, the choice in statistical approach dictated the conclusions that could be reached, and a failure to recognize this distinction contributed to continuing disagreements about the recovery status of species.

14.4 Chronic effects

With any large environmental disturbance, some effects may emerge only later. This could be due to indirect effects, time lags, or low-level, chronic exposure (Ford et al., 1982; Wiese and Robertson, 2004). In PWS, some effects only appeared to be delayed because some studies did not begin until several years after the spill, as the focus shifted from documenting acute effects to assessing recovery and chronic effects (see Chapter 2). A series of studies on harlequin ducks, however, illustrates the difficulty of assessing chronic effects from an oil spill.

In studying harlequin ducks in 1995–97 in oiled (western) and unoiled (eastern) PWS, Rosenberg and Petrula (1998) concluded that there were no differences in population structure that would indicate continuing oil exposure. They found no differences in sex ratios, proportions of paired females and subadult males, the timing of molt, or reproductive recruitment. Nonetheless, they noted a trend of decreasing abundance in the oiled area, suggesting to them that harlequin ducks had not yet recovered from the spill, perhaps because of poor overwinter survival of females.

Esler et al. (2000c) studied overwinter survival by radio-tracking adult females during 1995–97 in the oiled part of PWS and at unoiled Montague Island. They found that mean winter survival was slightly lower in the oiled area (78%) than at Montague Island (84%). In 2000–03, however, Esler and Iverson (2010) found no differences in the overwinter survival of adult females, although the survival of first-year females was about 7% lower than that of adults in both areas. Iverson and Esler (2010) then used this information to model harlequin duck demography in oiled and unoiled parts of PWS, concluding that the cumulative mortality associated with chronic oil exposure, driven largely by reduced overwinter survival, exceeded the acute-phase mortality. According to the model, full population recovery (defined as "a return to the long-term average from survey counts") would take 24–32 years. Esler et al. (2000c) and Iverson and Esler (2010) suggested that the lower survival of adult females in previously oiled areas, the modeled long recovery times, and the elevated induction of cytochrome P450 1A (CYP1A) in harlequin ducks from the spill area (see Chapter 9) were all associated with continued exposure to persistent *Exxon Valdez* oil.

Assessing chronic effects requires a determination of the likelihood of hydrocarbon exposure. By using an ecological risk-assessment approach and a more comprehensive data set, Harwell et al. (2012) showed that the probabilities of such exposure are

---

[5] Murphy et al. did not use habitat covariates in their analyses.

vanishingly small because the only plausible pathways for harlequin ducks to be exposed to oil are no longer contaminated with spill hydrocarbons (Chapter 16). In the absence of documented linkages between CYP1A and biologically significant consequences, the conclusion that harlequin ducks "remained at risk of potential deleterious consequences of that exposure" (Esler et al., 2010, p. 1144) remains speculative (see Chapter 9).

14.5 Dealing with multiple species in a variable environment

14.5.1 Underlying variation in distribution

The patchy distribution of oil on shorelines (e.g., Chapter 1, Fig. 1.3) accounts for some of the variation in spill effects and species' short- and long-term responses to the spill. Inherent differences in the distribution and natural history of species also play a role. These factors are all part of the underlying background of variation that must be addressed when measuring species abundance, assessing patterns of distribution and habitat use, or drawing comparisons to determine how species might have responded to the oil spill.

To provide a glimpse of this reality, Figure 14.5 shows the variation in densities of five marine-bird species in 66 contiguous sections of the 453-km shoreline of Knight Island (Fig. 14.6) in four years. The five species – bald eagle, black oystercatcher, glaucous-winged gull, harlequin duck, and pigeon guillemot – are common in PWS but differ markedly in ecological and life-history characteristics (Box 14.1). Although they exhibited different patterns of effects and recovery from the oil spill (Tables 14.1 and 14.2), all were considered recovered when postspill data were collected in 2000, 2004, and 2008 (Wiens et al., 2004) and, obviously, none had been affected by the spill when Irons et al. (1988) conducted surveys on the same shoreline sections in 1984. Hence, the distributions and abundances depicted in Figure 14.5 represent natural variations in habitat occupancy that are uncomplicated by any effects of the oil spill.

There are some spatial patterns to the variations in abundance among shoreline sections, both within and among years. Bald eagles, for example, had a wide but highly variable distribution on Knight Island, both within and among years. In contrast, black oystercatchers showed interannual consistency in the parts of the island where they were most common. The spatial distribution of glaucous-winged gulls among shoreline sections was extremely variable, both within and among years. Harlequin ducks occurred in low numbers and were fairly evenly distributed around Knight Island in 1984; as their numbers increased in 2001, 2004, and 2008, large aggregations occurred on the eastern side of the island. The distribution of pigeon guillemots on Knight Island in 1984, when the species was still fairly abundant in PWS, differed substantially from the considerable variation in abundance among sections in 2001–08. Consistency in use of shoreline sections that contained large numbers of guillemots probably reflected locations of breeding colonies.

Such variations in distribution and abundance are not unique to Knight Island, nor are they unique to these species. Small-scale spatial variation in abundance that is inconsistent over time may be the norm in many systems, but it is usually not evident because surveys are often conducted in spatially separated subsamples and are not replicated over time. Consequently, the variations in Figure 14.5 should be examined for what factors might be contributing to the variation, rather than being regarded as noise or statistical variance around some overall mean. These factors can then be included as covariates in analyses of oil effects (see Chapter 10).

Patterns in the distribution and abundance of species are known to be scale-dependent (Wiens, 1989; Peterson and Parker, 1998). The fine-scale variations apparent in Figure 14.5 will inevitably be smoothed out when viewed at a broader spatial scale. To assess the issue of scale, we considered the distributions of the five bird species at the broader spatial scale of four geographically defined quadrants on Knight Island (Fig. 14.6). Because of Knight Island's orientation to open waters on the eastern side and the associated bathymetry, some parts of the island have greater exposure to waves and marine currents or are adjacent to more offshore islands and islets than are others. Shoreline substrates also differ among quadrants. At this broader scale of analysis, the jagged patterns of spatial variation of Figure 14.5 are smoothed out (Fig. 14.7). Bald eagles were generally most abundant in the southeastern quadrant, whereas black oystercatchers never occurred there and were most abundant in the northwestern quadrant. In contrast, glaucous-winged gulls were most abundant in the two southern quadrants in all years except 2001, when numbers were high in the northeastern quadrant. Numbers of harlequin ducks were consistently high in the northeastern quadrant, consistently low in the northwestern quadrant, and increased over time in the two southern quadrants. The long-term interannual declines in pigeon guillemot abundance were exhibited in different quadrants in different ways: densities declined steadily in the southwestern quadrant and guillemots nearly disappeared from the northwestern quadrant, but numbers were fairly stable in the two eastern quadrants from 2001 through 2008.

Spatial differences among these species indicate the potential importance of broad-scale spatial variation in habitat, and the interannual pattern for pigeon guillemots suggests that long-term population declines may be expressed differently in different habitats. Consequently, when and where sampling is conducted would almost certainly yield different results for BACI analyses to assess effects (e.g., Klosiewski and Laing, 1994; Murphy *et al.*, 1997) and trend analyses to assess recovery (e.g., McKnight *et al.*, 2008).

14.5.2 Multispecies analyses

The challenge of dealing with spatial and temporal variations in distribution among species is less daunting if a single species is studied. Studies of the effects of an oil spill on a single focal species may be warranted if the species has suffered extraordinary mortality (e.g., murres), has economic value (e.g., pink salmon, *Oncorhynchus gorbuscha*, Chapter 12; Pacific herring, *Clupea pallasii*, Chapter 13), or receives extraordinary public attention (e.g., sea otters, *Enhydra lutris*, Chapter 15). Sampling designs for single species can be customized to account for seasonal occurrence, breeding status, patterns of habitat use, or other factors.

Gauging the full scope of the effects of an oil spill such as the *Exxon Valdez* spill, however, still requires considering multiple species. Multispecies studies are typically more efficient, financially and logistically, and may have other scientific benefits that single-species studies lack. For example, Day *et al.* (1995, 1997b) used natural history and multispecies surveys to examine whether some species might have ecological barriers to recovery, such as restricted habitat preferences or delayed maturation. Wiens *et al.* (1996) used multispecies surveys to examine the effects of the *Exxon Valdez* spill on ecologically defined groupings of species (guilds) that included the rarer species that Day *et al.* (1995, 1997b) and Murphy *et al.* (1997) could not evaluate statistically.

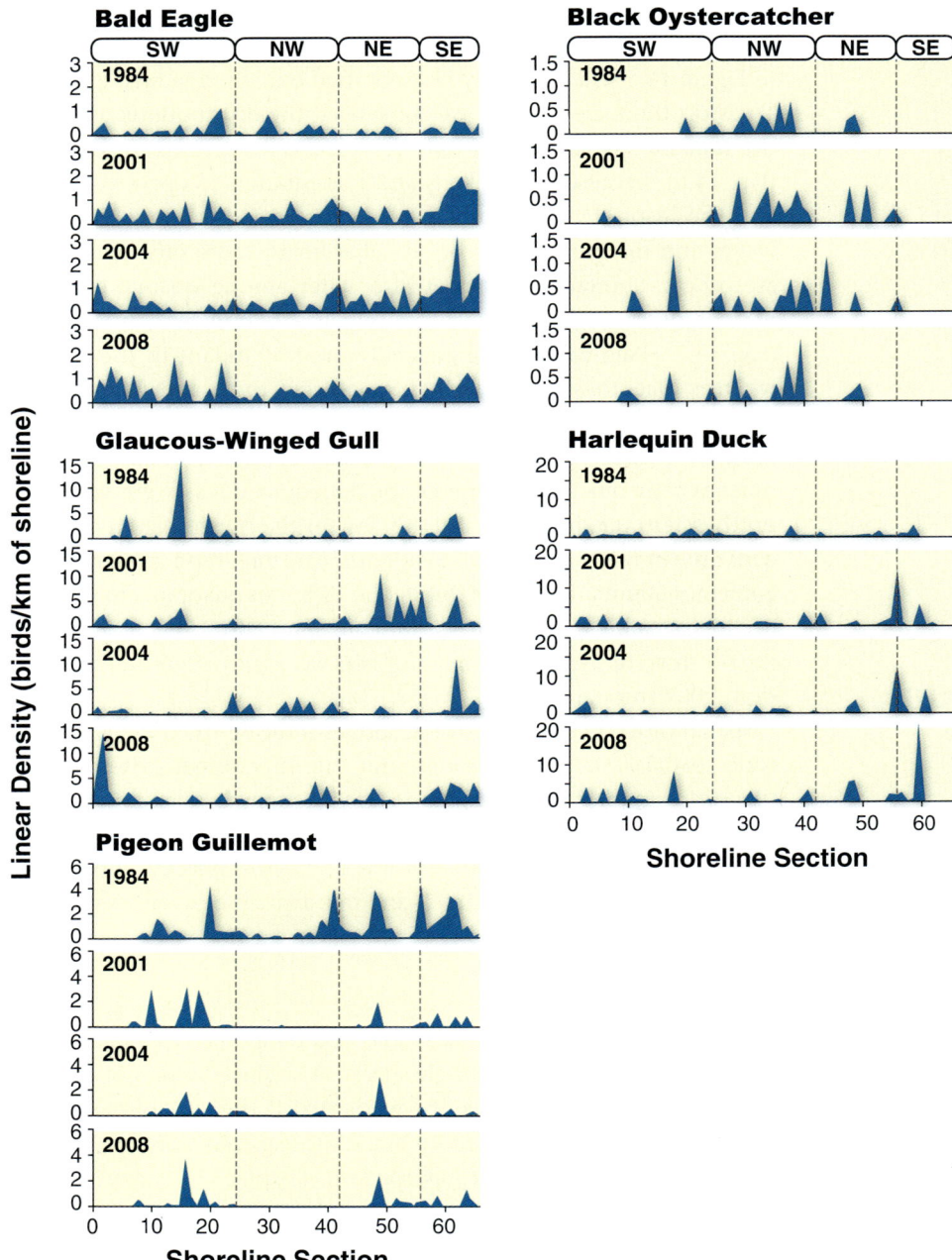

Figure 14.5 Densities of five marine-bird species recorded in sections of shoreline of Knight Island, 1984–2008. Because section boundaries were originally drawn based on the degree of exposure, they differed in length (1.6–23.7 km, mean = 6.9 km, SD = 4.6) and usually contained a mixture of habitat types. Surveys were conducted by driving a small skiff slowly along the shoreline and identifying and counting all birds seen on the water within 200 m of the shore, on the beach, up to 100 m inland, or flying over these areas (methods are described in detail by Day *et al*., 1995, 1997b). The 1984 data are from Irons *et al*. (1988). The scale on the *y*-axis is consistent within a species but differs among species. Shoreline sections are numbered sequentially moving clockwise around the island from Point Helen (Fig. 14.6).

Figure 14.6 Knight Island in PWS showing the locations of shoreline sections used to assess the fine-scale patterns of distribution and abundance of the focal species shown in Figure 14.5 (black lines) and the demarcations between quadrants of the island used to evaluate the broad-scale patterns shown in Figure 14.7 (red lines).

There are tradeoffs between the quality and quantity of information that can be collected and between the depth and breadth of understanding of spill effects in a single- versus a multispecies study. Multispecies studies may be sensitive to variation and confounding factors (which, correspondingly, diminish statistical power; see Chapter 10), but they may nonetheless be the most efficient option for broad surveys of disturbance effects. The decision about which approach is best, however, will generally be governed by whether stakeholders are most concerned about effects on particular species or potential damages to the broader biological communities or ecosystems.

Box 14.1 Five species, five dramatically different life histories.

For in-depth analyses, we selected five species to represent different ecological and life-history characteristics: bald eagles, black oystercatchers, glaucous-winged gulls, harlequin ducks, and pigeon guillemots (Fig. 14.1.1). Bald eagles are large raptors that are common breeders and residents along shorelines of PWS; they feed primarily on fish, are territorial while nesting, and have strong site fidelity (White *et al.*, 1995; Bernatowicz *et al.*, 1996). Black oystercatchers are shorebirds that commonly nest along shorelines throughout PWS (Murphy and Mabee, 2000); they feed intertidally on marine invertebrates and leave PWS during the coldest months of winter (Isleib and Kessel, 1973). Glaucous-winged gulls are the most common large gull of the northern GOA; they reside and breed throughout PWS (Isleib and Kessel, 1973) and feed on a variety of marine invertebrates, fishes, birds, mammals, and carrion. Harlequin ducks are small seaducks that breed in low numbers but are resident, with a large nonbreeding summer population and an even larger wintering population of residents and migrants; they typically occur along rocky shorelines and forage on intertidal invertebrates and seasonal foods such as herring eggs (Robertson and Goudie, 1999). Pigeon guillemots are medium-sized seabirds that occur in PWS primarily from spring through fall, nesting in cliff crevices and talus piles along shores; they feed primarily on nearshore fishes and have been undergoing a long-term population decline in the northern GOA that predates the *Exxon Valdez* spill (Ewins, 1993; Oakley and Kuletz, 1996).

Figure 14.1.1. The five focal bird species: (a) bald eagle; (b) black oystercatcher; (c) glaucous-winged gull; (d) harlequin duck; and (e) pigeon guillemot. (Photos: (a) Jennifer Boisvert; (b) John A. Wiens; (c, e) Robert H. Day; (d) Hugh Rose)

Bernatowicz, J.A., P.F. Schempf, and T.D. Bowman (1996). Bald eagle productivity in south-central Alaska in 1989 and 1990 after the *Exxon Valdez* oil spill. In *Proceedings of the* Exxon Valdez *Oil Spill Symposium*. S.D. Rice, R.B. Spies, D.A. Wolfe, and B.A. Wright, eds. Bethesda, MD, USA: American Fisheries Society; Symposium 18; ISBN-10: 0913235954; ISSN: 08922284; pp. 785–797.

Ewins, P.J. (1993). Pigeon guillemot (*Cepphus columba*). In *The Birds of North America*. F. Gill and A. Poole, eds. Philadelphia, PA, USA: Birds of North America, Inc.; No. 49.

Isleib, M.E. and B. Kessel (1973). Birds of the North Gulf Coast-Prince William Sound region, Alaska. *Biological Papers of the University of Alaska* **14**: 1–149.

Murphy, S.M. and T.J. Mabee (2000). Status of black oystercatchers in Prince William Sound, Alaska, nine years after the *Exxon Valdez* oil spill. *Waterbirds* **23**(2): 204–213.

Oakley, K.L. and K.J. Kuletz (1996). Population, reproduction, and foraging of pigeon guillemots at Naked Island, Alaska, before and after the *Exxon Valdez* oil spill. In *Proceedings of the* Exxon Valdez *Oil Spill Symposium*. S.D. Rice, R.B. Spies, D.A. Wolfe, and B.A. Wright, eds. Bethesda, MD, USA: American Fisheries Society; Symposium 18; ISBN-10: 0913235954; ISSN: 08922284; pp. 759–769.

Robertson, G.J. and R.I. Goudie (1999). Harlequin duck (*Histrionicus histrionicus*). In *The Birds of North America*. F. Gill and A. Poole, eds. Philadelphia, PA, USA: Birds of North America, Inc.; No. 466. [*The Birds of North America Online;* http://bna.birds.cornell.edu/bna/species/466]

White, C.M., R.J. Ritchie, and B.A. Cooper (1995). Density and productivity of bald eagles in Prince William Sound, Alaska, after the *Exxon Valdez* oil spill. In Exxon Valdez *Oil Spill: Fate and Effects in Alaskan Waters*. P.G. Wells, J.N. Butler, and J.S. Hughes, eds. Philadelphia, PA, USA: American Society for Testing and Materials; ASTM Special Technical Publication 1219; ISBN-10: 0803118961; pp. 762–779.

14.6 Evaluating cause–effect relationships

Ultimately, assessing the effects of an oil spill and subsequent recovery requires considering how the documented patterns (effects) are related to driving factors (causes). Some immediate causes and effects may be obvious. But it is often (incorrectly) assumed that all effects consistent with oiling are caused by oiling. If the abundance of a species is lower in oiled areas but then increases relative to unoiled areas (e.g., murrelets; Irons *et al.*, 2000), the species may be classified as "recovering." Other explanations for the increase (e.g., habitat availability or suitability, shifts in prey abundance) may be ignored. Differences among areas may be falsely attributed to a few isolated pockets of residual oil that are assumed to cause chronic exposure over a broad area (e.g., Esler *et al.*, 2010).

Pigeon guillemots illustrate the difficulty of establishing causality in a spatially and temporally variable environment. Oakley and Kuletz (1996) found that guillemot populations declined more in oiled than in unoiled areas in the year after the spill, and Golet *et al.* (2002) reported that adult guillemots had higher CYP1A levels in the spill area than in the reference area a decade after the spill, which they presumed indicated continued exposure to persistent *Exxon Valdez* oil. The *Exxon Valdez* Oil Spill Trustee Council (2010) later indicated that these differences had disappeared by 2004 and that exposure to lingering oil was "likely intermittent." Other analyses (Chapters 7, 9, 16) indicate that exposure was in fact highly unlikely.

Other things were not equal during this time period, however. As populations of lipid-rich schooling fishes declined after an oceanic regime shift in the GOA in 1977, populations of lipid-poor fishes increased; these changes persisted through a subsequent oceanic regime shift in 1989 (Springer, 1998; Hare and Mantua, 2000). The guillemots' food web was altered (Piatt and Anderson, 1996; Agler *et al.*, 1999; Anderson

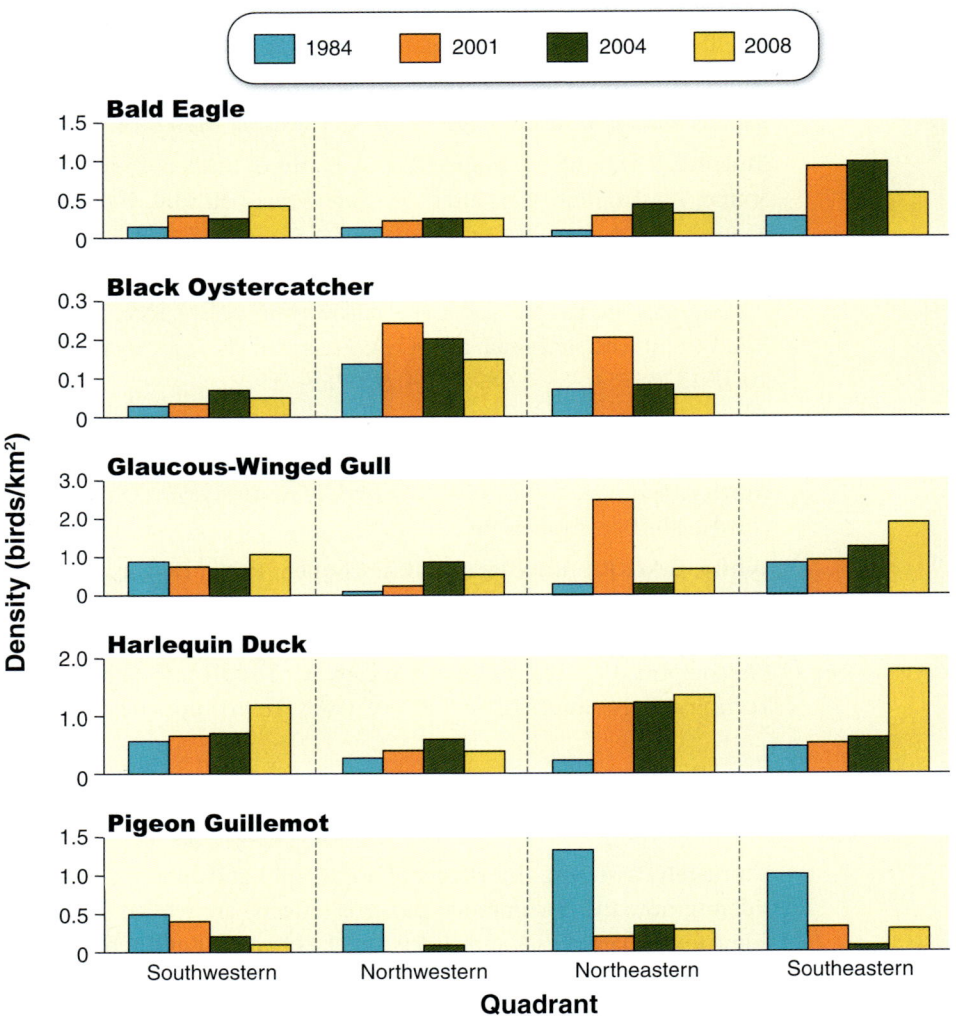

Figure 14.7. Densities of the five focal marine-bird species in the four quadrants of Knight Island (Fig. 14.6) in 4 years. Note the differences in scales on the *y*-axes.

and Piatt, 1999), and their recovery from the *Exxon Valdez* spill was limited by the availability of lipid-rich prey (Litzow et al., 2002).

Most of the guillemot studies focused on the Naked Island group, where mink (*Neovison vison*) first appeared in the 1980s (Hayes, 1995). By the early 1990s, mink were causing major egg and chick loss; by 2007–08, predation rates on nesting adults and chicks were extremely high (Bixler, 2010). Population declines over this period were substantially greater at Naked Island than at nearby, mink-free (but oiled) Smith Island. Because the Naked Island group supported about 25% of the PWS population of guillemots after the 1989 oil spill, predation was a major factor in the postspill decline of guillemots in the spill area. In the context of the long-term regional decline of guillemots and the effects of mink at the primary nesting area, the recovery goal of "sustained or increasing productivity" of guillemots (*Exxon Valdez* Oil Spill Trustee Council, 2010) seems unrealistic.[6]

---

[6] The mink on the Naked Island group are feral descendants of fur-farm stock (Chapter 5). A Trustee pigeon guillemot restoration project to remove the mink has been proposed (Irons and Roby, 2011).

Ultimately, attempts to establish cause–effect relationships must pass the test of biological plausibility. Doing so requires considering two things. First, are the proposed mechanisms linking cause to effect credible? It is usually difficult to measure the mechanisms by which oil affects individuals or populations directly, so indirect measures (e.g., CYP1A levels) are often used. This requires that potential exposure pathways be thoroughly evaluated. Concurrent studies of water chemistry, oil sequestration in shoreline sediments, polycyclic aromatic hydrocarbon (PAH) levels in forage species, and the like can help determine whether postulated exposure pathways actually exist (Chapter 9).

Second, are the potential effects biologically significant? Many acute effects, such as direct mortality, are undeniably significant to individuals, but the consequences may not be manifested at the population or community levels. Even small, long-term, chronic decreases in reproductive output or adult survival, however, may have population-level effects (Ford *et al.*, 1982; Iverson and Esler, 2010) – *if* all else is equal. In a variable environment, however, such small differences can be overwhelmed (or amplified) by numerous other factors, making it difficult to evaluate chronic effects on populations. Because oiling effects diminish over time and the remaining oil becomes localized and sequestered (Chapters 6 and 7), any lingering effects will be difficult to detect or to ascribe unambiguously to the spill and will be unlikely to affect the distribution, abundance, or biological functioning of marine bird populations.

14.7 Lessons learned

During more than two decades spent investigating the effects of the *Exxon Valdez* oil spill on marine birds, we learned that developing scientifically credible studies requires careful attention to a variety of issues. The primary lessons we learned are summarized below and in Table 14.3.

- Take time to think clearly about which potential causal pathways can lead to the observed effects.
- Stipulate clearly which pathways are being investigated and consider factors other than oil in appropriately framed alternative hypotheses that are then tested. Avoid rushing to conclusions about spill effects (or their absence). Cause–effect linkages must be biologically plausible, empirically documented, and associated with realistic exposure pathways.
- Measures of important habitat and environmental features should be included as covariates in analyses to disentangle potential oiling effects from those of habitat differences. Spatial variation in habitats or differences in other environmental features between "treatment" and "reference" areas, such as broadly defined or widely separated oiled and unoiled areas, may compromise statistical comparisons and lead to erroneous conclusions.
- Studies of chronic effects initiated long after a spill demand a rigorous study design and careful analysis of covariates to associate an effect with oil-related causes because the environment changes, the remaining oil diminishes and becomes localized, and the influences of confounding factors increase with time.

Table 14.3 Major issues, recommended solutions, and pros and cons related to those solutions for studies of marine birds following a major environmental perturbation such as an oil spill.

Issue	Recommended solution	Pros	Cons
Multi- versus single-species studies	Conduct multispecies studies – but depends on your objectives and amount of money available.	Single-species studies can be tailored to account for specific life-history attributes, such as seasonal abundance, patterns of habitat use, and breeding phenology, whereas multispecies studies have to be generalized. Multispecies studies are more efficient for addressing questions pertaining to the entire community.	Multispecies studies are confounded by differing phenologies, patterns of habitat use, and other life-history traits among species, which introduce additional variability. Conducting many single-species studies is less efficient (and more expensive) than conducting a multispecies study.
Short-term environmental variability	Conduct multiple-replicate sampling.	Accounts for environmental variability (e.g., tides, weather) and increases statistical power.	For certain types of surveys, decreases amount of time/money available for overall sampling effort, reducing the total amount of area that can be surveyed.
Medium-term environmental variability	Conduct multiple-season sampling.	Enables one to examine effects if there is seasonal turnover in avifauna or if effect may be manifested in different ways at different seasons.	Requires additional sampling efforts/year.
Long-term environmental variability	Conduct multiple-year sampling.	Enables one to assess both effects and recovery.	Requires long-term financial commitment. The oiling signal can be overwhelmed in time by long-term environmental variability.
Quantifying oiling levels	Use quantitative measure of oiling.	Much more accurate measure of independent variable than qualitative measure.	Requires high-resolution oiling mapping.
Quantifying levels of effect (oiling)	Use as fine a resolution as possible for quantifying oiling level.	Gives most precise information on responses of birds to oiling. Avoids gross stratification such as "oiled part of PWS," especially if you are examining a part within it that was never oiled. Provides greater flexibility in analyses.	Requires high-resolution oiling mapping, but that mapping is required by previous solution, so no additional cost.
Habitat variation	Quantify and factor in effects of habitat in studies of distribution and abundance.	Enables one to separate effects of habitat from those of oiling.	Requires high-resolution habitat mapping. Analyses become more complex.
Appropriate scale of habitat analysis	Use fairly short shoreline sections for sampling.	Habitats can be characterized for fairly small areas, providing greater resolution for habitat analyses.	Requires increased GIS effort.
Fuzziness of evidence	Use multiple lines of evidence for determination of effects and recovery.	Provides better overall view of how a species is doing than does single-line evaluation.	Requires additional levels of analysis and (sometimes) multiple data sets. Results sometimes conflict among different lines of evidence.
Uncertainty about effects	Develop diagram of effect pathways involving specific steps and linkages that can be evaluated.	Provides organizational flow-chart (e.g., Wiens et al., 2010; Harwell et al., 2012) for investigating factors that may produce effects.	May require an enormous number of steps and linkages that have to be evaluated.
Size of shoreline sections	Make sections similar (but not necessarily identical) length.	Reduces variability in density estimates, especially over section lengths that may vary by more than an order of magnitude. Lengths ≥ 2 km are appropriate for most species.	May make it impossible to use prespill data for comparison.

Table 14.3 (cont.)

Issue	Recommended solution	Pros	Cons
Layout of shoreline sections	Use a general layout with no stratification or effects of habitat on layout.	Provides flexibility in analysis; you also can apply habitat information later, during analyses.	May make it impossible to use prespill data for comparison if the sampling sections were laid out for specific reasons other than general population monitoring. Some species with specific habitat requirements may have biased density or population estimates.

- Uncertainty must be factored into study designs. Spatial and temporal variations in distribution and abundance differ among species and are expressed in different ways at different scales. It is important to recognize which sources of variation can be addressed through sampling and analytical design (e.g., replication, use of more permissive α-levels in statistical analyses, accepting results that have lower statistical power) and which must simply be understood and explicitly acknowledged.
- There are tradeoffs between conducting single- versus multispecies studies. Single-species studies provide greater insight into specific effects. Multispecies studies can better assess the overall scope of effects, but they suffer from the additional variation due to differences among species.
- The duration of a study should be determined by the objectives, characteristics of the species studied, and the risk of continuing exposure to petroleum hydrocarbons. Short-term studies may be sufficient to document acute spill effects, but longer studies are needed to determine chronic effects, particularly for wide-ranging, long-lived organisms such as birds.
- Documenting recovery requires an operational definition of "recovery" and appropriate statistical analyses. Exposure risk should be evaluated based on direct measurements of hydrocarbon levels in the environment, combined with risk analyses that calculate the probability that a target species will encounter the hydrocarbons. Studies can be prolonged unnecessarily when claims and counterclaims about effects and recovery are based on differing definitions and study designs that lead to different conclusions.

REFERENCES

Agler, B.A., S.J. Kendall, D.B. Irons, and S.P. Klosiewski (1999). Declines in marine bird populations in Prince William Sound, Alaska, coincident with a climatic regime shift. *Waterbirds* **22**(1): 98–103.

Ainley, D.G., R.G. Ford, E.D. Brown, R.M. Suryan, and D.B. Irons (2003). Prey resources, competition, and geographic structure of kittiwake colonies in Prince William Sound. *Ecology* **84**(3): 709–723.

Anderson, P.J. and J.F. Piatt (1999). Community reorganization in the Gulf of Alaska following ocean climate regime shift. *Marine Ecology Progress Series* **189**: 117–123.

Andres, B.A. (1997). The *Exxon Valdez* oil spill disrupted the breeding of black oystercatchers. *The Journal of Wildlife Management* **61**(4): 1322–1328.

Andres, B.A. (1999). Effects of persistent shoreline oil on breeding success and chick growth in black oystercatchers. *The Auk* **116**(3): 640–650.

Bernatowicz, J.A., P.F. Schempf, and T.D. Bowman (1996). Bald eagle productivity in south-central Alaska in 1989 and 1990 after the *Exxon Valdez* oil spill. In *Proceedings of the* Exxon Valdez *Oil Spill Symposium*. S.D. Rice, R.B. Spies, D.A. Wolfe, and B.A. Wright, eds. Bethesda, MD, USA: American Fisheries Society; Symposium 18; ISBN-10: 0913235954; ISSN 08922284; pp. 785–797.

Bibby, C.J. and C.S. Lloyd (1977). Experiments to determine the fate of dead birds at sea. *Biological Conservation* **12**(4): 295–309.

Bixler, K.S. (2010). *Why Aren't Pigeon Guillemots in Prince William Sound, Alaska, Recovering from the* Exxon Valdez *Oil Spill?* Corvallis, OR, USA: Oregon State University; M.S. Thesis.

Boersma, P.D., J.K. Parrish, and A.B. Kettle (1995). Common murre abundance, phenology, and productivity on the Barren Islands, Alaska: The *Exxon Valdez* oil spill and long-term environmental change. In Exxon Valdez *Oil Spill: Fate and Effects in Alaskan Waters*. P.G. Wells, J.N. Butler, and J.S. Hughes, eds. Philadelphia, PA, USA: American Society for Testing and Materials; ASTM Special Technical Publication 1219; ISBN-10: 0803118961; pp. 820–853.

Byrd, G.V., J.H. Reynolds, and P.L. Flint (2009). Persistence rates and detection probabilities of bird carcasses on beaches of Unalaska Island, Alaska, following the wreck of the M/V *Selendang Ayu*. *Marine Ornithology* **37**(3): 197–204.

Day, R.H., S.M. Murphy, J.A. Wiens, G.D. Hayward, E.J. Harner, and B.E. Lawhead (1997a). Effects of the *Exxon Valdez* oil spill on habitat use by birds along the Kenai Peninsula, Alaska. *The Condor* **99**(3): 728–742.

Day, R.H., S.M. Murphy, J.A. Wiens, G.D. Hayward, E.J. Harner, and L.N. Smith (1995). Use of oil-affected habitats by birds after the *Exxon Valdez* oil spill. In Exxon Valdez *Oil Spill: Fate and Effects in Alaskan Waters*. P.G. Wells, J.N. Butler, and J.S. Hughes, eds. Philadelphia, PA, USA: American Society for Testing and Materials; ASTM Special Technical Publication 1219; ISBN-10: 0803118961; pp. 726–761.

Day, R.H., S.M. Murphy, J.A. Wiens, G.D. Hayward, E.J. Harner, and L.N. Smith (1997b). Effects of the *Exxon Valdez* oil spill on habitat use by birds in Prince William Sound, Alaska. *Ecological Applications* **7**(2): 593–613.

Day, R.H., S.M. Murphy, J.A. Wiens, and K.R. Parker (2003). Changing habitat use by birds after the *Exxon Valdez* oil spill. In *Proceedings of the 2003 International Oil Spill Conference (Prevention, Preparedness, Response and Restoration-Perspectives for a Cleaner Environment)*, April 6–11, 2003, Vancouver, British Columbia, Canada. Washington DC, USA: American Petroleum Institute.

Dwyer, T.J., P. Isleib, D.A. Davenport, and J.L. Haddock (1976). *Marine Bird Populations in Prince William Sound, Alaska*. Anchorage, AK, USA: US Fish and Wildlife Service; unpublished report.

Erikson, D.E. (1995). Surveys of murre colony attendance in the Northern Gulf of Alaska following the *Exxon Valdez* oil spill. In Exxon Valdez *Oil Spill: Fate and Effects in Alaskan Waters*. P.G. Wells, J.N. Butler, and J.S. Hughes, eds. Philadelphia, PA, USA:

American Society for Testing and Materials; ASTM Special Technical Publication 1219; ISBN-10:0803118961; pp. 780–819.

Esler, D., T.D. Bowman, T.A. Dean, C.E. O'Clair, S.C. Jewett, and L.L. McDonald (2000a). Correlates of harlequin duck densities during winter in Prince William Sound, Alaska. *The Condor* **102**(4): 920–926.

Esler, D., T.D. Bowman, C.E. O'Clair, T.A. Dean, and L.L. McDonald (2000b). Densities of Barrow's goldeneyes during winter in Prince William Sound, Alaska, in relation to habitat, food and history of oil contamination. *Waterbirds* **23**(3): 423–429.

Esler, D. and S.A. Iverson (2010). Female harlequin duck winter survival 11 to 14 years after the *Exxon Valdez* oil spill. *The Journal of Wildlife Management* **74**(3): 471–478.

Esler, D., J.A. Schmutz, R.L. Jarvis, and D.M. Mulcahy (2000c). Winter survival of adult female harlequin ducks in relation to history of contamination by the *Exxon Valdez* oil spill. *The Journal of Wildlife Management* **64**(3): 839–847.

Esler, D., K.A. Trust, B.E. Ballachey, S.A. Iverson, T.L. Lewis, D.J. Rizzolo, D.M. Mulcahy, A.K. Miles, B.R. Woodin, J.J. Stegeman, J.D. Henderson, and B.W. Wilson (2010). Cytochrome P450 1A biomarker indication of oil exposure in harlequin ducks up to 20 years after the *Exxon Valdez* oil spill. *Environmental Toxicology and Chemistry* **29**(5): 1138–1145.

Exxon Valdez **Oil Spill Trustee Council** (2009). *2009 Status Report.* Anchorage, AK, USA: *Exxon Valdez* Oil Spill Trustee Council. [http://www.evostc.state.ak.us/Universal/Documents/Publications/AnnualStatus/2009AnnualReport.pdf]

Exxon Valdez **Oil Spill Trustee Council** (2010). *2010 Update on Injured Resources and Services.* Anchorage, AK, USA: *Exxon Valdez* Oil Spill Trustee Council. [http://www.evostc.state.ak.us/universal/documents/publications/2010IRSUpdate.pdf]

Ford, R.G., M.L. Bonnell, D.H. Varoujean, G.W. Page, H.R. Carter, B.E. Sharp, D. Heinemann, and J.L. Casey (1991). *Assessment of Direct Seabird Mortality in Prince William Sound and the Western Gulf of Alaska Resulting from the* Exxon Valdez *Oil Spill.* Portland, OR, USA: Ecological Consulting, Inc.; *Exxon Valdez* Oil Spill State/Federal Natural Resource Damage Assessment Bird Study Number 1; unpublished final report.

Ford, R.G., M.L. Bonnell, D.H. Varoujean, G.W. Page, H.R. Carter, B.E. Sharp, D. Heinemann, and J.L. Casey (1996). Total direct mortality of seabirds from the *Exxon Valdez* oil spill. In *Proceedings of the* Exxon Valdez *Oil Spill Symposium*. S.D. Rice, R.B. Spies, D.A. Wolfe, and B.A. Wright, eds. Bethesda, MD, USA: American Fisheries Society; Symposium 18; ISBN-10: 0913235954; ISSN 08922284; pp. 684–711.

Ford, R.G., J.A. Wiens, D. Heinemann, and G.L. Hunt (1982). Modelling the sensitivity of colonially breeding marine birds to oil spills: Guillemot and kittiwake populations on the Pribilof Islands, Bering Sea. *Journal of Applied Ecology* **19**(1): 1–31.

Fry, D.M. (1993). How do you fix the loss of half a million birds? In Exxon Valdez *Oil Spill Symposium Program and Abstracts.* Anchorage, AK, USA: *Exxon Valdez* Oil Spill Trustee Council; pp. 30–33.

Golet, G.H., P.E. Seiser, A.D. McGuire, D.D. Roby, J.B. Fischer, K.J. Kuletz, D.B. Irons, T.A. Dean, S.C. Jewett, and S.H. Newman (2002). Long-term direct and indirect effects of the *Exxon Valdez* oil spill on pigeon guillemots in Prince William Sound, Alaska. *Marine Ecology Progress Series* **241**: 287–304.

Hare, S.R. and N.J. Mantua (2000). Empirical evidence for North Pacific regime shifts in 1977 and 1989. *Progress in Oceanography* **47**(2–4): 103–145.

Harwell, M.A., J.H. Gentile, K.R. Parker, S.M. Murphy, R.H. Day, A.E. Bence, J.M. Neff, and J.A. Wiens (2012). Quantitative assessment of current risks to harlequin ducks in Prince William Sound, Alaska, from the *Exxon Valdez* oil spill. *Human and Ecological Risk Assessment* **18**(2): 261–328.

Hatch, S.A. and M.A. Hatch (1983). Populations and habitat use of marine birds in the Semidi Islands, Alaska. *The Murrelet* **64**(2): 39–46.

Hayes, D.L. (1995). *Recovery Monitoring of Pigeon Guillemot Populations in Prince William Sound, Alaska*. Anchorage, AK, USA: US Fish and Wildlife Service; unpublished report.

Integral Consulting, Inc. (2006). *Information Synthesis and Recovery Recommendations for Resources and Services Injured by the* Exxon Valdez *Oil Spill: Final Report*. Mercer Island, WA, USA: Integral Consulting, Inc.; *Exxon Valdez* Oil Spill Restoration Project 060783 Final Report. [http://www.evostc.state.ak.us/Files.cfm?doc=/Store/FinalReports/2006–060783-Final.pdf&]

Irons, D.B. (1996). Size and productivity of black-legged kittiwake colonies in Prince William Sound before and after the *Exxon Valdez* oil spill. In *Proceedings of the* Exxon Valdez *Oil Spill Symposium*. S.D. Rice, R.B. Spies, D.A. Wolfe, and B.A. Wright, eds. Bethesda, MD, USA: American Fisheries Society; Symposium 18; ISBN-10: 0913235954; ISSN: 08922284; pp. 738–747.

Irons, D.B., S.J. Kendall, W.P. Erickson, L.L. McDonald, and B.K. Lance (2000). Nine years after the *Exxon Valdez* oil spill: Effects on marine bird populations in Prince William Sound, Alaska. *The Condor* **102**(4): 723–737.

Irons, D.B., D.R. Nysewander, and J.L. Trapp (1988). *Prince William Sound Waterbird Distribution in Relation to Habitat Type*. Anchorage, AK, USA: US Fish and Wildlife Service; unpublished report.

Irons, D. and D. Roby (2011). *Pigeon Guillemot Restoration Research in Prince William Sound, Alaska*. Anchorage, AK, USA: *Exxon Valdez* Oil Spill Trustee Council, Restoration Project 11100853. [http://www.evostc.state.ak.us/Projects/ProjectInfo.cfm?project_id=2190]

Iverson, S.A. and D. Esler (2010). Harlequin duck population injury and recovery dynamics following the 1989 *Exxon Valdez* oil spill. *Ecological Applications* **20**(7): 1993–2006.

Klosiewski, S.P. and K.K. Laing (1994). *Marine Bird Populations of Prince William Sound, Alaska, before and after the* Exxon Valdez *Oil Spill: Final Report*. Anchorage, AK, USA: US Fish and Wildlife Service; *Exxon Valdez* Oil Spill State/Federal National Resource Damage Assessment Bird Study Number 2; Final Report. [http://www.evostc.state.ak.us/Files.cfm?doc=/Store/FinalReports/1989-B02-Final.pdf&]

Kuletz, K.J. (1996). Marbled murrelet abundance and breeding activity at Naked Island, Prince William Sound, and Kachemak Bay, Alaska, before and after the *Exxon Valdez* oil spill. In *Proceedings of the* Exxon Valdez *Oil Spill Symposium*. S.D. Rice, R.B. Spies,

D.A. Wolfe, and B.A. Wright, eds. Bethesda, MD, USA: American Fisheries Society; Symposium 18; ISBN-10: 0913235954; ISSN 08922284; pp. 770–784.

Lance, B.K., D.B. Irons, S.J. Kendall, and L.L. McDonald (2001). An evaluation of marine bird population trends following the *Exxon Valdez* oil spill, Prince William Sound, Alaska. *Marine Pollution Bulletin* **42**(4): 298–309.

Litzow, M.A., J.F. Piatt, A.K. Prichard, and D.D. Roby (2002). Response of pigeon guillemots to variable abundance of high-lipid and low-lipid prey. *Oecologia* **132**(2): 286–295.

McKnight, A., K.M. Sullivan, D.B. Irons, S.W. Stephensen, and S. Howlin (2008). *Prince William Sound Marine Bird Surveys, Synthesis and Restoration*. Anchorage, AK, USA: US Fish and Wildlife Service, Migratory Bird Management; *Exxon Valdez* Oil Spill Restoration Project 080751 Final Report. [http://www.evostc.state.ak.us/Files.cfm?doc=/Store/FinalReports/2008–080751-Final.pdf&]

Murphy, S.M., R.H. Day, J.A. Wiens, and K.R. Parker (1997). Effects of the *Exxon Valdez* oil spill on birds: Comparisons of pre- and post-spill surveys in Prince William Sound, Alaska. *The Condor* **99**(2): 299–313.

Murphy, S.M. and T.J. Mabee (2000). Status of black oystercatchers in Prince William Sound, Alaska, nine years after the *Exxon Valdez* oil spill. *Waterbirds* **23**(2): 204–213.

Nysewander, D.R., C.H. Dippel, G.V. Byrd, and E.P. Knudtson (1993). *Effects of the* Exxon Valdez *Oil Spill on Murres: A Perspective from Observations at Breeding Colonies*. Homer, AK, USA: US Fish and Wildlife Service, Alaska Maritime National Wildlife Refuge; *Exxon Valdez* Oil Spill State/Federal National Resource Damage Assessment Bird Study Number 3; Final Report. [http://www.evostc.state.ak.us/Files.cfm?doc=/Store/FinalReports/1992-B03-Final.pdf&]

Oakley, K.L. and K.J. Kuletz (1996). Population, reproduction, and foraging of pigeon guillemots at Naked Island, Alaska, before and after the *Exxon Valdez* oil spill. In *Proceedings of the* Exxon Valdez *Oil Spill Symposium*. S.D. Rice, R.B. Spies, D.A. Wolfe, and B.A. Wright, eds. Bethesda, MD, USA: American Fisheries Society; Symposium 18; ISBN-10: 0913235954; ISSN 08922284; pp. 759–769.

Parker, K.R. and J.A. Wiens (2005). Assessing recovery following environmental accidents: environmental variation, ecological assumptions, and strategies. *Ecological Applications* **15**(6): 2037–2051.

Patten, S.M., Jr., T. Crowe, R. Gustin, R. Hunter, P. Twait, and C. Hastings (2000). *Assessment of Injury to Sea Ducks from Hydrocarbon Uptake in Prince William Sound and the Kodiak Archipelago, Alaska, following the* Exxon Valdez *Oil Spill*. Anchorage, AK, USA: Alaska Department of Fish and Game, Division of Wildlife Conservation; *Exxon Valdez* Oil Spill State/Federal National Resource Damage Assessment Bird Study Number 11 Final Report; Volumes I and II. [http://www.evostc.state.ak.us/Files.cfm?doc=/Store/FinalReports/1992-B11-Final.pdf&]

Peterson, C.H. (2001). The "Exxon Valdez" oil spill in Alaska: Acute, indirect and chronic effects on the ecosystem. *Advances in Marine Biology* **39**: 1–103.

Peterson, D.L. and V.T. Parker, eds (1998). *Ecological Scale: Theory and Applications*. New York, NY, USA: Columbia University Press; ISBN-10: 0231105037; ISBN-13: 9780231105033.

Piatt, J.F. and P. Anderson (1996). Response of common murres to the *Exxon Valdez* oil spill and long-term changes in the Gulf of Alaska marine ecosystem. In *Proceedings of the* Exxon Valdez *Oil Spill Symposium*. S.D. Rice, R.B. Spies, D.A. Wolfe, and B.A. Wright, eds. Bethesda, MD, USA: American Fisheries Society; Symposium 18; ISBN-10: 0913235954; ISSN 08922284; pp. 720–737.

Piatt, J.F. and R.G. Ford (1996). How many seabirds were killed by the *Exxon Valdez* oil spill? In *Proceedings of the* Exxon Valdez *Oil Spill Symposium*. S.D. Rice, R.B. Spies, D.A. Wolfe, and B.A. Wright, eds. Bethesda, MD, USA: American Fisheries Society; Symposium 18; ISBN-10: 0913235954; ISSN 08922284; pp. 712–719.

Piatt, J.F. and C.J. Lensink (1989). *Exxon Valdez* bird toll. *Nature* **342**(6252): 865–866.

Piatt, J.F., C.J. Lensink, W. Butler, M. Kendziorek, and D.R. Nysewander (1990). Immediate impact of the *Exxon Valdez* oil spill on marine birds. *The Auk* **107**(2): 387–397.

Rice, S.D., J.W. Short, M.G. Carls, A. Moles, and R.B. Spies (2007). The *Exxon Valdez* oil spill. In *Long-Term Ecological Change in the Northern Gulf of Alaska*. R.B. Spies, ed. Amsterdam, The Netherlands: Elsevier; ISBN-10: 0444529608; ISBN-13: 9780444529602; pp. 419–520.

Rosenberg, D.H. and M.J. Petrula (1998). *Status of harlequin ducks in Prince William Sound, Alaska, after the* Exxon Valdez *oil spill, 1995–1997*. Anchorage, AK, USA: Alaska Department of Fish and Game, Division of Wildlife Conservation; *Exxon Valdez* Oil Spill Restoration Project 97427 Final Report. [http://www.evostc.state.ak.us/Files.cfm?doc=/Store/FinalReports/1997–97427-Final.pdf&]

Sharp, B.E., M. Cody, and R. Turner (1996). Effects of the *Exxon Valdez* oil spill on the black oystercatcher. In *Proceedings of the* Exxon Valdez *Oil Spill Symposium*. S.D. Rice, R.B. Spies, D.A. Wolfe, and B.A. Wright, eds. Bethesda, MD, USA: American Fisheries Society; Symposium 18; ISBN-10: 0913235954; ISSN 08922284; pp. 748–758.

Skalski, J.R., D.A. Coats, and A.K. Fukuyama (2001). Criteria for oil spill recovery: A case study of the intertidal community of Prince William Sound, Alaska, following the *Exxon Valdez* oil spill. *Environmental Management* **28**(1): 9–18.

Skalski, J.R. and D.H. McKenzie (1982). A design for aquatic monitoring programs. *Journal of Environmental Management* **14**(3): 237–251.

Sowls, A.L., S.A. Hatch, and C.J. Lensink (1978). *Catalog of Alaskan Seabird Colonies*. Anchorage, Alaska, USA: US Fish and Wildlife Service, Office of Biological Services; Project FWS/OBS 78/78.

Springer, A.M. (1998). Is it all climate change? Why marine bird and mammal populations fluctuate in the North Pacific. In *Biotic Impacts of Extratropical Climate Variability in the Pacific: Proceedings, 'Aha Huliko'a 10th Hawaiian Winter Workshop, January 25–29, 1998, University of Hawaii at Manoa, Hawaii*. G. Holloway, P. Müller, and D. Henderson, eds. Honolulu, HI, USA: University of Hawaii, Department of Oceanography and School of Ocean and Earth Science and Technology; pp. 109–119.

US Fish and Wildlife Service (2008). *North Pacific Seabird Colony Database*. Anchorage, AK, USA: US Fish and Wildlife Service, Migratory Bird Management. [http://alaska.fws.gov/mbsp/mbm/northpacificseabirds/colonies/default.htm]

White, C.M., R.J. Ritchie, and B.A. Cooper (1995). Density and productivity of bald eagles in Prince William Sound, Alaska, after the *Exxon Valdez* oil spill. In Exxon Valdez *Oil Spill: Fate and Effects in Alaskan Waters*. P.G. Wells, J.N. Butler, and J.S. Hughes, eds. Philadelphia, PA, USA: American Society for Testing and Materials; ASTM Special Technical Publication 1219; ISBN-10: 0803118961; pp. 762–779.

Wiens, J.A. (1989). Spatial scaling in ecology. *Functional Ecology* **3**(4): 385–397.

Wiens, J.A. (1995). Recovery of seabirds following the *Exxon Valdez* oil spill: An overview. In Exxon Valdez *Oil Spill: Fate and Effects in Alaskan Waters*. P.G. Wells, J.N. Butler, and J.S. Hughes, eds. Philadelphia, PA, USA: American Society for Testing and Materials; ASTM Special Technical Publication 1219; ISBN-10: 0803118961; pp. 854–893.

Wiens, J.A., T.O. Crist, R.H. Day, S.M. Murphy, and G.D. Hayward (1996). Effects of the *Exxon Valdez* oil spill on marine bird communities in Prince William Sound, Alaska. *Ecological Applications* **6**(3): 828–841.

Wiens, J.A., R.H. Day, S.M. Murphy, and M.A. Fraker (2010). Assessing cause-effect relationships in environmental accidents: Harlequin ducks and the *Exxon Valdez* oil spill. *Current Ornithology* **17**: 131–189.

Wiens, J.A., R.H. Day, S.M. Murphy, and K.R. Parker (2004). Changing habitat and habitat use by birds after the *Exxon Valdez* oil spill, 1989–2001. *Ecological Applications* **14**(6): 1806–1825.

Wiens, J.A. and K.R. Parker (1995). Analyzing the effects of accidental environmental impacts: Approaches and assumptions. *Ecological Applications* **5**(4): 1069–1083.

Wiese, F.K. and G.J. Robertson (2004). Assessing seabird mortality from chronic oil discharges at sea. *The Journal of Wildlife Management* **68**(3): 627–638.

Zabala, J., I. Zuberogoitia, J.A. Martínez-Climent, and J. Etxezarreta (2011). Do long lived seabirds reduce the negative effects of acute pollution on adult survival by skipping breeding? A study with European storm petrels (*Hydrobates pelagicus*) during the *Prestige* oil-spill. *Marine Pollution Bulletin* **62**(1): 109–115.

CHAPTER FIFTEEN

Sea otters: trying to see the forest for the trees since the *Exxon Valdez*

David L. Garshelis and Charles B. Johnson

15.1 Introduction

The *Exxon Valdez* oil spill generated enormous public and scientific attention on sea otters (*Enhydra lutris*). Photos of oil-covered sea otters hauled out on beaches or collected in boats frequently appeared in the media and in government reports, making it one of the most notable "poster species" of this spill (Batten, 1990). Rice *et al.* (2007, p. 450) commented, that "Perhaps our most persistent collective memory of the oil spill is the dead and dying sea otters." A major report, *Legacy of an Oil Spill 20 Years after* Exxon Valdez, featured sea otters on the cover and used this species as the predominant case study (*Exxon Valdez* Oil Spill Trustee Council, 2009).

Attention to sea otters was fueled by their charismatic nature and appearance, combined with the fact that no mammal suffered greater spill-related mortality. Given the large number of otters that died (or were not born) as an immediate or long-term result of the spill, the significance of this species in terms of natural resource damage assessment and public relations was enormous. It was argued, for example, that each otter killed in the spill was "worth" at least $80 000, the minimal cost to Exxon for each otter that was captured, cleaned, and rehabilitated (Estes, 1991).

At the time of the spill (March 1989), sea otters were abundant and thriving in Alaska, from the Aleutian Islands to Prince William Sound (PWS) and portions of Southeast Alaska. In the nineteenth century, fur harvests had nearly exterminated them (Box 15.1). But with legal protection in the early 1900s, a few remnant populations bounced back dramatically, even to the extent that in some areas sea otter foraging forced the closure of commercial shellfisheries (Garshelis *et al.*, 1986). Sea otters became the most compelling victim of the spill, not only because of their public appeal and high death toll, but also because they were misrepresented by the media as already rare before 1989 (Batten, 1990). Did this spill once again send this species to the brink?

Oil in the Environment: Legacies and Lessons of the Exxon Valdez *Oil Spill*, ed. J. A. Wiens. Published by Cambridge University Press. © Cambridge University Press 2013.

Box 15.1 Characteristics of sea otters relevant to the oil spill

As to the beauty of the animal, and particularly of its skin, this sea otter is alone incomparable, without a peer; it surpasses all other inhabitants of the vast ocean, and holds the first rank in point of beauty and softness of its fur.

That is how sea otters were first described to science (Steller, 1751, p. 76). It was later discovered that sea otters have the densest fur of any mammal. This luxuriant fur inevitably led to a commercial fur harvest that nearly drove this species to extinction. They were saved in 1911 by an international treaty to protect declining stocks of northern fur seals (*Callorhinus ursinus*) – the treaty included sea otters somewhat as an afterthought. At the time, fewer than 2000 otters in 13 isolated populations remained from a range that once stretched from Baja California, across southern Alaska and the Aleutian Islands, to northern Japan. One of those remnant populations was in western Prince William Sound (PWS). With protection, that population grew and eventually spread to eastern PWS by the late 1970s.

Some limited harvesting of sea otters resumed after Alaska attained statehood in 1959, but this changed in 1972 with passage of the US Marine Mammal Protection Act. This act allows harvests by native people for traditional uses but prohibits killing, capture, and harassment (close approach) by nonnatives without a permit.

As the PWS otter population increased in size and range, it eventually began to conflict with shellfisheries. Otters in PWS subsist mainly on clams and crabs, which they dig or gather from submerged bottom sediments. As otter foraging significantly reduced densities of clams and crabs, commercial shellfisheries were eventually closed throughout PWS.

Because otters must dive to the bottom to obtain food and then swim to the surface to consume it, they are limited to foraging in depths where they can complete a dive within 2 or 3 minutes. Therefore, foraging otters generally remain within the 40-m depth contour, so highest densities are found close to shore. When resting or swimming, however, they can cross much deeper water. The amount of time spent feeding varies from 9–13 hours per day, depending mainly on the availability of food.

Otters also spend a good deal of time grooming their fur. Unlike other marine mammals, they have no underlying layer of blubber, so fur with an entrapped layer of air provides their only insulation. When their fur gets matted, their skin gets wet and they can suffer hypothermia. Having an already high metabolic rate, they cannot compensate. They are able to groom only small quantities of oil off their fur, which makes them very vulnerable to floating oil.

Natural mortality among sea otters is heaviest in the first year of life, especially as pups become independent from their mother at about 6 months of age. At this age they are not capable of deep dives, so they are limited in where they can forage and may suffer malnutrition. In PWS, females at least 3 years old give birth to a single pup annually, typically in May–June. Female–pup pairs are easy to distinguish because females carry their pup on their chest (see Fig. 15.2). Pups become independent in early winter, the harshest time of year in terms of air and water temperatures and choppy seas, causing otters to expend more energy. Pup mortality varies widely (40–80%) among areas and years, related to food conditions and weather. Older otters have much lower rates of mortality (10–20%).

Otters feeling weak often come ashore (otherwise, in PWS, they rarely do); they may die onshore or get washed ashore afterwards. Most mortality occurs in winter so their carcasses or skeletons may be found by walking beaches along the highest tide line in spring. Following the *Exxon Valdez* spill, carcasses of otters that died from oiling in PWS, as well as those that died naturally the previous winter or in years past, were found during cleanup and salvage operations.

Steller, G.W. (1751). De bestiis marinis (The beasts of the sea). W. Miller and J.E. Miller, translators (1899). In *Fur Seals and Fur-Seal Islands of the North Pacific Ocean*. D.S. Jordan, ed. Washington DC, USA: US Government Printing Office; Part III; pp. 179–218.

More than two decades later, the sea otter remains one of the few species still not officially listed as recovered from the spill (*Exxon Valdez* Oil Spill Trustee Council, 2009). Here we review the principal findings from the many studies aimed at assessing the effects of the *Exxon Valdez* oil spill on the status of this species in PWS; we also evaluate factors that led to conflicting assessments over the extent and duration of these effects. We show that limited data from before the spill and a host of confounding factors since the spill made it nearly impossible to judge whether the affected population has returned to its prespill level of abundance; even more difficult was judging whether the population rose to where it would have been, absent the spill. These commonly used recovery targets proved to be vague and elusive, and hence controversial.

15.2 What was known about sea otter status in PWS before the spill?

The number of otters inhabiting the area of western PWS (WPWS) that was subsequently oiled (Fig. 15.1) was unknown immediately before the spill. Occasional surveys had been conducted in this area, but because corrections were not made for otters that were submerged or otherwise unobserved, we refer to these data as "counts" rather than abundance estimates. Nevertheless, because many of these counts were made during ideal conditions (calm seas, good lighting, and at times when otters tend to rest on the surface), they may reasonably approximate abundance for a few distinct areas and can provide a means for tracking trends through time.

Counts were conducted from a small boat around Green Island in WPWS one to four times per year during 1977–85 (Johnson, 1987; Fig. 15.1), providing the best available time series of otter numbers before the spill. Another boat-based count was conducted within 200 m of the entire shoreline of WPWS in 1984 (Irons et al., 1988). Prior to that, a helicopter-based count of WPWS was conducted in the summer and winter of 1973–74 (Pitcher, 1975).

Results of these surveys showed substantial spatial variation in otter numbers, which tended to persist through time as high- and low-density areas. A general pattern of population stability was perceived during the late 1970s through the mid-1980s, and reproduction, as judged from the percentage of pups in the population, also appeared to be fairly stable. This suggested that, by 1989, the WPWS population of sea otters was at carrying capacity (Johnson, 1987), stemming from an extended period of occupancy and legal protection (Box 15.1).

In portions of eastern PWS (EPWS), which was colonized much later than WPWS, food supplies remained relatively plentiful in the late 1980s, and sea otter density was apparently still increasing (Garshelis et al., 1986; Monnett and Rotterman, 1989). Otters in PWS subsisted mainly on clams, which they dug from submerged sediments, as well

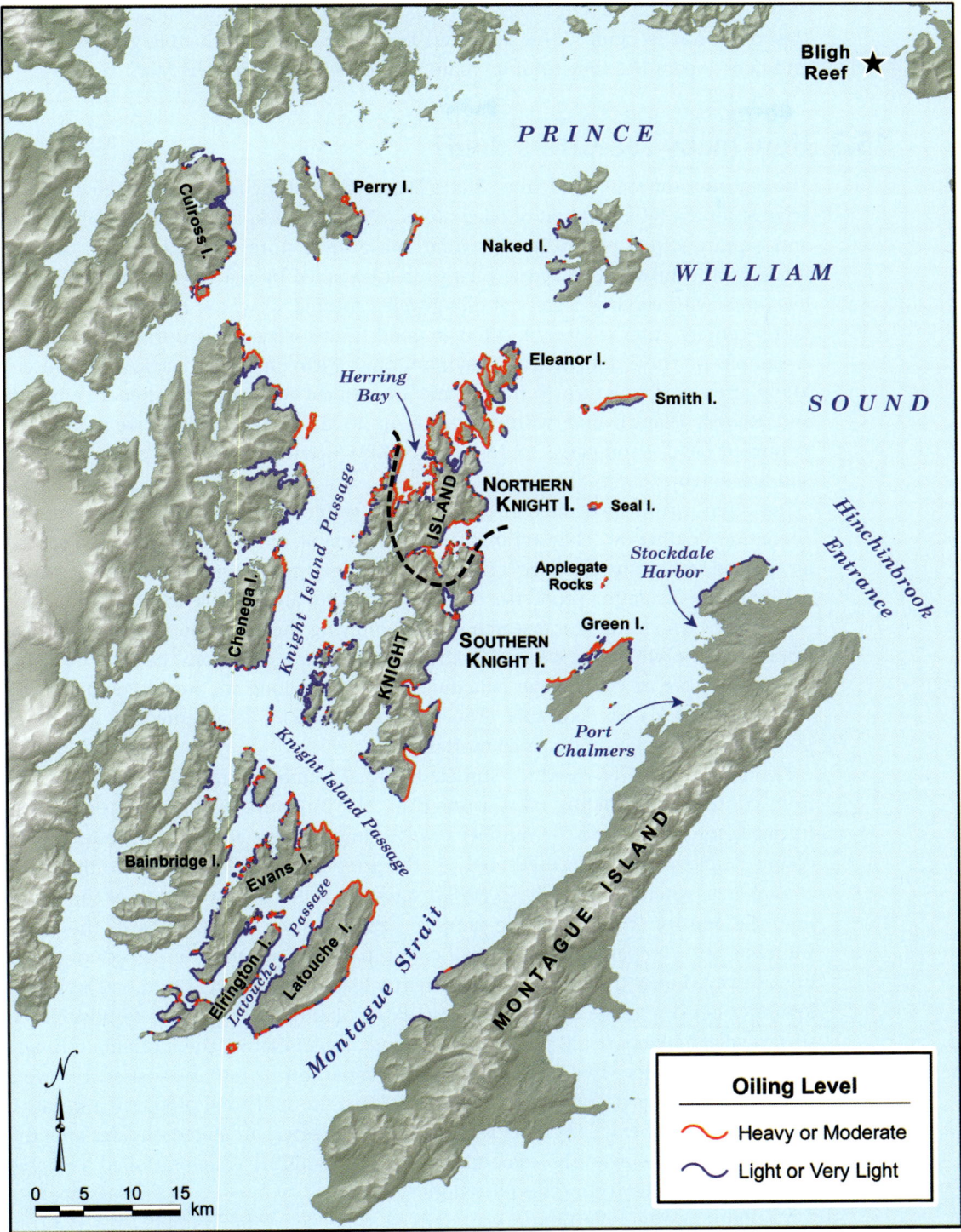

Figure 15.1 Principal sea otter study areas and maximum distribution of oil in western Prince William Sound, Alaska, following the *Exxon Valdez* oil spill. Source: Shoreline Cleanup Assessment Technique 2 (SCAT2) oiling map.

as some epibenthic invertebrates, such as crabs and mussels (Johnson, 1987). Prior to the spill it was recognized that sea otters feeding in EPWS obtained larger food items, and hence spent less time feeding, than otters in WPWS (Garshelis *et al.*, 1986).

15.3 How many sea otters died?

The day after the spill, the United States Department of the Interior directed Exxon to arrange for the rehabilitation of oiled otters. Within a week, facilities were constructed and capture efforts began. By this time, however, a storm had already pushed the floating oil southward, covering a large area occupied by sea otters. Ultimately, 344 live otters with varying degrees of oiling were captured, of which 128 died at rehabilitation centers. After treatment, otters in good health were released in unoiled EPWS. A sample of these individuals was monitored with implanted radio transmitters. Many attempted to return to their home areas (most were from the Kenai Peninsula and Kodiak Island), and within a year up to 55% had died or were lost and presumed dead – compared to 15–17% for EPWS resident control groups (Monnett and Rotterman, 1995a).

Sea otters spend most of their time on the water surface, where they are apt to encounter floating oil. They are particularly vulnerable to the oil because they rely on air trapped in their fur ("pelage") for insulation. They groom themselves frequently to maintain the insulating properties of their pelage, but are not able to remove oil.

During beach cleaning and search efforts following the spill, 866 sea otter carcasses were found, of which 602 were oiled and judged to have died from the spill. Of these, 391 (65%) were in WPWS; the remainder were from along the Kenai Peninsula and Kodiak Island–Alaska Peninsula. The retrieved carcasses represented an unknown portion of the overall spill-related mortality.

Two approaches were used to estimate the total number of otters that died directly from contact with floating oil. Garrott *et al.* (1993) compared the number existing 5 years before the spill to the number counted immediately afterward. Although both counts employed the same methodology, the prespill survey (1984) covered the entire coastal area, whereas the postspill (1989) survey sampled only a portion of the shoreline; the density observed in the sampled area was then extrapolated to the entire shoreline and compared to the prespill count. It turned out that the sample of shorelines surveyed after the spill was inadvertently biased toward areas that had relatively low otter density in the earlier survey, so extrapolating these samples to unsurveyed sites yielded a low postspill population estimate, thus inflating the mortality estimate (Garshelis and Estes, 1997). Restricting the comparison to just those sites that were surveyed in both pre- and postspill periods within the spill zone produced a negative value for the number killed (Table 15.1); that is, more otters inhabited this area after the spill than before. Obviously, some unforeseen confounding issue precluded a reliable estimate of mortality using this procedure.

A second mortality estimate was derived from the total number of recovered carcasses. DeGange *et al.* (1994) marked and released 25 frozen carcasses near Kodiak Island during cleanup operations, 20% of which were later found on beaches. The total number of carcasses actually recovered after the spill was therefore assumed to be 20% of the total mortality (Table 15.1). However, this experiment considered only one route of carcass recovery – a dead otter washing ashore.

Table 15.1 Estimates of the total number of sea otters that died directly from the *Exxon Valdez* oil spill within Prince William Sound.

Method	Key points	Mortality estimate	Source
Comparison of 1989 versus 1984 living population	Surveyed 25% of the spill area in 1989 and extrapolated density to whole area (whole area was surveyed in 1984). Included area north of the spill path (not oiled). Assumed the population in WPWS grew by 12.7% from 1984 to 1989, as observed in EPWS.	2650	Garrott et al. (1993)
Comparison of 1989 versus 1984 living population (reanalysis)	Compared only the 25% of WPWS that was surveyed in both years. Excluded the area north of the spill path. Included and excluded the 12.7% assumed growth from 1984 to 1989.	−540 to −940	Garshelis and Estes (1997)
Experimental release and recovery of carcasses postspill	Released 25 marked carcasses at Kodiak Island; recovery crews found 5 (20%) on the beach.	~2000	DeGange et al. (1994)
Estimation of carcass recovery rate from existing literature	Accounted for moribund otters that hauled out on shore and carcasses collected at sea. Accounted for higher search effort on beaches of PWS than on Kodiak.	750	Garshelis (1997)

Notes: Negative estimates by Garshelis and Estes (1997) were generated because the areas sampled in 1989 had more otters remaining after the spill than were counted in the same areas, using the same technique, in 1984. The authors did not derive an actual mortality estimate, but presented these values to demonstrate that the estimation procedure used by Garrott et al. (1993) was flawed. DeGange et al. (1994) only provided an estimate for the entire spill area, including PWS, Kenai Peninsula, Kodiak Island, and Alaska Peninsula, and used an incorrect figure for the total number of spill-related carcasses found in this area. The number of presumed spill-related carcasses found in PWS was 391, so this divided by 20% and added to the number of otters that were captured alive and eventually died yields the total estimate for PWS.

After the spill, many otter carcasses were also recovered in the water, and many live otters that became oiled hauled out on beaches where they eventually died. Moreover, by the time this experiment was conducted, the search effort for carcasses had waned, so the retrieval rate was no longer directly comparable to that of the extensive effort shortly after the spill. Garshelis (1997) reassessed these data by trying to account for the various ways that carcasses were recovered: using a plausible range of values to make up for lacking empirical data, he estimated a significantly higher carcass-retrieval rate (44–78%) and obtained correspondingly lower estimates of mortality (Table 15.1).

Quantifying how many otters died in the spill informs the question of recovery, as our notion of recovery is often based on whether the immediate death toll has been matched by subsequent population growth. For example, Bodkin et al. (2011) believed that sea otter recovery was still not complete two decades after the spill because the difference between the latest counts and counts made shortly after the spill still fell short of Garrott et al.'s (1993) mortality estimate. Furthermore, they took this as evidence that lower mortality estimates (Garshelis, 1997) must have been wrong. This conclusion presumes that the population would increase just to its immediate prespill level and not surpass that – in other words, that all postspill growth reflects recovery. Of course, this does not account for the conundrum that otter counts just after the spill already exceeded those from 5 years before the spill (Table 15.1), indicating that population growth was occurring before spill-related recovery.

15.4 Obstacles to postspill studies

In the first 4 years after the spill, over 20 scientists were involved in a wide range of sea otter research, mainly in WPWS (Ballachey *et al.*, 1994). Since then, an enormous effort has continued to ascertain whether this species recovered from the initial effects of the spill through normal population growth, or was plagued by chronic effects of residual oil for many years afterward.

To varying degrees, these studies were hindered by the following factors.

15.4.1 Insufficient prespill data

Most prespill data on sea otter numbers and ecology were from Green Island, which became oiled on one half but not the other. The oiled and unoiled halves of the island, however, supported different otter densities before the spill. The heaviest oiling occurred at Knight Island (Fig. 15.1), where few prespill counts and no corresponding ecological information on sea otters were available. The northern half of this island received the brunt of the oil, but like Green Island, the two halves of Knight Island, with different extents of oiling, appeared to have had different sea otter densities before the spill.

15.4.2 Lack of valid reference sites

Various postspill studies used unoiled portions of Green or Knight Island, Montague Island (Fig. 15.1), or EPWS as reference sites; but differences in habitat, occupation history, or prespill otter densities confounded comparisons with oiled sites. Moreover, otters at Green and Knight islands easily moved among portions of shoreline that had been subjected to different degrees of oiling (from heavy to none). Montague Island was essentially unoiled, but in the two bays where counts had been conducted before the spill (Fig. 15.1), otter densities were the highest known in WPWS. All of EPWS was unoiled, but before the spill, otter numbers and distribution in EPWS were much less stable than in WPWS.

15.4.3 Inadequate understanding of natural population dynamics

One of the greatest unknowns hampering postspill studies was the extent of natural variation in this system. Before the spill, it was recognized that otter numbers in local areas could surge or decline in association with periodic redistributions in a matter of days, but the underlying causes were not well understood (Garshelis and Garshelis, 1984; Johnson, 1987). Thus, scientists were faced with assessing long-term responses and recovery from a spill where the only certainty was that a large (but unknown) number of otters died. That otter numbers appeared to be higher after the spill than before in some of the most severely affected portions of WPWS (Johnson and Garshelis, 1995; Garshelis and Johnson, 2001) demonstrated how little was really known about the population dynamics of otters in this area.

15.5 Design of postspill studies

Postspill studies of sea otters were designed to investigate a range of possible effects, from the level of organ damage to general body condition, rates and causes of mortality, abundance, and population growth. Limited prespill information and inadequate

Figure 15.2 Some postspill surveys of sea otters followed the techniques used to survey otters in western PWS during the early–mid-1980s in order to ensure that pre- and postspill counts were comparable. Two observers sat or stood in the bow of a small boat that paralleled the shoreline. Surveys were conducted only during calm weather conditions when otters were easily visible on the water surface (inset shows female carrying her pup). Sea otters use the water surface for resting, swimming, and consuming food. Otters were not visible when foraging underwater, but the boat traveled slowly, and observers looked front and backward to spot otters that surfaced from foraging dives. (Photos: David L. Garshelis)

postspill reference sites necessitated a weight-of-evidence approach, rather than rigorous before-versus-after or control-versus-impact comparisons.

Many of the general methods for studying sea otters were well established before the spill, including capture and handling techniques and implanting radio transmitters for tracking individuals. There was no general agreement, however, on the best methods for counting otters. Some investigators followed prespill counting methods to maintain consistency (Johnson and Garshelis, 1995; Fig. 15.2). Others developed new methods, including surveys from airplanes (Bodkin and Udevitz, 1999), which yielded results not directly comparable to boat-based prespill counts.

Because otters are a federally protected species, study design was also influenced by permit restrictions (Box 15.1). Government-funded scientists were allowed to "take" otters (i.e., capture, handle, collect samples from live or dead specimens), whereas nongovernment scientists would have had difficulty acquiring the necessary permits. Consequently, government scientists conducted a wider range of research, including studies of blood parameters, liver function, polycyclic aromatic hydrocarbon (PAH) metabolites, body condition, movements, and survival of marked otters (e.g., Ballachey *et al.*, 1994, 2003; Bodkin *et al.*, 2002). Nongovernment scientists were much more constrained in the types of studies they could do. Marking, tracking with telemetry, evaluating body condition and organ function, collecting parts of dead otters, and even closely approaching live otters were prohibited, so nongovernment scientists designed projects using observational methods and publicly available data (e.g., Garshelis, 1997; Garshelis and Johnson, 2001). This dichotomy led to very different research questions and study designs, which in turn fueled debates when findings conflicted.

15.6 Undisputed findings from scientific studies

There was no dispute among scientists that sea otters had suffered significant mortality, even though investigators did not agree on the actual death toll (Table 15.1). There was also no disputing that a large number of sea otters survived. This was because oil did not completely engulf the area (see Chapter 1, Fig. 1.3): large stretches of shoreline inside bays or coves, hidden behind islands, or facing south (opposite to the movement of the oil), were protected. Sea otters in those areas did not encounter oil unless they swam out into it.

Another area of (eventual) agreement was that more otters inhabited WPWS 1–2 years after the spill than were thought to have lived there 5 years before the spill. Initially, Burn (1994) reported a 35% decline in otters from the last prespill count in 1984 to counts made a few months after the spill, and then a further decline in 1990. A re-examination of these data several years later, when an electronic database of Burn's counts was made available, revealed that the declines initially reported arose by combining data from the spill zone with an unoiled swath in northern PWS (NPWS). It is unknown what caused the precipitous drop in otters in portions of NPWS, but when data from this unoiled region were separated from WPWS, otters in WPWS showed an increase from 1984 to 1989 (Garshelis and Johnson, 2001). This finding was consistent with the results of Johnson and Garshelis (1995).

These results presented an obvious paradox: if as many otters were living in the spill zone in 1989 and 1990 as were counted there 5 years before the spill, the population must have increased during the period before the spill when no counts were conducted, and the spill removed this previous increase (Johnson and Garshelis, 1995; Garshelis and Johnson, 2001). The prespill increase may have resulted from enhanced reproduction, reduced mortality, and/or increased immigration from NPWS. None of this was recognized at the time of the spill, making the initial postspill counts enigmatic. This situation made recovery more ambiguous because, immediately after the spill, the population was already at or above the level that would normally serve as the de facto recovery target, absent population-trend information.

15.7 Conflicting or equivocal findings from scientific studies

A year after the spill, oil was rarely detected on or in the water in WPWS (Taft et al., 1995; Chapter 3), but concerns remained that otters could suffer from exposure to oil residues buried in sediments or in tainted prey, or from decreased abundance of prey. Because some oil remained in the environment, effects were expected to continue for a few more years in some areas. With time and the continued weathering of remaining residues, it was expected that baseline conditions for otters would soon resume.

In assessing recovery of a species, the chief variable of interest is abundance, so measures of changing abundance or parameters directly affecting abundance (reproduction and survival) were the primary focus. No study detected any spill-related effects on sea otter reproduction, which was easily observed and quantified (Garshelis and Johnson, 2001; Bodkin et al., 2002). Survival, though, was much more difficult to measure.

15.7.1 Changes in otter survival in WPWS

Mortality is highest in young otters, just after weaning (approximately 6 months old), so it made sense to direct attention to this age group as potentially the most sensitive to spill effects (Box 15.1). Two studies (Rotterman and Monnett, 1995; Ballachey et al., 2003) surgically implanted radio transmitters in sea otter pups and monitored their survival for the year immediately postweaning (weanling survival). Pups captured in WPWS 2–4 years after the spill were less likely to survive the year than those captured in EPWS. This comparison was initially designed to examine the chronic effects of oiling in WPWS, but different food conditions in the two areas, owing to differences in duration of occupancy by otters (Garshelis et al., 1986), confounded the results. No comparisons were made between pups raised in oiled versus unoiled portions of WPWS. In fact, most of the WPWS pups in the two telemetry studies were not from oiled areas, and no observed mortalities were directly attributable to oil.

Sea otter carcasses (generally skeletons) collected on beaches during the spring, after the normal winter die-off (Box 15.1), provided another means for examining changes in mortality over time. Age-at-death was assessed from growth layers in the teeth. Prespill data on the age structure of dead otters at Green Island existed from carcass collections made during 1976–85 (Johnson, 1987). Carcasses were also collected across a wide area of WPWS following the spill, and age-at-death was determined from those judged to have died naturally before the spill. Systematic postspill collections were made at Green Island from 1990–97, and then within a larger oiled area in 1998 (Monson et al., 2000).

The observed age composition of dead animals changed over time, and a mathematical model was developed to explain this change. Holding reproduction, immigration, and emigration constant, age-specific survival was altered in the model to produce results that best matched observed ages at death (Monson et al., 2000). The results indicated that young animals had higher rates of mortality immediately after the spill than before, but their survival returned to normal or even above normal within a few years. Conversely, survival declined for middle-aged and older otters. These results were interpreted as indicative of a prolonged spill-related effect on survival even for otters born after the spill, with gradual recovery due to the eventual loss of these older-aged, debilitated cohorts (Monson et al., 2000; Bodkin et al., 2002).

A major incongruity existed, though, between the results of the modeling and the number of otters actually observed. Carcasses collected after the spill were primarily from Green Island, where counts of otters had been stable or increasing since 1990 and were equal to or greater than prespill levels (Johnson and Garshelis, 1995; Garshelis and Johnson, 2001). If survival of adult animals had been declining through time, it must have been compensated for by increased reproduction or immigration in order for total numbers to be so high. However, such an increase in reproduction or immigration violates the assumptions of the model; in other words, the model could not explain both the carcass age distribution and the number of otters living at Green Island.

Annual carcass collections were continued over a wide area of WPWS from 1999 to 2008, and the observed age structure continued to change in a way that suggested prolonged negative effects on survival (Monson *et al.*, 2011). Whereas the proportion of pups remained fairly stable, the proportion of 2–8 year olds ("prime age") increased while the proportion of older otters declined. In an attempt to explain this continued apparent depression of survival in prime-age otters in the face of continuing overall increase in the WPWS population, Monson *et al.* (2011) developed a more complex "source–sink" model in which otter numbers in one portion of the population could be increasing (as observed), while emigrants from that source area supported a population sink (an area of population loss that is maintained by immigrants from elsewhere; Liu *et al.*, 2011).

Monson *et al.*'s model used data from an unoiled site on Montague Island as a source population, and a large portion of WPWS, with variable degrees of past oiling (from none to heavy; e.g., Green Island, Fig. 15.1) as the presumed sink. The model predicted an unchanging "sink population" of about 900 otters during 1990–2009, supported by a continually growing source population; as a result, the hypothesized sink portion declined from $>40\%$ to $<25\%$ of the total WPWS otter population. Model results indicated no improvement of survival in the sink for at least two decades postspill, yielding a cumulative loss over this period of ~800 animals beyond what was expected had baseline survival not changed; these deaths were considered to be chronic effects of the oil spill (Monson *et al.*, 2011).

Models such as this are an attempt to make sense of a complex array of data, and may do so elegantly – but that does not guarantee that the model correctly portrays reality. The credibility of the model results may be assessed by examining the validity of its key assumptions and predictions:

- The model assumed emigration of juvenile otters, mainly males, from Montague Island to various sites throughout WPWS. Although emigration of young males has been observed, there is no direct evidence from tagging and telemetry studies of young males from Montague settling elsewhere in WPWS.
- If dispersing male otters from unoiled areas did preferentially settle in oiled areas (e.g., because of lower densities) and lived into adulthood (as the model predicted), the sex ratio and hence reproductive output would change, a violation of model assumptions and counter to observations.
- With high rates of male immigration in the modeled sink area, mortality would have to be male-biased to maintain the female-biased sex ratio of otters in WPWS. Thus, most carcasses should have been male. However, the sex of carcasses could not be reliably determined, so sexes were lumped together and used in the model as the distribution of female ages at death.
- Model results indicated that "otters born within the sink population faced chronic, low level exposures [to oil] during development (as pups and juveniles), which presumably could lead to decreased survival rates in adulthood" (Monson *et al.*, 2011, p. 2929). Otters that dispersed into such areas from other (unoiled) areas as older juveniles should therefore suffer less mortality than individuals born in sink areas. In the model, however, male dispersers died at a high rate when they were older.
- The model used population trends at one small area on northern Knight Island (NKI; discussed in more detail below) to represent the trends in the entire hypothesized sink area, which encompassed more than 10 times the number of otters at NKI.

However, the NKI area is unique: no other area with the same population trend has been identified, despite over two decades of sound-wide surveys.

- The source–sink model attempted to overcome the issue encountered in the original model, where stable or increasing counts were observed in the very areas where modeled age-at-death data predicted declines. However, this same conundrum plagued the more complex model. The carcasses that provided the age-at-death data for the sink population were obtained from a broad region of WPWS where otter numbers were increasing.

- Model predictions did not match observed trends. The model predicted a continued decline of the sink population, but counts at NKI during 2007–09 showed a sizeable increase (discussed further below).

This model has been held out as the chief evidence of continuing elevated mortality of otters attributable to the spill (Bodkin et al., 2012). Like any model, however, there are assumptions, and in this case there appear to be inconsistencies in assumed conditions and predicted outcomes.

15.7.2 Trends in otter abundance in WPWS

Scientists generally agreed that otter numbers increased in WPWS following the oil spill. Boat surveys from an area of WPWS with both oiled and unoiled sites indicated a population growth rate of 2.5% per year from 1991 to 1996, but slower for the next several years (Fig. 15.3) (Garshelis and Johnson, 2001). Aerial surveys in 1993–2009

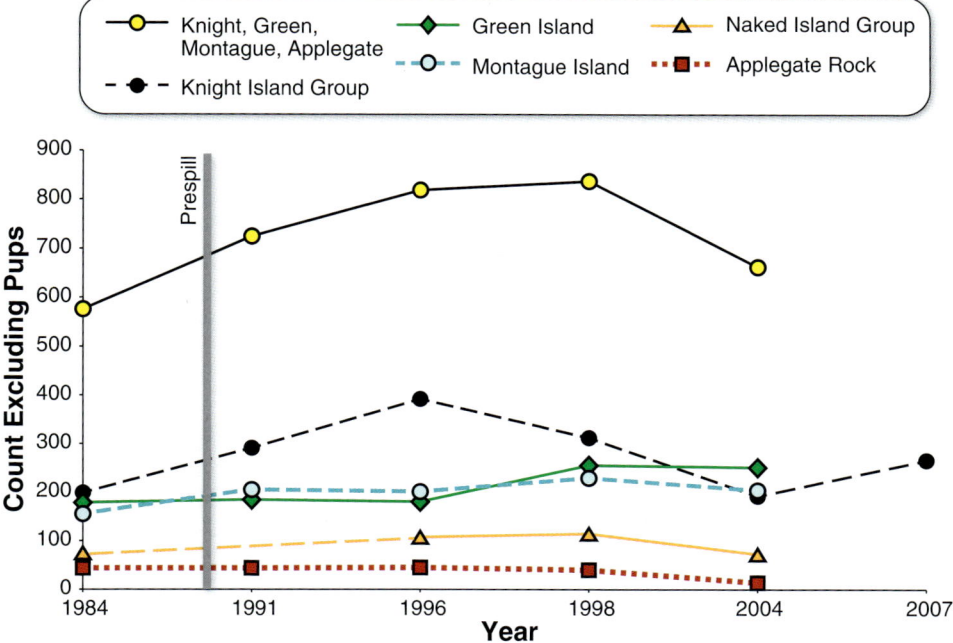

Figure 15.3 Counts of sea otters, excluding dependent pups, obtained during boat-based surveys of portions of western PWS (site locations shown in Fig. 15.1). Prespill data are from Johnson (1987) and Irons et al. (1988), and postspill data from Garshelis and Johnson (2001; and unpublished data). Postspill survey methods followed those of prespill investigators. Yearly points are the means of 1–8 surveys per site. Only years with survey data are shown on x-axis.

across a wider area in WPWS indicated that otter numbers continued to grow at an average of 2.6% per year; in fact, the population virtually doubled, increasing by nearly 2000 otters over this period (Bodkin et al., 2011). (McKnight et al.'s (2006) boat survey recorded no trend in otter numbers, but variability in population estimates derived from this survey was extreme, possibly due in part to movements of otters in and out of the small sampling units.)

One area that appeared to stand out as an exception to this general increasing trend was the northern half of Knight Island (including Disk, Ingot, and Eleanor islands). The north-facing shorelines of this island group captured one of the first major landings of oil following the grounding of the *Exxon Valdez*, resulting in heavy oiling (Fig. 15.1). Consequently, NKI became a focal point of extensive cleanup and of postspill recovery studies for many species. There appeared to be a significant contrast between the flat trend in otter numbers at NKI and the surge in numbers observed at unoiled Montague Island (indicated by aerial but not boat-based counts; Fig. 15.4), suggesting to some that the oil had a depressing effect on population growth (Bodkin et al., 2002).

Counts of otters sharply declined across a broad region of WPWS, including unoiled Montague Island, from 2001 to 2002 (Fig. 15.4a). At NKI, numbers did not fully rebound until 2007 (Fig. 15.4b). No other areas were investigated as intensely as NKI, so it is unknown whether the more prolonged depression in numbers there was an anomaly or part of a wider trend.

This raises the general issue of whether it was preferable to examine otter abundance in relatively small but heavily oiled sites like NKI, looking for discrepancies from a reference site; or to examine variation across a broader spatial and temporal scale and attempt to discern whether outliers matched places that had significant oiling. The first approach creates more Type I errors (detecting oiling effects that are not real), whereas the latter is more prone to Type II errors (not finding oiling effects that are present). Postspill studies of sea otters were made more difficult by the fact that potential reference sites were not only ecologically different from oiled sites, but otter numbers at reference sites were changing.

Although a population increase was expected across WPWS as otters recovered from the spill, such an increase was not expected for unoiled sites such as Montague Island. Historically, numbers of otters at Montague varied widely, but no trend was apparent in nine counts made from 1959 to 1984 (Lensink, 1962; Pitcher, 1975; Johnson, 1987). A postspill surge in otters at this site, followed by a series of large fluctuations (Bodkin et al., 2011; Fig. 15.4a), was as anomalous as the lack of an increase at other sites (Fig. 15.3), some of which were oiled. Up-and-down swings in otter populations at various sites in PWS, associated with periodic redistributions, predated the spill (Garshelis and Garshelis, 1984; Johnson, 1987) but are still not well understood.

Arguably a better comparative site with NKI is the southern half of Knight Island (SKI). This area was oiled, but far less than NKI, and areas with high otter densities at SKI received very little oil (Fig. 15.1, Table 15.2). Eight boat-based surveys conducted during 1991–2007 around Knight Island found a strikingly parallel trend in otter numbers between the northern and southern halves of the island (Fig. 15.5a), even though reproduction (measured as the proportion of otters with pups) was consistently higher at SKI (Fig. 15.5b). Notably, a change in distribution of otters around Knight Island was evident prior to the spill: otter numbers declined precipitously at NKI from 1973 to 1984 while increasing at SKI (Table 15.3). It is not wise to infer trends from only two

Table 15.2 Extent of shoreline area along the northern and southern portions of Knight Island, Prince William Sound, covered by three categories of oil as recorded during two time periods.

Portion of Knight Island	Date	Percentage of shoreline in oiling category			
		Heavy	Moderate	Light	No Impact
Northern	Summer 1989	29%	21%	27%	24%
	Fall 1989	29%	21%	39%	10%
Southern	Summer 1989	8%	5%	7%	80%
	Fall 1989	9%	5%	28%	58%

Notes: Data for summer represent the cumulative extent of oiling from the date of the spill in late March through late August, before cleanup was suspended for the winter. Some areas that were not directly impacted became lightly oiled by fall (Sept.–Nov.) 1989 from cleanup operations that dispersed some of the oil. Digital data were obtained from the Alaska Department of Natural Resources and analyzed using ArcMap 9.1. Oiling categories are described by Neff *et al*. (1995) and Chapter 4, Table 4.2. Total shoreline lengths were 136 km and 299 km in northern and southern portions of the island, respectively. Oil distribution was patchy (Fig. 15.1).

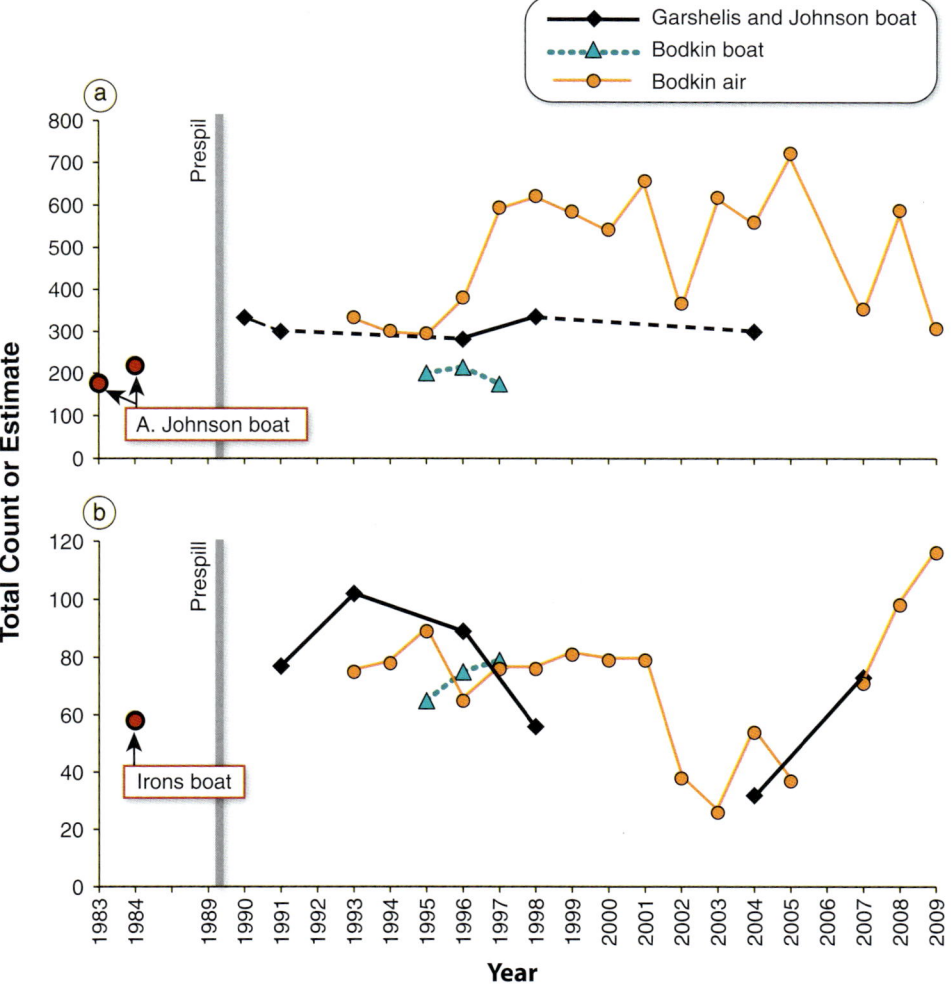

Figure 15.4 Trends in total sea otter numbers at (a) northwestern Montague Island, and (b) northern Knight Island, PWS. Boat surveys were counts along the complete shoreline, unadjusted for detection rate. Postspill counting methods followed prespill methods at each site. Aerial counts were made along sample transects, corrected for detection and extrapolated to the entire shoreline. Aerial counts at Montague Island included a portion of Green Island, so are somewhat higher than boat counts. Pups could not be differentiated during aerial surveys, so all data include pups. Dashed lines denote significant gaps between boat surveys at Montague. Data sources: Johnson (1987), Irons *et al*. (1988), Johnson and Garshelis (1995), Garshelis and Johnson (2001; and unpublished data), and Bodkin *et al*. (2002, 2011).

Table 15.3 Counts of sea otters made 16 years to 5 years before the *Exxon Valdez* oil spill (1973–84) at Southern Knight Island (SKI), Northern Knight Island (NKI), and Herring Bay (part of NKI), Prince William Sound, Alaska, compared to counts made 2 years and 18 years after the spill, using the same survey methodology as used in 1984.

Survey			Count				
Year	Month	Platform	SKI	NKI	% Knight Is. Otters in NKI	Herring Bay	% NKI Otters in Herring Bay
1973	June	Helicopter	103	105	50%	22	21%
1974	March	Helicopter	50	27	35%	9	33%
1984	June–Aug.	Boat	200	58	22%	7	12%
1991	July–Aug.	Boat	344	77	18%	11	15%
2007	June	Boat	299	73	20%	15	21%

Notes: Although counts from different survey platforms are not directly comparable, the counts highlight changes in otter distribution around Knight Island over time. Surveys in 1973 and 1974 were conducted by Pitcher (1975), 1984 survey by Irons *et al*. (1988), and 1991 and 2007 surveys by Garshelis and Johnson (unpublished data). Values shown for 1991 are the mean of three surveys. All counts include pups; pups were distinguished during boat-based counts but not during helicopter counts. Counts are not adjusted for imperfect detection.

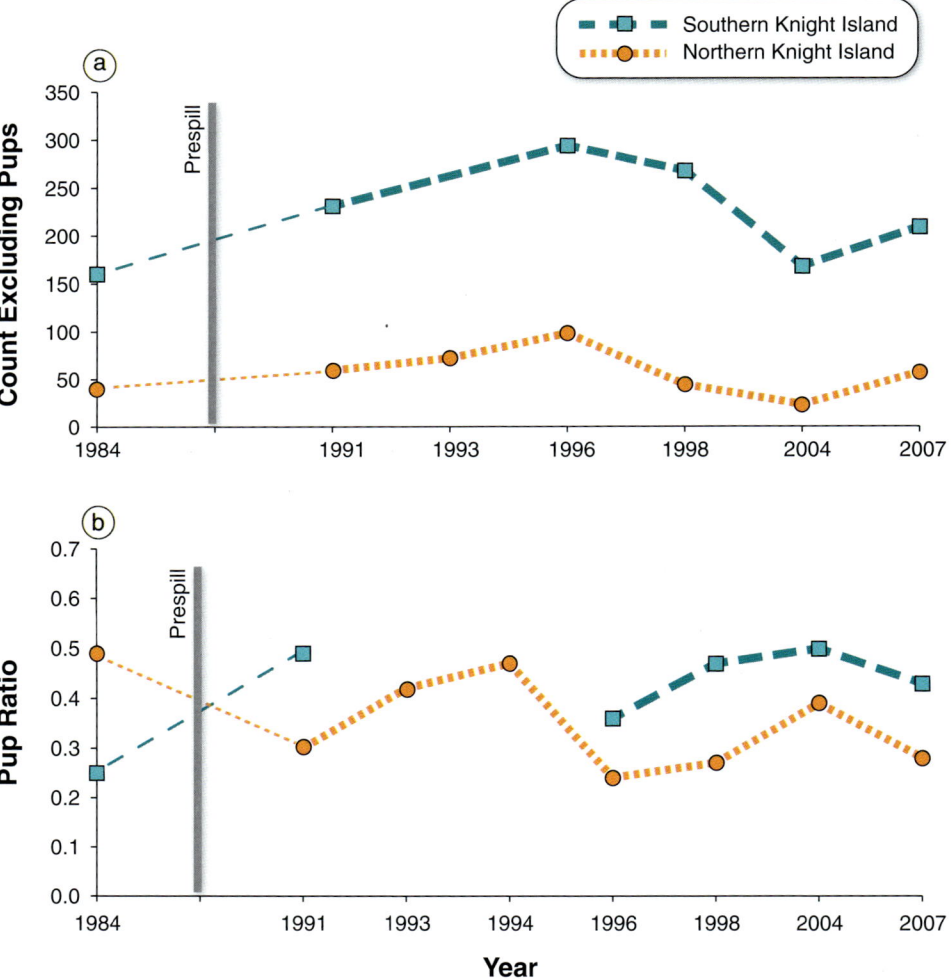

Figure 15.5 (a) Counts of sea otters (excluding pups), and (b) ratios of dependent pups: independent otters tallied during boat-based surveys conducted during 1991–2007 (some years missing) at Knight Island, PWS (Garshelis and Johnson, 2001; and unpublished data) versus a single prespill survey using the same technique (Irons *et al*., 1988). Only years with survey data are shown on *x*-axis.

points in time, but that is all that was available in this case, and the counts changed dramatically over this 11-year period.

The unexplained dynamics in the sea otter population around Knight Island (Fig. 15.5), at Montague Island (Fig. 15.4a), and at other sites in WPWS (Fig. 15.3) complicated interpretations of population trends at NKI. There is little empirical or conceptual basis for claims about what the population trajectory at individual sites in WPWS should have been in the period since the oil spill.

15.7.3 Prespill versus postspill abundance at northern Knight Island

Over time, as oil continued disappearing from shorelines, concerns about sea otter recovery focused on a smaller and smaller portion of WPWS, eventually centering solely on NKI. Some scientists believed that sea otters at NKI had not recovered nearly two decades after the spill, based on purportedly low otter abundance compared to prespill estimates (Rice et al., 2007; Bodkin et al., 2012). There is considerable uncertainty and disagreement, however, as to the number of otters that occupied NKI before the spill (Box 15.2).

Dean et al. (2000, 2002) derived an estimate of prespill abundance at NKI from a count made by Pitcher (1975) 16 years before the spill. Pitcher surveyed all of PWS from a helicopter during June 1973 and again in March 1974. At NKI, these two counts differed by nearly four-fold (Table 15.3). To assess the proportion of otters missed, Pitcher compared the March helicopter counts to counts made by boat. Overall, boat counts were 73% higher than helicopter counts, although at Knight Island the difference was 205%. Applying this range of correction factors to the March 1974 helicopter count at NKI yielded estimates of 47–82 otters. Pitcher did not compare helicopter to boat counts during summer. However, because of better lighting (higher sun angle) and less wind, summer aerial counts tend to be more accurate than in winter. Given that the uncorrected summer helicopter count at NKI (105 otters) was higher than the corrected winter count, it seems likely that significantly fewer otters were missed during the summer. Dean et al. (2000, 2002) applied an unexplained correction factor of 230% to Pitcher's summer count to derive their estimate (237) of the number of otters present at NKI during the summer of 1973, which they assumed was a good approximation to the number there just before the spill in March 1989.

Dean et al. (2000) also used a second approach to estimate prespill numbers of otters at NKI. They reasoned that the number of dead and moribund otters collected shortly after the spill provided a minimum estimate of the number of otters that lived there at the time of the spill. Adjusting for carcasses that were not recovered (per Garshelis, 1997), they estimated that 165 otters were present at NKI at the time of the spill. This estimate assumes that the number of floating carcasses that drifted into the area after the spill equaled the number that drifted out. This assumption is unlikely to be true because prevailing currents flowed *into* this area from the origin of the spill. Moreover, shortly after the spill, PWS was struck by a large storm with northerly winds that pushed the floating oil southward, heavily oiling the north-facing shorelines (Galt and Payton, 1990). Given that oil landed disproportionately at NKI (Fig. 15.1), it follows that dead and moribund otters drifted on a similar course (Hill et al., 1990). With these currents and wind conditions, the number of sea otter carcasses collected at NKI would have been substantially higher than the number living there at the time of the spill.

Box 15.2 The elusive baseline

The number of animals in the area impacted by the spill at the instant before the spill is considered the baseline. If this is known with some degree of certainty, recovery from the loss may be evaluated as a return to this baseline value. However, baseline abundance is an elusive target because it is confounded by population trends prior to the spill, population changes in nonimpacted areas after the spill, redistributions, and seasonal fluctuations. Moreover, the number of animals occupying the impacted area just before the mortality-causing event is rarely known, so disparate baseline estimates are likely to arise.

In the case of sea otters at northern Knight Island (NKI), at least seven potential baseline estimates are available in the literature. These were generated from counts made during 1973–84 and carcasses collected just after the spill. Values varied by nearly an order of magnitude. Selection of a baseline value (or correction of a past estimate to derive a baseline value – e.g., estimate 2 in Table 15.2.1) thus yielded vastly different perceptions of recovery when compared to postspill estimates. Many of the discrepancies among perceptions of the recovery of otters at NKI related directly to varying baseline estimates. Postspill estimates of abundance were more consistent, even though they were generated with different techniques.

Table 15.2.1 Multiple ways of estimating baseline abundance of sea otters in northern Knight Island, PWS.

Prespill estimate	Description	Comments	Source
(1)	1973 summer helicopter count	NKI summer population declined during 1970s–80s, so this 16-year-old estimate probably did not represent the 1989 population	Pitcher (1975)
(2)	1973 summer count corrected for unobserved otters	Correction factor not based on empirical data	Dean et al. (2002)
(3)	1974 March helicopter count (same month as spill)	Uncorrected count underestimated true abundance, especially in winter when low sun makes otters hard to see	Pitcher (1975)
(4)	1974 March count corrected based on boat count	No comparable winter counts made postspill	Pitcher (1975)
(5)	1984 summer boat count	Count concentrated near shoreline	Irons et al. (1988)
(6)	1989 carcasses collected	Some carcasses not found; includes carcasses that washed into NKI from farther north	Dean et al. (2000)
(7)	1989 carcasses collected, corrected for those not found	Includes carcasses that washed into NKI from farther north	Dean et al. (2000)
Postspill estimate			
(A)	1993[a] boat count	Same technique as 1984 count, so should be comparable, even though neither was corrected for imperfect detection and unsurveyed areas far from shore	Garshelis and Johnson (2001; and unpublished data)
(B)	1993 airplane count	Area sampled with series of straight transects; estimate included large correction factor (~5×) to account for incomplete coverage and imperfect detection	Bodkin et al. (2002, 2011)

[a] 1993 was the first year after the spill when both a boat-based and airplane-based estimate of NKI were available.

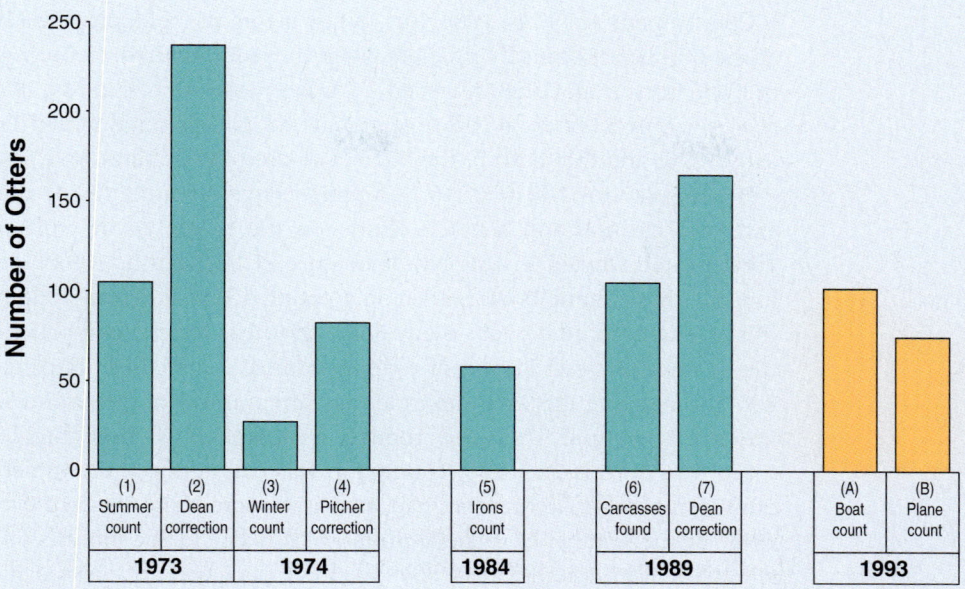

Figure 15.2.1 Various baseline estimates of sea otter numbers at northern Knight Island, PWS. See Table 15.2.1 for details.

Bodkin, J.L., B.E. Ballachey, T.A. Dean, A.K. Fukuyama, S.C. Jewett, L. McDonald, D.H. Monson, C.E. O'Clair, and G.R. VanBlaricom (2002). Sea otter population status and the process of recovery from the 1989 *Exxon Valdez* oil spill. *Marine Ecology Progress Series* **241**: 237–253.

Bodkin, J.L., B.E. Ballachey, and G.G. Esslinger (2011). *Synthesis of Nearshore Recovery Following the 1989* Exxon Valdez *Oil Spill: Trends in Sea Otter Population Abundance in Western Prince William Sound*. Anchorage, AK, USA: US Geological Survey, Alaska Science Center; *Exxon Valdez* Oil Spill Restoration Project 070808, 070808A, and 090808 Final Report. [http://www.evostc.state.ak.us/Files.cfm?doc=/Store/FinalReports/2009-090808-Final.pdf&]

Dean, T.A., J.L. Bodkin, A.K. Fukuyama, S.C. Jewett, D.H. Monson, C.E. O'Clair, and G.R. VanBlaricom (2002). Food limitation and the recovery of sea otters following the *Exxon Valdez* oil spill. *Marine Ecology Progress Series* **241**: 255–270.

Dean, T.A., J.L. Bodkin, S.C. Jewett, D.H. Monson, and D. Jung (2000). Changes in sea urchins and kelp following a reduction in sea otter density as a result of the *Exxon Valdez* oil spill. *Marine Ecology Progress Series* **199**: 281–291.

Garshelis, D.L. and C.B. Johnson (2001). Sea otter population dynamics and the *Exxon Valdez* oil spill: Disentangling the confounding effects. *Journal of Applied Ecology* **38**(1): 19–35.

Irons, D.B., D.R. Nysewander, and J.L. Trapp (1988). *Prince William Sound Sea Otter Distribution in Relation to Population Growth and Habitat Type*. Anchorage, AK, USA: US Fish and Wildlife Service; unpublished report.

Pitcher, K.W. (1975). Distribution and abundance of sea otters, Steller sea lions, and harbor seals in Prince William Sound, Alaska. In *Distribution and Abundance of Marine Mammals in the Gulf of Alaska*. D.G. Calkins, K.W. Pitcher, and K. Schneider, eds. Anchorage, AK, USA: Alaska Department of Fish and Game, Division of Game; Appendix A.

Counts made at NKI in 1984, just 5 years before the spill, provide a better indication of the number of otters living there when the spill occurred. Counting over the whole of PWS, Irons et al. (1988) recorded 2.5 times more otters than did Pitcher. That Irons et al. saw only 58 otters at NKI suggests that the NKI population had declined since the early 1970s and that both of the Dean et al. prespill estimates were unrealistic.

Herring Bay on NKI (Fig. 15.1) captured large amounts of oil and thus attracted extensive cleanup and research efforts. At the height of the spill response, nearly 1000 people worked in this bay (Hooten and Highsmith, 1996). Anecdotal reports indicate that cleanup boats herded oil, floating debris, and some wildlife carcasses into this area in order to contain them. Some scientists have drawn particular attention to the 38 otter carcasses known or estimated to have come from Herring Bay (Rice et al., 2007), suggesting this as a minimum baseline number of otters present in that bay at the time of the spill. This value, though, is more than five times higher than the seven otters that Irons et al. (1988) counted in this bay during the summer of 1984. Either otter numbers in Herring Bay had increased more than five-fold between 1984 and 1989, or the number of carcasses found in this bay in the months following the spill represented an accumulation of individuals that died and drifted in from the large expanse of water to the north.

Depending on which prespill values are used, otter numbers in Herring Bay and other parts of NKI could have been above baseline as early as 1991, or could have been below baseline for two decades postspill (Table 15.3, Fig. 15.4b). The problem of defining a reliable baseline severely hindered attempts to assess recovery (Box 15.2). Nevertheless, the contention that abundance of otters at NKI remained below prespill levels prompted studies into the mechanisms that might explain possible persistent oiling effects.

15.8 Evaluating mechanisms for long-term oiling effects

15.8.1 Demographic conditions

Various demographic conditions were proposed to explain the lack of growth in the NKI sea otter population during the 1990s. These included higher than normal mortality and emigration from NKI (Bodkin et al., 2002); higher rates of loss, despite immigration from source populations (Monson et al., 2011); and immigration into WPWS from EPWS, but avoiding NKI (Rice et al., 2007). This assortment of proposals reflects the lack of direct evidence that mortality, immigration, or emigration were aberrant at this site compared to elsewhere in WPWS. Moreover, suggestions of unbalanced rates of immigration and emigration at heavily oiled sites such as NKI were contrary to empirical telemetry data, which showed no indication that immigration or emigration was affected by degree of shoreline oiling (Monnett and Rotterman, 1995b).

Long-term effects on sea otter health were not detectable anywhere in the spill zone. Monson (2009) reported improving body condition for WPWS otters beginning about 2 years after the spill and lasting about a decade. Likewise, 7–9 years after the spill, Dean et al. (2002) found that 1–4-year-old otters at NKI had better body condition than those at Montague Island, where otter density was higher and food availability per otter estimated to be lower.

15.8.2 Biomarkers

Elevated levels of cytochrome P450 1A (CYP1A), an enzyme system involved in metabolizing planar aromatic hydrocarbons (Chapter 9), were higher for otters at NKI than at Montague; this was considered evidence of continued exposure to oil at NKI (Ballachey *et al.*, 2002). However, it was found that the reverse transcriptase–polymerase chain reaction (RT-PCR) used to measure CYP1A messenger RNA levels was based on a genetic sequence not related to sea otter CYP1A (Chapter 9; Hook *et al.*, 2008). Miles *et al.* (2012) investigated another potential biomarker of oil exposure based on immune-function gene transcripts, which differed between WPWS otters and a control site. But within WPWS, those same transcription profiles occurred at unoiled Montague, indicating that the "anomaly" was not linked to oil (Garshelis and Johnson, 2013).

15.8.3 Subsurface oil residues

The idea that otters might still be exposed to oil two decades after the spill stemmed in part from what appeared to be a plausible pathway of exposure to oil residues that remained buried beneath the surface of some shorelines in WPWS, most notably at NKI. Short *et al.* (2006) investigated the distribution of subsurface oil residues (SSOR) along NKI shorelines in 2003 and suggested that otters digging for clams in this region would "encounter lingering *Exxon Valdez* oil repeatedly during the course of a year" (Short *et al.*, 2006, p. 3728), perhaps at least once every 2 months. They asserted that this frequency of encounter would be sufficient to affect the health of otters and thus hamper their population growth.

Neff *et al.* (2011), however, pointed out that Short *et al.*'s estimate was based on the assumptions that otters have an equal likelihood of digging for clams at all tidal heights and in all sediment types along the shoreline and that SSOR is distributed evenly across all shoreline substrates – none of which is correct. Otters dig for clams in perpetually wet mixed sand/gravel in subtidal and lower intertidal zones, whereas the remaining SSOR occurred mostly in small patches in poorly irrigated, middle- and upper-tidal sediments behind boulders or under a boulder/cobble surface armor, where it was protected from wave action, a substrate unsuitable for large clams (Chapters 6, 7; Neff *et al.*, 2011). Indeed, the protection afforded by this rocky substrate is the very reason the oil persisted. Clams rarely are found in this type of habitat, and otters do not (and cannot) dig there.

When otters dig for clams, they leave pits in the substrate, which may last for many months and are readily visible along shorelines at low tide (Fig. 15.6). Boehm *et al.* (2007, 2011) and Neff *et al.* (2011) found that foraging pits along NKI shorelines in 2006 were distinctly separated by habitat and tidal zone from pockets of SSOR, suggesting that foraging otters would rarely encounter SSOR. These results spurred a further investigation by Bodkin *et al.* (2012), who searched soft-sediment beaches in 2008 and found more otter pits in the middle-intertidal zone than Boehm *et al.* did along all shoreline types in NKI. Bodkin *et al.* also found traces of oil in or near some otter pits, suggesting a higher rate of potential encounter with SSOR than projected by Boehm *et al.* (2007, 2011).

15.8.4 Quantitative ecological risk assessment

Are the disparities between the findings of Bodkin *et al.* (2012) and Boehm *et al.* (2007, 2011) regarding the extent of overlap between foraging otters and SSOR likely to be important in terms of effects on individuals? To address this issue, Harwell *et al.* (2010; Chapter 16) developed an ecological risk-assessment model to quantify potential sea otter exposure to SSOR at NKI and its toxicological effects. The model included a range

Figure 15.6 Sea otters in PWS feed primarily on clams that they dig from bottom sediments. They feed in the intertidal zone only when the area is submerged. At lower tidal levels, exposed areas where otters have fed show characteristic foraging pits and scattered clam shells with one side bitten and cracked off (inset). This easily visible sign was used to map intertidal otter foraging areas at northern Knight Island and their relationship to residual buried oil, which could be exposed by such foraging. The extent to which otters encountered buried oil via their foraging has been debated in the literature, but quantitative ecological-risk modeling indicated that even the highest estimated rates of exposure to SSOR from foraging pits have a very low probability of harming individual otters. Demographic consequences of this exposure are even less likely. (Photos: David L. Garshelis)

of SSOR-encounter frequencies that well exceeded the higher estimates of Bodkin *et al.* (2012). According to this model, oil-encounter rates for the maximally exposed individuals would have to be > 30 times higher than predicted to reach the minimum dose that would cause chronic effects. With fewer than 100 otters at NKI, it would take on average 10 years for any individual otter to obtain this minimum dose. Moreover, sensitivity analyses conducted using the risk-assessment model (Harwell *et al.*, 2010, 2012) indicated that, for toxicological effects to occur, maximally exposed otters would need to dig 2–10 pits intersecting SSOR per *day* over several months. This far exceeds Bodkin *et al.*'s (2012) estimated SSOR encounter rate of 2–24 pits per *year*. The conclusion from this modeling was that no plausible toxicological risk from remnant oil existed for individual otters, much less for the population of otters at NKI.

15.9 Confounding factors affecting sea otter population growth

15.9.1 Predation by killer whales

Following the *Exxon Valdez* oil spill, sea otter numbers began to plummet in the western Aleutian Islands, even though *Exxon Valdez* oil never reached the Aleutians and was never implicated in the otter decline there. Based on several lines of evidence,

Estes *et al.* (1998) concluded that killer whale (*Orcinus orca*) predation was the most plausible cause. Garshelis and Johnson (1999) found that there was equally compelling evidence for killer whale predation on otters around Knight Island. Common to both Knight Island and the Aleutians, there was no indication of reduced birthing or pup survival, few dead otters washed ashore (as they would have in cases of disease, malnutrition, or contamination), and body condition of otters indicated that food supplies were adequate (Dean *et al.*, 2000, 2002; Laidre *et al.*, 2006).

Few instances of killer whale predation on otters have been observed in either the Aleutians or WPWS. In the Aleutians, only six killer whale attacks were observed, and among these only three of the otters died (Hatfield *et al.*, 1998). Given the low probability of actually witnessing such brief events in this huge area, the three confirmed mortalities were extrapolated to an estimated 40 000 otters consumed by killer whales (Estes *et al.*, 1998). It is now widely believed that killer whale predation reduced the Aleutian Islands' otter population by more than 95% (Estes *et al.*, 2005). Doroff *et al.* (2003, p. 55) called it "one of the most widespread and precipitous population declines for a mammalian carnivore in recorded history." Despite the rather scant observational evidence of the cause for this decline, when southwestern Alaska sea otters were proposed as a threatened population under the US Endangered Species Act, killer whale predation was considered the sole threat (US Fish and Wildlife Service, 2004).

Although studies of both otters and killer whales have been conducted in PWS since the early 1980s, the first attack was not witnessed until 1992 – by coincidence, shortly after the spill. All three observed killer whale attacks since then occurred at Knight Island, two of which were in Herring Bay, NKI (Hatfield *et al.*, 1998). Additionally, in 2003 a killer whale was found dead in Latouche Passage, south of Knight Island, with five sea otters in its stomach. This whale was identified as part of a pod whose range was centered in the Knight Island area (Vos *et al.*, 2006). Its stomach contents and the three observed attacks on sea otters, all near Knight Island, far exceed the killer whale predation rate observed anywhere in the Aleutians.

Killer whales could not only consume several otters per day at Knight Island, but the risk of predation could drive otters to safer areas. Accordingly, it seems that killer whales should be foremost on a list of potential factors affecting population trends of sea otters at Knight Island. The reasons for increased killer whale predation on otters remain uncertain, but it may be linked to widespread declines of other marine mammals, which are preyed upon preferentially by some pods of killer whales (Estes *et al.*, 2009). Dramatic declines of harbor seals (*Phoca vitulina*) in WPWS (Frost *et al.*, 1999) may have spurred some killer whale pods to consume more sea otters (Saulitis *et al.*, 2000).

15.9.2 Subsistence harvest by Alaska natives

Alaska natives are legally permitted to harvest sea otters for subsistence or handicrafts, and these harvests may have affected population trends in portions of WPWS. In parts of southeast Alaska, the reported harvest rate (up to 8% of the total otter population per year) was sufficient to limit or depress otter numbers (Esslinger and Bodkin, 2009). After the *Exxon Valdez* spill, at least 139 otters were harvested from the spill zone (US Fish and Wildlife Service, unpublished data, 1990–2009). Harvests were especially high at Knight Island: in 2000 and 2003, natives took 5–10% of the 200–300 otters living there (data were inadequate to trace losses to the northern or southern halves of the island). That these harvests exceeded the highest population growth rate observed in other portions of WPWS suggests that they caused a population decline at Knight Island. By contrast, since 1998 only two otters were harvested from Montague Island, which harbors a larger sea otter population than Knight Island (note: Fig. 15.3 shows otter numbers for only Stockdale Harbor and Port Chalmers, see Fig. 15.1, not all of Montague Island).

During 2005–09, only two sea otters were reported harvested at Knight Island – this coincides with a period of otter population growth at NKI (Fig. 15.4b). Although the effects of subsistence harvests on otter numbers at NKI remain equivocal, they cannot be discounted as a factor that affected the dynamics of the otter population in this area.

15.9.3 Disturbance from human activity

Ironically, one of the largest impacts to PWS following the *Exxon Valdez* spill – aside from the oil itself – was a substantial increase in human activity directed at assessing impacts in the most heavily oiled areas. NKI was not only the target of extensive cleanup operations, but also of numerous postspill studies of shoreline chemistry, kelp, invertebrates, fish, birds, and otters. Investigators established campsites and ran boats in and out of this area for two decades. In other parts of PWS, otters tended to leave areas with high boat traffic (Garshelis and Garshelis, 1984). The sea otter studies at NKI also involved the capture and handling of over 200 individuals (out of a population averaging fewer than 80 individuals, but with substantial individual turnover), adding more disturbance. Bodkin *et al.* (2011) suggested that the disturbance from a new fishery contributed to many otters leaving the Montague Island survey areas in 2009 (Fig. 15.4a).

15.9.4 Natural environmental changes

Large natural perturbations also significantly affected the ecosystem of WPWS. In 1964, the area was struck by the largest earthquake on record in North America (Chapter 1), severely affecting habitat and food availability for sea otters over an extended period (Garshelis and Johnson, 2001). Beginning in the mid-1970s, abrupt, large-scale changes in atmospheric conditions and ocean currents caused higher water temperatures in the northern Gulf of Alaska (GOA), including PWS, which altered the physical and biological processes of this region on a massive scale (Spies, 2007). This "regime shift" has been linked to marked changes in the abundance of a number of marine species, including some affected by the oil spill (Agler *et al.*, 1999; Chapter 13). Changes in oceanic conditions are still taking place, including a minor regime shift during the year of the spill, which affected various biota in the region (Hare and Mantua, 2000). Recent, large-scale declines in marine mammals have been witnessed from the GOA through the Aleutian Islands, although the cause is uncertain (Springer *et al.*, 2003; Trites *et al.*, 2007).

In the face of all this ecosystem "noise," it is probably impossible to discern an unambiguous signal from an oil spill that occurred more than two decades in the past in an area such as NKI, with fewer than 100 sea otters.

15.10 Measuring recovery

Whether or not sea otters have recovered from the spill depends on how recovery is defined. At least three definitions of recovery have been posed for sea otters:

1. Abundance equals what it was prespill.
2. Abundance equals what it would have been had there been no spill.
3. All conditions returned to what they would have been in the absence of the spill.

Regarding condition (1), it was clear that within about 2 years after the spill, otter numbers in most areas (Fig. 15.3), including heavily oiled NKI (Table 15.3), equaled or exceeded what they had been 5 years before the spill. This, though, does not mean that otters had recovered to their prespill 1989 level. One approach to judging whether numbers rebounded to prespill levels would be to compare the postspill increase in abundance to estimated spill-related mortality. This entails numerous difficulties, however: (a) total WPWS mortality estimates varied more than three-fold (Table 15.1); (b) spill-related mortality is impossible to accurately estimate for small areas like NKI; and (c) unoiled areas, such as Montague Island, increased postspill, and it is unclear whether that growth should be included as part of recovery of the WPWS spill zone.

Recovery target (2) incorporates both prespill abundance and population trend, absent the spill. That the number of otters counted immediately after the spill was similar to 1984, despite the large number that died directly from the oil spill, suggests that otters in WPWS were generally increasing from 1984 to 1989. However, this might not have been the case at NKI, where otter numbers declined dramatically prior to 1984 (Table 15.3). The different trajectory of population growth at NKI since the spill could therefore reflect conditions present before the spill. But even if the NKI population was also increasing before the spill, otter numbers were so low that a very small additional source of mortality or emigration could stifle population growth. Bodkin et al. (2002) noted that, with an average of fewer than 80 otters at NKI, an extrinsic factor that caused an additional loss of only three otters per year would offset the "recovering" annual population growth rate of 4% per year observed elsewhere in WPWS. One killer whale could easily consume this number of otters in just one day (and still not satisfy its daily caloric requirements; Williams et al., 2004). It would be virtually impossible to ascertain the cause of such small numbers of losses and also determine whether they differed from what occurred before the spill.

Recovery target (3) involves assumptions about prespill conditions combined with forecasts of what would have occurred over the ensuing decades in terms of otter abundance and distribution, absent the spill but with changing environmental conditions. Bodkin et al. (2012) argued that a failure to return to prespill levels combined with evidence of potential continuing exposure is sufficient to conclude nonrecovery, given the difficulty of directly attributing demographic changes to a specific cause. The problem, of course, is that if a species faces a number of changing potential risks (as was the case for sea otters), none of which can be empirically linked to demographic harm, then gauging "recovery" becomes a matter of subjective judgment. Such unmeasurable recovery targets provoke continuing debate, with little chance of resolution.

15.11 Lessons learned

- An event of the nature and magnitude of the *Exxon Valdez* oil spill inevitably leads to disagreements about short- and long-term effects. In the case of sea otters, scientists with differing perspectives and with different regulatory and funding constraints posed questions differently, designed studies differently, and interpreted similar sets of data differently – resulting in different conclusions.

- Conflicting conclusions are likely to arise from contrasting approaches to discerning effects. Some sea otter studies scrutinized data for possible indications of lingering harm, while others explored a broader suite of factors impacting the study sites looking for evidence of recovery. The latter strategy might be perceived as an industry-friendly search for feasible alternatives to an oiling effect. But had the catastrophe not been human-caused, and thus free of legal and monetary repercussions, the "explore all alternatives" approach would be the most scientifically justified.

- If the pre-event dynamics are not well understood, before-versus-after study designs will not yield reliable results. Baseline conditions that are erroneously assumed to be static can skew perceptions of recovery unless the signal from the impact is particularly strong. Environmental and demographic conditions for sea otters were not static prior to the spill.

- The complexity of confounding factors makes understanding ecological catastrophes extremely challenging (Chapter 10). With large background variation, a comparison of impacted versus reference sites requires too many replicates to be feasible. Moreover, each site must be sufficiently large to contain a demographically meaningful population – this was the difficulty highlighted by the small "population" of sea otters at NKI.

- Resolution of opposing conclusions may be aided by insights from other disciplines. While sea otter biologists struggled with determining whether a small population at NKI deviated significantly from an ambiguous prespill baseline or an ecologically different reference site, shoreline ecologists and specialists in ecological risk assessment investigated the likelihood of continuing effects from oil exposure. Their finding that encounter rates with SSOR by pit-digging sea otters had infinitesimally small toxicological risks provided what appeared to be the most conclusive result among the various studies aimed at investigating the potential long-term effects of the spill on this species.

- Recovery must be gauged against a measureable target. A particularly vivid lesson from the sea otter studies is that assessment of recovery depends on how recovery is defined. Vague and unmeasurable recovery targets, based on scant data from the past, provide fertile ground for conflicting views about whether recovery has occurred.

- Perspectives are often clouded when catastrophic events are human-caused (Estes, 1991, 1999). The sea otter case was especially muddled and prolonged, in part because it involved such a high-profile, charismatic species. Claims of nonrecovery were essentially reduced to only three otters, the missing "expected" annual growth at NKI (Bodkin et al., 2002), within a total population of about 12 000 in all of PWS (US Fish and Wildlife Service, 2008). This small deviation, insignificant in terms of the overall demographics of sea otters in PWS and unmeasurable for almost any species, still spawns new studies and continuing controversy. Meanwhile, an even larger "natural" event (apparently related to increased killer whale predation) has been devastating sea otter populations in southwestern Alaska. Otter numbers there have declined by more than 100 000 since the late 1980s and that population is now threatened with extirpation (US Fish and Wildlife Service, 2005). However, natural events, even of such extraordinary magnitude, often do not attract nearly as much public attention as human-caused catastrophes.

REFERENCES

Agler, B.A., S.J. Kendall, D.B. Irons, and S.P. Klosiewski (1999). Declines in marine bird populations in Prince William Sound, Alaska, coincident with a climatic regime shift. *Waterbirds* **22**(1): 98–103.

Ballachey, B.E., J.L. Bodkin, and A.R. DeGange (1994). An overview of sea otter studies. In *Marine Mammals and the* Exxon Valdez. T.R. Loughlin, ed. San Diego, CA, USA: Academic Press; ISBN-10: 0124561608; pp. 47–59.

Ballachey, B.E., J.L. Bodkin, S. Howlin, A.M. Doroff, and A.H. Rebar (2003). Correlates to survival of juvenile sea otters in Prince William Sound, Alaska, 1992–1993. *Canadian Journal of Zoology* **81**(9): 1494–1510.

Ballachey, B.E., J.J. Stegeman, P.W. Snyder, G.M. Blundell, J.L. Bodkin, T.A. Dean, L. Duffy, D. Esler, G. Golet, S. Jewett, L. Holland-Bartels, A.H. Rebar, P.E. Seiser, and K.A. Trust (2002). Oil exposure and health of nearshore vertebrate predators in Prince William Sound following the 1989 *Exxon Valdez* oil spill. In *Mechanisms of Impact and Potential Recovery of Nearshore Vertebrate Predators Following the 1989* Exxon Valdez *Oil Spill, Vol. 1*. L.E. Holland-Bartels, ed. Anchorage, AK, USA: US Geological Survey, Alaska Biological Science Center; *Exxon Valdez* Oil Spill Restoration Project 99025 Final Report. [http://www.evostc.state.ak.us/Files.cfm?doc=/Store/FinalReports/1999-99025-Final.pdf&]

Batten, B.T. (1990). Press interest in sea otters affected by the *T/V Exxon Valdez* oil spill: a star is born. In *Sea Otter Symposium: Proceedings of a Symposium to Evaluate the Response Effort on Behalf of Sea Otters after the* T/V Exxon Valdez *Oil Spill into Prince William Sound, April 17–19, 1990, Anchorage, Alaska*. K. Bayha and J. Kormendy, eds. Anchorage, AK, USA: US Fish and Wildlife Service; *Biological Report* **90**(12): 32–40.

Bodkin, J.L., B.E. Ballachey, H.A. Coletti, G.G. Esslinger, K.A. Kloecker, S.D. Rice, J.A. Reed, and D.H. Monson (2012). Long-term effects of the *Exxon Valdez* oil spill: sea otter foraging in the intertidal as a pathway of exposure to lingering oil. *Marine Ecology Progress Series* **447**: 273–287.

Bodkin, J.L., B.E. Ballachey, T.A. Dean, A.K. Fukuyama, S.C. Jewett, L. McDonald, D.H. Monson, C.E. O'Clair, and G.R. VanBlaricom (2002). Sea otter population status and the process of recovery from the 1989 *Exxon Valdez* oil spill. *Marine Ecology Progress Series* **241**: 237–253.

Bodkin, J.L., B.E. Ballachey, and G.G. Esslinger (2011). *Synthesis of Nearshore Recovery Following the 1989* Exxon Valdez *Oil Spill: Trends in Sea Otter Population Abundance in Western Prince William Sound*. Anchorage, AK, USA: US Geological Survey, Alaska Science Center; *Exxon Valdez* Oil Spill Restoration Project 070808, 070808A, and 090808 Final Report. [http://www.evostc.state.ak.us/Files.cfm?doc=/Store/FinalReports/2009-090808-Final.pdf&]

Bodkin, J.L. and M.S. Udevitz (1999). An aerial survey method to estimate sea otter abundance. In *Marine Mammal Survey and Assessment Methods*. G.W. Garner, S.C. Amstrup, J.L. Laake, B.F.J. Manly, L.L. McDonald, and D.G. Robertson, eds. Rotterdam, The Netherlands: A.A. Balkema; ISBN-10: 9058090434; pp. 13–26.

Boehm, P.D., D.S. Page, J.M. Neff, and J.S. Brown (2011). Are sea otters being exposed to subsurface intertidal oil residues from the *Exxon Valdez* oil spill? *Marine Pollution Bulletin* **62**(3): 581–589.

Boehm, P.D., D.S. Page, J.M. Neff, and C.B. Johnson (2007). Potential for sea otter exposure to remnants of buried oil from the *Exxon Valdez* oil spill. *Environmental Science & Technology* **41**(19): 6860–6867.

Burn, D.M. (1994). Boat-based population surveys of sea otters in Prince William Sound. In *Marine Mammals and the* Exxon Valdez. T.R. Loughlin, ed. San Diego, CA, USA: Academic Press; ISBN-10: 0124561608; pp. 61–80.

Dean, T.A., J.L. Bodkin, A.K. Fukuyama, S.C. Jewett, D.H. Monson, C.E. O'Clair, and G.R. VanBlaricom (2002). Food limitation and the recovery of sea otters following the *Exxon Valdez* oil spill. *Marine Ecology Progress Series* **241**: 255–270.

Dean, T.A., J.L. Bodkin, S.C. Jewett, D.H. Monson, and D. Jung (2000). Changes in sea urchins and kelp following a reduction in sea otter density as a result of the *Exxon Valdez* oil spill. *Marine Ecology Progress Series* **199**: 281–291.

DeGange, A.R., A.M. Doroff, and D.H. Monson (1994). Experimental recovery of sea otter carcasses at Kodiak Island, Alaska, following the *Exxon Valdez* oil spill. *Marine Mammal Science* **10**(4): 492–496.

Doroff, A.M., J.A. Estes, M.T. Tinker, D.M. Burn, and T.J. Evans (2003). Sea otter population declines in the Aleutian archipelago. *Journal of Mammalogy* **84**(10): 55–64.

Esslinger, G.G. and J.L. Bodkin (2009). *Status and Trends of Sea Otter Populations in Southeast Alaska, 1969–2003*. Reston, VA, USA: US Department of the Interior, US Geological Survey; Scientific Investigations Report 2009–5045. [http://pubs.usgs.gov/sir/2009/5045/pdf/sir20095045.pdf]

Estes, J.A. (1991). Catastrophes and conservation: Lessons from sea otters and the Exxon Valdez. *Science* **254**(5038): 1596.

Estes, J.A. (1999). Otter-eating orcas: response. *Science* **283**(5399): 177.

Estes, J.A., D.F. Doak, A.M. Springer, and T.M. Williams (2009). Causes and consequences of marine mammal population declines in southwest Alaska: a food-web perspective. *Philosophical Transactions of the Royal Society B* **364**: 1647–1658.

Estes, J.A., M.T. Tinker, A.M. Doroff, and D.M. Burn (2005). Continuing sea otter population declines in the Aleutian archipelago. *Marine Mammal Science* **21**(1): 169–172.

Estes, J.A., M.T. Tinker, T.M. Williams, and D.F. Doak (1998). Killer whale predation on sea otters linking oceanic and nearshore ecosystems. *Science* **282**(5388): 473–476.

***Exxon Valdez* Oil Spill Trustee Council** (2009). *2009 Status Report*. Anchorage, AK, USA: *Exxon Valdez* Oil Spill Trustee Council. [http://www.evostc.state.ak.us/Universal/Documents/Publications/AnnualStatus/2009AnnualReport.pdf]

Frost, K.J., L.F. Lowry, and J.M. ver Hoef (1999). Monitoring the trend of harbor seals in Prince William Sound, Alaska, after the *Exxon Valdez* oil spill. *Marine Mammal Science* **15**(2): 494–506.

Galt, J.A. and D.L. Payton (1990). Movement of oil spilled from the *T/V Exxon Valdez*. In *Sea Otter Symposium: Proceedings of a Symposium to Evaluate the Response Effort on Behalf of Sea Otters after the* T/V Exxon Valdez *Oil Spill into Prince William Sound, April 17–19, 1990, Anchorage, Alaska*. K. Bayha and J. Kormendy, eds. Anchorage, AK, USA: US Fish and Wildlife Service; Biological Report 90(12); pp. 4–17.[http://archive.org/details/seaottersymposiu00bayh]

Garrott, R.A., L.L. Eberhardt, and D.M. Burn (1993). Mortality of sea otters in Prince William Sound following the *Exxon Valdez* oil spill. *Marine Mammal Science* **9**(4): 343–359.

Garshelis, D.L. (1997). Sea otter mortality estimated from carcasses collected after the *Exxon Valdez* oil spill. *Conservation Biology* **11**(4): 905–916.

Garshelis, D.L. and J.A. Estes (1997). Sea otter mortality from the *Exxon Valdez* oil spill: Evaluation of an estimate from boat-based surveys. *Marine Mammal Science* **13**(2): 341–351.

Garshelis, D.L. and J.A. Garshelis (1984). Movements and management of sea otters in Alaska. *Journal of Wildlife Management* **48**(3): 665–678.

Garshelis, D.L., J.A. Garshelis, and A.T. Kimker (1986). Sea otter time budgets and prey relationships in Alaska. *Journal of Wildlife Management* **50**(4): 637–647.

Garshelis, D.L. and C.B. Johnson (1999). Otter-eating orcas. *Science* **283**(5399): 176–177.

Garshelis, D.L. and C.B. Johnson (2001). Sea otter population dynamics and the *Exxon Valdez* oil spill: Disentangling the confounding effects. *Journal of Applied Ecology* **38**(1): 19–35.

Garshelis, D.L. and C.B. Johnson (2013). Prolonged recovery of sea otters from the *Exxon Valdez* oil spill? A re-examination of the evidence. *Marine Pollution Bulletin*. DOI: 10.1016/j.marpolbul.2013.3.027.

Hare, S.R. and N.J. Mantua (2000). Empirical evidence for North Pacific regime shifts in 1977 and 1989. *Progress in Oceanography* **47**(2–4): 103–145.

Harwell, M.A., J.H. Gentile, C.B. Johnson, D.L. Garshelis, and K.R. Parker (2010). A quantitative ecological risk assessment of the toxicological risks from *Exxon Valdez* subsurface oil residues to sea otters at northern Knight Island, Prince William Sound, Alaska. *Human and Ecological Risk Assessment* **16**(4): 727–761.

Harwell, M.A., J.H. Gentile, K.R. Parker, S.M. Murphy, R.H. Day, A.E. Bence, J.M. Neff, and J.A. Wiens (2012). Quantitative assessment of current risks to harlequin ducks in Prince William Sound, Alaska, from the *Exxon Valdez* oil spill. *Human and Ecological Risk Assessment* **18**(2): 261–328.

Hatfield, B.B., D. Marks, M.T. Tinker, K. Nolan, and J. Peirce (1998). Attacks on sea otters by killer whales. *Marine Mammal Science* **14**(4): 888–894.

Hill, K.A., F. Weltz, T.P. Monahan, and R.W. Davis (1990). Capture operations. In *Sea Otter Rehabilitation Program: 1989* Exxon Valdez *Oil Spill*. T.M. Williams and R.W. Davis, eds. Galveston, TX, USA: International Wildlife Research; pp. 59–81. [www.valdezsciences.com]

Hook, S.E., M.E. Cobb, J.T. Oris, and J.W. Anderson (2008). Gene sequences for cytochromes p450 1A1 and 1A2: The need for biomarker development in sea otters (*Enhydra lutris*). *Comparative Biochemistry and Physiology, Part B* **151**(3): 336–348.

Hooten, A.J. and R.C. Highsmith (1996). Impacts on selected intertidal invertebrates in Herring Bay, Prince William Sound, after the *Exxon Valdez* oil spill. In *Proceedings of the* Exxon Valdez *Oil Spill Symposium*. S.D. Rice, R.B. Spies, D.A. Wolfe, and B.A. Wright, eds. Bethesda, MD, USA: American Fisheries Society; Symposium 18; ISBN-10: 0913235954; ISBN: 08922284; pp. 249–270.

Irons, D.B., D.R. Nysewander, and J.L. Trapp (1988). *Prince William Sound Sea Otter Distribution in Relation to Population Growth and Habitat Type*. Anchorage, AK, USA: US Fish and Wildlife Service; unpublished report.

Johnson, A.M. (1987). *Sea Otters of Prince William Sound, Alaska*. Anchorage, AK, USA: US Fish and Wildlife Service; unpublished report.

Johnson, C.B. and D.L. Garshelis (1995). Sea otter abundance, distribution, and pup production in Prince William Sound following the *Exxon Valdez* oil spill. In Exxon Valdez *Oil Spill: Fate and Effects in Alaskan Waters*. P.G. Wells, J.N. Butler, and J.S. Hughes, eds. Philadelphia, PA, USA: American Society for Testing and Materials; ASTM Special Technical Publication 1219; ISBN-10: 0803118961; pp. 894–929.

Laidre, K.L., J.A. Estes, M.T. Tinker, J. Bodkin, D. Monson, and K. Schneider (2006). Patterns of growth and body condition in sea otters from the Aleutian archipelago before and after the recent population decline. *Journal of Animal Ecology* **75**(4): 978–989.

Lensink, C.J. (1962). *The History and Status of Sea Otters in Alaska*. West Lafayette, IN, USA: Purdue University; Ph.D. Dissertation; UMI (ProQuest) Publication Order No. 6203469. [http://disexpress.umi.com/dxweb]

Liu, J., V. Hull, A.T. Morzillo, and J.A. Wiens (2011). *Sources, Sinks and Sustainability*. Cambridge, UK: Cambridge University Press; ISBN-10: 0521145961; ISBN-13 9780521145961.

McKnight, A., K.M. Sullivan, D.B. Irons, S.W. Stephensen, and S. Howlin (2006). *Marine Bird and Sea Otter Population Abundance of Prince William Sound, Alaska: Trends Following the* T/V Exxon Valdez *Oil Spill, 1989–2005*. Anchorage, AK, USA: US Fish and Wildlife Service, Migratory Bird Management; *Exxon Valdez* Oil Spill Restoration Projects 040159/050751 Final Report. [http://www.evostc.state.ak.us/Files.cfm?doc=/Store/FinalReports/2005-050751-Final.pdf&]

Miles, A.K., L. Bowen, B. Ballachey, J.L. Bodkin, M. Murray, J.L. Estes, R.A. Keister, and J.L. Stott (2012). Variation of transcript profiles between sea otters *Enhydra lutris* from Prince William Sound, Alaska, and clinically normal reference otters. *Marine Ecology Progress Series* **451**: 201–212.

Monnett, C. and L.M. Rotterman (1989). *Distribution and Abundance of Sea Otters in Southeastern Prince William Sound*. Anchorage, AK, USA: US Fish and Wildlife Service, Marine Mammals Management; unpublished report.

Monnett, C. and L.M. Rotterman (1995a). *Mortality and Reproduction of Sea Otters Oiled and Treated as a Result of the* Exxon Valdez *Oil Spill*. Anchorage, AK, USA: US Fish and Wildlife Service, Alaska Fish and Wildlife Research Center; *Exxon Valdez* State/Federal Natural Resource Damage Assessment; Marine Mammal Study 6–14 Final Report. [http://www.evostc.state.ak.us/Files.cfm?doc=/Store/FinalReports/1992-MM0614-Final.pdf&]

Monnett, C. and L.M. Rotterman (1995b). *Movements of Weanling and Adult Female Sea Otters in Prince William Sound, Alaska, After the* T/V Exxon Valdez *Oil Spill*. Anchorage, AK, USA: US Fish and Wildlife Service, Alaska Fish and Wildlife Research Center; *Exxon Valdez* State/Federal Natural Resource Damage Assessment; Marine Mammal Study 6–12 Final Report. [http://www.evostc.state.ak.us/Files.cfm?doc=/Store/FinalReports/1992-MM0612-Final.pdf&]

Monson, D.H. (2009). *Sea Otters (*Enhydra lutris*) and Steller Sea Lions (*Eumetopias jubatus*) in the North Pacific: Evaluating Mortality Patterns and Assessing Population Status at Multiple Time Scales*. Santa Cruz, CA, USA: University of California at Santa Cruz; Ph.D. Dissertation; UMI (ProQuest) Publication Order No. 3351047. [http://disexpress.umi.com/dxweb]

Monson, D.H., D.F. Doak, B.E. Ballachey, and J.L. Bodkin (2011). Could residual oil from the *Exxon Valdez* spill create a long-term population "sink" for sea otters in Alaska? *Ecological Applications* **21**(8): 2917–2932.

Monson, D.H., D.F. Doak, B.E. Ballachey, A. Johnson, and J.L. Bodkin (2000). Long-term impacts of the *Exxon Valdez* oil spill on sea otters, assessed through age-dependent mortality patterns. *Proceedings of the National Academy of Sciences of the United States* **97**(12): 6562–6567.

Neff, J.M., E.H. Owens, S.W. Stoker, and D.M. McCormick (1995). Shoreline oiling conditions in Prince William Sound following the *Exxon Valdez* oil spill. In Exxon Valdez *Oil Spill: Fate and Effects in Alaskan Waters*. P.G. Wells, J.N. Butler, and J.S. Hughes, eds. Philadelphia, PA, USA: American Society for Testing and Materials; ASTM Special Technical Publication 1219; ISBN-10: 0803118961; pp. 312–346.

Neff, J.M., D.S. Page, and P.D. Boehm (2011). Exposure of sea otters and harlequin ducks in Prince William Sound, Alaska, USA, to shoreline oil residues 20 years after the *Exxon Valdez* oil spill. *Environmental Toxicology and Chemistry* **30**(3): 659–672.

Pitcher, K.W. (1975). Distribution and abundance of sea otters, Steller sea lions, and harbor seals in Prince William Sound, Alaska. In *Distribution and Abundance of Marine Mammals in the Gulf of Alaska*. D.G. Calkins, K.W. Pitcher, and K. Schneider, eds. Anchorage, AK, USA: Alaska Department of Fish and Game, Division of Game, Appendix A.

Rice, S.D., J.W. Short, M.G. Carls, A. Moles, and R.B. Spies (2007). The *Exxon Valdez* oil spill. In *Long-term Ecological Change in the Northern Gulf of Alaska*. R.B. Spies, ed. Amsterdam, The Netherlands: Elsevier; ISBN-10: 0444529608; ISBN-13: 9780444529602; pp. 419–520.

Rotterman, L.M. and C. Monnett (1995). *Mortality of Sea Otter Weanlings in Eastern and Western Prince William Sound, Alaska, During the Winter of 1990–91*. Anchorage, AK, USA: US Fish and Wildlife Service, Alaska Fish and Wildlife Research Center; *Exxon Valdez* State/Federal Natural Resource Damage Assessment; Marine Mammal Study 6–18 Final Report. [http://www.evostc.state.ak.us/Files.cfm?doc=/Store/FinalReports/1992-MM0618-Final.pdf]

Saulitis, E., C. Matkin, L. Barrett-Lennard, K. Heise, and G. Ellis (2000). Foraging strategies of sympatric killer whale (*Orcinus orca*) populations in Prince William Sound, Alaska. *Marine Mammal Science* **16**(1): 94–109.

Short, J.W., J.M. Maselko, M.R. Lindeberg, P.M. Harris, and S.D. Rice (2006). Vertical distribution and probability of encountering intertidal *Exxon Valdez* oil on shorelines of three embayments within Prince William Sound, Alaska. *Environmental Science & Technology* **40**(12): 3723–3729.

Spies, R.B., ed. (2007). *Long-term Ecological Change in the Northern Gulf of Alaska*. Amsterdam, The Netherlands: Elsevier; ISBN-10: 0444529608; ISBN-13: 9780444529602.

Springer, A.M., J.A. Estes, G.B. van Vliet, T.M. Williams, D.F. Doak, E.M. Danner, K.A. Forney, and B. Pfister (2003). Sequential megafaunal collapse in the North Pacific Ocean: an ongoing legacy of industrial whaling? *Proceedings of the National Academy of Sciences of the United States of America* **100**(21): 12223–12228.

Taft, D.G., D.E. Egging, and H.A. Kuhn (1995). Sheen surveillance: An experimental monitoring program subsequent to the 1989 *Exxon Valdez* shoreline cleanup. In Exxon Valdez *Oil Spill: Fate and Effects in Alaskan Waters.* P.G. Wells, J.N. Butler, and J.S. Hughes, eds. Philadelphia, PA, USA: American Society for Testing and Materials; ASTM Special Technical Publication 1219; ISBN-10: 0803118961; pp. 215–238.

Trites, A.W., A.J. Miller, H.D.G. Maschner, M.A. Alexander, S.J. Bograd, J.A. Calder, A. Capatondi, K.O. Coyle, E. Di Lorenzo, B.P. Finney, E.J. Gregr, C.E. Grosch, S.R. Hare, G.L. Hunt, Jr., J. Jahncke, N.B. Kachel, H-J. Kim, C. Ladd, N.J. Mantua, C. Marzban, W. Maslowski, R. Mendelssohn, D.J. Neilson, S.R. Okkonen, J.E. Overland, K.L. Reedy-Maschner, T.C. Royer, F.B. Schwing, J.X.L. Wang, and A.J. Winship (2007). Bottom-up forcing and the decline of Steller sea lions (*Eumetopias jubatus*) in Alaska: Assessing the ocean climate hypothesis. *Fisheries Oceanography* **16**: 46–67.

US Fish and Wildlife Service (2004). Endangered and threatened wildlife and plants: Listing the southwest Alaska distinct population segment of Northern Sea Otter (*Enhydra lutris kenyoni*) as threatened. *Federal Register* **69**(28): 6600–6621. [http://www.gpo.gov/fdsys/pkg/FR-2004-02-11/pdf/04-2844.pdf]

US Fish and Wildlife Service (2005). Endangered and threatened wildlife and plants: Determination of threatened status for the Southwest Alaska distinct population segment of the northern sea otter (*Enhydra lutris kenyoni*). *Federal Register* **70**(152): 46366–46386. [http://www.gpo.gov/fdsys/pkg/FR-2005-08-09/pdf/05-15718.pdf]

US Fish and Wildlife Service (2008). *Northern Sea Otter (*Enhydra lutris kenyoni*): Southcentral Alaska Stock*. Anchorage, AK, USA: US Fish and Wildlife Service, Marine Mammals Management Office; Sea Otter Stock Assessment Report. [http://alaska.fws.gov/fisheries/mmm/stock/FinalsouthcentralAlaskaseaotterSAR01AUG2008.pdf]

Vos, D.J., L.T. Quakenbush, and B.A. Mahoney (2006). Documentation of sea otters and birds as prey for killer whales. *Marine Mammal Science* **22**(1): 201–205.

Williams, T.M., J.A. Estes, D.F. Doak, and A.M. Springer (2004). Killer appetites: assessing the role of predators in ecological communities. *Ecology* **12**: 3373–3384.

PART IV

ASSESSING OIL SPILL EFFECTS AND ECOLOGICAL RECOVERY

INTRODUCTION

The chapters in the previous section provide a detailed picture of how studies were designed and conducted, what they assumed, what they found, and what was learned in the process. Putting it all together to assess the real effects of the oil spill and ecologically significant, long-term consequences requires a broader, more synthetic approach. This is the topic of Chapter 16. Mark Harwell, John Gentile, and Keith Parker develop the conceptual framework and operational models for formalized ecological risk assessments, in which all plausible pathways of exposure to oil hydrocarbons are evaluated to determine the likelihood of exposure and its potential consequences. Using the rich lode of data and insights developed in the studies reviewed in the rest of the book, they apply the approach to two species: harlequin ducks and sea otters. Because this risk-assessment framework was not available when the *Exxon Valdez* spill occurred in 1989, the analyses were not conducted until much later. If applied in the early phases of a spill, however, the risk-assessment approach can help to direct studies to the exposure pathways posing the greatest potential risk.

CHAPTER SIXTEEN

Characterizing ecological risks, significance, and recovery

Mark A. Harwell, John H. Gentile, and Keith R. Parker

16.1 Introduction

When the *Exxon Valdez* oil spill occurred in 1989, there was no accepted framework for conducting ecological risk assessments. The primary guidance for assessing effects from oil spills was *Oil in the Sea* (National Research Council, 1985), which emphasized individual species but provided little information on communities or ecosystem processes. More importantly, it did not provide a systematic, risk-based methodology for assessing relationships between exposures to oil and their effects. The United States Environmental Protection Agency's (EPA) risk-assessment framework had been developed specifically for determining human cancer risks from chemical exposures (National Research Council, 1983). Assessing ecological risks, however, presents significant additional challenges: diverse ecological systems are subjected to multiple natural and anthropogenic stressors, each potentially causing many different effects on many different ecological attributes. Consequently, EPA developed a new framework and guidance designed for conducting ecological risk assessments (ERAs) (US Environmental Protection Agency, 1992, 1998), but this did not occur until several years after the *Exxon Valdez* oil spill.

Another decade passed before we applied this ERA framework to evaluate the ecological significance of any remaining effects of the *Exxon Valdez* spill. Now, 20 years after the ERA framework was developed, the process has matured considerably into an integrated environmental-assessment framework (Cormier and Suter, 2008). The integrated framework no longer simply assesses exposures and effects, but now provides a more comprehensive approach to thinking about ecological incidents. This expanded construct focuses on identifying the system attributes that matter ecologically or societally and provides a means to assess both immediate and long-term effects on those important attributes. It also places spill risks in the context of other natural and anthropogenic stressors and

---

Oil in the Environment: Legacies and Lessons of the Exxon Valdez *Oil Spill*, ed. J. A. Wiens. Published by Cambridge University Press. © Cambridge University Press 2013.

incorporates a process to account for uncertainty. It ultimately provides a means to characterize recovery in light of natural variability over time.

Studies of the *Exxon Valdez* spill provided important information and hypotheses that we have used to illustrate and advance the application of the integrated assessment approach. In this chapter, we review the application of this comprehensive framework and illustrate how it provides a systematic and objective process for evaluating scientific information and quantifying population risks and uncertainty, while helping to focus research and monitoring on what really matters ecologically, rather than on things that do not. Most importantly, we describe how it can be applied to future oil spills and other natural resource damage assessment (NRDA) activities.

16.2 Risk-based environmental-assessment frameworks

Like many other regional-scale systems, Prince William Sound (PWS) is a complex, diverse, and dynamic ecosystem comprising many different habitat types (Chapter 1). Within the ecosystem are hundreds of species, hierarchically organized into interactions among one another and driven by physical, chemical, and biological processes.

In order to address this complexity, the ecological risk framework is structured into three sequential phases: problem formulation, analysis, and risk characterization (US Environmental Protection Agency, 1992). Each of these phases focuses on the two essential components of risk: (1) *characterization of the stressor (exposure) regime*, where "stressor" is defined as any physical, chemical, or biological agent that could affect an ecological system; and (2) *characterization of ecological consequences from environmental stressors*, evaluated as effects on ecological assessment endpoints, or *valued ecosystem components* (VECs) (Canadian Council of Ministers of the Environment, 1996; Hegmann *et al.*, 1999).

The objective of the ERA is to understand the magnitude and likelihood of adverse ecological effects from human activities and natural processes, while considering uncertainties, recovery potential, and natural variability. The risk framework is a powerful tool for analyzing causal relationships between exposure and effects and for designing more effective research and monitoring programs to support informed decisions. Implementing a risk-based, integrated assessment process immediately following a major oil or chemical spill can provide a systematic and comprehensive roadmap for characterizing injury, risk, remediation, recovery, and restoration and bringing sound science to bear on environmental decision-making.

When we began our risk-based assessments a decade after the spill, two overarching issues guided our studies:

- Are there residual ecological risks from the spill?
- Have those species or populations defined by the *Exxon Valdez* Oil Spill Trustee Council (hereafter, "Trustees") as injured actually recovered?

Using these two points of departure, we developed questions concerning ecological risks and ecological recovery. Although these questions were addressed separately, the integrated assessment framework provides many points of commonality and is applicable to both oil-spill and NRDA situations (as well as a broader range of environmental disruptions).

16.2.1 Problem formulation

Problem formulation is the critical first phase of integrated assessments. It should be implemented as soon after the spill as feasible. If it is done correctly and as early in the process as the situation warrants, the rest of the oil spill (or NRDA) process can unfold systematically and reach resolution more effectively and efficiently.

Problem formulation is designed to articulate the purpose of the assessment, define the issues, plan for analyzing and characterizing risk, and select the VECs for which effects, risks, and recovery will be evaluated (US Environmental Protection Agency, 1992, 1998). The objective for selecting VECs is to identify a parsimonious set of populations, habitats, or ecological processes that have particular significance to the structure and health of the ecosystem and/or particular importance for their services to society. Problem formulation also identifies the at-risk ecosystem components by acquiring information on the oil spill, other stressors on the ecosystem from human activities and natural processes, and observed ecological effects. A key product of problem formulation is the development of *conceptual ecosystem models* (CEMs) that form the core of integrated assessments, guiding assessments of risks in the context of natural variability and the other natural and anthropogenic stressors.

There are many different types of CEMs. Our *Exxon Valdez* studies employed two particular classes of CEMs, qualitative and ecotoxicological, that are derived from the ERA framework (US Environmental Protection Agency, 1992, 1998).

Qualitative CEMs (Section 16.3) capture the natural and anthropogenic forces and associated stressors that can affect an ecosystem, defining the risk regime by graphically illustrating pathways by which a stressor affects ecosystem attributes and providing the foundation for formulating risk hypotheses (Gentile et al., 2001). The power and value of these CEMs are in defining the relationships that control ecosystem structure and functioning and providing a systematic framework for ranking the strength of these relationships. These qualitative CEMs can be used to identify the types of information needed to conduct an integrated assessment and indicate data gaps to focus research and monitoring. We used this qualitative class of CEM both to assess the ecological significance of exposures and effects (Harwell and Gentile, 2006) and to assess the relative risks of oil-spill stressors compared with other stressors affecting PWS (Harwell et al., 2010a).

Ecotoxicological CEMs (Section 16.4) were developed for sea otters (*Enhydra lutris*) and harlequin ducks (*Histrionicus histrionicus*), each capturing all of the plausible pathways of exposure to polycyclic aromatic hydrocarbons (PAH). These ecotoxicological CEMs provided the bases for developing simulation models to quantitatively characterize population-level risks and to attribute risks to specific causal factors (Harwell et al., 2010b, 2012a, b).

16.3 Qualitative description of ecological significance, risks, and recovery

16.3.1 Framework for assessing ecological significance

A key issue in oil-spill or NRDA assessments is determining what constitutes an ecologically significant adverse effect, which in the context of the regulations controlling oil and chemical spills is termed *injury*. There are two types of significance: statistical and ecological. Statistically significant results are not necessarily ecologically meaningful in the sense that a measured change may not make any ecological difference. Virtually any change can be statistically significant with sufficient sampling effort: whereas a small sampling effort may only detect large changes, intensive sampling could find even extremely small changes statistically significant, even if ecologically irrelevant (National Research Council, 1990; Chapter 10). Here we focus on ecological significance. In essence, the issue is distinguishing environmental responses that matter from those that do not.

Gentile and Harwell (1998) explored concepts and procedures for assessing ecological significance. One important issue is differentiating a *biological response* from an *ecological effect*. A biological response can be defined as any change in an exposed organism's biochemistry, physiology, or behavior. However, such biological responses will only result in an ecological effect if they affect survival, reproductive ability, or other characteristics of the population, and thus cause population-, community-, or ecosystem-level effects. In the absence of such higher-level effects, a biological response is not ecologically significant.

In the context of risk, if a biochemical response (also called a biomarker; see Chapter 9) is diagnostic of a particular chemical, then it may be appropriate to use it as an indicator of exposure because it indicates bioavailability of the chemical and it integrates exposure over time and space. However, if a biochemical or individual-level response occurs without concomitant changes at the population level or higher, then that biomarker is not an appropriate indicator of ecological effects. Even when both biochemical responses and observed population-level effects occur simultaneously, this does not necessarily mean there is a causal relationship between the two, as other stressors may be responsible for the observed effects. Such situations require a careful assessment of attributable risk, as discussed below.

Quantifying adverse effects (injuries) may involve comparisons to regulatory benchmarks (e.g., water-quality criteria, fishery advisories) for which injury characterization is clear-cut – if the criterion is exceeded, then injury has occurred. Another approach to assessing injury is to characterize the significance of an ecological effect on the population or community (Gentile and Harwell, 1998). This approach is particularly appropriate in complex, multistressor situations or in the absence of regulatory benchmarks (Gentile and Harwell, 2001). Both conditions applied to the *Exxon Valdez* spill in PWS.

Finally, it is not unusual in regional-scale risk assessments that there are insufficient data to conduct detailed, quantitative analyses. In many such cases, however, qualitative assessments are quite sufficient to reach sound conclusions concerning ecological significance and relative risks. In those cases, we have proposed using reasonable expert judgment, uniform criteria, and simple ranking schemes to assess the ecological significance of effects, rather than applying rigorous or quantitative metrics. Ideally, this approach is best used with an independent panel of experts who apply a systematic framework and scientific principles to objectively analyze, evaluate, and interpret the data and provide written rationale and documentation for their conclusions. The value of this approach is that it oftentimes can readily separate the wheat from the chaff, i.e., focus risk and restoration efforts only on what is truly important to ecosystem health, to the exclusion of things that are unimportant. Even though this qualitative approach cannot distinguish degrees of importance among different types of wheat, separating the chaff is often sufficient and invaluable, as we demonstrate below.

16.3.2 Assessing ecological significance of the *Exxon Valdez* oil spill

To characterize ecological significance, we developed a working definition, a set of evaluation criteria, and a process for applying those criteria to actual environmental problems (Gentile and Harwell, 1998). The ecological significance framework has two components: (1) selecting VECs with ecological and/or societal relevance and importance to ecosystem health and sustainability; and (2) distinguishing whether a change in a VEC exceeds its natural variability and is of sufficient type, intensity, spatial extent, and duration to constitute an injury. Additional criteria include stressor persistence, reversibility, and time for recovery of VECs after stressor removal.

Using this framework, we examined the extensive postspill literature on *Exxon Valdez* to assess the ecological significance of exposures and effects in immediate and long-term periods – at the time, 15 years after the spill (Harwell and Gentile, 2006). Typical of oil spills, *Exxon Valdez* produced four stressors:

1. Volatile organic compounds (e.g., benzene, xylenes), which dissipated within a few days but may have posed early inhalation risks to some species;
2. Oiling, which caused most of the immediate effects through loss of thermoregulation and resulting hypothermia;
3. PAH, which posed longer-term toxicological risks; and
4. Stressors associated with intertidal and shoreline cleanup activities, e.g., physical disturbance and extensive human presence.

Since the magnitude and extent of these stressors in the early aftermath of the spill were so large, no further analysis was required to assess the ecological significance of exposures and effects in that time period.

Because PWS is a very dynamic system, with extreme storms and high-energy wave and tidal regimes, residual *Exxon Valdez* oil was largely eliminated naturally from shorelines in the initial months to few years after the spill, assisted by cleanup during the springs and summers of 1989–92 (Chapters 4 and 6). Thus, the only potential residual or long-term risk to the PWS ecosystem was from PAH.

Assessing ecological significance of long-term exposure to PAH involved multiple lines of evidence: the nature of sources of PAH; their spatial and temporal extent, heterogeneity, and concentrations; source persistence; bioavailability; and potential causal relationships between oil residues and observed or potential ecological effects (Harwell and Gentile, 2006). The key for assessing ecological significance of effects, and more generally assessing ecosystem risks, was selecting an appropriate set of VECs representative of the trophic structure and critical structural/functional attributes of the PWS ecosystem. The Trustees had previously developed a list of endpoints for assessing recovery (*Exxon Valdez* Oil Spill Trustee Council, 2002) that focused on species-level attributes, without including higher levels of ecological organization or ecological processes. Consequently, we expanded upon this list to develop a more comprehensive suite of VECs to characterize PWS ecosystem health, distributed across the ecological hierarchy (Fig. 16.1) (Harwell and Gentile, 2006).

Applying the Gentile and Harwell (1998) criteria for assessing ecological significance, we concluded that shortly after the spill, most VECs showed high- or medium-level initial ecological significance (Fig. 16.1a). These effects differed in the likely causes of observed effects – for example, direct oiling caused most of the mortality to seabirds and sea otters; volatile organic compounds potentially affected marine mammals; and

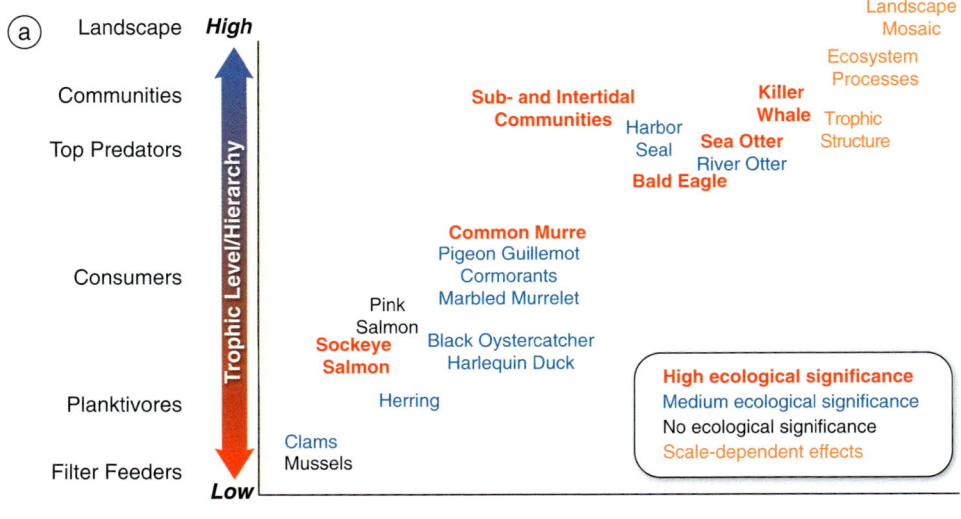
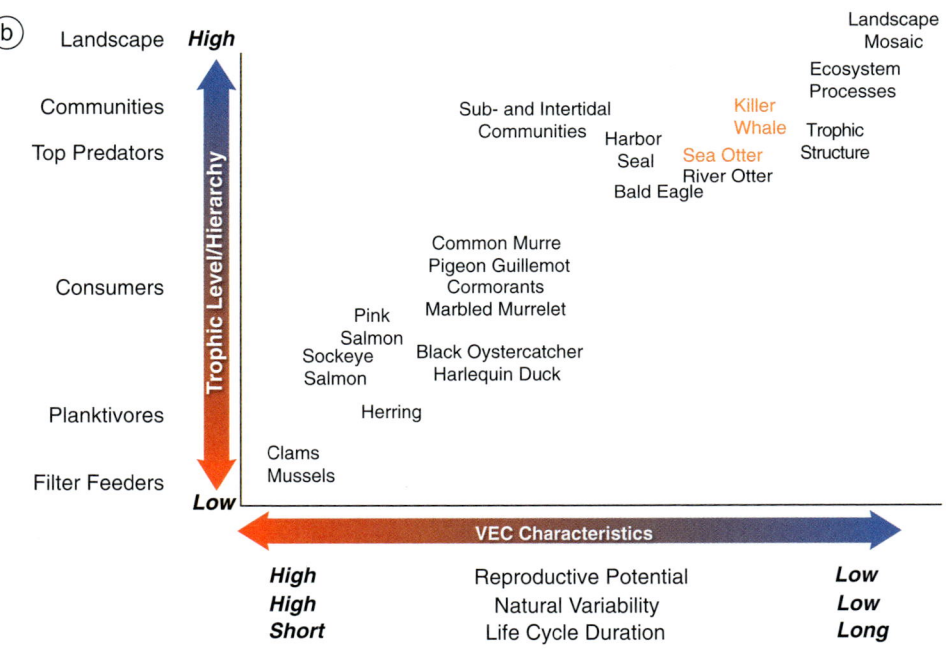

Figure 16.1 Ecological significance of effects on PWS valued ecosystem components (VECs) from the *Exxon Valdez* oil spill. Each VEC of the PWS ecosystem was assessed for the ecological significance of the effects of the *Exxon Valdez* oil spill, following the qualitative approach detailed in Harwell and Gentile (2006). VECs are shown by trophic-level position and arrayed along a gradient of VEC characteristics. Ecological significance is characterized as high, medium, or none. Scale-dependent effects means the ecological significance of effects depends on the scale of observation. (a) Initial effects. Effects from all stressors from the oil spill and the cleanup activities were evaluated. (b). Fifteen years after oil spill. Long-term stressors from the oil spill were essentially limited to toxicological risks from PAH exposure as all of the other stressors had virtually been eliminated by then. All high- and medium-level ecologically significant effects had dissipated by the time of the assessment, and only some potential scale-dependent effects may remain. (Data from Harwell and Gentile, 2006)

cleanup activities adversely affected subtidal and intertidal communities. Ecologically significant effects on sockeye salmon (*Oncorhynchus nerka*) were caused by the fishery closure rather than physical stressors from the spill (Harwell and Gentile, 2006).

In later years after the spill (Fig. 16.1b), no VEC showed ecologically significant population- or higher-level effects, perhaps with the exception of lingering effects on a subpopulation of sea otters, discussed in detail below, and potentially one pod of killer whales (*Orcinus orca*) (Harwell and Gentile, 2006; Integral Consulting, Inc., 2006, reached similar conclusions for most endpoints).

16.3.3 Assessing relative risks to the ecosystem

An important aspect of understanding long-term risks is to assess the relative importance of oil spills compared to the other stressors affecting the ecosystem. Assessing relative risks is essential both in order to attribute any lingering effects on VECs to oil residues (i.e., determine causality) and to assess ecosystem recovery. Consequently, we drew upon the extensive literature concerning PWS and the Gulf of Alaska (GOA) and the judgment of a group of experts on climate, oceanography, fisheries, and ecology to develop risk-based CEMs of PWS and GOA ecosystems (Fig. 16.2) (Harwell *et al.*, 2010a). Using the

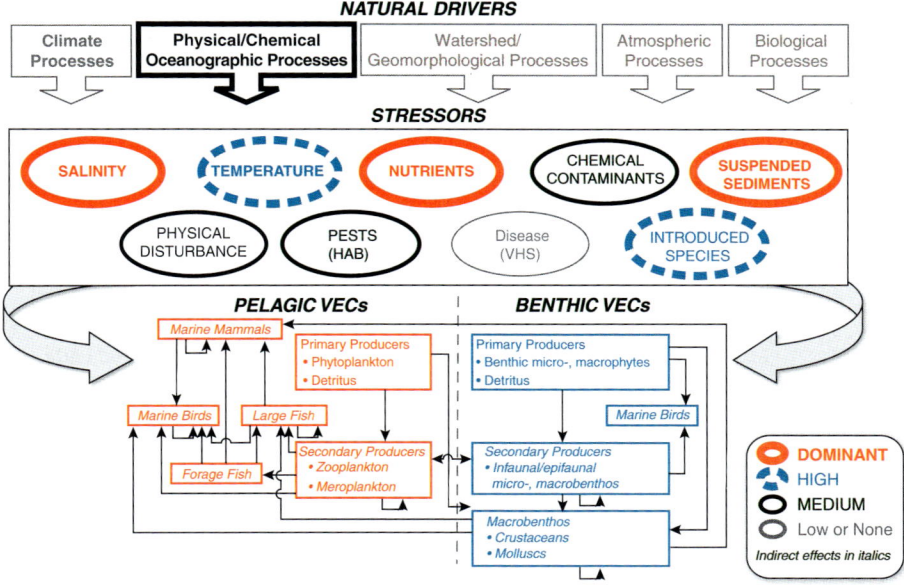

Figure 16.2 Selected graphical conceptual ecosystem models (CEMs) for the PWS–Gulf of Alaska ecosystem. The top tier (highlighted rectangular box) indicates the specific natural or anthropogenic driver for the CEM. The middle tier (ovals) identifies the stressors associated with the natural drivers. For a particular driver, the resulting stressor that has a dominant role in causing ecological effects is highlighted in bold red; a stressor that has a high role in causing effects is indicated in dotted blue; medium stressors are identified in black; stressors with low or no effects are shown in background coloring. The third tier is the trophodynamical model, showing dominant (red), high (blue), medium (black), or no (background) trophic-structure-mediated effects for the particular driver. Direct pathways are shown in normal font; indirect pathways are shown in italics. Separate figures are shown for each natural or anthropogenic driver: (a) Physical/chemical oceanographic processes; (b) Oil spill and cleanup, immediate period; (c) Oil spill and cleanup, long-term period. (Reproduced from Harwell *et al.*, 2010a, with permission of Taylor & Francis Ltd., http://www.tandf.co.uk/journals.)

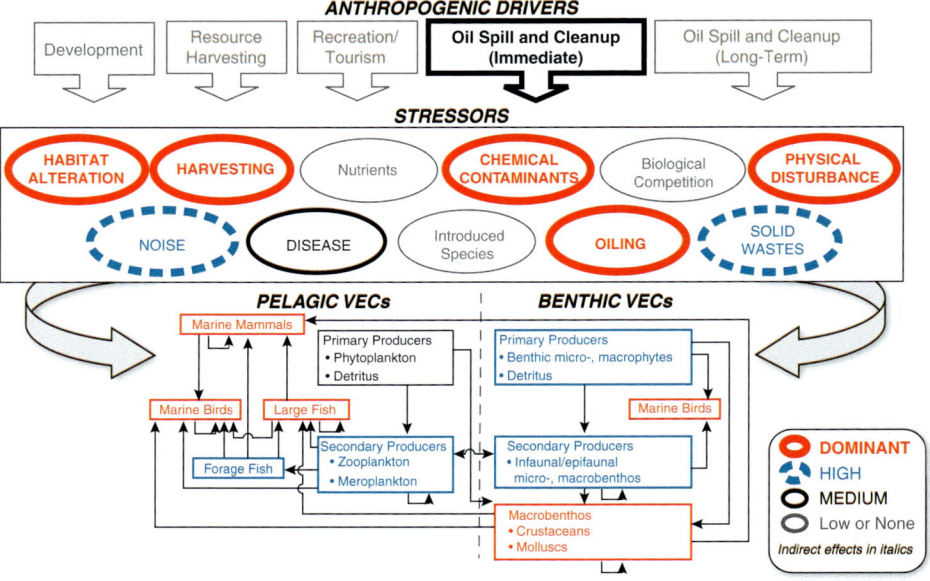

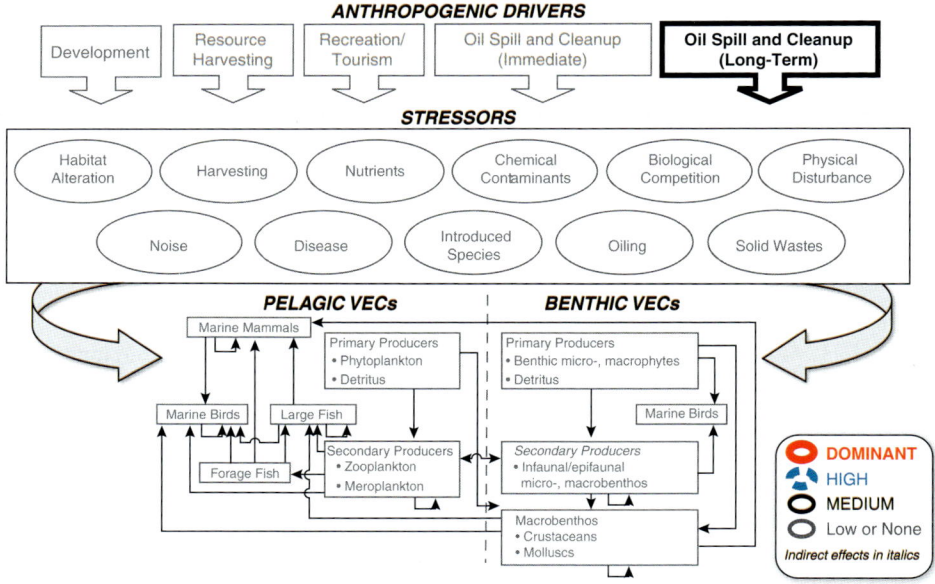

Figure 16.2 (cont.)

risk-based CEM-development methodology discussed in Gentile *et al.* (2001) and US Environmental Protection Agency (2002), we identified the natural and anthropogenic drivers, associated environmental stressors, and VECs of PWS and GOA and ranked the relative strengths of relationships among these components. From these relationships we developed the graphical qualitative CEMs of the PWS-GOA ecosystem (Fig. 16.2).

These CEMs provide a synoptic view that differentiates: (1) those factors that truly drive the ecosystem's structure and functioning, particularly climatic and oceanographic processes; (2) those factors that are important to the ecosystem but not dominant; and (3) those factors that are not at all important to ecosystem condition (Fig. 16.3). Thus, the qualitative CEM informs assessments of ecological recovery and provides an effective communications tool for scientists and nonscientists, particularly decision-makers, to place long-term relative risks from oil spills in the context of the overall stressor regime of the impacted ecosystem.

The couplings of physical and biological processes and of pelagic and coastal systems are fundamental features of PWS-GOA that are central to the integration of the trophic and physiochemical dynamics of the Sound. As a result, we integrated a trophodynamic model into the qualitative CEM construct to illustrate how trophic relationships mediate the dominating influence of climatic and oceanographic processes (Fig. 16.3). This integrated CEM shows how direct stressor effects on VECs not only cause indirect

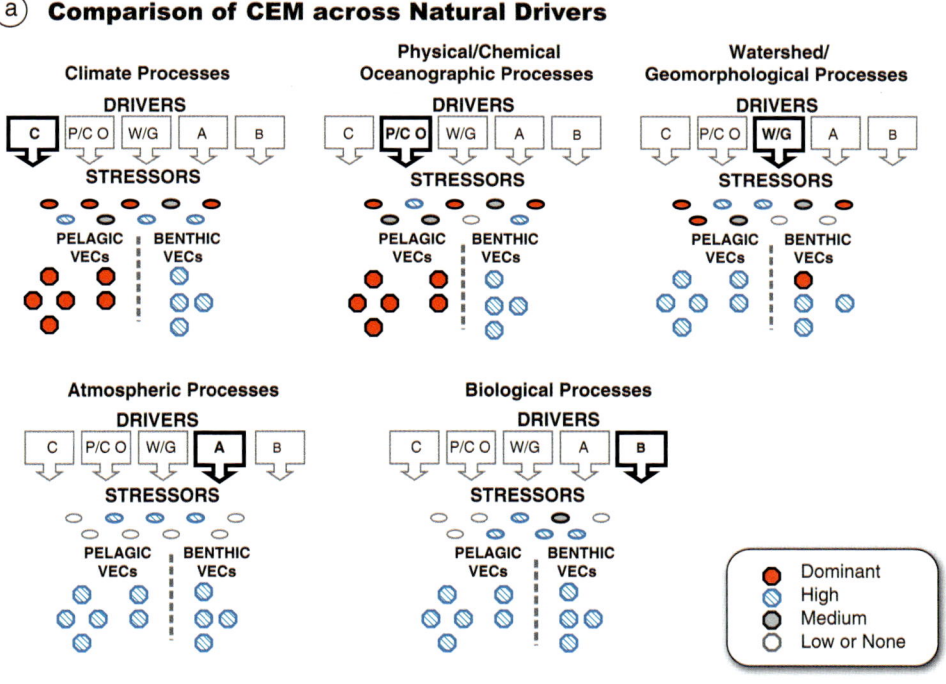

Figure 16.3 Comparisons of the CEMs across natural and anthropogenic drivers. Each of the natural or anthropogenic driver-specific CEMs is shown schematically, summarizing the information shown in the individual CEM. The same construct is followed as in the CEMs, with the top tier of each graphic showing the driver, middle tier reflecting the stressors for that driver, and lower tier the components of the trophodynamical model. The location of each stressor and trophodynamical component is identical to the associated CEM. For example, in (a) Physical/Chemical oceanographic processes (identified by the bold rectangle labeled "P/CO"), the top left stressor is salinity, the next stressor to the right is temperature, etc. (derived from associated Figure 16.2a). Stressor and trophodynamical component symbols having dominant roles are filled with red; high stressors or trophodynamical components are filled with striped blue; medium stressors or components are filled with gray; and low or no are open symbols. (a) Natural drivers; (b) Anthropogenic drivers. (Reproduced from Harwell et al., 2010a, with permission of Taylor & Francis Ltd., http://www.tandf.co.uk/journals.)

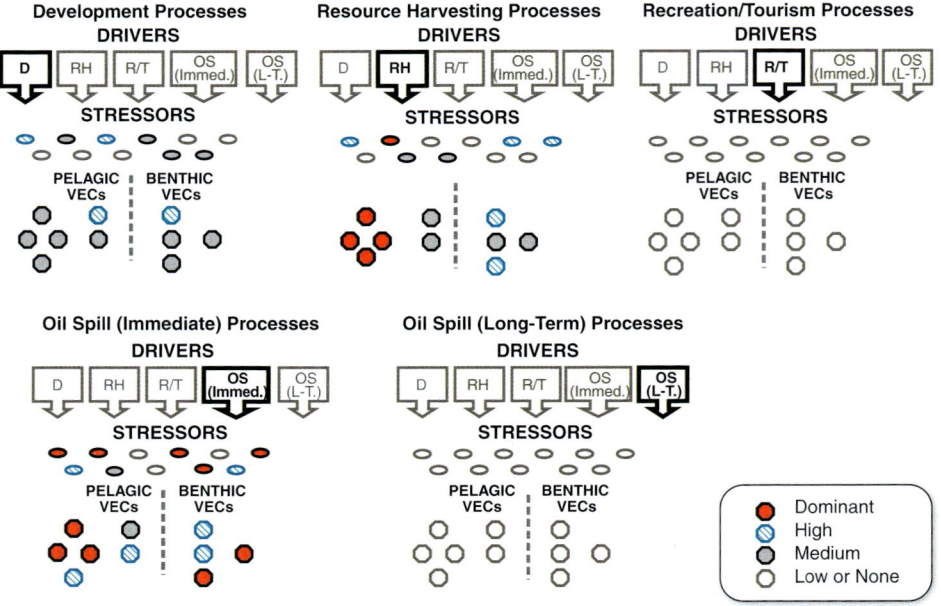

Figure 16.3 (cont.)

effects on other ecosystem attributes but also potentially cause cascading effects, i.e., major changes at one level that propagate throughout the trophic structure. Cascading effects in PWS are largely driven by bottom-up processes: climatic–oceanographic variability causes large changes to the energetic base of the trophic structure (phytoplankton productivity), which in turn propagate throughout the Sound via major impacts on planktivorous forage-fish populations (e.g., Pacific herring, *Clupea pallasii*) and their predators, particularly large fish, marine birds, and marine mammals. This dynamic is captured graphically in our CEM (Harwell et al., 2010a) and discussed more fully in Mundy (2005) and elsewhere (e.g., Finney et al., 2000; Miller et al., 2005; Lees et al., 2006). Because the foundation of the trophic structure in PWS is so strongly controlled by coupled climate–oceanographic processes, the magnitude of this bottom-up cascading effect exceeds the top-down cascades often seen in other ecosystems caused by loss of a top predator (e.g., Paine, 1980).

Our relative risk assessment included examination of the ecological effects of the Great Alaska Earthquake of 1964 on the PWS–GOA ecosystem. The qualitative CEMs showed that the initial impacts from that earthquake and from the *Exxon Valdez* oil spill were for a period of time as important to the ecosystem as the natural, biophysically coupled climate–oceanographic processes. For example, the extensive vertical displacements of land by the earthquake (Chapter 1, Fig. 1.2) caused massive habitat alteration of coastal zones throughout PWS, affecting virtually the entire intertidal zone. Previously submerged or upland habitats suddenly became intertidal. Consequently, the coastal habitats at local scales were permanently changed. At the broad scale, a new regime emerged after several years to a decade, when the overall distribution of habitats in PWS essentially recovered. For a similar period of time after the *Exxon Valdez* spill, the anthropogenic stressors from the oil spill dominated the ecosystem, but PWS subsequently recovered, even at the local scale. However, the climate regime shifts occurring over the ensuing decades altered PWS–GOA so much that any residual spill effects have become inconsequential (Harwell *et al.*, 2010a).

16.3.4 Assessing ecological recovery

Assessing recovery is a recurrent and important theme throughout this book (see Chapters 1, 10, and 17). The particular perspective on recovery explored in the present chapter derives from the integrated assessment approach and its applicability to oil and chemical spills. As discussed in Chapter 1, the NRDA (National Oceanic and Atmospheric Administration, 1996, 2010) and Oil Pollution Act of 1990[1] (OPA 90) regulations define ecological recovery as a return of injured resources and services to baseline, i.e., the condition that would have existed had no spill occurred. Unfortunately, this regulatory definition leaves many critical issues unresolved, such as: What constitutes baseline? What attributes should or should not be monitored to assess recovery? How do natural variability and other nonspill-related stressors affect recovery? How should recovery objectives change over time?

Conceptually, ecological recovery relates to whether or not and how quickly an ecosystem returns to some predisturbance condition once the stressor is removed (Holling, 1973). Although ecological recovery has been examined theoretically using linear or linearized models (e.g., DeAngelis, 1980; DeAngelis *et al.*, 1989a, b), such models have failed to incorporate recovery-determining factors such as nonlinearities, effects thresholds, alternate system states, differential responses across ecosystem attributes, spatiotemporal variability, and cascading effects – each of which is directly relevant to recovery processes. These limitations have reinforced misperceptions that ecosystems have steady-state, balance-of-nature endpoints, and that recovery consists of returning to a precise, deterministic preperturbation state, exactly where the system would have been in the absence of the stressor. We use a more ecologically relevant and practical approach that remains consistent with the OPA 90 definition but is risk-based and focuses on assessing ecological recovery in a complex and changing ecosystem.

Assessing ecological recovery from oil or chemical spills is complicated by the lack of steady-state endpoints that results because of succession and because of other changes that may occur in the ecosystem over time and space that are caused by other anthropogenic and natural stressors. Many studies of regional-scale ecosystem restoration and recovery across a broad diversity of ecosystem and stressor types illustrate this point, including: the restorations of the Everglades (e.g., Ogden *et al.*, 2002; US Army Corp of Engineers, 1999), Missouri River (National Research Council, 2002), and Upper Mississippi River (Upper Mississippi River Basin Association, 1995); re-establishment of threatened/endangered species (e.g., National Marine Fisheries Service, 2010); and recovery from catastrophic natural events, such as Hurricane Andrew (e.g., Lirman and Fong, 1995; Lovelace and McPherson, 1996) and the eruption of Mount St. Helens (Dale *et al.*, 2005a, b). Key lessons from these examples are: (1) the need to focus recovery assessments on carefully selected VECs that capture the diversity of the

---

[1] 33 United States Code 2701 *et seq.* [//www.law.cornell.edu/uscode/html/uscode33/usc_sup_01_33_10_40_20_I.html]

ecosystem; (2) the importance of understanding the complexities of ecological stress–response relationships; and (3) the central roles of spatial and temporal variability and cumulative and multiple stressors on ecosystem recovery. Each of these factors can confound the determination of ecosystem recovery – and each was critical to appropriately assessing the recovery of PWS from the *Exxon Valdez* spill.

Recovery from oil spills contrasts with other regional-scale problems, such as the Everglades restoration, in which the ecosystems have not recovered in part because the important stressors (e.g., hydrological and habitat alteration) continue to dominate the structure and functioning of the ecosystem. As a result, restoration must first focus on stressor removal, management, or mitigation before recovery can proceed. By contrast, in an oil spill such as the *Exxon Valdez*, all four oil-spill-caused stressors are essentially eliminated from the system within a few years and do not impede ecological recovery thereafter. Once the stressors are removed, it remains for the ecosystem to undergo recovery through processes that differ across ecosystem components, stressor characteristics, population characteristics (e.g., species' demographics, social structures, and longevity), and ecosystem resiliency.

The issues of baseline and scale (especially heterogeneity across space and time) are critical to assessing recovery. For example, the population of coastal Alaska sea otters was rapidly increasing and expanding for decades prior to the spill, recovering from near-extinction in the early twentieth century (Chapter 15). As a result, many areas, including parts of PWS, may not have been at carrying capacity at the time of the spill, meaning that the baseline was continually changing over time. Additionally, the sea otter population initially expanded into western PWS in the 1960s, reaching Knight Island by 1970, and continuing on through eastern PWS thereafter (Johnson and Garshelis, 1995). Consequently, the baseline at the time of the spill may have varied considerably in different areas within PWS. In the past decade or so, most of the southwestern Alaska sea otter population (west of Cook Inlet but not including the subpopulation within PWS) has declined precipitously (US Fish and Wildlife Service, 2005), likely because climate regime shifts altered food and predation regimes (Estes *et al.*, 1998; Burn and Doroff, 2005). This illustrates the uncertainty and limitations of basing recovery assessments on a localized area (e.g., northern Knight Island) and on the subpopulation there, which represents but a small fraction of the overall PWS sea otter population.

Natural variability across time is also important. For example, the large interannual variability in populations of many PWS species, such as forage fish, is driven by variability in dominant climatic and oceanographic factors controlling the ecosystem. This, too, makes determining baseline conditions very difficult and greatly affects the ability to detect ecologically significant effects and establish goals for recovery. This issue only becomes more problematic over time as uncertainty from natural variability remains unabated while the signal of the effects of the spill continuously diminishes. Eventually the point is reached where any residual effects from the oil spill are lost in the noise of natural variability, and risks can no longer be attributed to the spill.

Given these issues, we suggest that in the context of the integrated risk framework, recovery of an ecosystem can be defined as the absence of any ecologically significant adverse effects or ecologically significant continuing risks on the VECs. If the suite of VECs has been properly selected, then the health of the ecosystem is captured by the condition of the VECs. If an injured VEC returns to its baseline, given these issues of temporal and spatial heterogeneity, then it has recovered, and if all VECs have done so, then the ecosystem has recovered. This requires that there are no remaining ecologically significant risks to any VECs from stressors caused by the spill, or else future injuries could occur.

Under this perspective, we assessed ecosystem recovery from the *Exxon Valdez* oil spill by qualitatively evaluating whether any ecologically significant effects or risks remain from the oil spill, as discussed in the preceding sections. The other risk-based component of assessing recovery, discussed in the following section, consisted of quantitatively assessing whether any residual stressors from the oil spill *could* cause ecologically significant effects on individuals or populations.

16.4 Quantitative characterization of risks and uncertainty

16.4.1 Individual-based modeling in ecological risk assessments

The practice of ecological risk assessment has evolved considerably since the guidelines were issued (US Environmental Protection Agency, 1998), yet there remains a notable lack of quantitative, population-level risk assessments. Although risk management is primarily concerned with populations, most risk assessments solely assess individual-level effects. This is partly because population-level data are sparse, contrasting with the wealth of suborganismal- and organismal-level toxicity data, and partly because of the lack of readily accessible models to quantitatively project population-level effects given natural variability and natural and anthropogenic stressors. Irrespective of the causes, there is an unmet need for a closer relationship between the decision-making focus of risk management and the current methodologies of risk assessment (Stern and Fineberg, 1996; Barnthouse et al., 2007).

Our approach to conducting quantitative ecological risk assessments for the oil spill involved development of simulation models. Munns *et al.* (2007) provide a useful typology of classes of population-level models, discussing attributes, limitations, and applications of each. One class, *individual-based models* (IBMs), offers particular promise for realistic, quantitative risk characterization. IBMs have been successfully developed for a wide range of species, but they have not been widely used in ERAs (Munns et al., 2007; Forbes *et al.*, 2010). Traditional analytical population models directly model population-level attributes (e.g., birth and death rates, predator–prey interactions) as solutions of differential or difference equations. In contrast, IBMs simulate individual-level attributes, mimicking the hierarchical relationships of the environment (DeAngelis and Gross, 1992; DeAngelis and Mooji, 2005), in which higher-level properties emerge from the collective activities at lower levels (von Bertalanffy, 1968).

IBMs have the unique ability to readily simulate alternative behaviors, spatial and temporal heterogeneity, thresholds of actions or processes, activities in which an individual chooses among alternatives, and other nonlinear processes, each of which can greatly affect recovery processes. Because none of these factors can be addressed readily in traditional population-level models, the factors that are most important to accurate predictions of effects or recovery processes previously had to be either ignored, linearized, assumed away, or otherwise not explicitly addressed. In contrast, an adequately parameterized IBM can realistically simulate activities, processes, and environmental conditions as they are experienced by individuals in the real world. For example, in quantifying complex and often probabilistic toxicological-risk pathways, an IBM can explicitly model exposure regimes that differ from one individual to another, one day to another, or one location to another, just as an individual organism would encounter variability of exposures in the naturally heterogeneous environment. This characteristic makes IBMs ideal for quantifying toxicological risks and their uncertainties to individuals and populations.

Consequently, we used IBMs to quantitatively assess whether residual oil deposits continued to pose a risk to populations of sea otters and harlequin ducks (Harwell et al., 2010b, 2012a, b), as had been suggested.

16.4.2 Quantitative ecological risk assessment of sea otters

Sea otters are found in Alaskan coastal waters, where they feed on benthic invertebrates living either on top of sediments (epifauna) or within sediments (infauna) (Bodkin and Ballachey, 1997). Although the initial acute effects of the oil spill on sea otters in PWS have long since disappeared, an issue has been raised concerning the subpopulation of sea otters at northern Knight Island (NKI), whose numbers have remained essentially unchanged for several years but are below one estimate of prespill levels (Chapter 15). Bodkin et al. (2002) and Dean et al. (2002) attributed these putative effects to subsurface oil residues (SSOR). No current effects are posited for sea otters elsewhere in PWS, even though SSOR are not limited to NKI. As discussed in Chapters 6 and 7, SSOR exists in patches, primarily in boulder/cobble/gravel shorelines in the middle- to upper-intertidal zones. Short et al. (2006) suggested that, when digging in sediments for food, NKI sea otters may encounter SSOR with sufficient frequency and quantity to affect their health.

To judge the plausibility of this suggestion, we quantitatively assessed current risks to NKI sea otters associated with oil residues (Harwell et al., 2010b). We developed an ecotoxicological CEM (Fig. 16.4) presenting all plausible pathways for PAH exposures from *Exxon Valdez*-derived sources, such as SSOR, and from non-*Exxon Valdez* sources, such as hydrocarbon-combustion (pyrogenic) sources. This CEM was converted into a quantitative IBM by simulating each exposure pathway to sea otters living in initially oiled areas and to those living in reference (unoiled) areas. This IBM was developed using the simulation language Stella™ (ver. 9.1.2; copyright isee systems, inc.; www.iseesystems.com), an object-oriented software. One individual sea otter is simulated at a time. During each time step (1-hour period), each exposure pathway is simulated for each of 40 individual PAH analytes; assimilated PAH are then accumulated to generate a daily dose. Because PAH are readily metabolized and consequently do not biomagnify through the food web, the sea otter IBM could accurately simulate assimilated doses of PAH on a daily basis, permitting quantification of chronic average daily-assimilated doses to a population of sea otters.

The model is stochastic, with variability built into each component of the exposure pathways. This allows quantification of doses to a large population of sea otters as they would experience actual exposure variability in the environment. The model was parameterized using the most recent empirical data on concentrations and relative frequencies of PAH as measured in PWS, along with an extensive database on sea otter characteristics developed by the co-authors of Harwell et al. (2010b) or derived from the literature.

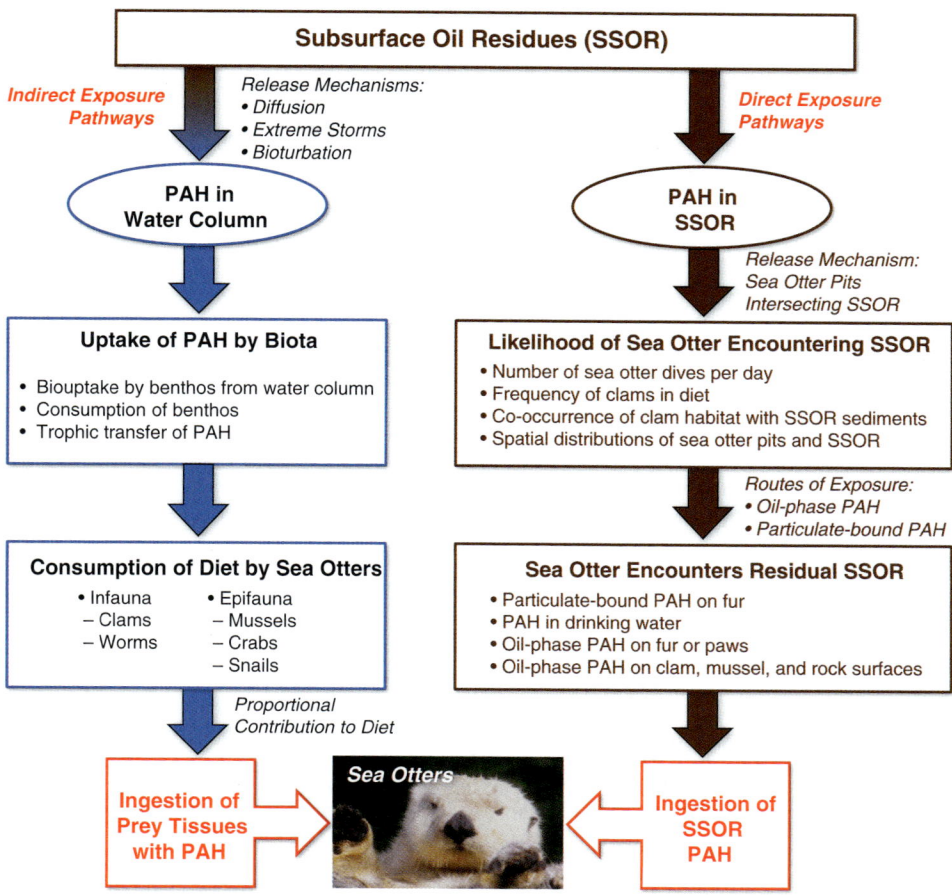

Figure 16.4 Conceptual model of exposure pathways for sea otters at northern Knight Island from PAH in SSOR. Two pathways are shown: direct exposure to SSOR through a sea otter pit intercepting an SSOR deposit (right side of diagram), and indirect exposure to PAH in the environment (left side of diagram). Each element in the direct exposure pathways was modeled quantitatively; for the indirect pathways, ingestion of PAH in prey tissues, sediments, and water was quantitatively modeled based on measured concentrations in the environment. (Modified from Harwell et al., 2010b, with permission of Taylor & Francis Ltd., http://www.tandf.co.uk/journals.)

When a PWS sea otter feeds on infauna (primarily clams), it excavates a pit several centimeters into sediments until clams are encountered (Chapter 15, Fig. 15.6). The highest-risk exposure pathway occurs when a sea otter inadvertently excavates a pit in the intertidal zone that directly intersects a patch of SSOR. Because unoiled areas by definition have no SSOR, the reference sea otters could not be exposed via this pathway. The other potential pathways were also modeled for both oiled and reference sea otters using data on existing PAH concentrations in their respective areas.

Four factors affect the likelihood of a sea otter's digging a pit that intersects SSOR:

1. The distribution of SSOR patches across tidal zones and shoreline subdivisions;
2. The distribution of clam habitat across tidal zones;
3. The co-occurrence of clam habitat and SSOR; and
4. The number of pits excavated by each sea otter per day in the intertidal zone.

The probability of a sea otter pit intersecting SSOR is low in part because only about one-tenth of the SSOR occurs within the lower intertidal zone, where sea otters feed (Short et al., 2006), and almost all SSOR exists under a surface covering of stable armor composed of sheltered bedrock/rubble and boulder/cobble/gravel (Hayes and Michel, 1999; Taylor and Reimer, 2008; Chapters 6 and 7), a habitat in which sea otters do not dig for clams (Dean et al., 2002).

To simulate the co-occurrence of a sea otter pit and SSOR, the model randomly makes the following choices during each time step:

1. Did the sea otter feed?
2. If so, did foraging occur in the intertidal zone?
3. If so, did it excavate a pit to harvest infauna?
4. If so, what was the size of the excavation pit?
5. Where was the pit excavated with respect to the distribution of SSOR?
6. If there was co-occurrence, what category of SSOR was hit (e.g., heavy, moderate, or low oil residues)?
7. Given the category selected, what was the total PAH (TPAH) (chosen from the lognormal distribution of measured TPAH in the environment)?
8. What was the specific distribution of PAH analytes within the TPAH?

By the end of a time step, if a sea otter dug a pit that intersected SSOR, then the quantity of SSOR-derived PAH for each of 40 analytes was assigned and allocated among the various exposure pathways identified in the conceptual model (Fig. 16.4). If no SSOR intersection occurred during the time step, or if the sea otter was in an unoiled area, the sea otter could still assimilate PAH from prey, sediments, and seawater through the indirect exposure pathways in the food web.

In addition to empirical data on SSOR and habitat distributions, the model required data on sea otter behavior (e.g., diet, frequency of dives per tidal zone, number of pits per hour), which were derived from the extensive observational databases of Garshelis and Johnson reported in Harwell et al. (2010b). Other necessary data (e.g., sea otter energetics, body size and mass, prey energy–density) were derived from the literature. The PAH distributions in prey, sediments, and seawater were derived from analyses of samples from oiled and unoiled areas.

Considerable variability exists in PAH exposures because of spatial heterogeneity and differing concentrations and relative proportions of PAH in each exposure pathway. To capture that exposure variability, the model simulated chronic doses to 500 000 individuals, randomly assigning a specific value, sampled from the PAH distributions as measured in the environment, for each stochastic parameter for each time step and for each sea otter. Thus, the simulated population captured the full range of exposures to PAH that could occur to sea otters at NKI- which has a subpopulation size of about 75 individuals. To provide highly conservative estimates of potential effects from the exposures, we rank-ordered the assimilated doses and focused on the 1-in-1000th most-exposed individuals (i.e., 99.9% quantile) for effects assessments. Exposure values were calculated for other quantiles, thereby allowing inferences to be drawn concerning population-level effects (discussed below). Sensitivity analyses were conducted to explore effects of different model parameters, model structures, and sources of uncertainty on the results. Altogether, over 1 billion sea otter hours were simulated to capture environmental, SSOR, and sea otter variability, demonstrating the exceptional ability of a well-designed IBM to thoroughly explore variability and uncertainty in exposures to individuals and populations. Details of the model and its parameterization are provided in Harwell *et al.* (2010b).

The effects component of the risk assessment used a standard, EPA-approved approach to establishing *toxicity reference values* (TRVs) for PAH exposures to sea otters (US Environmental Protection Agency, 2007). The TRV is defined as the dose from chronic exposures above which ecologically relevant effects might occur to wildlife species and below which it is reasonable to expect that such effects would not occur (US Environmental Protection Agency, 2003). The ratio of the model-simulated dose to the TRV is the *hazard quotient* (HQ), where an HQ ≥ 1 indicates that exposure could lead to chronic effects and HQ < 1 indicates that no ecologically relevant adverse chronic effect would occur. The TRVs used in this study were derived from the EPA-approved set of experimental data in its Eco-SSL database (US Environmental Protection Agency, 2007). We extracted the mammal-relevant no-observed-adverse-effects level (NOAEL; the highest administered dose that did not cause an effect) and lowest-observed-adverse-effects level (LOAEL; the lowest administered dose that did cause an effect), based on the population-relevant endpoints of mortality, growth, or reproductive effects. To accommodate interspecies uncertainty in the toxicity tests, geometric 95% lower confidence limits were calculated for use as the TRVs. Following US Environmental Protection Agency (2007), TRVs were separately established for the low-molecular weight (2- and 3-ring) PAH, high-molecular weight (4-, 5-, or 6-ring) PAH, and TPAH.

The resulting HQs for the most-exposed individual sea otters for high molecular weight PAH were about 30–125 times lower than the NOAEL TRV threshold, and about 75–310 times lower than the LOAEL threshold. Figure 16.5 is an example of the rank-ordered frequency distribution of assimilated doses of PAH, indicating NOAEL and LOAEL TRVs and doses to various quantile sea otters. The sensitivity analyses showed that these results are robust (Harwell *et al.*, 2010b).

Bodkin *et al.* (2012) subsequently reported studies on the frequency of sea otter pits potentially intersecting SSOR. They monitored the diving patterns of sea otters near northern Knight Island using 19 recovered time-depth recorders during 2003–08. Additionally, in summer 2008, they surveyed the intertidal zone of soft-sediment beaches for the presence and location of sea otter pits. Their resulting calculations of the frequency of sea otters' encountering SSOR were 2–24 times per year for females and 2–4 times per year for males. By comparison, the estimates of Harwell *et al.* (2010b; Table 2) were about 2–7 times per year, depending on the sea otter class; consequently, assimilated doses using the new SSOR–sea otter pit co-occurrence estimates from Bodkin *et al.* (2012) would remain at least an order of magnitude below the no-effects level for the 1-in-1000th most-exposed individuals, and our results remain valid. Moreover, as shown in Section 16.4.5, the sensitivity analyses reported in Harwell *et al.* (2012b) indicate that it would require 4–10 SSOR-intersecting pits per *day* for there to be effects on the 99.9% quantile most-exposed individuals, putting the new Bodkin *et al.* (2012) estimate of 2–24 SSOR-intersecting pits per *year* in context.

We conclude that there is currently no plausible risk to any individual sea otter at NKI from *Exxon Valdez* oil residues.

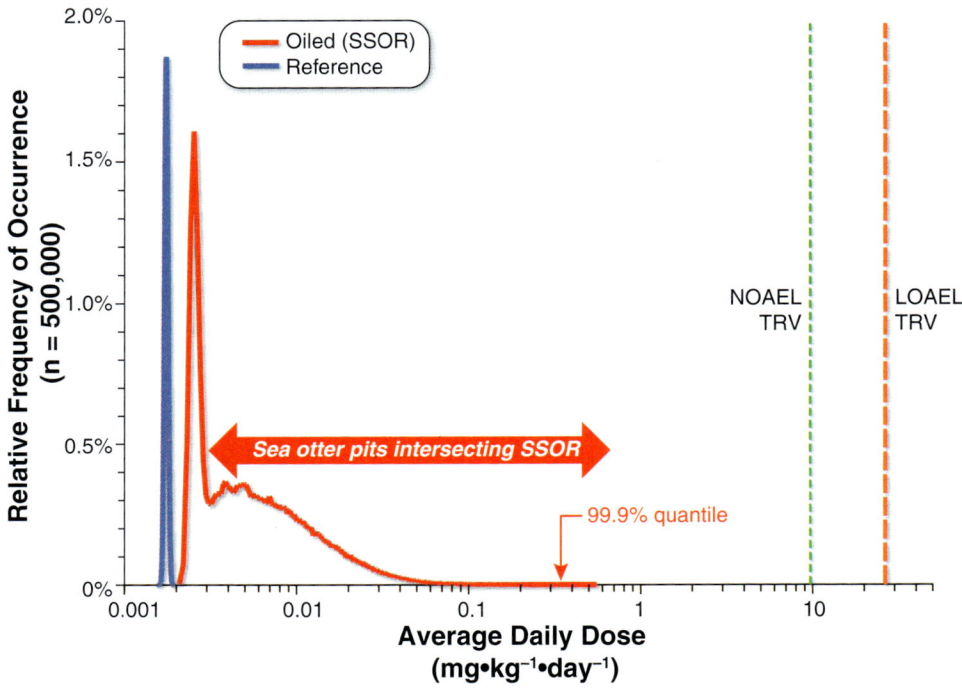

Figure 16.5 Relative frequency distributions of modeled average daily assimilated doses (mg/kg/day) of 4–6-ring polycyclic aromatic hydrocarbons (PAH) for sea otters under oiled and reference conditions. Base model results are shown for the sea otter pup class (n = 500 000); similar results occurred for all other modeled classes of sea otters. The magnitudes of the relative frequencies are a function of the number of bins used in each histogram. Because the ranges of assimilated doses are so different between the oiled and reference sea otter classes, it is necessary to scale the number of bins in order to illustrate clearly the shape of each distribution – in this case, 200 bins were used for reference data and 5000 bins for oiled data. This does not affect the relevant aspect of each distribution of assimilated doses represented on the x-axis. Also shown are the 99.9% quantile of assimilated doses (maximum-exposed individuals) and the No Observed Adverse Effects Level (NOAEL) and Lowest Observed Adverse Effects Level (LOAEL) 4–6-ring PAH toxicity reference values (TRVs). The blue line represents the assimilated doses to the sea otters in the reference sites. The red line represents the assimilated doses to two sets of sea otters foraging in the initially oiled sites: those that did not intersect SSOR (left peak) and those that did intersect SSOR while excavating sediments for infauna (portion of the curve to the right under the horizontal arrow). The TRVs are more than an order of magnitude above the highest assimilated dose in the population, indicating that there is essentially no risk to PWS sea otters from SSOR. (Data based on Harwell et al., 2010b.)

16.4.3 Quantitative ecological risk assessment of harlequin ducks

In response to suggestions of continuing toxicological effects on harlequin ducks (Bodkin et al., 2002; Esler et al., 2002; Short et al., 2006), we used a similar approach to assess current risks from the oil spill to this species in oiled and unoiled areas of PWS (Harwell et al., 2012a). We developed an ecotoxicological CEM (Fig. 16.6) describing all potential pathways of PAH exposure from residual *Exxon Valdez* and non-*Exxon Valdez* sources: (1) direct encounter with surface oil residues (SOR); (2) direct encounter with SSOR; (3) consumption of prey tissues containing PAH; (4) consumption of sediments containing PAH; and (5) consumption of seawater containing PAH.

By 2002, PWS had very little remaining SOR (< 0.2% by area of surveyed sites), and what remained had weathered to small patches of hard, highly weathered asphaltic "pavement," primarily in middle- to upper-intertidal or supratidal zones on beaches having

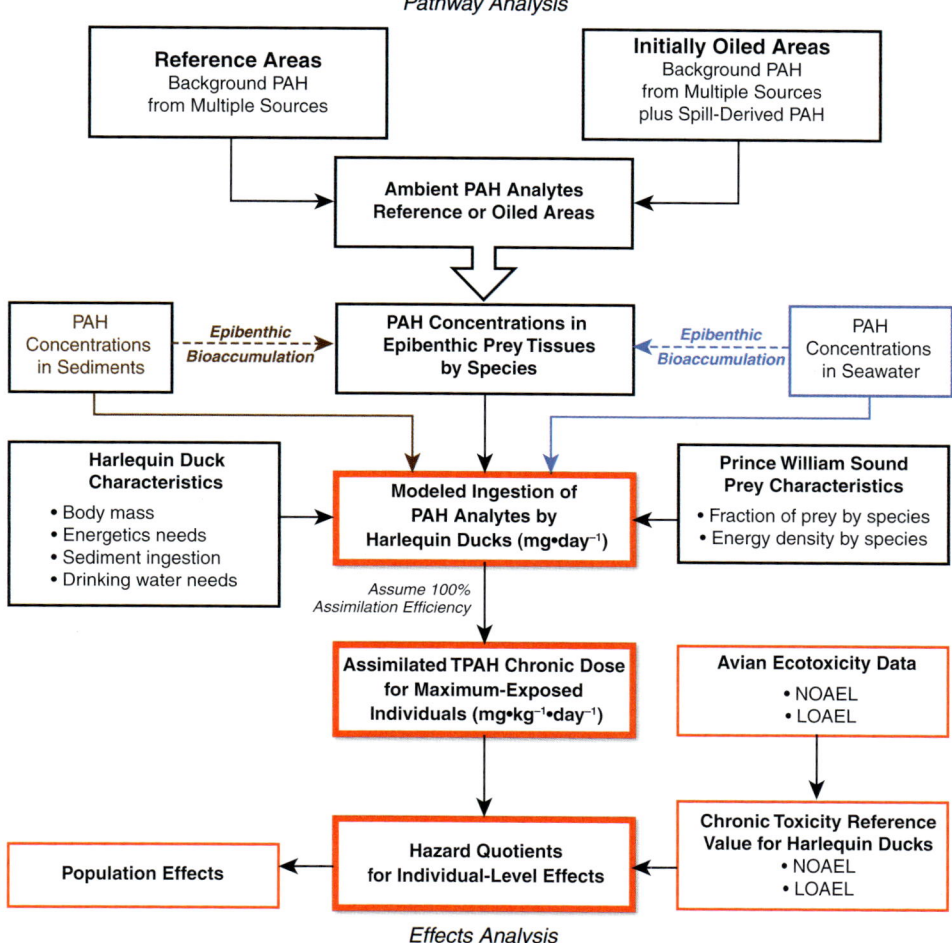

Figure 16.6 Harlequin duck risk assessment framework. Pathways conceptual model showing: the sources of PAH in the PWS environment in reference and initially oiled sites; concentrations of PAH in sediments, seawater, and prey tissues; risk assessment model inputs, including characteristics of harlequin ducks in PWS and their prey; elements of the risk assessment model; generation of assimilated doses; development of chronic TRVs for seaducks; and assessment of individual- and population-level effects. (Reproduced from Harwell *et al*., 2012a, with permission of Taylor & Francis Ltd., http://www.tandf.co.uk/journals.)

cobble, boulder, and pebble cover (Taylor and Reimer, 2008). Harlequin ducks feed primarily on epibenthic invertebrates in the lower intertidal zone (Robertson and Goudie, 1999), so the asphaltic SOR is a very unlikely target for feeding. Because the SOR is so highly weathered, its PAH are essentially not bioavailable (Michel *et al*., 2006; Integral Consulting, Inc., 2006). Consequently, consumption of SOR is not a plausible pathway for PAH exposure to harlequins and was not modeled.

Similarly, the possibility that a harlequin duck would directly excavate SSOR, given its depth and location under surface-armored sediments in the middle- and upper-intertidal zones, is extremely unlikely because harlequin ducks do not have the physical capability to dig to those depths or under such armoring (Robertson and Goudie, 1999). As a result, the pathway in which a harlequin duck excavated sediments to feed on infauna and in the process intersected SSOR does not warrant modeling.

Whether in oiled or unoiled areas, however, harlequin ducks may be routinely exposed to environmental PAH in prey tissues, sediments, and seawater, providing the three potential routes of exposure that warranted quantitative assessment in an IBM. The model was parameterized with empirical data for each exposure pathway, including: (1) PAH concentrations in PWS prey, sediments, and seawater; (2) dietary information on PWS harlequin ducks; (3) other literature-derived data about seaduck behavior and physiology; and (4) energy densities of each prey category (sources of data are described in Harwell *et al.*, 2012a).

The PWS sediment- and prey-PAH data showed complex patterns of heterogeneity with respect to both spatial distributions and sources of PAH; these patterns are relevant to understanding current risks. In general, five sources can be identified in the PWS samples (Harwell *et al.*, 2012a): petrogenic PAH from natural petroleum background, diesel fuel (e.g., from boats), or *Exxon Valdez* oil; biogenic PAH; and pyrogenic PAH. In spite of this diversity of sources, to provide an upper-bound estimate of risks we assumed that all doses to oiled harlequin ducks exceeding reference doses were attributable to the oil spill.

Average daily assimilated doses for each PAH analyte and TPAH were calculated for a population of 500 000 individuals for each of eight age- and gender-based classes, capturing expected exposure variability within a population of harlequin ducks living in PWS. Sensitivity analyses explored model uncertainty, bringing the total number of seaducks modeled to 64 million. We conservatively emphasized the 99.9% quantile, considered here to be the most-exposed individuals.

For the effects component, none of the EcoSSL studies met criteria for establishing reliable avian TRVs for PAH (US Army, 2000; US Environmental Protection Agency, 2003), but toxicity studies by Stubblefield *et al.* (1995a, b) on mallard ducks (*Anas platyrhynchos*) using weathered *Exxon Valdez* crude oil provided a sound basis for assigning TRVs for harlequin ducks. Those authors measured a suite of endpoints, including the two population-relevant parameters used here, eggshell thickness and eggshell strength. We applied US Environmental Protection Agency (2003, 2007) protocols to these toxicity studies to derive chronic TRVs.

Results for the 99.9% quantile harlequin ducks show that assimilated doses in oiled areas are ~400 and ~4000 times lower than NOAEL and LOAEL TRVs, respectively. Frequency distributions of assimilated doses (Fig. 16.7) have much lower variance than for sea otters because harlequin ducks do not have the low-probability but high-consequence SSOR-intersecting pathway of exposure. The oiled-site PAH is only minimally higher than reference-site PAH, and both are essentially at background levels.

Based on these results and the series of sensitivity analyses (Harwell *et al.*, 2012a, b), we conclude that essentially no risk currently exists to harlequin ducks from the *Exxon Valdez* oil spill.

16.4.4 Quantitative assessment of uncertainties: sensitivity analyses

An important attribute of IBMs is the inherent ease of conducting sensitivity analyses in which a single model component is altered and results are compared with the outputs from the base model (i.e., the specific model formulation using best estimates for all parameters). Sensitivity analyses were conducted on our models to assess the effects on risks of the following: data uncertainties; alternate model assumptions about exposure

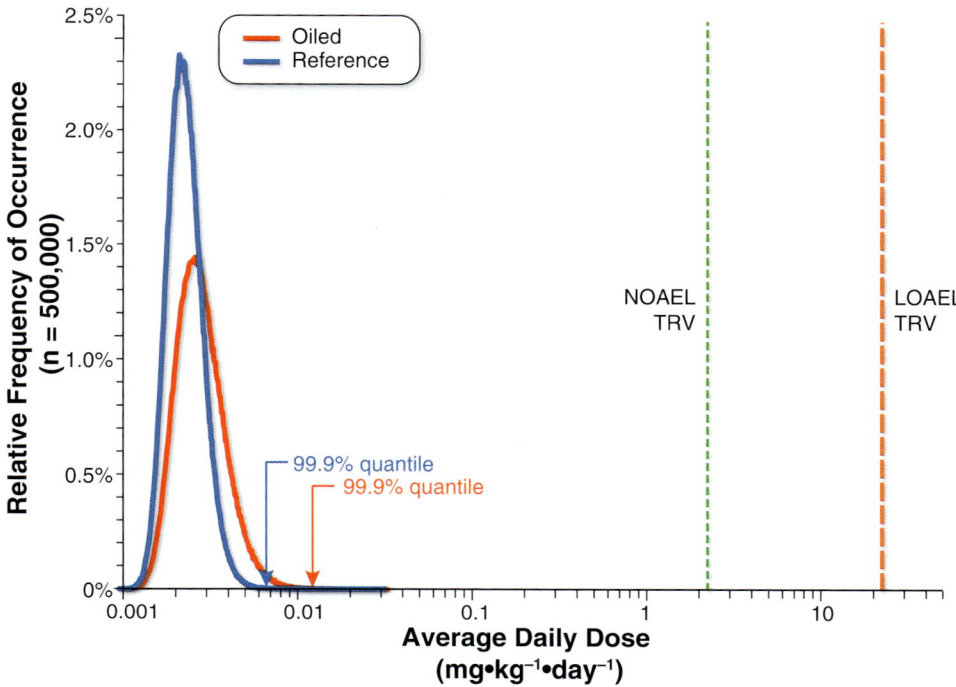

Figure 16.7 Relative frequency distributions of modeled average daily assimilated doses of total polycyclic aromatic hydrocarbons (TPAH) for harlequin ducks under oiled and reference conditions. Base model results are shown for adult female harlequin ducks in winter (n = 500 000); very similar results occurred for all other modeled classes of seaducks. The magnitudes of the relative frequencies are affected by the number of bins used in each histogram. Because the ranges of assimilated doses were similar for the oiled and reference harlequin ducks, each distribution could be based on the same number of bins, in this case 1000. Also shown are the 99.9% quantile of assimilated doses (maximum-exposed individuals) and the No Observed Adverse Effects Level (NOAEL) and Lowest Observed Adverse Effects Level (LOAEL) TPAH toxicity reference values (TRVs). The TRVs are more than two to three orders of magnitude above the highest assimilated dose in the population, indicating that there is essentially no risk to PWS harlequin ducks. (Data based on Harwell *et al.*, 2012a.)

pathways; alternate model assumptions about animal behavior, diet, and physiology; alternative sources of data; alternative classes of individuals (e.g., different age- and gender-specific parameters); alternative exposure regimes; and environmental heterogeneity.

Because our models capture variance in risks as it exists in the environment, sensitivity analyses can show the consequences to not just a single model output but to the full range of risk relationships, thereby significantly enhancing the utility for risk management. For example, sensitivity analyses can define an upper-bound range of risks under plausible assumptions, indicating how much change in the environment or exposure regime would be necessary to reach a specified threshold of effects. Sensitivity analyses that explore alternative model constructs, assumptions, or data sources can also answer the "so what?" question. They are often able to demonstrate whether some controversial or contentious issue even matters to the overall conclusions about risks.

Structural-sensitivity analyses examine alternate model constructs. For the sea otter model (Harwell *et al.*, 2010b), one such set directly assigned probabilities of co-occurrence of a sea otter pit with SSOR rather than using the base model's spatially explicit algorithm discussed previously. This set included one analysis that – incorrectly – assumed that sea otters can dig pits throughout the intertidal zone (based on the assumptions in Short *et al.*, 2006) instead of reflecting their actual foraging, which occurs only in the lower intertidal and subtidal zones. Even under such implausible conditions, our analyses showed that risks remained an order of magnitude below no-effects thresholds. Another structural-sensitivity analysis used sea otter diving data derived from Ballachey and Bodkin (2006) instead of our observational data, with little effect on the results.

Parameter-sensitivity analyses examine variations in individual parameters. For the sea otter model, this included altering pit size, PAH-assimilation efficiency, grooming efficiency, number of dives per pit, and thickness of oil-phase coatings on clams and paws. In all cases, the results were quite similar to those of the base model.

For the harlequin duck model, parameter-sensitivity analyses examined effects of spatial heterogeneity in PAH distributions, seaduck metabolic rates, alternative diets (e.g., 100% mussel diet), and increased sediment ingestion. The sensitivity analysis that came closest to having effects remained more than 2–3 orders of magnitude below effects thresholds. This occurred when the duck was assigned solely to a highly contaminated peat habitat, which was unrealistic because the peat site is extreme with regard to residual PAH concentrations in PWS and because harlequin ducks do not feed in peat habitats.

The set of harlequin duck sensitivity analyses that had the greatest relative effect on risk estimates involved method detection limits (MDLs) for chemical analyses. In the base model, those PAH in field samples that measured "0" (also called nondetects) were assigned the value of 0.5 MDL (following US Environmental Protection Agency, 1991), but the sensitivity analyses assigned values of either "0" or the full MDL. Sensitivity to MDLs occurs because the prey pathway contributes ~90% of the total harlequin duck dose. PAH in prey tissues were essentially identical to background, and most samples in both oiled and reference sites were nondetects. Thus, this sensitivity analysis indicates that the environmental concentrations in both oiled and unoiled areas were essentially indistinguishable from background.

16.4.5 Quantifying population-level risks

By taking advantage of the stochasticity of our IBMs, we could go beyond assessing risks to individuals to quantitatively assess risks at the population level. As shown previously, there are essentially no residual risks for sea otters and harlequin ducks, even for the most-exposed individuals. Consequently, no population-level risk could exist, since that would require a significant fraction of all individuals in the population to be affected. Nevertheless, to demonstrate the utility of IBMs for assessing population-level risks, and to assess what would be required to cause population-level effects on sea otters and seaducks, we created hypothetical exposure regimes of sufficient magnitude to force effects to occur.

Because sea otter risks are dominated by the low-probability, high-consequence intersection of SSOR with a sea otter pit, we artificially forced the number of pits intersecting SSOR per day to unrealistically high levels. We bypassed all probabilistic algorithms for determining whether an SSOR-intersecting pit was dug by a sea otter during a time step and forced one SSOR-intersecting pit to occur every day while keeping all other stochastic processes intact. Outputs from this series remained well below NOAEL HQ and LOAEL HQ levels (NHQ and LHQ, respectively). Assimilated doses from this fixed pit-per-day model are linear with increasing number of pits, resulting in predictions of effects on maximally exposed individuals occurring at ~4 and ~10 pits per day, for NHQ and LHQ, respectively. This rate contrasts with base model predictions of one SSOR-intersecting pit occurring, on average, about once every 50 to 180 days, depending on the sea otter class (Harwell *et al.*, 2010b).

We next extended this analysis to assess population-level effects. If the 99.9% quantile sea otter did actually dig 4–10 SSOR-intersecting pits per day, then there still would be only 1 sea otter in 1000 showing effects, clearly an incidence rate neither detectable nor of consequence to a population. The question then became, what percentage of individuals (quantile level) would need to be affected to have a detectable population-level effect? We reasoned that the median level (50% quantile), in which half of the individuals in the population had effects, would likely be detectable. Precisely where between the 99.9% and 50% quantiles a population-level effect would become detectable is in part a function of the population's natural variability. For example, a 10% change in a population with an interannual variability of 10-fold, as in some forage fish populations, would not likely be detectable, whereas a 10% change in a population with virtually no interannual variability, as in large marine mammals, might be detectable. Rather than selecting a single quantile, we generated a family of curves representing a spectrum of quantile levels so that risk managers and others could choose population-level effects thresholds.

These outputs, shown in Figure 16.8 for the most sensitive of the sea otter classes (juveniles), illustrate the consequences of selecting alternate quantile levels and using NOAEL or LOAEL thresholds. The shape of these curves illustrates that the quantile level for reaching effects thresholds is essentially linear and has low sensitivity over the

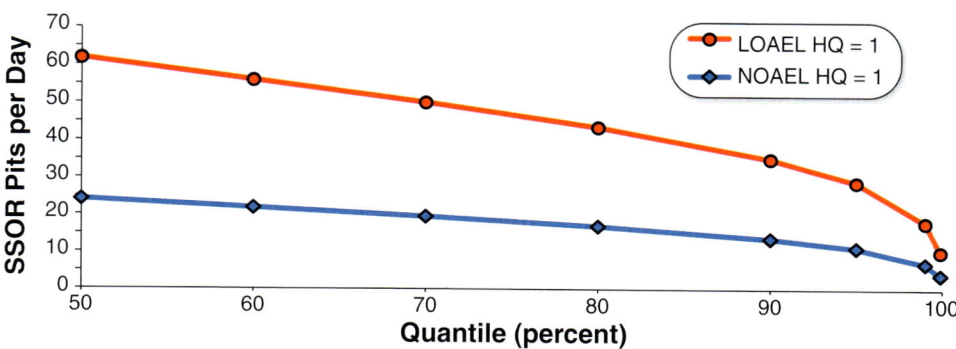

Figure 16.8 Effects thresholds across quantiles for juvenile sea otters in sensitivity analyses of hypothetical elevated exposure to SSOR. Sensitivity analyses quantified how many SSOR-intersecting pits would be required to cause effects, based on 4–6-ring-PAH TRVs (the most toxic components of TPAH) at LOAEL and NOAEL thresholds. Quantiles based on model simulations of a population of 500 000 individuals in the most sensitive (juvenile) class of sea otters for hypothetical multiples of the one-SSOR-intersecting-pit-per-day scenario. Graphs show the number of SSOR-intersecting sea otter pits per day required to reach the hazard quotient (HQ = 1) threshold for each quantile level. For example, for the median-exposed juvenile sea otter (50% quantile) about 62 SSOR-intersecting pits per day would result in reaching the LOAEL hazard quotient (LHQ = 1) threshold, but for the 90% quantile-exposed juvenile sea otter, about 35 SSOR-intersecting pits per day would suffice; this contrasts with the actual rate, as projected in the base model, of one SSOR-intersecting pit occurring on average once every 65 days for juvenile sea otters. (Data based on Harwell *et al.*, 2012b.)

range of 50–90% quantiles (less than two-fold difference over that range), but is more sensitive and nonlinear above 90%. However, because the latter applies only to infrequently higher-exposed individuals, it is not relevant to detectable population-level effects. Additional results reported in Harwell *et al.* (2012a) showed about a two-fold range across sea otter classes, with the nonadults (pups, juveniles, and subadults) and females with pups being similarly sensitive and adult females without pups the least sensitive class.

Finally, the population-level risk regime emerges using particular assumptions about risk thresholds. We assumed that: (1) no population-level effect would be detectable for exposures with $\leq$ 20% of the population reaching NOAEL (i.e., below the 80% quantile NHQ) (a quite conservative assumption, since even if 100% reached NOAEL levels, there would, by definition, still be "no effects") and (2) detectable population-level effects would be expected where more than half of the population exceeded LHQ thresholds (i.e., 50% quantile level). Thus, depending on sea otter class, anything below ~18–25 SSOR-intersecting pits occurring every day would not result in a detectable population-level effect, whereas anything above ~60–100 SSOR-intersecting pits every day would likely be detectable as a population-level effect. The actual exposure level at which population-level effects would become detectable is within the ranges shown in Figure 16.9. These values are orders of magnitude greater than actual estimated rates in PWS, in which about one SSOR intersection occurs every 50–180 days (2–7 SSOR-intersecting pits per year), depending on the sea otter class, according to our studies (Harwell *et al.* 2012a, b), or about 2–24 SSOR-intersecting pits per year, according to the estimates of Bodkin *et al.* (2012).

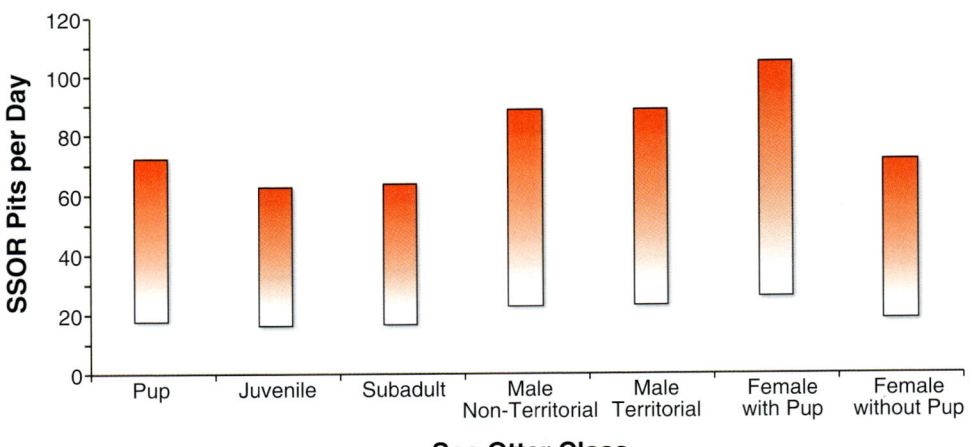

Figure 16.9 Thresholds for detecting population-level effects from exposure to 4–6-ring PAH by seven classes of sea otters. Suggested bounds of detecting population-level effects are: no population-level effects would be detectable at exposures below the NOAEL hazard quotient (NHQ) 80% quantile (i.e., $\leq$ 20% of individuals reaching NOAEL exposures) and population-level effects are expected to be detected at exposures above the LOAEL hazard quotient (LHQ) 50% quantile (i.e., $\geq$ 50% of individual reaching LOAEL effects levels). While alternate bounds could be chosen by risk managers and others, this illustrates the process to bound population-level risk thresholds. In this example, no population effects would occur below about 20 SSOR-intersecting pits per day, and population-level effects would likely be detectable above about 60–100 SSOR-intersecting pits per day, depending on the sea otter class. (Data based on Harwell *et al.*, 2012b.)

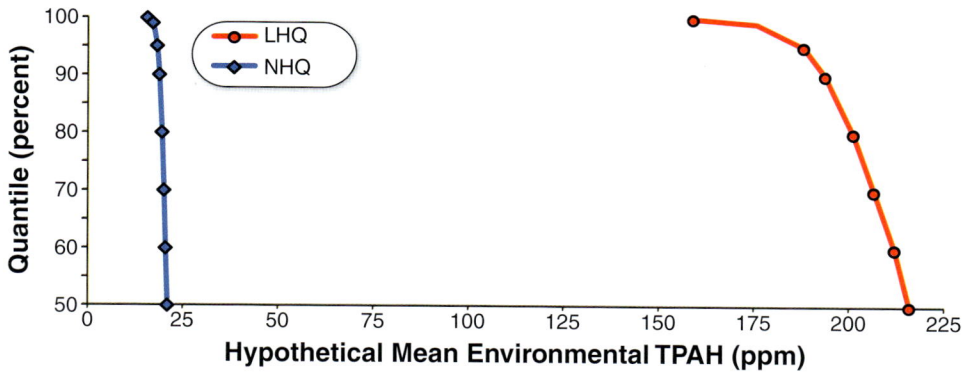

Figure 16.10 Hypothetical mean environmental PAH concentrations needed to reach HQ = 1 thresholds across exposure quantiles for adult female harlequin ducks in winter. Sensitivity analyses quantified how high the mean environmental TPAH concentrations would have to be to cause effects, using TPAH TRVs. For example, at the 80% quantile-exposure level, about 200 ppm mean environmental TPAH would be sufficient to reach the LOAEL TRV. Very similar results occurred for all other classes of harlequin ducks. Shown are the NOAEL HQ = 1 series (NHQ) and LOAEL HQ = 1 series (LHQ). (Data based on Harwell et al., 2012b).

Because risks to harlequin ducks are controlled by concentrations of PAH in their environment (there are no SSOR exposure pathways to harlequin ducks), we artificially forced those concentrations to unrealistically high levels (Harwell et al., 2012a, b). To accomplish this, we heuristically increased the environmental concentrations of TPAH in the model, bracketing HQ = 1 thresholds but keeping relative proportional contributions across PAH sources unchanged. Figure 16.10 reflects hypothetical mean environmental TPAH concentrations needed for various exposure quantiles to reach HQ = 1 thresholds for NOAEL and LOAEL. We concluded that mean environmental TPAH concentrations would have to be ~20 ppm to reach NHQ = 1 and ~200 ppm to reach LHQ = 1, values not seen in PWS since 1990, an order of magnitude greater than those measured since the cleanup ended in 1992, and three orders of magnitude greater than existed in PWS in 2008.

We used the same reasoning as applied to sea otters to assess where population-level effects would be detectable in harlequin ducks, bounded by the 80% quantile NOAEL (~20 ppm) and 50% quantile LOAEL TRV thresholds (~215 ppm). One may choose alternate thresholds for inferring detectable population-level effects, but selecting a particular set of thresholds is not a sensitive parameter – using an 80% quantile LHQ value instead of 50% quantile results in population-level effects being detectable at ~200 ppm rather than ~215 ppm.

To put these environmental TPAH concentrations in perspective, we examined PAH concentrations measured soon after the spill as reported in the *Exxon Valdez* Trustees Database (Nelson et al., 1999) and the Oil Spill Health Task Force database (Bence and Burns, 1995). We concluded that even as early as 90 days postspill, mean measured TPAH levels in the PWS environment were well below these population-level-effects thresholds for harlequin ducks. This confirms that the likelihood of population-level effects occurring now, over 20 years after the spill, is essentially nonexistent.

Using this approach, either for a case in which actual environmental exposures are in the range of toxicity thresholds or in cases like this, in which environmental exposures have to be hypothetically increased in order to reach toxicity thresholds, risk managers could select levels of population detectability for the endpoint or VEC using the appropriate model-output curve. Further, reporting the range of values across quantiles provides a comparative context for assessing the magnitude of risks, the potential for individual-level effects, and, importantly, the potential for population-level effects. This approach provides a quantitative understanding of how population-level risks can emerge from the distributions of outputs from an IBM.

16.4.6 Quantifying sources of toxicological risks

A unique capability provided by IBMs is the ability to attribute risks of PAH to particular sources in the environment. In our model, harlequin ducks act as integrators of PAH from all environmental sources, and modeled distributions of the 40 assimilated PAH analytes provide diagnostic information on the sources and exposure pathways of PAH as they are ingested by seaducks. As detailed in Harwell *et al.* (2012a), we characterized the relative distributions of PAH as assimilated by the harlequin ducks and examined those distributions to identify characteristic "fingerprints" of the various sources of PAH in PWS. From this we could attribute risks to specific sources.

The seaduck model was designed to distinguish exposures from prey and sediments pathways. Prey-pathway doses, constituting ~90% of the total dose, were essentially identical to background. Further, the close similarity between oiled and unoiled (reference) distributions indicated that oiled-site prey cannot be distinguished from reference-site prey. Thus, there was little if any remaining signal from oil residues in prey tissues by the time samples were collected (2002–08). Moreover, the small differences between oiled- and reference-site distributions that existed did not reflect significant amounts of additional petrogenic PAH in oiled sites, whether from oil-spill residues or other petrogenic sources. Rather, prey-based doses were dominated by pyrogenic PAH, as evidenced by elevated concentrations of parent compounds relative to their alkylated homologs (Bence *et al.*, 2007).

The unoiled-site sediments pathway contained a relatively large fraction of perylene, a biogenic PAH. The oiled-sediment pathway had both petrogenic and pyrogenic components. The petrogenic component, which occurred in only ~10% of oiled sediment samples (Harwell *et al.*, 2012a), was heavily weathered and showed diagnostic PAH ratios of *Exxon Valdez* spill-derived residues (Page *et al.*, 1995). However, sediment-derived PAH contributed < 10% of the total assimilated dose to harlequin ducks, resulting in exposures three and four orders of magnitude below NHQs and LHQs, respectively. Thus, the continuing oil-spill signal seen in some sediment PAH is not ecologically significant.

16.5 Applying the integrated assessment framework to future spills

The integrated studies described here were not initiated until more than a decade after the *Exxon Valdez* oil spill. What might have been the benefits had this risk-based approach been available and applied soon after the spill? Had that occurred, we believe it could have significantly improved the planning, analysis, synthesis, and decision processes. To illustrate how the integrated assessment process could be applied in new situations, we distinguish three postspill time periods, discussed below, that demonstrate how assessments should evolve over time as risks change and the ecosystem recovers.[2]

As a general rule, the amount of site- and spill-specific information available increases over time, allowing analyses to become more quantitative and more predictive. Nevertheless, much valuable information exists even early in the postspill period that can be used to gain insights on injury, risks, and recovery characteristics. Clearly, each spill situation will require spill-specific information concerning the nature, magnitude, spatial extent, and toxicity of the spill components. If the spill-stressor regime is not understood in at least its broad outlines, then any assessments of injury, risk, or recovery become problematic. Beyond that base level of information, however, much may already be known about the candidate VECs of an oil or chemical spill in a marine environment. In PWS, for example, prespill information on many species was sufficient such that an exposure pathway CEM and a preliminary IBM could have been developed soon after the spill, based on existing literature-derived information on the species' characteristics. Such a preliminary model would have been more than sufficient to bound the risks and, through sensitivity analyses, explore the potential importance (or lack thereof) of factors in the environment. Consequently, even early after the spill, implausible risk hypotheses could have been eliminated.

Later in the postspill period, as illustrated by our analyses, a richer set of data specific to the system would enable the development of a more accurate and quantitative characterization of the risk regime as it unfolds over time while recovery proceeds. Using the risk-based, integrated assessment framework, scientists and decision-makers could take advantage of new information as it becomes available. They could use preliminary and interim analyses to inform additional studies, identify research areas not worth pursuing, and discern the most efficient pathways for achieving ecosystem recovery.

16.5.1 Immediate postspill period

During and immediately after an oil spill (or other chemical release), primary emphasis is on containment and removal of the spilled material to minimize the spatial extent of contamination, limit exposure of natural resources, and mitigate acute ecological effects. This immediate postspill period overlaps with Phases 1 and 2 described in Chapter 2. During this time period, problem formulation should begin to assess the magnitude and extent of initial injury and risks. Attention should be on characterizing the physical and chemical characteristics of the spill; assessing fate-and-transport processes and the resulting spatial extent and pathways of exposures; and evaluating initial effects on natural resources. Initially, a qualitative CEM should be developed to capture those elements and to identify other natural and anthropogenic stressors that may affect the ecosystem. These steps can result in a carefully crafted CEM that identifies all plausible routes of exposure to the VECs at-risk, setting the stage for formulating risk and recovery hypotheses that will guide future research, analysis, and monitoring. In addition, these CEMs can be used for an initial, qualitative relative-risk ranking study based on existing information and collective scientific expertise, even before extensive data from research and monitoring programs are available.

---

[2] Note that these time periods overlay but do not directly correspond to the three phases of an oil spill described in Chapter 2.

In retrospect, we believe that had this been done soon after the *Exxon Valdez* oil spill, greater consideration could have been given to the selection of appropriate baselines and endpoints for assessing injury and recovery. This illustrates how a properly developed problem formulation could avoid unnecessary expenditure of resources on research and monitoring of inappropriate endpoints, and instead specifically target studies to assess risks and reduce uncertainties on those VECs of importance. Concomitant with this is a better recognition of the roles of natural variability and other stressors on the recovery dynamics of species and the ecosystem as captured in the VECs. Applying the integrated assessment approach soon after a spill can also foster a more consistent application of scientific principles for identifying recovery objectives and metrics and establishing a more comprehensive and integrated database for use in assessing injury, ongoing risks, and recovery.

16.5.2 Intermediate postspill period

Once the field studies are underway, but before extensive data have been acquired or risk analyses conducted, there is a continuing role for integrated assessments. This intermediate assessment period corresponds primarily to the early part of Phase 3 but may also overlap Phase 2 (Chapter 2). During this time period, quantitative analysis tools should be developed for selected VECs, as exemplified by our sea otter and harlequin duck IBMs. Initial versions of such IBMs could be derived from previous models and from existing information in the literature, along with insights from the qualitative CEMs on the plausible exposure pathways.

Although more extensive data (such as those available to us during our quantitative risk assessments) likely would not yet exist, preliminary models nevertheless could be used very effectively for conducting scenario–consequence and sensitivity analyses. Such analyses could be used to identify the most important uncertainties and the specific pathways contributing the most to individual- and population-level risks. These analyses could also define the outer bounds of risk, thereby indicating what is and is not plausible, and what exposures or pathways are or are not capable of causing ecologically significant effects. Such information could inform remediation activities by examining the efficacy of alternative remedial, recovery, and restoration options.

Finally, during this time period, analyses that couple the versatility and power of IBMs with qualitative CEM-based relative-risk-ranking analyses could help to identify areas of continuing risks while eliminating other factors as incapable of causing further effects. This approach can begin to place the risks from the spill-caused stressors in the context of other impacts on the ecosystem.

16.5.3 Long-term postspill period

This is the period of extensive quantitative analysis and synthesis of information and data derived from ongoing research, monitoring, and modeling programs such as those described in this chapter; it corresponds to the later part of Phase 3 (Chapter 2). During this period, the types of quantitative analyses and risk-based studies presented here and in the associated publications provide a template for future integrated assessment-based studies. This is the period in which quantitative analyses can be conducted to determine residual and attributable risks to species and populations. These assessments can identify and reduce critical uncertainties and guide the achievement of remediation and recovery goals.

16.6 Lessons learned

Our studies have demonstrated that by applying the risk-based integrated environmental assessment framework throughout the postspill time periods, the assessment, remediation, and recovery objectives could be met efficiently and effectively, thereby achieving the overarching goal of making the environment and public whole for injuries to natural resources and services. Additional lessons include the following:

- The integrated environmental assessment process has particular utility and value because it is comprehensive, robust, systematic, risk-based, transparent, and based on sound ecosystem science.
- The integrated environmental assessment framework can identify what is at risk, define how risks should be characterized and quantified, and suggest how information should be interpreted and used in decision-making in the presence of uncertainty.
- Problem formulation constitutes the essential first step in an integrated assessment, in which the nature and scope of the problem are defined, questions regarding what is at risk are articulated, essential data are identified, research is initiated, and science-based plans are laid for remediation, assessments, recovery, and restoration.
- Early application of problem formulation and development of an initial conceptual ecosystem model incorporating appropriate VECs to guide the assessment process could have prevented later misperceptions that residual effects attributable to the oil spill continued to affect PWS natural resources, even when the stressors of concern had long since returned to background levels.
- The individual-based models developed in our studies were used to quantify both individual- and population-level risks. These models provided a comprehensive understanding of the risk regime following a disturbance; they were effective tools to evaluate risk hypotheses, and gave unique insights into attributable risks.
- Qualitative conceptual ecosystem models provide a systematic, weight-of-evidence methodology that applies objective criteria to compare natural and anthropogenic stressors and assign relative risks in a multistressor environment. Our qualitative CEMs showed that climatic and oceanographic processes fundamentally control the PWS–GOA ecosystems, leading to cascading trophic effects with Sound-wide implications, and indicated that long-term effects from the oil spill were inconsequential compared to these natural processes.
- Quantitative models can be used to reach definitive conclusions on residual risks, ecological significance, recovery, and uncertainties. Our quantitative analyses showed clearly that there could not be a causal link between the oil spill and current risks to species of concern, and if any effects on those species do exist, they must derive from stressors other than those caused by the oil spill.
- To assure the validity of our modeling studies, we examined all plausible pathways for PAH exposure and parameterized the models with the most-relevant empirical data from PWS. Building a series of conservatisms into the models and analyses ensured that predicted risks represent upper-bound estimates.
- By modeling stochasticity, driven by the actual variance in the environment, we captured the complete distributions of exposure variability, including rare but consequential exposures.

- Characterizing model uncertainty is an important component of all risk assessments, especially when complex quantitative and probabilistic models are employed. We were able to assess model uncertainty by disaggregating the models into their constituents and quantifying the uncertainties associated with each component through a series of sensitivity analyses. These uncertainty analyses resulted in high confidence in the conclusions derived from the modeling studies.
- A careful quality assurance plan is also essential to ensure the validity of modeling studies; ours required that every simulation be archived, each parameter characterized and referenced, and each of the simulated exposures could be precisely repeated.
- Variability across space and time makes determining baseline conditions very difficult and greatly affects the ability to detect ecologically significant effects and establish goals for recovery. This issue only becomes more problematic over time as natural variability remains unabated while the remnant exposures and associated effects of the spill continue to diminish. Eventually the point is reached where the risks from the oil spill are lost in the noise of the variability of natural processes, and ecological risks can no longer be attributed to the spill.

Collectively, our studies applied the integrated assessment process, relying heavily on CEMs to guide our analyses. We used quantitative models to reach definitive conclusions on residual risks, ecological significance, and recovery, and to characterize uncertainties. These analyses assured the robustness of our causal pathway analyses and placed our results within the context of natural stressors. Using the multiple-lines-of-evidence approach, we fully characterized residual risks by showing that current exposure to remnant *Exxon Valdez* oil is so far below the concentrations needed for effects to occur that we can confidently assert that no ecologically significant risks to PWS populations remain from the oil spill.

REFERENCES

Ballachey, B.E. and J.L. Bodkin (2006). *Lingering Oil and Sea Otters: Pathways of Exposure and Recovery Status*. Anchorage, AK, USA: *Exxon Valdez* Oil Spill Trustee Council Restoration Project //620 Draft Final Report; April 2006.

Barnthouse, L.W., W.R. Munns, Jr., and M.T. Sorensen, eds (2007). *Population-Level Ecological Risk Assessment*. Boca Raton, FL, USA: CRC Press; ISBN-10: 1420053329; ISBN-13: 9781420053326.

Bence, A.E. and W.A. Burns (1995). Fingerprinting hydrocarbons in the biological resources of the *Exxon Valdez* spill area. In Exxon Valdez *Oil Spill: Fate and Effects in Alaskan Waters*. P.G. Wells, J.N. Butler, and J.S. Hughes, eds. Philadelphia, PA, USA: American Society for Testing and Materials; ASTM Special Technical Publication 1219; ISBN-10: 0803118961; pp. 84–140.

Bence, A.E., D.S. Page, and P.D. Boehm (2007). Advances in forensic techniques for petroleum hydrocarbons: the *Exxon Valdez* experience. In *Oil Spill Environmental Forensics: Fingerprinting and Source Identification*. Z. Wang and S.A. Stout, eds. San Diego, CA, USA: Academic Press; ISBN-10: 0123695236; ISBN-13: 9780123695239; pp. 449–487.

Bodkin, J.L. and B.E. Ballachey (1997). *Restoration Notebook: Sea Otter* (Enhydra lutris). Anchorage, AK, USA: *Exxon Valdez* Oil Spill Trustee Council. [http://www.evostc.state.ak.us/Universal/Documents/Publications/RestorationNotebook/RN_seaotter.pdf]

Bodkin, J.L., B.E. Ballachey, T.A. Dean, A.K. Fukuyama, S.C. Jewett, L. McDonald, D.H. Monson, C.E. O'Clair, and G.R. VanBlaricom (2002). Sea otter population status and the process of recovery from the 1989 *Exxon Valdez* oil spill. *Marine Ecology Progress Series* **241**: 237–253.

Bodkin, J.L., B.E. Ballachey, H.A. Coletti, G.C. Esslinger, K.A. Kloecker, S.D. Rice, J.A. Reed, and D.H. Monson (2012). Long-term effects of the *Exxon Valdez* oil spill: sea otter foraging in the intertidal as a pathway of exposure to lingering oil. *Marine Ecology Progress Series* **447**: 273–287.

Burn, D.M. and A.M. Doroff (2005). Decline in sea otter (*Enhydra lutris*) populations along the Alaska Peninsula, 1986–2001. *Fishery Bulletin* **103**(2): 270–279.

Canadian Council of Ministers of the Environment (1996). *A Framework for Ecological Risk: General Guidance*. Winnipeg, MB, Canada: The National Contaminated Sites Remediation Program; PN 1195; ISBN-10: 0662242463. [http://www.ccme.ca/assets/pdf/pn_1195_e.pdf]

Cormier, S.M. and G.W. Suter, II (2008). A framework for fully integrating environmental assessment. *Environmental Management* **42**(4): 543–556.

Dale, V.H., F.J. Swanson, and C.M. Crisafulli, eds (2005a). *Ecological Responses to the 1980 Eruption of Mount St. Helens*. New York, NY, USA: Springer; ISBN-13: 9780387238685.

Dale, V.H., F.J. Swanson, and C.M. Crisafulli (2005b). Ecological perspectives on management of the Mount St. Helens landscape. In *Ecological Responses to the 1980 Eruption of Mount St. Helens*. V.H. Dale, F.J. Swanson, and C.M. Crisafulli, eds. New York, NY, USA: Springer; ISBN-13: 9780387238685; pp. 277–286.

Dean, T.A., J.L. Bodkin, A.K. Fukuyama, S.C. Jewett, D.H. Monson, C.E. O'Clair, and G.R. VanBlaricom (2002). Food limitation and the recovery of sea otters following the *Exxon Valdez* oil spill. *Marine Ecology Progress Series* **241**: 255–270.

DeAngelis, D.L. (1980). Energy flow, nutrient cycling, and ecosystem resilience. *Ecology* **61**(4): 764–771.

DeAngelis, D.L., S.M. Bartell, and A.L. Brenkert (1989a). Effects of nutrient recycling and food-chain length on resilience. *American Naturalist* **134**(5): 778–805.

DeAngelis, D.L. and L.J. Gross (1992). *Individual-Based Models and Approaches in Ecology: Populations, Communities and Ecosystems*. New York, NY, USA: Springer; ISBN-10: 041203171X; ISBN-13: 978–0412031717. Also, London, UK: Chapman and Hall; ISBN-10: 0412031612, ISBN-13: 9780412031618.

DeAngelis, D.L. and W.M. Mooij (2005). Individual-based modeling of ecological and evolutionary processes. *Annual Review of Ecology, Evolution, and Systematics* **36**: 147–168.

DeAngelis, D.L., P.J. Mulholland, A.V. Palumbo, A.D. Steinman, M.A. Huston, and J.W. Elwood (1989b). Nutrient dynamics and food-web stability. *Annual Review of Ecology, Evolution, and Systematics* **20**: 71–95.

Esler, D., T.D. Bowman, K.A. Trust, B.E. Ballachey, T.A. Dean, S.C. Jewett, and C.E. O'Clair (2002). Harlequin duck population recovery following the *Exxon Valdez* oil spill: progress, process and constraints. *Marine Ecology Progress Series* **241**: 271–286.

Estes, J.A., M.T. Tinker, T.M. Williams, and D.F. Doak (1998). Killer whale predation on sea otters linking oceanic and nearshore ecosystems. *Science* **282**(5388): 473–476.

Exxon Valdez **Oil Spill Trustee Council** (2002). Exxon Valdez *Oil Spill Restoration Plan Update on Injured Resources and Services, April 10, 2002*. Anchorage, AK, USA: *Exxon Valdez* Oil Spill Trustee Council. [http://www.evostc.state.ak.us/Universal/Documents/Publications/2002IRSUpdate.pdf]

Finney, B.P., I. Gregory-Eaves, J. Sweetman, M.S.V. Douglas, and J.P. Smol (2000). Impacts of climatic change and fishing on Pacific salmon abundance over the past 300 years. *Science* **290**(5492): 795–799.

Forbes, V.E., P. Calow, V. Grimm, T. Hayashi, T. Jager, A. Palmqvist, R. Pastorok, D. Salvito, R. Sibly, J. Spromberg, J. Stark, and R.A. Stillman (2010). Integrating population modeling into ecological risk assessment. *Integrated Environmental Assessment and Management* **6**(1): 191–193.

Gentile, J.H. and M.A. Harwell (1998). The issue of significance in ecological risk assessments. *Human and Ecological Risk Assessment* **4**(4): 815–828.

Gentile, J.H. and M.A. Harwell (2001). Strategies for assessing cumulative ecological risks. *Human and Ecological Risk Assessment* **7**(2): 239–246.

Gentile, J.H., M.A. Harwell, W.P. Cropper, Jr., C.C. Harwell, D. DeAngelis, S. Davis, J.C. Ogden, and D. Lirman (2001). Ecological conceptual models: a framework and case study on ecosystem management for South Florida sustainability. *Journal of Science and the Total Environment* **274**(1–3): 231–253.

Harwell, M.A. and J.H. Gentile (2006). Ecological significance of residual exposures and effects from the *Exxon Valdez* oil spill. *Integrated Environmental Assessment and Management* **2**(3): 204–246.

Harwell, M.A., J.H. Gentile, K.W. Cummins, R.C. Highsmith, R. Hilborn, C.P. McRoy, J. Parrish, and T. Weingartner (2010a). A conceptual model of natural and anthropogenic drivers and their influence on the Prince William Sound, Alaska, ecosystem. *Human and Ecological Risk Assessment* **16**(4): 672–726.

Harwell, M.A., J.H. Gentile, C.B. Johnson, D.L. Garshelis, and K.R. Parker (2010b). A quantitative ecological risk assessment of the toxicological risks from *Exxon Valdez* subsurface oil residues to sea otters at Northern Knight Island, Prince William Sound, Alaska. *Human and Ecological Risk Assessment* **16**(4): 727–761.

Harwell, M.A., J.H. Gentile, K.R. Parker, S.M. Murphy, R.H. Day, A.E. Bence, J.M. Neff, and J.A. Wiens (2012a). Quantitative assessment of current risks to harlequin ducks in Prince William Sound, Alaska, from the *Exxon Valdez* oil spill. *Human and Ecological Risk Assessment* **18**(2): 261–328.

Harwell, M.A., J.H. Gentile, and K.R. Parker (2012b). Quantifying population-level risks using an individual-based model: Sea otters, harlequin ducks, and the *Exxon Valdez* oil spill. *Integrated Environmental Assessment and Management* **8**(3): 503–522.

Hayes, M.O. and J. Michel (1999). Factors determining the long-term persistence of *Exxon Valdez* oil in gravel beaches. *Marine Pollution Bulletin* **38**(2): 92–101.

Hegmann, G., C. Cocklin, R. Creasey, S. Dupuis, A. Kennedy, L. Kingsley, W. Ross, H. Spaling, and D. Stalker (1999). *Cumulative Effects Assessment Practitioners Guide.* Hull, QC, Canada: Canadian Environmental Assessment Agency. [http://publications.gc.ca/collections/Collection/En106-44-1999E.pdf]

Holling, C.S. (1973). Resilience and stability of ecological systems. *Annual Review of Ecology, Evolution, and Systematics* **4**: 1–23.

Integral Consulting, Inc. (2006). *Information Synthesis and Recovery Recommendations for Resources and Services Injured by the* Exxon Valdez *Oil Spill.* Mercer Island, WA, USA: Integral Consulting, Inc.; *Exxon Valdez* Oil Spill Restoration Project 060783 Final Report. [http://www.evostc.state.ak.us/Files.cfm?doc=/Store/FinalReports/2006-060783-Final.pdf&]

Johnson, C.B. and D.J. Garshelis (1995). Sea otter abundance, distribution, and pup production in Prince William Sound following the *Exxon Valdez* oil spill. In Exxon Valdez *Oil Spill: Fate and Effects in Alaskan Waters.* P.G. Wells, J.N. Butler, and J.S. Hughes, eds. Philadelphia, PA, USA: American Society for Testing and Materials; ASTM Special Technical Publication 1219; ISBN-10: 0803118961; pp. 894–929.

Lees, K., S. Pitois, C. Scott, C. Frid, and S. Mackinson (2006). Characterizing regime shifts in the marine environment. *Fish and Fisheries* **7**(2): 104–127.

Lirman, D. and P. Fong (1995). The effects of Hurricane Andrew and Tropical Storm Gordon on Florida reefs. *Coral Reefs* **14**(3): 172.

Lovelace, J.K. and B.F. McPherson (1996). *Restoration, Creation, and Recovery: Effects of Hurricane Andrew (1992) on Wetlands in Southern Florida and Louisiana.* Washington DC, USA: United States Geological Survey; National Water Summary on Wetland Resources; USGS Water Supply Paper 2425. [http://water.usgs.gov/nwsum/WSP2425/andrew.html]

Michel, J., Z. Nixon, and L. Cotsapas (2006). *Evaluation of Oil Remediation Technologies for Lingering Oil from the* Exxon Valdez *Oil Spill in Prince William Sound, Alaska.* Juneau, AK, USA: National Oceanic and Atmospheric Administration, National Marine Fisheries Service; *Exxon Valdez* Oil Spill Trustee Council Restoration Project 050778 Final. [http://www.evostc.state.ak.us/Files.cfm?doc=/Store/FinalReports/2005-050778-Final.pdf&]

Miller, A.J., E. Di Lorenzo, D.J. Neilson, H-J. Kim, A. Capotondi, M.A. Alexander, S.J. Bograd, F.B. Schwing, R. Mendelssohn, K. Hedstrom, and D.L. Musgrave (2005). Interdecadal changes in mesoscale eddy variance in the Gulf of Alaska circulation: possible implications for the Steller sea lion decline. *Atmosphere-Ocean* **43**(3): 231–240.

Mundy, P.R., ed. (2005). *The Gulf of Alaska: Biology and Oceanography.* Fairbanks, AK, USA: Alaska Sea Grant College Program, University of Alaska; Publication AK-SG-05-01; ISBN-10: 156612090x.

Munns, W.R., Jr., J. Gervais, A.A. Hoffman, U. Hommen, D.R. Nacci, M. Nakamaru, R. Sibly, and C.J. Topping (2007). Modeling approaches to population-level risk assessment. In *Population-Level Ecological Risk Assessment.* L.W. Barnthouse, W.R.

Munns, Jr., and M.T. Sorensen, eds. Boca Raton, FL, USA: CRC Press; ISBN-10: 1420053329; ISBN-13: 9781420053326; pp. 179–210.

National Marine Fisheries Service (2010). *Interim Endangered and Threatened Species Recovery Planning Guidance. Version 1.3. June 2010.* Silver Spring, MD, USA: National Oceanic and Atmospheric Administration, National Marine Fisheries Service. [http://www.nmfs.noaa.gov/pr/pdfs/recovery/guidance.pdf]

National Oceanic and Atmospheric Administration (1996). *Guidance Document for Natural Resource Damage Assessment under the Oil Pollution Act of 1990, August 1996.* Washington DC, USA: General Counsel for Natural Resources, Office of Habitat Conservation, Office of Response and Restoration, Damage Assessment Remediation and Restoration Program. [http://www.darrp.noaa.gov/library/1_d.html]

National Oceanic and Atmospheric Administration (2010). *Guidance Document for Natural Resource Damage Assessment under the Oil Pollution Act of 1990, Revised July 2010.* Washington DC, USA: General Counsel for Natural Resources, Office of Habitat Conservation, Office of Response and Restoration, Damage Assessment Remediation and Restoration Program. [http://www.darrp.noaa.gov/library/1_d.html]

National Research Council (1983). *Risk Assessment in the Federal Government: Managing the Process.* Washington DC, USA: National Academy Press.

National Research Council (1985). *Oil in the Sea: Inputs, Fates, and Effects.* Washington DC, USA: National Academy Press; ISBN-10: 0309034795.

National Research Council (1990). *Managing Troubled Waters: The Role of Marine Environmental Monitoring.* Washington DC, USA: National Academy Press; ISBN-10: 0309074789; ISBN-13: 9780309074780.

National Research Council (2002). *The Missouri River Ecosystem. Exploring the Prospects for Recovery.* Washington DC, USA: National Academy Press; ISBN-10: 0309083141; ISBN-13: 9780309083140.

Nelson, B.D., J.D. Short, and S.D. Rice (1999). *Hydrocarbon Data Analysis, Interpretation and Database Maintenance for Restoration and NRDA Environmental Samples Associated with the* Exxon Valdez *Oil Spill.* Juneau, AK, USA: National Oceanic and Atmospheric Administration, National Marine Fisheries Service; *Exxon Valdez* Oil Spill Restoration Project 98290 Annual Report. [http://www.evostc.state.ak.us/Files.cfm?doc=/Store/AnnualReports/1998-98290-Annual.pdf&]

Ogden, J.C., S.M. Davis, and L.A. Brandt (2002). Science strategy for a regional ecosystem monitoring and assessment program: the Florida Everglades example. In *Monitoring Ecosystems: Interdisciplinary Approaches for Evaluating Ecoregional Initiatives.* D. Busch and J.C. Trexler, eds. Washington DC, USA: Island Press; ISBN-10: 1559638516; ISBN-13: 9781559638500; pp. 135–166.

Page, D.S., P.D. Boehm, G.S. Douglas, and A.E. Bence (1995). Identification of hydrocarbon sources in the benthic sediments of Prince William Sound and the Gulf of Alaska following the *Exxon Valdez* oil spill. In Exxon Valdez *Oil Spill: Fate and Effects in Alaskan Waters.* P.G. Wells, J.N. Butler, and J.S. Hughes, eds. Philadelphia, PA, USA: American Society for Testing and Materials; ASTM Special Technical Publication 1219; ISBN-10: 0803118961; pp. 41–83.

Paine, R.T. (1980). Food webs: Linkage interaction strength and community infrastructure. *Journal of Animal Ecology* **49**(3): 666–685.

Robertson, G.J. and R.I. Goudie (1999). Harlequin duck (*Histrionicus histrionicus*). In *The Birds of North America*. F. Gill and A. Poole, eds. Philadelphia, PA, USA: Birds of North America, Inc.; No. 466. [*The Birds of North America Online;* http://bna.birds.cornell.edu/bna/species/466]

Short, J.W., J.M. Maselko, M.R. Lindeberg, P.M. Harris, and S.D. Rice (2006). Vertical distribution and probability of encountering intertidal *Exxon Valdez* oil on shorelines of three embayments within Prince William Sound, Alaska. *Environmental Science & Technology* **40**(12): 3723–3729.

Stern, P.C. and H.V. Fineberg, eds (1996). *Understanding Risk: Informing Decisions in a Democratic Society*. Washington DC, USA: National Research Council, National Academy Press; ISBN-10: 030905396X.

Stubblefield, W.A., G.A. Hancock, W.H. Ford, and R.K. Ringer (1995a). Acute and subchronic toxicity of naturally weathered *Exxon Valdez* crude oil in mallards and ferrets. *Environmental Toxicology and Chemistry* **14**(11): 1941–1950.

Stubblefield, W.A., G.A. Hancock, H.H. Prince, and R.K. Ringer (1995b). Effects of naturally weathered *Exxon Valdez* crude oil on mallard reproduction. *Environmental Toxicology and Chemistry* **14**(11): 1951–1960.

Taylor, E. and D. Reimer (2008). Oil persistence on beaches in Prince William Sound: a review of SCAT surveys conducted from 1989 to 2002. *Marine Pollution Bulletin* **56**(3): 458–474.

Upper Mississippi River Basin Association (1995). *Forging a New Framework for the Future: A Report to the Governors on State and Federal Management of the Upper Mississippi River*. St. Paul, MN, USA: Upper Mississippi River Basin Association. [http://www.umrba.org/publications/rivermgt/rptgov.pdf]

US Army (2000). *Standard Practice for Wildlife Toxicity Reference Values*. Washington DC, USA: US Army Center for Health Promotion and Preventive Medicine, Environmental Health Risk Assessment Program, Health Effects Research Program; Technical Guide Number 254. [http://usaphcapps.amedd.army.mil/erawg/tox/tg254(Oct00final).pdf]

US Army Corps of Engineers (1999). *Central and Southern Florida Project, Comprehensive Review Study, Final Integrated Feasibility Report and Programmatic Environmental Impact Statement, April 1999*. Washington DC, USA: Jacksonville District, South Florida Water Management District. [http://www.evergladesplan.org/docs/comp_plan_apr99/summary.pdf]

US Environmental Protection Agency (1991). *Chemical Concentration Data near the Detection Limit*. Philadelphia, PA, USA: Environmental Protection Agency Region III, Office of Superfund Hazardous Waste Management, Hazardous Waste Management Division; Region III Technical Guidance Manual, Risk Assessment; EPA/903/8–91/001. [http://www.epa.gov/nscep/index.html, search by title]

US Environmental Protection Agency (1992). *Framework for Ecological Risk Assessment*. Washington DC, USA: Risk Assessment Forum; EPA/630/R-92/001. [http://rais.ornl.gov/documents/FRMWRK_ERA.PDF]

US Environmental Protection Agency (1998). *Guidelines for Ecological Risk Assessment.* Washington DC, USA: Risk Assessment Forum; EPA/630/R-95/002F. [http://www.epa.gov/raf/publications/pdfs/ECOTXTBX.PDF]

US Environmental Protection Agency (2002). *Waquoit Bay Watershed Ecological Risk Assessment: The Effect of Land-Derived Nitrogen on Estuarine Eutrophication.* Washington DC, USA: National Center for Environmental Assessment, Office of Research and Development; EPA/600/R-02/079. [http://ofmpub.epa.gov/eims/eimscomm.getfile?p_download_id=36757]

US Environmental Protection Agency (2003). *Guidance for Developing Ecological Soil Screening Levels (Eco-SSL).* Washington DC, USA: Office of Solid Waste and Emergency Response; OSWER Directive 9285.7–55. [http://rais.ornl.gov/documents/ecossl.pdf]

US Environmental Protection Agency (2007). *Ecological Soil Screening Levels for Polycyclic Aromatic Hydrocarbons (PAHs).* Washington DC, USA: Office of Solid Waste and Emergency Response; Interim Final OSWER Directive 9285; pp. 7–78. [http://rais.ornl.gov/documents/eco-ssl_pah.pdf]

US Fish and Wildlife Service (2005). Endangered and threatened wildlife and plants; determination of threatened status and special rule for the southwest Alaska Distinct Population Segment of the northern sea otter (*Enhydra lutris kenyoni*). *Federal Register* **70**(152): 46365–46386; August 9, 2005; Final Rule and Proposed Rule. [http://www.fws.gov/policy/library/2005/05-15718.html]

von Bertalanffy, L. (1968). *General System Theory. Foundations, Development, Applications.* New York, NY, USA: George Braziller, Inc.

PART V

CONCLUSIONS

INTRODUCTION

When all is said and done, what is the legacy of the *Exxon Valdez* oil spill? The hundreds of scientific studies, many of which are discussed in the chapters of this book, enriched our knowledge of Prince William Sound (PWS) and its ecosystems and of how the elements of these ecosystems responded to oil in the environment. In the process, they also revealed much about the challenges of conducting high-level science in a harsh and variable environment. In this concluding chapter, John Wiens reviews the broad themes that have emerged from these studies, which provide cogent lessons for those who must grapple with assessing the consequences of other large environmental disruptions, whether caused by human accidents such as oil spills or by natural processes such as floods or forest fires.

The *Exxon Valdez* oil spill did more than unleash oil into PWS. It also spawned litigation that lasted for more than two decades, fueling competing agendas and controversy. These factors created the additional challenge of separating science from underlying agendas or preconceptions about spill effects. The quality of the studies summarized in this book testifies that it can be done, although it was not always easy.

CHAPTER SEVENTEEN

Science and oil spills: the broad picture

John A. Wiens

17.1 Introduction

The previous chapters have synthesized and evaluated the science that was brought to bear on the *Exxon Valdez* oil spill and its effects. Several overarching insights have emerged, including the importance of following a multidisciplinary, collaborative approach; of clearly defining one's objectives to design rigorous studies and identify data requirements; of recognizing the value of natural processes in facilitating restoration or recovery from spill effects; of assessing and documenting exposure of organisms to harmful oil constituents from all sources; and of evaluating possible avenues of spill exposure and effects through risk assessment. These insights have been central to developing a science-based understanding of the environmental effects of the *Exxon Valdez* spill (see Box 17.1). In this chapter, I highlight these lessons and describe how they provide a foundation for dealing with future oil spills or other large environmental disruptions.

I also explore several challenges that emerged during the *Exxon Valdez* studies. The confounding effects of environmental factors other than oil, natural variability of the environment, the attendant uncertainty in scientific data, and contradictory interpretations and disagreements among scientists present challenges in any high-profile and contentious situation, as major environmental accidents are likely to be. These factors confuse those looking to science for clear answers and straightforward guidance about what to do, and they foster public perceptions of conflicts between industry scientists and government scientists that diminish the credibility and value of the science itself. These challenges should be recognized, anticipated, and addressed. I offer some perspectives on how this might be accomplished, based on the collective experience of many scientists over two decades studying the effects of the *Exxon Valdez* spill.

Oil in the Environment: Legacies and Lessons of the Exxon Valdez *Oil Spill*, ed. J. A. Wiens. Published by Cambridge University Press. © Cambridge University Press 2013.

Box 17.1 What were the consequences of the *Exxon Valdez* oil spill? The key conclusions

The scientific studies of the *Exxon Valdez* oil spill described in the chapters of this book address three central questions about the environmental and biological consequences of the spill:

- What happened to the oil?
- Is there ongoing exposure to *Exxon Valdez* oil or its residues that could have continuing, chronic effects?
- Have biological resources recovered?

What happened to the oil?

When the *Exxon Valdez* grounded on Bligh Reef in March 1989, some 11 million gallons (40 million liters) of Alaska North Slope crude oil were released into the waters of Prince William Sound (PWS). What happened to all this oil? The simple answer is that little of the oil remained after a few years, and virtually none remains two decades later. Small amounts of oil residues persist today at a few isolated shoreline locations, although these deposits are sequestered in small lenses buried in armored shoreline sediments and are thus unavailable to affect intertidal organisms unless the sediments are disturbed by digging. It is now very difficult to find remnants of *Exxon Valdez* oil in PWS unless one knows exactly where to dig. Both the rapid, progressive disappearance of oil and its toxic components following the *Exxon Valdez* spill and the persistence of small amounts of oil residues in sheltered, low-permeability shoreline sediments match what has been observed in studies of other major oil spills elsewhere in the world (Chapter 6, Fig. 6.1).

More specifically:

- Approximately 20% of the oil, consisting of the more volatile and water-soluble hydrocarbons, dissipated within a few days of the spill (Chapter 3).
- Eventually, roughly half of the remaining oil was stranded on shorelines, ~40% in western PWS. Oiling levels varied substantially along those shorelines (Chapter 1, Fig. 1.3). Eastern PWS was not oiled at all.
- As a result of cleanup efforts and natural processes (e.g., biodegradation, Chapter 8; mineral particle-oil flocculation, Chapter 6, Box 6.1; and the highly dynamic tidal and storm regimes of PWS), nearly all of the visible oil was gone within 1–2 years. By 2001, only a very tiny percentage (0.12–0.28%) of the oil originally stranded on shorelines remained. This oil continued to dissipate at a rate of about 4% per year for several years after 2001 and has been dissipating at a progressively slower rate since then (Chapter 6).
- Some oil penetrated coarse sediments and formed subsurface oil residues (SSOR) in finer-grained sediments below the coarser surface layer. In 2008, patches of SSOR remained at a few isolated shoreline locations, primarily in the middle- and upper-intertidal zones (Chapter 6).
- Tidal energy is insufficient to wash the thin lenses of fine-grained subsurface sediments containing this SSOR; otherwise, they would have disappeared long ago. The concentration of toxic hydrocarbons measured in the water in contact with SSOR is extremely low. Fluid-flow models clearly show how these extremely low subsurface

values are then further reduced by dilution (Chapter 7). These SSOR are all in locations where Shoreline Cleanup Assessment Technique (SCAT) surveys conducted between 1989 and 1992 recorded residual oil (Chapter 4). Subsequent systematic searches failed to find additional SSOR in shoreline sediments not previously documented by the SCAT surveys.

- The SCAT surveys were remarkably effective and valuable in identifying both where oil needed to be cleaned from shorelines immediately after the spill and where it persisted (Chapter 6). In the process of conducting the surveys, SCAT teams also discovered and identified more than twice the number of previously known archaeological sites in the spill area (Chapter 5).

Is there a continuing exposure risk and are there chronic effects?

The short answers are no and no. Levels of toxic contaminants such as polycyclic aromatic hydrocarbons (PAH) from *Exxon Valdez* oil diminished rapidly after the spill. The remaining SSOR, some of which contain potentially toxic levels of PAH, are sequestered and release PAH extremely slowly unless shoreline sediments are disturbed, in which case any contaminants that are released are rapidly dispersed by tidal action. Sequestration is therefore both the reason that some oil residues remain and the reason that those residues are not bioavailable.

More specifically:

- Within one year of the spill, levels of total PAH (TPAH) in offshore water-column samples collected from the spill area in PWS were indistinguishable from background levels (Chapter 3).
- By 1993, TPAH levels attributable to *Exxon Valdez* oil were not elevated in sea-floor sediments except in a few shallow areas near some sheltered shorelines that were heavily oiled in 1989 and where shoreline cleanup during 1989–92 released oil back into the water. Otherwise, TPAH levels were at background levels (Chapter 6).
- TPAH in intertidal prey biota at sites containing SSOR were at background levels by 1998–2002 (Carls *et al.*, 2004; Neff *et al.*, 2006), as were PAH biomarkers of hydrocarbon exposure in intertidal fish (Huggett *et al.*, 2003).
- Biota in PWS may be exposed to hydrocarbons from multiple sources – normal boat traffic, leakage from abandoned mines and canneries, or sediments carried into PWS by currents from natural oil seeps, among others (Chapter 6). Measures of cytochrome P450 1A (CYP1A), which have been used to infer continuing exposure of several species to *Exxon Valdez* PAH, cannot by themselves be used to distinguish among sources of hydrocarbon exposure and are an unreliable indicator of exposure to *Exxon Valdez* hydrocarbons (Chapter 9).
- Quantitative ecological risk assessments showed that the likelihood of exposure to residual *Exxon Valdez* hydrocarbons for even the most-exposed sea otters (*Enhydra lutris*) or harlequin ducks (*Histrionicus histrionicus*) was more than an order of magnitude lower than any potential effects levels, and thus risks to populations are virtually nonexistent (Chapter 16).
- The ecological risk-based framework did not exist at the time of the oil spill. Had it been available and used at the outset, the selection of recovery goals and the design,

analysis, and interpretation of monitoring and research studies could have been done more efficiently (Chapter 16).

Have biological resources recovered?

The short answer is yes. However, the responses of organisms and populations to oil in the environment are determined by highly variable biological processes such as movement, habitat selection, feeding behavior, fidelity to breeding areas, reproductive rates, and recruitment. These processes contrast with the more predictable physical and chemical processes, both of which are affected by local circumstances and the properties of the spilled oil, which determine what happens to oil in the environment.[1] These factors, and the challenges of surveying mobile species for which adequate baseline or reference data are sparse or confounded by environmental variation, add uncertainty to assessments of biological recovery, which is further exacerbated by the use of different definitions and measures of "recovery."

Overall, the evidence reviewed in the previous chapters shows that virtually all of the species studied have recovered. No species was extirpated by the spill, as was initially feared; all species are present and reproducing where they were documented to be in PWS before the spill. Disagreements continue about the recovery status of a few species, but these relate to uncertainties in data or confounding effects from nonspill factors.

More specifically:

- Several studies showed that most taxa living on or in shoreline substrates in oiled areas had recovered by 1990–91 (Chapter 11). Some taxa, such as rockweed (*Fucus* spp.), took longer to achieve multi-age populations, but that period ended within a few years of the spill (Driskell *et al.*, 2001).

- Exposure of pink salmon (*Oncorhynchus gorbuscha*) eggs, embryos, and alevins to hydrocarbons in oiled streams occurred only at very low levels and had no effects on incubation success, timing of emergence, or populations overall. Only 2% of the salmon spawning streams in PWS were significantly oiled in 1989. Postspill returns of adult pink salmon to spawn in PWS were at or close to record high levels in 1990 and 1991 and in 8 of the 13 years between 1994 and 2007 (Chapter 12).

- Populations of Pacific herring (*Clupea pallasii*) in PWS were at record levels shortly after the spill. The collapse of herring populations in PWS in 1993 and their subsequent failure to recover reflect the effects of multiple environmental factors that are unrelated to the oil spill, including competition with hatchery-released pink salmon fry and predation by humpback whales (*Megaptera novaeangliae*) (Rice and Carls, 2007; Chapter 13).

- Among marine birds, murres (*Uria* spp.), which occur primarily in the Gulf of Alaska, accounted for nearly three-fourths of the overall mortality in the immediate aftermath of the oil spill. Attendance and reproduction at murre breeding colonies in the spill area were within the historical range within 2 years (Boersma *et al.*, 1995; Chapter 14).

- More than half of the 42 bird species analyzed were less abundant in oiled shoreline habitats than expected soon after the spill. However, a year later, habitat occupancy was as expected for 90% of the species, and recovery in habitat occupancy by all bird species was complete by 1996 (Chapter 14).

- As early as 2 years after the spill, and clearly by 1996, there were more adult and juvenile sea otters in the spill area in western PWS than had been recorded before the spill (Chapter 15). The population in western PWS has grown at an average annual rate of ~2.5% since 1991 (Chapter 15). There is continuing disagreement about the status of sea otter recovery in an area of northern Knight Island (Bodkin et al., 2012; Chapter 15), although the number of individuals potentially affected represents less than 3% of the otter population in western PWS.

Some components of the PWS ecosystem recovered rapidly from effects of the oil spill, whereas others took longer. In general, the ecological recovery processes followed the pattern expected for ecosystems subjected to major oil spills and most other episodic events, including the Great Alaska Earthquake of 1964. The few remaining disagreements about when (or whether) a resource could be considered to have recovered stem in large part from the various ways in which "recovery" is defined (Chapter 1, Box 1.3) and how studies were designed and hypotheses framed (Chapter 10), and they relate to only a few taxa in small areas of PWS.

The chapters in this book show that only a small amount of *Exxon Valdez* oil remains sequestered in shoreline sediments in PWS, potential exposure pathways are free of harmful contaminants from the spill, and risk analysis indicates that any remaining residues can have no meaningful effects on populations of biota that use the shorelines. Any remaining impediments to recovery are not tied to ongoing exposure to oil from the *Exxon Valdez* spill.

Bodkin, J.L., B.E. Ballachey, H.A. Coletti, G.G. Esslinger, K.A. Kloecker, S.D. Rice, J.A. Reed, and D.H. Monson (2012). Long-term effects of the *Exxon Valdez* oil spill: sea otter foraging in the intertidal as a pathway of exposure to lingering oil. *Marine Ecology Progress Series* **447**: 273–287.

Boersma, P.D., J.K. Parrish, and A.B. Kettle (1995). Common murre abundance, phenology, and productivity on the Barren Islands, Alaska: The *Exxon Valdez* oil spill and long-term environmental change. In Exxon Valdez *Oil Spill: Fate and Effects in Alaskan Waters*. P.G. Wells, J.N. Butler, and J.S. Hughes, eds. Philadelphia, PA, USA: American Society for Testing and Materials; ASTM Special Technical Publication 1219; ISBN-10: 0803118961; pp. 820–853.

Carls, M.G., P.M. Harris, and S.D. Rice (2004). Restoration of oiled mussel beds in Prince William Sound, Alaska. *Marine Environmental Research* **57**(5): 359–376.

Driskell, W.B., J.L. Ruesink, D.C. Lees, J.P. Houghton, and S.C. Lindstrom (2001). Long-term signal of disturbance: *Fucus gardneri* after the *Exxon Valdez* oil spill. *Ecological Applications* **11**(3): 815–827.

Huggett, R.J., J.J. Stegeman, D.S. Page, K.R. Parker, B. Woodin, and J.S. Brown (2003). Biomarkers in fish from Prince William Sound and the Gulf of Alaska: 1999–2000. *Environmental Science & Technology* **37**(18): 4043–4051.

Kingston, P.F. (2002). Long-term environmental impact of oil spills. *Spill Science & Technology Bulletin* **7**(1–2): 53–61.

Mielke, J.E. (1990). *Oil in the Ocean: The Short- and Long-Term Impacts of a Spill*. Washington DC, USA: The Library of Congress, Congressional Research Service; CRS Report for Congress 90-356 SPR.

Neff, J.M., A.E. Bence, K.R. Parker, D.S. Page, J.S. Brown, and P.D. Boehm (2006). Bioavailability of PAH from buried shoreline oil residues thirteen years after the *Exxon Valdez* oil spill: a multispecies assessment. *Environmental Toxicology and Chemistry* 25(4): 947–961.

Rice, S. and M.G. Carls (2007). Executive summary. In *Prince William Sound Herring: An Updated Synthesis of Population Declines and Lack of Recovery*. S. Rice and M.G. Carls, eds. Juneau, AK, USA: National Oceanic and Atmospheric Administration, National Marine Fisheries Service, Auke Bay Laboratory; *Exxon Valdez* Trustee Council Restoration Project 050794 Final Report; pp. 9–22. [http://www.evostc.state.ak.us/Files.cfm?doc=/Store/FinalReports/2005-050794-Final.pdf&]

Note

[1] This physical and chemical predictability is why what happened to spilled oil and the risks it might have posed following the *Exxon Valdez* spill were similar to those reported for other major oil spills (Mielke, 1990; Kingston, 2002).

17.2 What have we learned?

17.2.1 Objectives and study design are the foundations of spill assessment

With any scientific investigation, the critical first steps are formulating the problem and clearly stating objectives. The scientific objectives for assessing the consequences of an oil spill will be defined by the context of the incident – the environmental setting, the extent and nature of injuries to natural resources, legal requirements, public concerns, private losses, public policy, and economic and political constraints. The objectives must be framed in a manner that allows the tools of science to be used to resolve the unknowns and realize those objectives. Once the objectives are articulated, the study design and data requirements can be determined.

In the immediate aftermath of a major oil spill, however, the pressure to do something quickly is often overwhelming. Paying attention to the following points can help guide studies in the right directions.

17.2.1.1 Investigate what is relevant

In an oil spill, the immediate needs are to determine the oil's location to guide responses and provide an early assessment of the risk to critical resources, especially those of subsistence, commercial, and conservation importance. Within a week of the *Exxon Valdez* spill, chemistry teams began to systematically sample hydrocarbons in the water column within and outside the spill area to assess the risk to subsistence and commercial fisheries (Chapter 3). Shoreline Cleanup Assessment Technique (SCAT) teams initiated surveys to assess shoreline impacts, identify sensitive habitats, and guide cleanup activities (Chapter 4). Cultural resources teams documented the locations of known and newly discovered archaeological sites (Chapter 5). As the cleanup proceeded, the priority shifted to documenting the locations of any remaining oil residues and the status of species of particular concern.

Ideally, studies should focus on resources that are likely to be most at risk and locations where the potential consequences are likely to be greatest, although other factors may also influence the selection of study targets. For the *Exxon Valdez* spill, resources with commercial or subsistence value, such as pink salmon (*Oncorhynchus gorbuscha*)

(Chapter 12) or Pacific herring (*Clupea pallasii*) (Chapter 13), or that resonated with the public, such as seabirds (Chapter 14) and sea otters (*Enhydra lutris*) (Chapter 15), clearly merited particular attention. State and federal regulations required that archaeological sites and cultural artifacts be documented and protected (Chapter 5). Studies supported by Exxon, the state and federal governments, and the *Exxon Valdez* Oil Spill Trustee Council (hereafter, "Trustees") were conducted in the framework of natural resource damage assessment (NRDA) regulations existing at that time. Consequently, the focus was generally on particular species that were known or suspected to have suffered injuries. Several studies, however, conducted broad-based sampling of populations and communities of species that can be surveyed together, such as shoreline invertebrates (Chapter 11) or marine birds (Chapter 14). These studies helped to reveal the breadth of spill effects.

Whatever targets are selected for assessment of spill effects, data should be collected at a level sufficient for gauging significant effects. For biological systems, information on population or community attributes (e.g., abundance, habitat use, reproduction) will generally be more relevant and useful than information about the behavior or status of individuals.

It is also important to be attentive to what the data may *not* show. In particular, samples in which a species is absent or particular hydrocarbons are not detected can often be just as informative as samples in which they are present. The absence of a species, for example, may indicate that individuals were killed or left the area because of the oil spill, or they may be absent because the habitat is unsuitable, because they are not abundant in the region, or for some other reason. Such "zero values" in samples create challenges for statistical analyses (Chapter 10) and are often difficult to interpret, but they should not be ignored.

17.2.1.2 Determine a reference standard

Both federal regulations and scientific evaluations of potential oil-spill effects require a reference standard or baseline against which those effects can be compared. If the concern is with potential toxicological effects of petroleum hydrocarbons, standards that define toxicological threshold levels for specific hydrocarbons are available for many groups of organisms (although they may be difficult to apply in the environmental and ecological context of a spill). On the other hand, determining a baseline for population size, habitat occupancy, or rate of reproduction for a species is more difficult: there are no "standard" values. Instead, comparisons must be made to historical (prespill) data (which are often skimpy at best) or to samples from appropriate reference areas. Historical information can be useful in indicating what might be "normal" for locations in the spill area, although its value in establishing a baseline may be limited. Often such data were collected at a single time long before the spill, so they provide only an old snapshot taken under different conditions. The accuracy and precision of the data are usually unknown, hindering reliable statistical analyses. If the data were collected using different procedures, the values are probably not comparable to those from postspill samples. Historical data are best used as a general, qualitative baseline (e.g., presence versus absence in an area) rather than for numerically precise comparisons (e.g., densities in an area).

Another way to establish a baseline is through comparisons with appropriate reference areas. The key word is "appropriate." Defining broad areas as "oiled" and "unoiled," for example, ignores the patchiness of oil deposition and weathering on shorelines (Chapter 1, Fig. 1.3) and assumes that there are no systematic differences other than oiling history between the areas. Such broad-scale comparisons permit only general inferences about possible spill effects. Quantitative, statistical comparisons, while more useful, require that reference areas be closely matched with impacted areas in all respects except their oiling history. Testing the assumption that, aside from the presence of oil, all else is equal among the areas is facilitated if other factors that might covary with oiling level, such as shoreline substrate, wave exposure, depth of water, or shoreline orientation, are also measured (Chapters 10 and 14).

Assessing potential spill effects requires recognition that an oil spill is not the only potential source of hydrocarbons in the environment. Other natural and anthropogenic sources may contribute to a background level of hydrocarbons, on which the spilled oil is superimposed (Chapter 6). These background hydrocarbon sources may be relatively unimportant immediately following a spill when spill-derived hydrocarbons predominate, but they become increasingly relevant over time as the amount of spilled oil in the environment diminishes. Rather than attributing any and all hydrocarbons in the spill area to the spill itself, it is important to document background concentrations of hydrocarbons and determine which are natural, which are anthropogenic, and which are derived from the spill. Because hydrocarbons from different sources can have distinctive chemical compositions, "fingerprinting" these chemical profiles can be a powerful tool in discriminating spilled oil from background hydrocarbons (Chapter 1, Box 1.1).

Establishing baseline or background levels for populations of organisms or hydrocarbons is also essential to evaluating recovery. There are various ways to define recovery (Chapter 1, Box 1.3). Some studies of the *Exxon Valdez* spill measured recovery against prespill conditions. Others defined recovery as the absence of a previously documented, statistically significant difference between oiled and reference samples over time. Differences in how scientists defined recovery or selected appropriate reference conditions contributed to ongoing debates about the status of injured resources (Chapters 12–15).

17.2.1.3 Use direct rather than indirect measures

Whenever possible, the levels and composition of oil in the environment or the exposure of organisms to harmful hydrocarbons should be measured directly rather than inferred indirectly from measurements of something else (Chapter 9). The SCAT surveys following the *Exxon Valdez* spill successfully directed cleanup activities and identified vulnerable populations because they incorporated direct evaluation and mapping of where oil had actually been deposited on shorelines (Chapter 4). This information was subsequently used to develop an index of shoreline oiling that, when used to analyze habitat use by marine birds, pink salmon, sea otters, or spawning herring, led to different conclusions about spill effects and recovery than those derived from broad, regional categorizations of areas as "oiled" or "unoiled" based only on the general outline of the spill path (Chapters 12–15).

Direct measures are especially important in evaluating the risk of exposure of organisms to toxic hydrocarbons. Biomarkers such as cytochrome P450 1A (CYP1A) have been used as indirect measures of hydrocarbon exposure in several studies of the *Exxon Valdez* and other oil spills. Although CYP1A can indicate exposure to some hydrocarbons (as well as other naturally occurring and anthropogenic chemicals; Chapter 9), it does not discriminate among potential sources of those hydrocarbons. Consequently, when used alone, it is an unreliable indicator of exposure to spilled oil in an environment containing hydrocarbons from multiple sources. A more direct measure of exposure to hydrocarbons in the environment may be obtained from chemical analyses of the tissues of organisms (e.g., mussels) in proximity to or consumed by the species of concern. Translating such measures into estimates of exposure of species of concern, however, requires consideration of differences among species in the rates at which they metabolize hydrocarbons such as polycyclic aromatic hydrocarbons (PAH). Direct measures of PAH in sediments and water samples are necessary to determine the sources, locations, concentrations, and toxicity of oil as it changes following initial release, weathering, and sequestration as subsurface oil residues (SSOR) (Chapters 3, 4, 6, and 7). None of the measures of exposure, whether direct or indirect, tells one whether the exposure to hydrocarbons has had ecologically significant effects on individuals or populations (Chapter 16).

17.2.1.4 Pay attention to study design

As with all scientific analysis, how an oil spill study is designed will determine whether the results will withstand scrutiny and inform decisions and restoration actions, or whether they will be discounted and disregarded. Because there is no single "best" design for studying the consequences of an oil spill, study designs must be tailored to the questions asked, information needed, subjects studied, and accuracy and precision required. This all must be accomplished within the very real constraints of conducting research under difficult environmental conditions in a frequently contentious atmosphere. The following guidelines may prove useful in such situations.

First, *know what assumptions are being made in a particular study and state those assumptions explicitly*. Studies that define recovery as a return to prespill conditions, for example, assume that the system is in steady-state equilibrium, whereas studies that compare impacted with reference areas often assume spatial equilibrium (i.e., the areas are equivalent in all respects except their oiling history) (Chapter 10). Assuming that the spill had only detrimental effects, instead of being open to finding negative, positive, or no effects, influences the form of the hypotheses to be tested and the structure and power of statistical tests. Whatever the assumptions, they should be recognized and stated explicitly and transparently so the results can be interpreted appropriately.

Second, *consider how to conduct sampling*. There is an unavoidable tradeoff between collecting several samples of the same thing in the same place at the same time (i.e., replication) versus collecting the same number of samples over a broader area, with fewer (usually only one) samples per location and time. Logistical constraints come into play: in a large area, such as Prince William Sound (PWS), it may be easier to collect many samples at a few locations, especially if the time when samples must be taken is limited (e.g., during low tides, during salmon or herring spawning). If this is done, however, care should be taken to avoid sampling impacted (i.e., oiled) sites together at one time and reference (i.e., unoiled) sites together at another time, since time can then become a confounding factor in the analyses (Chapter 12). If sampling is too frequent or too intrusive, the act of sampling may itself affect the presence, abundance, or reproductive success of some species.

In general, broad-scale sampling can provide information about the geographic scope of spill effects or recovery, but at the expense of reduced statistical power and strength of conclusions. Replicate sampling increases power to detect differences, so the statistical conclusions are more robust, but they are also more limited in scope and applicability. The ideal may be a sequentially structured design, in which broad-scale sampling is first used to define particular areas of interest, followed by more intensive, replicated sampling at fewer sites, particularly if the sites are arrayed along a gradient of potential exposure. As time passed following the *Exxon Valdez* spill, for example, sampling of SSOR was increasingly restricted to the places where residual oil was known to remain. Because all areas later found to contain SSOR had been identified during the initial SCAT surveys (Chapters 4 and 6), the restriction of sampling was justified.

Third, *allow the types of scientific studies and their design to change over time*. This is necessary because the distribution and composition of oil in the environment, its effects on populations, and the recovery of populations all change over time. Considering studies in the context of the phases of a spill (Chapter 2) can help to anticipate when and what types of studies may be needed. But simply anticipating that hydrocarbon concentrations in the water column will diminish, oil will disappear from shorelines, or recovery will occur is insufficient. Quantitative measures are necessary to determine when and where hydrocarbons are no longer present at harmful levels or populations have or have not reached specified recovery goals. This means that some studies should be continuous, with samples collected using a standardized protocol at regular intervals. The trick is in knowing when to stop. Statistical analyses can help: if recovery is defined as the disappearance of a previously significant negative relationship with PAH or oiling history of habitats, for example, a failure to find statistically significant results in tests some time after the spill may indicate that further data collection is unnecessary. Even where spilled oil remains in the environment (e.g., as SSOR), it may not be necessary to continue to conduct studies looking for effects if it can be demonstrated that the risk from exposure is negligible (Chapters 7 and 16). The bottom line is that both continuity and flexibility in study design are needed; studies must be tailored to suit the circumstances as time passes following a spill.

17.2.1.5 Results are of little value unless they are communicated

After an oil spill, the need for communication goes far beyond the normal scientific imperative to publish papers. Those who need to know – spill responders, cleanup planners, on-site coordinators, agency personnel, those whose livelihoods depend on resources, landowners, other scientists, the concerned public, journalists, and lawyers – need to know quickly. In the weeks or months immediately following a spill, cleanup is being planned and implemented, and assessments of hydrocarbon levels, exposure risks, and injuries to biological resources are well underway. Making the scientific information and data understandable and available can help direct resources to meet the most urgent needs, reassure the public about what is being done to address the situation, and inform decision-makers and stakeholders about effects on important resources and the status of ecosystem recovery. Because the scientific studies in later phases of a spill build on the results of earlier studies, the results of the initial studies must be communicated quickly and effectively. Otherwise, speculation flourishes.

If the results being communicated are based on data and analyses rather than impressions and conjectures, the data – and not just the results – should be made available for thorough review and analysis. In this way, errors or misinterpretations can be detected before they become cast in stone through publication or media accounts. With all of the data available, analyses can be conducted to test alternative hypotheses, broadening the array of potential causal pathways examined. When the *Exxon Valdez* spill occurred, sophisticated data-management systems were in their infancy. This is no longer the case. Part of planning and implementation of a spill response should be the creation of data-management systems that enable researchers to enter data promptly and others to retrieve the data almost as quickly. This goal may be difficult to achieve in situations in which pending or threatened litigation precludes or delays the open exchange of the findings of scientific studies. In this case, data sharing, while scientifically desirable, may be difficult.

17.2.1.6 Be prepared

Much of what has been said here about data and study design is obvious in hindsight. It can easily be forgotten, however, in the emotional atmosphere and urgency that follow a major oil spill. Unless attention is given to the details of study design and data collection at the outset, subsequent investigations may be flawed or inadequate. Once started down that path, it becomes difficult to change course. The main message is to be prepared.

17.2.2 Natural recovery is often the best restoration strategy

It is important to recognize that many environments, particularly in coastal, high-latitude ecosystems such as PWS, are subjected to frequent natural disturbances from a variety of sources. Once the effects of a disturbance diminish or disappear, biological systems often have a remarkable capacity to bounce back. Recovery following a disturbance – be it natural or anthropogenic, small or large – is a natural process. Recovery of the PWS ecosystem from the effects of the *Exxon Valdez* spill began immediately, with some components exhibiting recovery within months of the spill and others taking several years (Integral Consulting, Inc., 2006). By some accounts, however, recovery is still deemed incomplete (*Exxon Valdez* Oil Spill Trustee Council, 2010). I will return to consider differing conclusions about the recovery status of biological resources in Section 17.3.3.1.

Restoration efforts aim to accelerate or augment the natural recovery process. Despite the huge investment of time and money in scientific research, however, restoration options following the *Exxon Valdez* spill were limited. For the most part, recovery of biological resources occurred without active restoration actions directed at any particular species or habitat (although considerable funds were invested in protecting forested watersheds bordering PWS). Once hydrocarbons in the water column reached background levels and shoreline cleanup was completed, the most immediate sources of stress on the biota were reduced or eliminated. Attention then turned to determining what happened to the remaining oil, when biological resources recovered, and whether there were any lingering or delayed effects of residual oil.

Does the fact that natural processes did most of the restoration work mean that the huge investment of time and money on scientific studies was unnecessary? Not at all! It was essential to document what happened to the oil, which resources were affected, and which were not. Even though these studies generally did not produce results that could be directly applied to improving or accelerating recovery, they did provide critical information about how (or whether) recovery was occurring and the rates at which a high-latitude system would recover from such a large disturbance. In the absence of this information, speculation and conjecture, rather than science, would drive restoration efforts.

17.2.3 Determining exposure risk underlies evaluations of potential injuries

In the rush to mobilize spill responses and identify injured resources immediately after a spill, it is easy to believe that any detrimental changes in the environment or biota are caused by the oil. Beyond the NRDA regulations, the demands of good science mandate that the linkage of apparent injuries to the oil spill be documented empirically rather than simply inferred. The painstaking work of documenting exposure and collecting data on exposure pathways cannot be avoided.

There are multiple pathways by which oil in the environment may affect resources (e.g., individuals, species, cultural sites, services; see Chapter 14, Fig. 14.1; Chapter 16, Figs. 16.4, 16.6). Some of these pathways are indirect: reductions in prey populations as a result of their exposure to oil, for example, may have cascading effects on prey consumption by predators, potentially affecting the predators' reproduction or distribution. Oil on shorelines or disturbance associated with cleanup may cause mobile species to move elsewhere without having contacted oil directly. Determining the contribution of oiling to such indirect pathways becomes more difficult the more indirect the potential effects and the more time that passes following the event. Do the apparent effects of changes in fish populations on their consumers reflect the effects of an oil spill on the prey, or are the changes due to regional factors such as oceanographic changes or overfishing (e.g., Piatt and Anderson, 1996; Harwell *et al.*, 2010a)? Is the reduced abundance of a seabird species on shorelines that were impacted by a spill a result of direct exposure to the oil or its residues (e.g., mortality from oiling, ingestion of contaminated prey), a consequence of cleanup activities on shorelines during critical times of the year (e.g., effects of disturbance on breeding productivity), or a reflection of underlying factors that reduce the suitability of habitat on those shorelines (e.g., effects of oil on habitat use)? There are ways to disentangle the effects of oil from those of other factors in such indirect pathways, as by including covariates in statistical analyses (Chapter 10) or by developing quantitative conceptual models of relative risks (Chapter 16). Without a careful consideration of a full array of potential causes, however, attributing a change in a biological resource to an indirect effect of oiling on something else that then affects the resource can easily lead to untestable speculations.

Documenting direct exposure pathways is more straightforward, and direct exposure to oil or its constituents is usually of greater concern than indirect pathways. In the days or weeks following a spill, when oil is floating on the water or deposited on shorelines, determining exposure is often a matter of direct observation: where there is oil and there are organisms, exposure is likely. As oil weathers on a shoreline or dissipates in the water column, determining exposure turns more to chemical analyses: what is the form of the oil, what are the toxic fractions, where are they, and are they bioavailable (i.e., able to move into tissues from the exposure medium)? As spill hydrocarbon concentrations return to background levels through natural weathering processes and oil residues become highly localized and sequestered, it becomes more difficult, but more important, to evaluate exposure risk carefully.

Direct analysis of PAH levels in tissues is a proven method of assessing exposure. In some instances, indirect measures of exposure (e.g., biomarkers; Chapter 9) may be useful. In those instances, measurements of PAH levels in the potential exposure pathways (e.g., prey tissue, water column, shoreline and seafloor sediments), combined with associated field observations, are required to identify the source of the inferred exposure (Neff *et al.*, 2011). Indirect measurements alone are usually insufficient to assess exposure (Chapter 9).

Showing that important biological resources are exposed directly to oil or its constituents is only the first step in exposure assessment. To have biologically significant effects, the oil must foul or be taken up by individuals. Fouling occurs by direct encounter with oil (e.g., oiling of feathers, pelage, or gills) or ingestion (e.g., preening of oiled feathers or fur; ingesting oil droplets from contaminated prey, sediment, or water). If contaminated prey are eaten, the contaminants must be assimilated and metabolized. If the hydrocarbons are bioavailable, they may be taken up into the tissues of the organism, primarily through the gills or digestive tract, where they may cause biochemical changes or cell damage (toxic responses). Contaminants must accumulate in tissues in sufficiently high concentrations to cause tissue damage that impairs individual well-being (e.g., survival, reproduction, development, growth). If tissue PAH levels attributable to the spill are below those known to cause harm, claims of continuing exposure-related injury are unfounded and further studies of possible harmful effects are probably unnecessary. Finally, the effects must involve a sufficient number of individuals to translate into population-level consequences.

Charting an exposure pathway requires considering all of these steps: breaking the chain at any point means that the presence of oil or its constituents in the environment is, by itself, unlikely to have important biological consequences.

17.2.4 The likelihood of injuries can be evaluated using ecological risk assessment

The multiple pathways of potential exposure risk can present a large number of ways in which oil in the environment might affect biological resources. Conducting scientific studies to evaluate all of these possibilities can be both challenging and expensive. Not all pathways, however, are equally likely, and some may entail a series of steps that, taken together, render a biologically significant effect implausible. Ecological risk assessment and modeling provide useful tools for integrating information from multiple sources to derive an overall picture of exposure pathways and risks (Chapter 16).

Ecological risk assessment is a powerful way of making explicit the linkages in a potential exposure pathway to injury of a resource and the probabilities associated with each step. It addresses the question: how can one get there (biologically significant consequences to resources of interest) from here (oil or its constituents in the environment)? Using individual-based models and sensitivity analyses, the approach allows one to assess the effects of environmental variation or of uncertainties in the data. It can be used to evaluate the plausibility of claims of oil-related injuries and to identify which steps or linkages between cause (oil) and effect (injuries) are most likely to be important, and therefore most critical to investigate scientifically.

In general, having more data on more variables is better, but accumulating data takes time. In the case of the *Exxon Valdez* spill, formalized risk assessment did not begin until a decade after the spill (largely because the tools were not fully developed earlier) and produced the most useful results after more than two decades, when multiple studies were able to provide the detailed quantitative data needed to populate the models (Chapter 16).

This does not mean, however, that risk assessment concepts and guidelines cannot be applied earlier in the chronology of a spill. Much of the value of the approach is in the initial clear formulation of the problem(s) to be addressed and the development of conceptual models that link the essential elements of exposure pathways with biological resources. These steps can be taken without quantitative information, using past experience, expert opinion, and qualitative observations. For example, Wiens (2007) used a variety of sources of information in an initial risk assessment to evaluate the likelihood that harlequin ducks (*Histrionicus histrionicus*) were suffering from exposure to residual hydrocarbons from the *Exxon Valdez* spill – as Harwell *et al.* (2012) later demonstrated (Chapter 16), the likelihood was vanishingly small. Even when based on preliminary data, sensitivity analyses and scenario assessments may provide insights that help to focus research on the issues that matter most for understanding injury and recovery.

17.2.5 Insights from multiple disciplines provide broad understanding

Oil spills are complex events, with many moving parts playing in an involved drama on a stage that is already complex and dynamic. The normal approach of science to such a problem, stemming from its reductionist roots, is to partition the components into manageable pieces, which are then studied in isolation. Carried to an extreme, this approach can lead to rigid disciplinary silos, in which studies in one discipline fail to inform or benefit from complementary information being generated by studies in other disciplines.

A consistent message from the studies in this book is that perspectives from many different disciplines and scientists are necessary to understand what happens after an oil spill. The SCAT teams that surveyed shorelines following the spill each included a geologist, a biologist, and a cultural resources specialist, enabling the team to catalog not only where the oil was, but also the characteristics of the shoreline substrate and the plants, animals, and archeological sites potentially at risk (Chapters 4 and 5). The combined data were invaluable in directing shoreline cleanup and subsequent scientific studies. An understanding of fluid flows in porous media helped explain why dyes injected into banks of salmon streams during ebb tide were rapidly distributed as the tide rose (Carls *et al.*, 2003) – and why the results had little bearing on the sequestration of SSOR in highly impermeable sediments (Chapter 12). The suggestion that harlequin ducks might be exposed to harmful levels of hydrocarbons by feeding close to sea otters, whose excavation of pits could release SSOR into the water column during their foraging for clams, was dispelled by conducting a formal ecological risk assessment based on combining information about the distribution of SSOR in shoreline sediments along with sea otter and harlequin duck distributions and foraging behaviors (Chapters 6, 14, 15, and 16).

Developing the necessary multidisciplinary perspectives and collaborations does not just happen, however. It takes planning and effort. Those responsible for managing spill responses and investigations should identify the complementary skills needed and engage experts who can work together to address problems that are beyond the scope of a single area of expertise. It is also essential that the people working in different disciplines communicate with one another, early and often. The close association of researchers with differing expertise has value well beyond logistical savings; such association helps build an appreciation for the views and abilities of individual scientists outside of their familiar disciplines. Following the *Exxon Valdez* spill, this was facilitated because several scientists often needed to share a single ship during their fieldwork. Researchers sampling oil residues in shoreline sediments, surveying intertidal invertebrates, conducting counts of marine birds, documenting the distribution of sea otters, and even the statisticians who would be analyzing the data could share their observations, perspectives, and problems with one another in real time.

Multidisciplinary collaboration has added value if it occurs among different organizations. For example, both Exxon and governmental personnel staffed SCAT survey teams (Chapter 4). Exxon's work on cultural resources in the spill area was closely coordinated with the State of Alaska, federal agencies, and Alaska Native organizations (Chapter 5). Investigations of the effectiveness of bioremediation in removing stranded oil from shorelines were conducted jointly by Exxon and the US Environmental Protection Agency (EPA). And, while not discussed in this book, a joint Trustee and Exxon collaboration demonstrated the safety of subsistence foods following the spill (Field et al., 1999). Such cooperation between Exxon and governmental agencies helped establish public trust. On the other hand, legal constraints on collaboration and communication between scientists supported by Exxon and those working for the Trustees meant that their studies were conducted largely separately and that the underlying data were withheld from public and peer review for several years. This delay sowed the seeds of disagreements that eroded public trust in the science and findings.

17.3 What are the challenges?

Most of the lessons learned from the *Exxon Valdez* spill lead to guidance about what to do or what not to do in the event of a large oil spill. Several issues, however, appeared and reappeared during the scientific research on the spill that do not lend themselves to easy solutions. Unrecognized, they can distort or compromise the findings of otherwise useful studies.

17.3.1 Environmental variation and confounding variables

Scientists conducting studies of oil spills often assume that the effects are so large that they overwhelm other factors in the environment. However, most environments vary substantially over time and from place to place, and short-term or local variations are often superimposed on long-term and/or regional differences, trends, or cycles. Thus, efforts to link causes with effects are complicated by differences in the prespill baseline, inherent differences between oiled and reference sites unrelated to the spill, or differences in sampling times, species sensitivities, or exposure risk between oiled and reference sites. Consideration of multiple environmental variables in addition to oiling increases confidence that any effects detected are in fact due to oil rather than something else.

Variations in time or space are usually dealt with by randomizing sampling over multiple times and/or places. Randomization may not be feasible or practical, however, for situations that encompass numerous habitats and are constrained by logistical (if not financial) limitations on data gathering and analysis. By taking multiple samples in time or space (i.e., replication), variation can be identified and incorporated into statistical analyses. Including additional factors as covariates in sampling and analyses of sediments, habitats, or biota may help to reduce their confounding influences and uncertainties, clarifying the effects of oiling (Chapters 11 and 14). The problem with including covariates in analyses is in knowing which variables to include. Here, an understanding of the behavior of petroleum hydrocarbons in the spill environment or the natural history of species can provide guidance. Unfortunately, as more variables are considered, statistical power diminishes and uncertainty increases.

17.3.2 Coping with uncertainty

In a dynamic and variable environment full of confounding factors, the results of scientific studies will always contain an element of uncertainty. Yet the public, decision-makers, stakeholders, and lawyers want answers that are certain and expect scientists to provide them. In fact, as the severity of environmental problems increases, many people demand increasingly precise data to support their actions, particularly if litigation is involved (Holling, 1995; Bormann and Stankey, 2009). Answers such as "The results are ambiguous," "It all depends," "On the other hand," or "There are four possible contributing causes" from scientific experts open the door to doubt, and skeptics pounce on scientific uncertainties, nuances, and complexities to challenge or dismiss findings they do not like. In the face of these pressures, tentative conclusions may be presented as more certain than they really are. Consequently, the uncertainty that is part of science vanishes.[1]

Not all sciences are equally uncertain, however, and the differences are important to consider when evaluating what we know about the *Exxon Valdez* spill and its consequences. It should come as no surprise that the most certain results came from chemical analyses of the fate of the spilled oil. Concentrations of hydrocarbons in water, sediments, and the tissues of organisms can be measured with great accuracy and precision; the constituents of crude oils are well known, and fingerprinting can document the composition of samples in detail (Chapter 1, Box 1.1). Standards for analytical methods, detection limits, and toxicological thresholds are well-established. Tens of thousands of samples were collected for chemical analysis by Trustee and Exxon scientists in PWS following rigorous study designs, reducing statistical uncertainty.

The level of accuracy and precision changes when biology enters the picture. The composition of shoreline invertebrate communities in PWS, for example, is determined in part by chance events (when and where disturbance occurred, which species colonized first). The numbers of sea otters in an area can be affected by the actions of a single killer whale (*Orcinus orca*) (Chapter 15). Because animals such as salmon or marine birds may move over great distances in a short time, what appears in surveys can be affected by what happens to be in an area when observers are there to record data. An individual predator's behavior influences the likelihood of consuming contaminated prey, and the dynamics of populations are determined by factors that operate over multiple scales of space and time.

Thus, although many attributes of the biological resources in PWS were measured or counted with considerable accuracy following the spill, there were inherent sources of variation and uncertainty that could not be controlled. Moreover, whereas investigations of hydrocarbon chemistry documented potential causes (e.g., the distribution, composition, and toxicity of petroleum PAH), biological studies focused on effects (e.g., reduced abundance, overwinter survival, habitat use, reproduction). Debates about SSOR, for example, were less about how much was there or whether the residues were from the *Exxon Valdez* spill than they were about how or whether SSOR might have been accessible to or actually affected sea otters, harlequin ducks, and other biota.

---

[1] See Pollack (2003) for a discussion of scientific uncertainty and its implications.

There are also no agreed-upon standards for measuring such things as the normal annual reproductive output of a colony of murres (*Uria* spp.), the number of sea otters around northern Knight Island, or the diversity of natural shoreline communities. This lack of standards leads to a search for elusive baselines or criteria for recovery (Chapter 1, Box 1.3; Chapter 15, Box 15.2). These criteria may be specified in different ways, contributing additional uncertainty about whether recovery or restoration has been achieved.

Some sources of uncertainty can be reduced through increasing replication, careful design of sampling to account for environmental variation, or statistical analyses that include confounding covariates. Some of the uncertainty must simply be accepted, however. Scientists, and those who use science to guide their actions and decisions, should consider how much certainty can be achieved, and at what cost, for different types of data and studies. Scientists will always argue for more and better data, of course, but there is a point of diminishing returns where additional information and further reduction in uncertainty will not provide a different or better answer to a question. If a confidence level of 80% is deemed adequate to assess spill effects, for example, collecting the additional samples to increase confidence to 95% would not change the conclusion, although it might decrease the uncertainty in the assessment (Chapters 10 and 14). Determining the acceptable degree of uncertainty involves assessing the inherent variability of the subject of study, the precision and accuracy required, the effort and cost of obtaining samples, the feasibility of replication, the potential uses of the data, and the urgency of taking immediate action. Efforts to reduce uncertainty also depend upon the objectives being pursued. It is important to distinguish questions and pursuits that are scientifically interesting for their own sake from those that may have direct relevance to reducing spill-related injuries or guiding future restoration.

17.3.3 Conflicting results

17.3.3.1 Disagreements and their sources

At the time of the *Exxon Valdez* oil spill, there was a naïve expectation that scientists studying the same thing in much the same way would reach similar conclusions about the short- and long-term effects of the spill. This was not to be. Hydrocarbon chemists disagreed about the source(s) of natural background hydrocarbons and the extent of the weathering of oil residues. Biologists disagreed about the effects of the spill on the population dynamics of salmon and Pacific herring and the recovery status of marine birds. Disagreements and debates are part of the scientific process, of course, but in some instances they became intense and acrimonious, leading to a polarization and hardening of positions.[2] Why?

The studies of both Trustee and Exxon scientists were conducted in support of litigation. Given their legal positions, the initial emphasis of Trustee-sponsored studies was on documenting injuries to biological resources so appropriate restitution could be established. The provisions of the Clean Water Act and other federal and state statutes, for example, require estimates of environmental harm to determine the damages to be paid by the responsible party – in this case, Exxon. Exxon's interests, on the other hand, were in determining the overall effects of the spill to guide cleanup efforts and defend

---

[2] Chapman (2012) has discussed several scientific "sins of omission and commission," many of which appeared during debates about the effects of the *Exxon Valdez* spill and contributed to the intensity of disagreements, diminishing the credibility of the science and scientists.

themselves against claims. Consequently, the Exxon-sponsored studies considered a broad suite of factors ranging from assessing the fate of the spilled oil and its effects on the environment and biota to evidence of recovery.

The Reopener provision of the legal settlement between the state and federal governments and Exxon concerning environmental damage (Chapter 1, Box 1.4) led much of the Trustee-sponsored research after 1991 to focus on documenting previously undetected or continuing injuries from the spill rather than on restoration of injured resources. In response, several Exxon-supported studies also investigated the potential for any residual *Exxon Valdez* oil to affect biota, particularly fish, marine birds, and sea otters (see Chapters 12–16). Much of this research was conducted after the trials over private-party claims were completed in 1994, so the information and insights that were gained did not figure in those deliberations.

The high stakes of the litigation over the *Exxon Valdez* spill placed the scientists and their work under the lens of intense scrutiny by the public, media, lawyers, and other scientists. As a result, data were shielded from public disclosure during the first several years of the litigation by all parties. Individual scientists retained as expert witnesses were progressively separated into opposing camps. The differing perspectives and regulatory and funding constraints of scientists led them to pose questions differently, design studies differently, conduct analyses differently, and interpret similar sets of data differently. It is little wonder that they reached different conclusions.

The different emphases might not matter, except that they reinforced the assumptions either that the spill had detrimental effects that must be documented or, alternatively, that it might or might not have had negative (or even positive) consequences, which should be investigated. These positions are fundamentally different, and their effects permeate all aspects of study design, analysis, and interpretation. For example, differences in study designs, such as whether covariates were included or whether sampling locations were based on the actual level of shoreline oiling rather than their location somewhere within the spill zone, stemmed from initial presumptions about spill impacts and effects. When scientists then adopted different criteria for defining recovery and used different baselines or reference areas for comparisons, the seeds for disagreement were well-sown. The disagreements, in turn, created the perception of what Spies (2007) called "dueling scientists" – scientists working on different sides of an issue, with different agendas, who conduct science by claim and counter-claim.

Spies (2007, p. 426) cast the conflict as one between industries pursuing their own best interests versus governments protecting "the greater public interest." Certainly there have been instances in which industry (or government, for that matter) has subverted the scientific process to advance an agenda or strengthen a position (Oreskes and Conway, 2010). This was not one of them. The scientific findings determined the positions to be taken by both Exxon and the governments in the litigation, not the other way around. The scientists conducted their studies to produce honest and reliable findings about what happened to the oil from the *Exxon Valdez*, what effects it had, and how long those effects persisted. These were interests common to both Exxon and the Trustees.

The metaphor of dueling scientists paints a misleading picture and does a grave disservice to the many scientists on both sides of the debates who did their best to document what happened under logistically difficult and emotionally challenging circumstances. It is the nature of science that disagreements will emerge, particularly when different scientists take different approaches to the same problem. The disagreements often spur more research, resolving some issues and clarifying the basis for disagreements about others. In the case of the *Exxon Valdez* spill, this process, coupled with the natural recovery of the ecosystem over time, led to a growing convergence of conclusions about the fate of the oil and effects on the biota (see Box 17.1).

17.3.3.2 A cooperative approach?

Disagreements might be avoided if there were greater cooperation and collaboration among scientists from the outset. Although a cooperative science program was suggested in the aftermath of the *Exxon Valdez* spill, the governments and Exxon were unable to agree and instead supported separate science programs.

During the first phase of the spill, however, there were several cooperative efforts of the state and federal governments and Exxon. These efforts succeeded in identifying environmental sensitivities and acting on them (e.g., protecting salmon hatcheries and marine-mammal breeding areas), conducting thorough shoreline surveys, and guiding cleanup in a safe and effective manner. These achievements were all the more impressive in view of the challenges presented by the remoteness of PWS and the extent of the spill zone. The PWS water-quality monitoring program (Chapter 3) was also effective in documenting the extent to which oil had or had not contaminated the water body that was critical to herring and salmon, both of which released fry into nearshore waters shortly after the spill and then migrated through PWS during the summer of 1989. The preliminary results of this water-monitoring program were communicated by its lead scientist in January 1990 (Neff, 1990, 1991), helping to allay public concerns about potential long-term effects on fisheries.

The focus of science then turned from the analysis of hydrocarbons in the water column to evaluating the acute effects on other components of the environment and their recovery and on assessing chronic effects associated with potential long-term exposure of species to oil that remained in the environment. These investigations went on for years after the spill and involved scientists sponsored separately by the Trustees and by Exxon. Several examples illustrate how these separate studies built upon one another in an iterative fashion to produce a better understanding of the oil spill and its effects.

Pink salmon experts were engaged by both parties to assess the potential injury to salmon in oiled streams used as breeding areas (Chapter 12). Both groups of experts adopted similar study designs to assess the risk to salmon eggs resulting from oil in their natal streams. One of the research groups found a correlation between oiling and increased toxicity to eggs, whereas the other group found no such correlation. These conflicting results led to a further evaluation of the protocol for sampling the streams in which toxicity had been observed. As documented in Chapter 12, the elevated egg mortality was found to be the result of the timing of sampling rather than an effect of oil.

Another group of researchers conducted surveys in 2001 and 2003 to determine the extent of oil remaining on shorelines. That work led to the conclusion that "oil was surprisingly persistent and often in a relatively unweathered state" (Short *et al.*, 2001, p. 1), although a consideration of previous oil spills would have shown that small amounts of oil often persist for years (e.g., Chapter 6, Fig. 6.1). Of greater concern, however, was the suggestion that there might be ongoing harm to biota as a result of the remaining oil. This conclusion prompted subsequent surveys that showed that the remaining oil was in locations where it was known to have existed at the end of the cleanup in 1992, albeit in a much reduced and more weathered state (Chapter 6). Other scientists then used field measurements of permeability contrasts and well-established principles that define the physics and chemistry of oil in porous media to demonstrate that the concentration of PAH components from subsurface oil residues deposits dissolved into the groundwater is extremely low and that these already extremely low concentrations are then further diluted to background concentrations by tidal action (Chapter 7).

Other studies of sea otters in the northern Knight Island area some years after the spill led some researchers to conclude that otters were not as numerous as they should be, prompting the conjecture that the otters were being exposed to remnant oil in sediments as they dug pits in search of food. To assess this suggestion, an expert in otter behavior and an expert on the form and location of remnant oil collaborated in a study in the northern Knight Island area that showed that there was little overlap between otter foraging areas and the locations of remnant oil (Chapter 15). More recently, other researchers have suggested that overlap in the locations of otter foraging pits and oil might nonetheless be sufficient for exposure to hydrocarbons to have harmful effects (Bodkin et al., 2012), although other analyses (Harwell et al., 2010b; Chapter 16) indicate that harmful effects are not plausible, even if individuals were to be exposed to hydrocarbons from residual oil. Disagreements continue, as does the research.

The point of these examples, and others covered in this book, is to illustrate that knowledge and understanding are improved when capable scientists closely evaluate and build on one another's work. In retrospect, one can argue that, for all the disagreements, the separate scientific efforts eventually produced greater scientific resolution of impacts, effects, and recovery from the *Exxon Valdez* spill than would have any single investigation alone.

17.4 What can be done?

When disagreements among scientists escalate from scholarly debates in scientific meetings and journals into publicized attacks about the validity of one another's findings or assertions about scientific integrity, the credibility of science as a process and its relevance in guiding decisions suffer. In the decades following the spill, nearly all areas of disagreement among scientists have been whittled away. But there were enormous costs of the scientific disagreements, in terms of missed opportunities for findings to be applied to spill remediation, eroding public perceptions of science, and the escalating expense of continuing research well past the point at which issues were resolved simply for the sake of rebuttal.

How can science be brought to bear more effectively on assessing the consequences of large environmental disturbances such as oil spills? The lessons we have detailed in this book provide considerable guidance, particularly for dealing with specific issues and needs. But there are also some broader messages. Ideally, the science that is done should be more open and transparent. Instead of engaging in defensive debates about findings only after research has been completed and results published, study designs could be shared at the outset of the research. In this way, hidden assumptions might be uncovered, methods could be evaluated, and definitions and standards could be proposed, if not agreed upon. Many of the roots of potential disagreements would be laid bare. These disagreements can advance understanding if they are framed as hypotheses to be tested rather than assumptions to be justified.

It is important also to recognize that science conducted in a litigation context is often driven by the overarching legal issues. This should not mean that the science is conducted any differently than the best basic science – only that the objectives and questions asked may be different. Science to inform litigation must answer the precise questions presented by legal proceedings, while also meeting the requirements of good science. Scientific standards, however, cannot be sacrificed in the interests of advocacy for a particular legal position.

These suggestions may strike some as idealistic and naïve. The reality is that for large environmental issues, governmental and private-stakeholder interests will often differ and will be aggressively, but separately, pursued. The only rational approach, amid the conflicting demands and expectations, is to do the science right, discuss it at conferences and symposia, and submit it to the scrutiny of peer review and publication. The scientific process of evaluating the design and underlying assumptions of studies, exposing the conclusions to critical scrutiny, and discussing areas of disagreement will inevitably reduce uncertainty and lead to a better understanding of the extent and duration of injuries from large environmental disruptions such as oil spills. The sooner after an event this process is implemented, the sooner the understanding gained will be available to guide responses and restoration efforts.

REFERENCES

Bodkin, J.L., B.E. Ballachey, H.A. Coletti, G.G. Esslinger, K.A. Kloecker, S.D. Rice, J.A. Reed, and D.H. Monson (2012). Long-term effects of the *Exxon Valdez* oil spill: sea otter foraging in the intertidal as a pathway of exposure to lingering oil. *Marine Ecology Progress Series* **447**: 273–287.

Bormann, B.T. and G.H. Stankey (2009). Crisis as a positive role in implementing adaptive management after the Biscuit fire, Pacific Northwest, USA. In *Adaptive Environmental Management: A Practitioner's Guide*. C. Allen and G.H. Stankey, eds. New York, NY, USA: Springer; ISBN-13: 9789048127108; pp. 143–167.

Carls, M.G., R.E. Thomas, M.R. Lilly, and S.D. Rice (2003). Mechanism for transport of oil-contaminated groundwater into pink salmon redds. *Marine Ecology Progress Series* **248**: 245–255.

Chapman, P.M. (2012). Crossing the scientific line: Sins of omission and commission. *Marine Pollution Bulletin* **64**(3): 457–458.

Driskell, W.B., J.L. Ruesink, D.C. Lees, J.P. Houghton, and S.C. Lindstrom (2001). Long-term signal of disturbance: *Fucus gardneri* after the *Exxon Valdez* oil spill. *Ecological Applications* **11**(3): 815–827.

***Exxon Valdez* Oil Spill Trustee Council** (2010). *Exxon Valdez Oil Spill Restoration Plan: 2010 Update Injured Resources and Services*. Anchorage, AK, USA: *Exxon Valdez* Oil Spill Trustee Council. [http://www.evostc.state.ak.us/Universal/Documents/Publications/2010IRSUpdate.pdf]

Field, L.J., J.A. Fall, T.S. Nighswander, N. Peacock, and U. Varanasi, eds (1999). *Evaluating and Communicating Subsistence Seafood Safety in a Cross-Cultural Context:*

Lessons Learned from the Exxon Valdez *Oil Spill*. Pensacola, FL, USA: Society of Environmental Toxicology and Chemistry, SETAC Press; ISBN-10: 1880611295.

Harwell, M.A., J.H. Gentile, K.W. Cummins, R.C. Highsmith, R. Hilborn, C.P. McRoy, J. Parrish, and T. Weingartner (2010a). A conceptual model of natural and anthropogenic drivers and their influence on the Prince William Sound, Alaska, ecosystem. *Human and Ecological Risk Assessment* **16**(4): 672–726.

Harwell, M.A., J.H. Gentile, C.B. Johnson, D.L. Garshelis, and K.R. Parker (2010b). A quantitative ecological risk assessment of the toxicological risks from *Exxon Valdez* subsurface oil residues to sea otters at northern Knight Island, Prince William Sound, Alaska. *Human and Ecological Risk Assessment* **16**(4): 727–761.

Harwell, M.A., J.H. Gentile, K.R. Parker, S.M. Murphy, R.H. Day, A.E. Bence, J.M. Neff, and J.A. Wiens (2012). Quantitative assessment of current risks to harlequin ducks in Prince William Sound, Alaska, from the *Exxon Valdez* oil spill. *Human and Ecological Risk Assessment* **18**(2): 261–328.

Holling, C.S. (1995). What barriers? What bridges? In *Barriers and Bridges to the Renewal of Ecosystems and Institutions*. L.H. Gunderson, C.S. Holling, and S.S. Light, eds. New York, NY, USA: Columbia University Press; ISBN-10: 0231101023; ISBN-13: 9780231101028; pp. 3–34.

Integral Consulting, Inc. (2006). *Information Synthesis and Recovery Recommendations for Resources and Services Injured by the* Exxon Valdez *Oil Spill*. Mercer Island, WA, USA: Integral Consulting, Inc.; *Exxon Valdez* Oil Spill Restoration Project 060783 Final Report. [http://www.evostc.state.ak.us/Files.cfm?doc=/Store/FinalReports/2006-060783-Final.pdf&]

Neff, J.M. (1990). *Water Quality in Prince William Sound*. Houston, TX, USA: Exxon Company, USA. [http://www.valdezsciences.com]

Neff, J.M. (1991). *Water Quality in Prince William Sound and the Gulf of Alaska*. Houston, TX, USA: Exxon Company, USA. [http://www.valdezsciences.com]

Neff, J.M., D.S. Page, and P.D. Boehm (2011). Potential for exposure of sea otters and harlequin ducks in the Northern Knight Island area of Prince William Sound, Alaska, to shoreline oil residues 20 years after the *Exxon Valdez* oil spill. *Environmental Toxicology and Chemistry* **30**(3): 659–672.

Oreskes, N. and E.M. Conway (2010). *Merchants of Doubt: How a Handful of Scientists Obscured the Truth on Issues from Tobacco Smoke to Global Warming*. New York, NY, USA: Bloomsbury Press; ISBN-10: 1596916109; ISBN-13: 9781596916104.

Piatt, J.F. and P. Anderson (1996). Response of common murres to the *Exxon Valdez* oil spill and long-term changes in the Gulf of Alaska marine ecosystem. In *Proceedings of the* Exxon Valdez *Oil Spill Symposium*. S.D. Rice, R.B. Spies, D.A. Wolfe, and B.A. Wright, eds. Bethesda, MD, USA: American Fisheries Society; Symposium 18; ISBN-10: 0913235954; ISSN: 08922284; pp. 720–737.

Pollack, H.N. (2003). *Uncertain Science ... Uncertain World*. Cambridge, UK: Cambridge University Press; ISBN-10: 0521781884; ISBN-13: 9780521781886.

Short, J., S. Rice, and M. Lindberg (2001). *The Exxon Valdez Oil Spill: How Much Oil Remains?* Juneau, AK, USA: National Oceanic and Atmospheric Administration, National Marine Fisheries Service; Alaska Fisheries Science Center Quarterly Research

Reports; July–September 2001. [http://www.afsc.noaa.gov/Quarterly/jas2001/feature_jas01.htm]

Spies, R.B. (2007). Dueling scientists. In *Long-Term Ecological Change in the Northern Gulf of Alaska*. R.B. Spies, ed. Amsterdam, The Netherlands: Elsevier; ISBN10: 0444529608; ISBN-13: 9780444529602, Box 5.1, pp. 426–428.

Wiens, J.A. (2007). Applying ecological risk assessment to environmental accidents: Harlequin ducks and the *Exxon Valdez* oil spill. *BioScience* **57**(9): 769–777.

INDEX

Locators in **bold** refer to figures, tables or boxed material

A Program postcleanup survey   88
abundance   254
   algae   **254**
   herring   **305**
   marine birds   320–327, 332–333
   sea otters   348, 359–366
   shoreline biota   246
accidents, natural   220–222, 235
   see also earthquakes
accuracy, data   73
activation pathways, cytochrome P450 1A   207–208
adverse effects. *see* injury
aerial reconnaissance   **81**, 81–83, **82**, 87
age structured assessment (ASA), herring   299, 306–309
Alaska Department of Environmental Conservation (ADEC)   81, 83, **92**, 185–186
Alaska Department of Fish and Game (ADFG)   272–275
Alaska Heritage Resources Survey   102
Alaska Implementation Guidelines (ARRT)   99, 110
Alaska North Slope crude oil   9–13, **10–12**
Alaska Office of History and Archaeology   102
Alaska Regional Response Team   99, 110
algae   187, 246, 250–252, **253**, **254**, 255
Alutiiq region   8, 99, 105–107,
ambiguity. *see* uncertainty
Amoco Cadiz spill   **117–119**
analytical models   159–162, **161**
ancient sites. *see* cultural resource protection
anthropogenic effects. *see* humans
Archaeological Index Site Monitoring program   106, 110
archeology, Prince William Sound   108.
   see also cultural resources
archives, data   102
armor layer   119–121, **121**, 133, **134**
aromatic hydrocarbon receptor (AHR)   206.
   see also polycyclic aromatic hydrocarbons
aromatic hydrocarbon response elements (AHREs)   207
Arrow spill   117
artefacts, cultural   99, 106, 108, 110
   see also cultural resources
asphalt   **12**, 39, 119, 192
assessment. *see* environmental effects assessment; monitoring
assumptions   222–225, 431
August Shoreline Assessment Program (ASAP)   91, **92**

B Program postcleanup survey   88
BACI. *see* Before-After-Control-Impact
bacteria   178, 192
balance of nature   225, 393. *see also* baseline data; endpoint criteria
bald eagles (*Haliaeetus leucocephalus*)   **336**, **323–325**, 336
baseline data   122–123, 429–430
   ecological effects   229–230, 393–395, 412
   marine birds   326
   oil, pre-spill   5, **17**, 19, 46, 122–123, 212, 430
   sea otters   348, 350–352, **351**, 354, **364**, **368**, 372
   water column   63–64, 71, 73
baseline services, definition   22–26, **23–25**
beach profiling   **90**
Beachwalk program   87
bedrock   **120**, 164–165, **166**, 168, **169**
Before-After-Control-Impact (BACI)   229–230, 322, **323–325**
benzene   59, 387. *see also* BTEX
Bering, Vitus Jonassen   8
bioaccessibility   128
bioavailability   128, 189
biochemistry   202, 386. *see also* biomarkers
biodegradation/bioremediation   56, 87, 176–193, **184**, **186**, **187**, 190, **191**
   consultation exercizes   **185**
   mathematical models   166–168
   measures   **180**
   shoreline oil   **124**, **125**, 124–137
biodiversity, shoreline biota   246
biological effects   199, 386. *see also* biological significance
Biological Monitoring Study (BMS) 241–242, 247–248, **248**, 256
   results   **249**, 252–255, **254**
   shoreline oil   128, **131–132**, 134–135
biological plausibility. *see* plausibility
biological resources   57, 60, 62
biological responses, definition   386
biological samples   43–44, 46
biological significance   209, 232–234, 339, 385–389, **388**
biomarkers   199, 202–204, 430
   biodegradation/bioremediation   **180**
   evidence for harm   212
   future role   212–213
   interpretation   210–212
   salmon   **204**
   sea otters   355, 367

TPAH **190**
toxicology **204**, 204–205
see also cytochrome P4501A
bioremediation index 190.
    see also biodegradation/bioremediation
biota, shoreline 241–242, **243**, 255–258
  recovery **426**
  spill-related studies 243, **244**.
    see also Biological Monitoring Study;
    Coastal Habitat Injury Assessment;
    Shoreline Ecology Program
birds, marine. *see* seabirds
black oystercatchers. *see* oystercatchers
Block Island 17
blue mussels *see* mussels
BMS. *see* Biological Monitoring Study
booms, cleanup 4
bottom-up processes 392
boulder beaches 119–121, **120**, 133, **134**, **154**, 154–156, **155**
boundary conditions **148–149**
breeding success. *see* reproduction
BTEX (benzene, toluene, ethylbenzene, xylenes) 59, 66, 387
burning oil 16

Cabin Bay 298
caged fish 210–211
cancer 203–204, 207, 383
capillary forces, models 147, **151**
carbon dating 102
cascade effects 392
cause–effect relationships 221, 227, 337–339, 412
cellular biomarkers 202
CEMs. *see* conceptual ecosystem models
change, natural. *see* variation
checklists, SCAT 81, 82
chemical analysis 42, 73
chemistry, oil **10–12**, 59, **60**, 116, 119
chronic effects. *see* long-term effects
CHIA. *see* Coastal Habitat Injury Assessment
clam species 255, 397–399
clay-oil flocculation 89
cleanup methods 13–17, **63**, 86–87, **87**
  see also cold-water washing; flooding; flushing; hot-water shoreline cleanup; rakes; skimmers
cleanup phase 40, **41–42**, 44–46
  ecological risk assessments 387
  herring 294–299, **296**, **298**, **299**, **300**
  SCAT support 86–87
  shoreline oil 116, **121**, 123–128
  water sampling **63**, 63

clustering, natural 231
Coastal Habitat Injury Assessment 241–242, 246–247, **247**, 256
  results **249**, 251–252, **253**, 256
cobble beaches 119–121, **120**, 133, **134**, **154**, 154–156, **155**
cold-water washing **14**, 16
collaboration. *see* cooperative approaches; multidisciplinary approaches
communication 111, 432–433
community biomarkers 202
community-level analyses 246
complexity, ecosystems 393
Comprehensive Environmental Response, Compensation and Liability Act (CERCLA) **20–21**, 22
conceptual models 146–159, **150**, **151**, **154**, **155**, **156**, **159**
conceptual ecosystem models (CEMs) 385, **389**, 389–393, **391**, 412
condition index, salmon 268, 281
conductivity–temperature–depth 64
confidentiality 102–103, 111
conflicting results 439–442.
  see also uncertainty
confounding factors
  ecological effects analysis 225–227, **226**, 229–230, 236
  science-based approaches 437–438
  sea otters 357, 368–370, 372
consequences, oil spill **424–427**
consultation exercizes **185**
Cook, Captain James 8
cooperative approaches 100–101
  bioremediation 177, **185**, 185–186, 192
  science-based 441–442
  see also multidisciplinary approaches
copper mining 8, 108, **109**
Copper River sediments 5
corroborating evidence 232–234
counts 25, 350–352, **355**, **361**, **362**
  carcass 321,
  fish 268
  marine birds 231, 321, 331, 437
  sea otters 332, 339, 350
covariate analysis
  ecological effects analysis 227
  herring 297–299, 308
  marine birds 332, 339
crude oil 9–13, **10–12**
  carbon dating 102
  characteristics **39**
  water column analysis 59, **60**
cultural resources 55, 98–101, **100**, 111
  archeology/history 108, **109**

cultural resources (cont.)
  effects on resources   105–106
  methodology   101–103
  postcleanup monitoring   109–111
  results   103–105, **104**, 104, 105
Cultural Technical Advisory Group (CTAG)   101
cumulative stressors   393
currents, ocean   6, 122, 370
Customblen   179, 183, **187** see also fertilizers
CYP gene battery   207
cytochrome P450 1A biomarkers   199, 201, **205**, 205–209, 213–214, 430
  harlequin ducks   331
  marine birds   337
  sea otters   367

data collection   **41–42**, **68**, 68–71, **69**
  accuracy   73
  archives   102
  cleanup phase   37
  cultural resource protection   99, 101, 110–111
  model development   145–146
  preimpact data   229–230
  recovery phase   46
  release/immediate response   43–44, 60–61
  replicate sampling   230–231
  SCAT   **80**, 80
  water column analysis   64–65
databases
  SCAT   82
  Trustees   68
Deepwater Horizon oil spill   57, 72
definitions
  biological responses   386
  biomarkers   202
  ecological risk assessments   235
  exposure pathways   **20**
  injury   **24**, 242
  oil residues   153
  recovery see below
  segments   101
  stressors   384
definitions of recovery   22–26, **23–25**, **24**
  and biological effects   200
  and ecological risk assessments   393–395
  marine birds   341
  science-based approach   431, 440
  sea otters   370–371
  shoreline biota   248, 252, 256
deformities, developmental   267, 275
demographic conditions, sea otters   366
density parameters   **148–149**
design, study   72, 200, 325, 354–356, **365**, 431–432
detoxification pathways   207–208
developmental deformities, salmon   267, 275
diesel   **12**, **39**, **60**, 65, 122, 402

direct effects   **319**
direct measures   430–431
direct water sampling   65
disagreement, scientific   271–275, 281–283, 356–359, 439–442
disasters, anthropogenic   372
diseases, herring   301, **302–303**, 308
Disk Island   186
dispersants   15
dissolved fraction   65
disturbance, human   370, 387
documentation
  exposure pathways   **20–21**, 20–22
  SCAT   85
dolphins   211
drift experiments   321
ducks, harlequin. see harlequin ducks
dueling scientists metaphor   440–441. see also disagreement

eagles. see bald eagles
earthquakes   6–7, **7**, 121, 370, 392
ecological effects   199, 386. see also biological significance
ecological effects analysis   220–225, **223**, **224**. see also ecological risk assessments
  confounding factors   225–227, **226**, 229–230, 236
  covariate analysis   227
  herring   222
  hypotheses, framing/testing   227–228
  interpretation, data   232–234
  Knight Island   **226**
  multidisciplinary approaches   234, 236
  multiple-lines-of-evidence   233
  multiple-year studies   231–232
  non-statistical   232
  pre-impact data   229–230
  qualitative   385–395, 411
  replicate sampling   230–231, 236
  salmon   **224**, 224
  statistical power   229
ecological monitoring. see monitoring
ecological risk assessments (ERAs)   **20**, 235–236, 383–385, 411–412
  application to future spills   408–410
  chronic effects   **425–426**
  models   144, 149. see also conceptual ecosystems model
  quantitative   395–408, **397**, **400**, **401**, **403**, **405**, **406**, **407**, 411
  science-based approach   435–436
  sea otters   367–368, 387–389, 396–399, **397**, **400**
ecological significance   209, 232–234, 385–389, **388**
ecological toxicology   **204**, 204–205

economic activity, Alaska 8–9
ecosystems
    biomarkers 202
    complexity 393
    effects. *see* environmental effects assessment
    variation 370
ecotoxicological conceptual ecosystem
    models 385, 396–402, **397, 400, 401, 403**
educational programs 99, 103
effects 23
    biomarkers 203–204, 212
    cultural resource protection 105–106
    definitions **23–25**, 26
    ecological 199, 386. *see also* biological significance
    mechanisms **362**, 366–368. *see also* plausibility
    thresholds **405**
    see also oil spill phases/effects
egg incubation
    herring 292–293, 298–299
    salmon 232, 272–277, **275**
Eleanor Island **154**, 154–155, **156**, 165, **167**
El Niño events 6
El Niño Southern Oscillation 6
Elrington Island 186
Endangered Species Act, U.S. 369
endpoint criteria 297, **298**, 308, 393–395.
    see also definitions of recovery
environmental processes 61. *see also* weathering
Environmental Protection Agency (EPA) 146, 182, 185–186, 383
Environmental Sensitivity Indexes
    43–44, 62, 119–121, **120**. *see also* sensitivity analyses
equilibria 222–225
    dynamic 222, **223**, 225
    spacial 222, **223**, 225
    steady state 222, **223**, 225
ERAs. *see* ecological risk assessments
EROD (ethoxyresorufin-O-deethylase) 208–209
erosion **100**, 105–106
escapement, salmon **224**, 224
ethylbenzene 59. *see also* BTEX
eutrophication 187
evidence, corroborating 232–234
experiments, vs. observations 220–222, 235. *see also* laboratory testing
expert advisors 386
exposure biomarkers 202–203, 210–212
exposure pathways 39, 200
    documentation **20–21**, 20–22
    harlequin ducks 400–402, **401**, **403**
    herring 308
    PAH 411
    salmon 266–268
    science-based approach 434–435

sea otters **397**, 397–399
Exxon Shoreline Ecology Program.
    see Shoreline Ecology Program
Exxon Valdez Cultural Resource
    Program 99–101, **100**
Exxon Valdez oil spill. *see* oil spill, *Exxon Valdez*
Exxon Valdez Oil Spill Trustee Council 22–26, 68, 91, **328–329**, 439–441

Fast Assessment Shoreline Survey Team
    (FASST) 90, **92**
feeding. *see* nutrition
fertilizers 177–178, **179**, **184**, **187**
    field trials 179
    guidelines 182
    and oxygen demand **184**
    toxicity 178
field trials 179–182, 185–186, **186**
filtration 65
Final Shoreline Assessment Program
    (FINSAP) 91, **92**
fingerprinting hydrocarbons **19**, 213, 430
finite-element/difference methods 163
fires, oil 60
fish, caged 210–211 *see also* herring; salmon threespine stickleback
fishing industry 8, 57. *see also* herring; salmon
fixed site (FS) sampling method 241, 246, **249**, 250–251, **251**, 256
flexibility, observational studies 234
flocculation, oil **124**, **125**, 124–137
flooding cleanup methods 14, 86–87, **87**
fluid saturation **148–149**
fluorometry 64
flushing cleanup methods 14, 86–87, **87**
food sources, shorelines as 78
footprints, oil 40
four-dimensional water sampling 62
Fucus spp. *see* rockweed
fuel oil 39
fur farming 108
future spills, risk assessment 408–410

gasoline 39
genetic effects, salmon 267, 276–277
geographical information systems **80**
geography/geology, GOA 4–6, 162
geomorphology
    cultural resource protection 109
    oil persistence 116–117, 119–121, **120**, **120**, 127, 133–135, **134**, **135**
    Prince William Sound 242–243
geostatistics 157
glacial flour 123, **124–137**
glaucous-winged gulls (*Larus glaucescens*)
    323–325, **328–329**, 329, **336**, 336

goldeneye (*Bucephala* spp.)   323–325, 327, 328–329
Graduated Severity Index (GSI)   297
gravel beaches   119–121, **120**, 133, **134**, **154**, 154–156, **155**
Great Alaska Earthquake   6–7, **7**, 121, 370, 392
Green Island   17, 354, 357–359
ground surveys   83–85
groundwater flow   **148–149**, 164, **165**, **166**, 168, **169** *see also* hydraulic-conductivity distributions
growth, salmon   268, 278–280, **279**, **280**
guidelines
    biodegradation/bioremediation   182
    Guidance for Conducting Remedial Investigations and Feasibility Studies Under CERCLA (US Environmental Protection Agency)   146
    Oil in the Sea   383
guillemots, pigeon (*Cepphus columba*)   209, 323–325, 326, **336**, 336, 337–339. *see also* murres
Gulf of Alaska (GOA)   4–9, 39, 55–56
    cultural resources   **104**, 104–105
    ecological risk assessments   **389**
    SCAT   81
    shorelines   **5**
gull species   231, 323–325, 328–329, 329, **336** *see also* glaucous-winged gulls

habitats
    environmental assessment   222
    marine birds   326–327, **340–341**
    sensitive   81
harlequin ducks (*Histrionicus histrionicus*)   233, 323–325, 328–329, **336**, 436
    chronic effects   331–332
    ecological risk assessments   400–402, **401**, **403**
    habitat occupancy   327
    models   321
    population level effects   **407**, 407–408
    science-based approach   436
    sensitivity analyses   404
    toxicological risk   408
harm. *see* injury
hazard quotient (HQ)   399
Hazardous Materials (HAZMAT). *see* Biological Monitoring Study
health, human   44
helicopter reconnaissance   81–83
herring (*Clupea* spp.)   9, 43–44, 108, 292–294, **293**, 308–309, 428
    cleanup phase   294–299, **296**, **298**, **299**, **300**
    ecological effects analysis   222

fertilizer toxicity testing   187
    recovery   294, 301–307, **302–303**, **305**, **426**
    release/immediate response phase   294
heterogeneity, modeling   162–163
history   8–9, 99, 108, **109**
homogeneity-of-slopes test   331
hot-water shoreline cleanup treatment   242, 247–248, 252–255, **254**, 257
Howes, Don   94
HQ (hazard quotient)   399
human(s)
    disasters   372
    disturbance   370, 387
    health   44
    history   8–9
    pollution, non-spill related   212
    predation of sea otters   369–370
hydraulic-conductivity   **148–149**, 157–159, 164, **165**
hydrocarbons
    mussel exposure   129–130
    salmon exposure   270–272
    water analysis   66
    see also specific hydrocarbons by name
hydrogeologic processes   151. *see also* wave action
hypotheses, framing/testing   227–228, 235–236, **302–303**, 308, 431

IBM. *see* individual-based modeling
Ichthyophonus hoferi   **302–303**, 304, 306
immediate response phase. *see* release/immediate response phase
immunohistochemical staining   208
impact   23
    definitions   23–25, 26
    phases   **38**, **39**. *see also* oil spill phases
Incident Command System (ICS)   93
individual-based modeling (IBMs)   246, 395–399, **397**, **400**, 411
Inipol®   **179**, 183, **184**. *see also* fertilizers
injury   23
    biological resources   57, 60
    biomarkers   212
    cultural resource protection   105–106
    definitions   **23–25**, **24**, 26, 242
    ecological risk assessments   385
    endpoint criteria   297, **298**, 308
    exposure pathways   434–435
    quantifying   386
inspections, postcleanup   87–92, **88**, **89**, **90**, 109–111. *see also* monitoring
Integral Consulting   87–92, **88**, **89**, **90**, 109–111, 328
integrated assessment framework. *see* ecological risk assessments

Interagency Shoreline Cleanup Committee (ISCC) 83, 86, 101
interdisciplinary approaches.
    see multidisciplinary approaches
intergenerational effects, salmon 267, 276–277
interpretation, data 199
    biomarkers 208–212
    ecological effects analysis 232–234
intertidal zone 91, 127, 387
invertebrates 253. see under individual species by name
Ixtoc I spill 117

jet fuel 39

kayak artefacts 108
keystone species 292. see also herring
killer whales (*Orcinus orca*) 231, 368–369, 387–389
Knight Island 181–182
    ecological effects analysis 226
    field trials 185–186
    marine birds 333, **334**, **335**, **336**, 338
    see also sea otters (Knight Island)
Kodiak Island **100**
Kuroshima spill 110

laboratory testing
    bioremediation 178, 183–184, **184**
    biomarkers 202–203, 210–211
    salmon 267, 275–276, 283
landscape biomarkers 202
laws/legislation 22–26, **27–29**.
    see also litigation; regulations
learning curves 234
legal settlement, Exxon Corporation 26, 222–225
life cycles
    herring **293**, 308
    salmon 263–265, **264**, 283
light nonaqeous phase liquids (LNAPL) 151
limpets 250
lingering effects 128. see also long term effects
literature reviews 387
litigation 26, **27–29**, 39, 421, 433
    conflicting results 439–442
    science-based approach 443
long-term effects 128, **425–426**
    ecological risk assessments 410
    marine birds 331–332, 339
    sea otters 350, 357
    shoreline oil, fate 128–135
lowest-observed-adverse-effects level (LOAEL)
    harlequin ducks 402, **407**, 407–408
    sea otters 399, **400**, 404–406

mammals, marine 211, 387. *see also* sea otters; whales
maps/mapping
    postcleanup surveys 87
    protocols, SCAT 82
    study locations 152
    see also videotape mapping
marine birds 318–319, 339, **340–341**, 429
    acute effects 319, 319–327, **320**, 323–325
    cause–effect relationships 337–339
    ecological risk assessments 387
    recovery **327**, 327–332, **328–329**, **330**, 426
    variation, natural 332–335, **334**, **335**, **336**, **338**
Marry Map study 85
mathematical models 159–170, **161**, **165**, **166**, **167**, **169**, 357
May Shoreline Assessment Program (MAYSAP) 91, **92**, 127, 129, 132–133, 135, **136**
mergansers (*Mergus merganser*) 231, **323–325**, **328–329**
metabolomics 213
Metula spill 25, **117**
microorganisms 178, 192
mineral-oil flocculation 89
mink (*Neovison vison*) 211, 337–339
minnow, sheepshead (*Cyprinodon variegatus variegatus*) 271
mitigation strategies 394
models 55, 144–146, 170–171
    characterization parameters **148–149**
    marine birds 321
    mathematical 159–170, **161**, **165**, **166**, **167**, **169**, 357
    numerical **158**, 159, 162–164, 168–170
    quantitative/quantitative **150**, 385
    see also conceptual models
MODFLOW-SURFACT model 163–164
molecular biomarkers 202
molecular profiling 213
monitoring
    bioremediation 183–184, **184**, 187
    postcleanup 87–92, **88**, **89**, **90**
Montague Island **226**, 226
mortality
    herring 292–293, 298–299, 301–304
    marine birds **320**, 320–325
    salmon 232, 267, 272–277, **273**, **274**, **274**, **275**
    sea otters 352–353, **353**, 357–359, 366, 348–350
mousse, oil 17, 122, 188
multidisciplinary approaches 26, 171
    cultural resource program 100–101
    ecological effects analysis 234, 236

multidisciplinary approaches (cont.)
  models   145
  science-based approach   436–437
  SCAT   83
  see also cooperative approaches
multiphase fluid flow   147, **148–149**, **151**, 160, 163
multiple-lines-of-evidence   199
  ecological effects analysis   233
  ecological risk assessments   412
  herring decline   309
  salmon   283
  science-based   437–438
multispecies analyses   333–335, **340–341**, 341
multivariate correspondence analyses   246
murrelets (*Brachyramphus* spp.)   **323–325**, 326, **328–329**
murres (*Uria* spp.)   320, 321–322, **323–325**, 325–326, **328–329**
mussels (*Mytilus* spp.)   45, 70–71, **70–72**, **71**, 73, 128
  Biological Monitoring Study   255
  fertilizer toxicity testing   187
  as passive samplers   65
  Shoreline Ecology Program   245, 250
  systematic hydrocarbon analysis   129–130
mutations, salmon   267, 276–277
mysid shrimp (*Mysidopsis bahia*)   267, 270, 276–277,

Naked Island   337–339
National Historic Preservation Act (NHPA)   100
National Oceanic and Atmospheric Administration (NOAA)   127–129, 188, 393 *see also* Biological Monitoring Study
National Research Council   383
natural recovery   130–132, 433
  see also recovery
natural resource damage assessment (NRDA)   **20–21**, 22–26, 39, 128, 384
  see also ecological risk assessments
natural variation. *see* variation
Net Environmental Benefit Analysis   90
NOAA. *see* National Oceanic and Atmospheric Administration
no-observed-adverse-effects level (NOAEL)
  harlequin ducks   402, **407**, 407–408
  sea otters   **400**, 404–406, 399
Northwest Passage   8
null hypotheses   227–228, 283
numerical models   **158**, 159, 162–164, 168–170
nutrients, biodegradation   176
nutrition
  cause–effect relationships   337–339

herring   301, **302–303**
salmon   268, 281
sea otters   357, **362**
objectives, stating   428–433
objectivity. *see* science-based approach
observational studies   222, 234
  see also ecological risk assessments
observations vs. experiments   220–222, 235
Occam's Razor   232, 236
ocean currents   6, 122, 370
oceanographic characterization   64, **302–303**, 306
oil migration   153–159
oil mousse   17, 122, 188
Oil Pollution Act (OPA)   22, 393
oil pollution, background   5, **17**, 19, 46, 122–123, 212, 430. *see also* baseline data
oil residues
  definitions   153
  ecological risk assessments   144
  lenses   165, **167**
  science-based approach   **424–425**
  see also persistence; residual oil; subsurface oil residues
oil samples   42–44
oil sequestration models   157
oil slicks, surface   60
oil spill, *Exxon Valdez* xxii, 9–19, **10–12**, **20–21**, **23–25**, **27–29**
  archeology   99
  consequences   **424–427**
  distribution   13–19
  shoreline oil, fate   121–135
  subsurface oil residues   135, **136**
  water column analysis   66–72, **67**, **68**, **69**, **70–72**, **71**
Oil Spill Health Task Force   263
oil spill phases   37–40, **38**, **41–42**, 47
  see also cleanup; recovery; release/immediate response phases
oilfields, Prince William Sound   9
omics   213
opinion, scientific differences   439–442
organ level biomarkers   202
organic sediments   121
organism level biomarkers   202
organisms, shoreline. *see* biota
oxygen
  and biodegradation   176, 183
  cytochrome P450 1A   **205**, 205
  and fertilizers   **184**
oystercatchers, black (*Haematopus bachman*)   323–325, 327, **328–329**, **336**, 336
oysters (various species) 187

Pacific Decadal Oscillation 6
Pacific herring. *see* herring
Pacific salmon. *see* salmon
PAH. *see* polycyclic aromatic hydrocarbons
parallelism test 252–254, **254**, 331
parameter-sensitivity analyses 404
partition coefficients 157, 162
Passage Cove 181–182
passive water column samplers 65
patch dynamics **243**, 243, 332, 429
peat deposits 127
pebble beaches 119–121, **120**, 133, **134**, **154**, 154–156, **155**
penetration, shoreline oil **117**, 118–119
pericardial edema, herring 297
permeability 121, 146, **148–149**, 165. *see also* porosity
persistence, oil residues **16**, 17–19, 60, 117
   and geomorphology 116–117, 119–121, 127, 133–135, **134**, **135**
   postcleanup surveys **88**, 88
   science-based approach 441–442
petroleum chemistry **10–12**, 59, **60**
petroleum seeps, natural 5, 19, 122, 212, 430. *see also* baseline data
phases, oil spill 37–40, **38**, **41–42**, 47 *see also* cleanup; recovery; release/immediate response phases
physical properties of oil **116**, **117**
pink salmon. *see* salmon
pit-sediment sampling 130, **136**
   bioremediation **190**
   biomarkers 211
   conceptual models 156–157
   postcleanup surveys **89**
   sea otters 367–368
   subsurface oil residues **135**
plausibility, biological 309, 339, **362**, 366–368
Plexiglas® columns 183
policy 22–26, 111 *see also* laws/legislation
polycyclic aromatic hydrocarbons **11**, 18, **19**, 68, 68–71, **69**, 73
   biodegradation/bioremediation 176
   Biological Monitoring Study 255
   chemistry **10–13**, 119
   chronic effects **425–426**
   conceptual models 152
   cytochrome P450 1A 207
   direct measures 431
   ecological risk assessments 232, 387
   exposure pathways 411, 434
   harlequin ducks 400–402, **401**
   herring 297, 306

   mathematical models 164–168
   mussels 129–130
   quantifying 408
   salmon 263, 271–272
   sea otters 396–399, **397**, **400**
   Smith Island **191**
   solubility 156
   spatial distributions **159**
   stream sediments 271–272
   water column analysis 59, 66
   see also total polycyclic aromatic hydrocarbons
pollution, background 5, **17**, 19, 46, 122–123, 212, 430 *see also* baseline data
population dynamics
   herring **299**, **300**, 301–307, **302–303**
   sea otters 354, **355**, **359**, 359–366, **361**, **362**, 372
population level effects
   biomarkers 202
   ecological risk assessments 395–396
   herring 298–299, **299**, **300**
   quantitative analyses 404–408, **405**, **406**, **407**
   salmon 283
   sea otters **406**
porosity **117**, 121, 146, **148–149**
   see also permeability
Port Valdez 8
Post Shoreline Assessment Program (PostSAP) **92**, 92
postcleanup inspections 87–92, **88**, **89**, **90**, 109–111 *see also* monitoring
power, statistical 229
precision
   data 73
   replicate sampling 230–231
predation
   cause–effect relationships 337–339
   herring **302–303**, 306
   sea otters 368–369
Prediction Map postcleanup survey 87
predictions, shoreline oil fate 132–133, **136**
pre-impact data. *see* baseline data
Prince William Sound (PWS) 4–9
   cultural resources **104**, 104–105, 108, **109**
   earthquake 6–7, **7**, 121, 370, 392
   ecological risk assessments 389
   shoreline 5, 242–243, **243**
Principal Component Analysis, shorelines **226**
productivity, marine 4
profiling
   beach **90**

profiling (cont.)
  molecular   213
proteomics   213
Prudhoe Bay   9
public perceptions   71–72
pyrogenic sources, oil   212. *see also* oil pollution, background

qualitative ecological risk assessments   385–395, **388, 389, 391,** 411
qualitative models   **150,** 385
quality assurance planning   412
quality control, biomarkers   208–209
quantitative ecological risk assessments   367–368, 395–408, **397, 400, 401, 403, 405, 406, 407,** 411
quantitative PCR   208–209
quantitative models   **150**

randomization   437–438
rate of natural oil loss   **129,** 129
rate-limited mass transfer   159
realtime PCR   208–209
recommendations. *see* guidelines
recording forms, standardized   83, **84,** 93, 102
recovery   23, 40, **41–42,** 46–47, **223**
  bioremediation   188–192, **190**
  cultural resource protection survey   109–111
  definitions. *see* definitions of recovery
  ecological risk assessments   384, 393–395
  herring   294, 301–307, **302–303,** 305
  marine birds   **327,** 327–332, **328–329,** 330
  measures   **228,** 370–371
  monitoring   87–92, **88, 89,** 90
  natural   130–132, 433
  science-based approach   **426–427**
  sea otters   350
  shoreline oil, fate   128–135
  targets   372
recreation, shoreline use   78
reference areas   225–227
  herring   295
  salmon   **224,** 274
  sea otters   354
  set-aside sites   258
  see also baseline data
regime shifts, oceanic   6, 370, 392–393
regulations   22–26, 111. *see also* laws/legislation
relative risk   **389,** 389–393, **391**
release/immediate response phase   40, **41–42,** 42–44
  data collection urgency   72
  ecological risk assessments   409–410
  herring   294

marine birds   **319,** 319–327, **320, 323–325**
  shoreline oil, fate   123–128
  water column analysis   60–61
  water sampling   61–63
release scenarios   **58**
relevance, spill-related studies   428–429
repetitive water sampling   63
replicate sampling   230–231, 236
reproduction
  herring   292–293, 298–299
  marine birds   325–326
  salmon   **232,** 267, 272–277, **275**
Research Planning Inc (RPI) surveys   131–132
residual oil
  biodegradation   188–192, **190**
  ecological risk assessments   384, 387
  shoreline   **117**
  see also oil residues
resins   **12,** 192
responses, biochemical   386. *see also* biomarkers
responses, biological, definition   386
restoration   433. *see also* bioremediation
reverse transcriptase-polymerase chain reactions (RT-PCR)   208–209
risk assessments. *see* ecological risk assessments
rockweed (*Fucus* spp.)   250–252, 255

salmon (*Onchorhynchus* spp.)   9, 43–44, 108, 263–265, **264,** 281–283, 428
  biomarkers   **204**
  caged   210–211
  ecological effects analysis   **224,** 224
  ecological risk assessments   387
  effects on herring recovery   **302–303,** 306–307
  egg mortality   **232**
  fertilizer toxicity testing   187
  life cycle   263–265, **264**
  multiple approaches   283
  recovery   **426**
  research protocols   266–270, **269**
  science-based approach   **441**
  studies   270–281, **273, 274, 275, 279, 280**
  susceptibility   265, **266**
sampling   72, 200
  data collection   43–44
  replicate sampling   230–231
  science-based approach   431–432
  see also fixed site sampling; stratified random sampling
saturated hydrocarbons   **11**

scale-dependence
  bird surveys 340–341, **334**, **335**, 333
  ecological risk assessments 393–394
  spatial 412
SCAT. *see* shoreline cleanup assessment technique
science-based approach 3, 73, 220, 423, **424–427**, 442–443
  challenges 437–442
  ecological risk assessments 435–436
  exposure pathways 434–435
  multidisciplinary approaches 436–437
  objectives, stating 428–433
  recovery, natural 433
scenarios, release 58
sea otters (*Enhydra lutris*) 43–44, 233, 348–350, 371–372, 429
  baseline data 348, 350–352, **351**, 354, **364**, **368**
  Biological Monitoring Study 255
  characteristics of 362
  confounding factors 368–370, 372
  counts 355
  cytochrome biomarkers 208–209
  ecological risk assessments 387–389, 394, 396–399, **397**, **400**, **406**
  effect mechanisms 362, 366–368
  mortality 348–350, 352–353, **353**, 357–359, 366
  population level effects 404–406, **405**, **406**
  recovery 370–371, **427**
  results 355, 356–366, **359**, **361**, **362**, **364**
  science-based approach 442
  sensitivity analyses 404
  studies **351**, 354–356, **365**
seabirds. *see* marine birds
seaducks 210, 408. *see also* harlequin ducks
seal species 211
sediment quality triad procedure 233, 245
sediments
  aerial reconnaissance 81
  Prince William Sound 5
  relocation 91
  salmon exposure 271–272
  subtidal 123–126
segments, shoreline **82**, 82–83
  definitions 101
  postcleanup surveys 87
Selendang Ayu spill 110
semipermeable membrane devices (SPMDs) 211
sensitive habitats 81, 428
sensitivity analyses 73
  harlequin ducks 402
  mathematical models 168
  salmon **273**

sea otters 398–399, **405**, 405–406
  see also Environmental Sensitivity Indexes
sentinel species 45
SEP. *see* Shoreline Ecology Program
set-aside sites 258. *see also* baseline data; reference areas
Shannon diversity index 246
shoreline cleanup assessment technique (SCAT) 55, 78–80, **80**, 86–87, 92–94
  oil distribution **10**, 127–128
  postcleanup surveys 81, 87–92
  process 80–87, 91
  teams 16, 79–80
  see also cultural resource protection
Shoreline Ecology Program (SEP) 128, **131–132**, 241–246, **246**
  results 248–251, **249**, **250**, **251**
  see also fixed site method; stratified random sampling
shoreline oil 55, 116–121, **117**, **136**, 136–137
  chemistry 116, 119
  cleanup phase **121**, 123–128
  distribution **10**
  Exxon Valdez experience 121–135
  geomorphology 116, 119–121, **120**
  oil flocculation **124**, **125**, 137
  recovery 128–135, **129**, **131–132**, **133**, **134**, **135**, **136**
Shoreline Oiling Summary (SOS) forms 83, **84**
Shoreline Restoration Plan 87
Shoreline Treatment Recommendation (STR) forms 93
shorelines
  Alaska, southcentral 79
  biota. *see* biota, shoreline
  Prince William Sound 5, 5–6, 242–243, **243**
  site surveys 130
  Smith Island 191
  types 133, **134**. *see also* geomorphology
  water column analysis 58
shrimp species **179**, 186–187, 270
significance. *see* biological significance; statistical significance
similarity analyses 254
simulations, ecological risk assessment 395–396. *see also* models
single-species perspectives 309
site surveys
  conceptual models 149–159
  mapping **152**
  model development 145–146
Sitka Sound 298
skimmers 15, 86–87
Smith Island 128, 130, 133, 149

Smith Island (cont.)
  models **155**, 164–165, 169
  photographs 5, **16**, **87**, **120**, **134**, **191**, **246**
  wildlife 338
snails, littorine 250
Snug Harbor 179
soil samplers 102
solubility **60**, 147–149, **148–149**, 156
SOR. *see* surface oil residues
SOS (Shoreline Oiling Summary) forms 83, **84**
source–sink models 358–359
spatial scale-dependence 393–394, 412
spawning
  herring 295–297, **296**, 299
  salmon **264**, 268, 281
  shoreline habitats 78
species composition
  multispecies analyses 333–335
  shoreline biota 254
species level modeling 246, 395–399, **397**, **400**, 411
species richness, shoreline biota 246
specific gravity, oil **12**, 59
specificity, data 73
Spill Impact Model Application Package (SIMAP) 60
spills, oil. *see* oil spills
Spring Oiling Prediction Maps 87
Spring Shoreline Assessment Team (SSAT) 90, **92**
SRS. *see* stratified random sampling
SSOR. *see* subsurface oil residues
standardized recording forms 102
standardized terminology **81**, 82, 93
standards
  measurement 439
  reference 429–430
statistical analysis 199
  herring 297–299
  hypotheses, framing/testing 227–228
  statistical power 229
statistical experiments 221
statistical power 229–231
statistical significance 209, 232–234, 385–386
Status and Trends Mussel Watch Program 129
steady state equilibrium 222, **223**, 225
Steller, Georg William 8
stickleback, threespine (*Gasterosteus aculeatus*) 187
stochastic models 411. *see also* ecological risk assessments
storms xxii, 387. *see also* weather
STR (Shoreline Treatment Recommendation) forms 93
stranded oil, aerial reconnaissance **81**, 82
stratified random sampling (SRS)
  method 241, 243–246, **245**, **246**, 255
  results 248–250, 256
stream sediments 271–272
stress
  characterization 384
  ecological consequences 384
  ecological risk assessments 387
  mitigation 394
  natural 297
structural sensitivity analyses 403
study design 72, 200, 221–232, 325, 354–356, 365, 431–432
subsurface oil residues (SSOR) 4, 118–121, **121**
  biodegradation 184, 188–192, **191**
  cleanup/release phase 126–128
  conceptual models 153–159
  Cultural Resource Program 102
  harlequin ducks 400–402, **403**
  mathematical models 164–168
  recovery phase **129**, 129–135, **131–132**, **133**, **134**, **135**, **136**
  science-based approach **424–425**
  sea otters 362, 367–368, 396–399, **397**, **400**, 404–406, **405**, 367–368
  shoreline **117**, **136**
  Smith Island **191**
subtidal sediments 123–126
surface deposits **117**, 118–119
surface oil residues (SOR)
  harlequin ducks 400–402
  recovery phase 129
  shoreline **117**
surface oil slicks 60
surface tension 147, **151**
surveys
  biodegradation/bioremediation 188
  ground 83–85
  shoreline oil, fate **131–132**, 135, **136**
  see also shoreline cleanup assessment technique
survival. *see* mortality
susceptibility
  biomarkers 204
  pink salmon 265, **266**
systematic hydrocarbon analysis 129–130
systematic shoreline mapping 82

targets 372, 428–433
teamwork 234, 236 *see also* multidisciplinary approaches
Technical Advisory Group (TAG) 90, 101
TEH (total extractable hydrocarbons) **154**, **156**, 156–157
terminology, standardized **81**, 82, 93
three-dimensional numerical models 163

weathering (cont.)
  geomorphology   119–121, **120**
  indexes   189
  shoreline oil, fate   116, 118
  subsurface oil residues   130–132
  see also environmental processes
websites
  cleanup methods   16, 86–87
  cultural resource protection   105
  individual-based modeling   396
  laws/legislation   **29**

MAYSAP   127
Trustee Council   68
weight-of-evidence approach   233
West Falmouth spill   **117**
wettability   **148–149**
whales/whaling   8, 211, 306
  humpback whales (*Megaptera novaeangliae*)   306. *see also* dolphins; killer whales

xylene   59. *see also* BTEX

zero values   429

time-series studies   129–130,
    231–232, 432
tines   **91**, 91
tissue analysis, pink salmon   268
tissue biomarkers. *see* biomarkers
toluene   59. *see also* BTEX
top-down processes   392
Torrey Canyon oil spill   256
total extractable hydrocarbons (TEH)   **154**,
    156, 156–157
total petroleum hydrocarbons (TPH)   59, 66
total polycyclic aromatic hydrocarbons
    (TPAH)
  analytical models   162
  baseline data   71
  biodegradation   **190**, 190
  Biological Monitoring Study   255
  biomarkers   **190**
  fixed site sampling   250–251
  harlequin ducks   **403**, **407**, 407–408
  mussels   70–71, **70–72**, **71**, 129–130
  numerical models   158
  oil spill data   **68**, 68–71, **69**, 128
  salmon species   266, 268, 270–272,
      275–280
  spatial distributions   157–159, **158**
  water analysis   66
  see also polycyclic aromatic hydrocarbons
tourism   78
toxicity/toxicology   65, 187
  biomarkers   **204**, 204–205
  bioremediation   176, 178, **179**
  quantifying   408
  weight-of-evidence approach   233
toxicity reference values (TRVs)   399
TPAH. *see* total polycyclic aromatic
    hydrocarbons
TPH (total petroleum hydrocarbons)
    59, 66
Trans Alaska Pipeline Act   9
transcription factors   207
transcriptomics   213
trophodynamic models   **391**, 392
Trustees *see Exxon Valdez* Oil Spill Trustee
    Council
tsunami   6–7
type A/B assessments, exposure
    pathways   **20–21**
TRVs (toxicity reference values)   399

uncertainty
  conceptual ecosystem models   402
  ecological risk assessments   384, 402, 412
  science-based approach   438–439
  sea otters   356–366, 372

studies   **340–341**, 341, 354
univariate modeling   246
upper shore oil residues   134–135
urgency   72, 433, 439

Valdez, Port   8
valued ecosystem components (VECs)   384
  future spills   409–410
  selection   385, 387–389, **388**, 393–395
  see also ecological risk assessments
vandalism   99, 103, 105–106, 111
variation between individuals   210–212
variation, natural   220, 222–225, **223**
  and cause–effect relationships   337–339
  ecological risk assessments   384
  marine birds   332–335, **334**, **335**, **336**, **338**,
      340–341
  replicate sampling   230–231
  salmon   283
  science-based approach   437–438
  spatial, permeability   146
  sea otters   370
  temporal   393–394, 412
very weathered oil (VWO)   275–276
videotape mapping   **81**, 81–83,
    82, 87
viral hemorrhagic septicemia (VHS)   **302–303**,
    308
viscosity, oil   **12**, 59, **148–149**
volatile aromatic hydrocarbons   59, 122
volatile organic compounds   387.
    see also polycyclic aromatic hydrocarbons;
    total polycyclic aromatic hydrocarbons
volatility, oil   **12**, **60**, 147–149

washing methods   **14**, 86–87, **87** *see also* cold-
    water washing; flooding; flushing;
    hot-water shoreline cleanup
water column analysis   55, 57–66, **60**, **63**,
    72–73
  Deepwater Horizon spill   72
  Exxon Valdez spill   66–72, **67**, **68**, **69**,
      70–72, **71**
water column exposure, salmon   270–271,
    277–281
water sampling   42, 58, 61–65
  data collection   43–44
  Exxon Valdez spill   66–67, **67**
water-in-oil emulsion   17, 122, 188
water-soluble hydrocarbons   122
wave action   6, 60, 91, 243, 387
wave shadow   119
weather   122. *see also* storms
weathering   10, **21**, 37, 57
  chemistry   119
  crude oil   13